AF597766

Transindividual Intelligence

Yige Liu · Hua Chai · Ding Wen 'Nic' Bao ·
Philip F. Yuan
Editors

Transindividual Intelligence

Proceedings of the 7th International Conference on Computational Design and Robotic Fabrication (CDRF 2025)

Editors
Yige Liu
Tongji University
Shanghai, China

Hua Chai
Tongji University
Shanghai, China

Ding Wen 'Nic' Bao
RMIT University
Melbourne, VIC, Australia

Philip F. Yuan
Tongji University
Shanghai, China

ISBN 978-981-92-0614-8 ISBN 978-981-92-0615-5 (eBook)
https://doi.org/10.1007/978-981-92-0615-5

This Springer imprint is published by the registered company Springer Nature Singapore Pte Ltd.
The registered company address is: 152 Beach Road, #21-01/04 Gateway East, Singapore 189721, Singapore

Contents

Computational Design and Generative Intelligence

ArchiLense: A Framework for Quantitative Analysis of Architectural Styles Based on Vision Large Language Models

Jing Zhong[1], Tianze Hao[1], Jun Yin[1](✉), Peilin Li[2], Ruolin Pan[3], Pengyu Zeng[1], Jingxuan Li[1], Hungcheng Tai[1], Miao Zang[1], and Shuai Lu[1]

[1] Shenzhen International Graduate School, Tsinghua University, Shenzhen, China
yinj24@tsinghua.edu.cn
[2] National University of Singapore, Clementi, Singapore
[3] South China University of Technology, Guangzhou, Guangdong, China

Abstract. Architectural cultures across regions are characterized by stylistic diversity, shaped by historical, social, and technological contexts in addition to geographical conditions. Understanding architectural styles requires the ability to describe and analyze the stylistic features of different architects from various regions through visual observations of architectural imagery. However, traditional studies of architectural culture have largely relied on subjective expert interpretations and historical literature reviews, often suffering from regional biases and limited explanatory scope.

To address these challenges, this study proposes three core contributions: (1) We construct a professional architectural style dataset named ArchDiffBench, which comprises 1,765 high-quality architectural images and their corresponding style annotations, collected from different regions and historical periods. (2) We propose ArchiLense, an analytical framework grounded in Vision-Language Models and constructed using the ArchDiffBench dataset. By integrating advanced computer vision techniques, deep learning, and machine learning algorithms, ArchiLense enables automatic recognition, comparison, and precise classification of architectural imagery, producing descriptive language outputs that articulate stylistic differences. (3) Extensive evaluations show that ArchiLense achieves strong performance in architectural style recognition, with a 92.4% consistency rate with expert annotations and 84.5% classification accuracy, effectively capturing stylistic distinctions across images.

The proposed approach transcends the subjectivity inherent in traditional analyses and offers a more objective and accurate perspective for comparative studies of architectural culture.

Keywords: Architectural Floor Plan Generation · Transformer Model · Convolutional Neural Networks (CNNs) · Hybrid Deep Learning Models

1 Introduction

See Fig. 1

Y. Liu et al. (Eds.): CDRF 2025, *Transindividual Intelligence*, pp. 3–13, 2026.
https://doi.org/10.1007/978-981-92-0615-5_1

Fig. 1. Graphic Abstract

Recent advances in machine learning have led to a wide range of studies in sustainable architecture [11, 13–15, 37, 46], energy-aware design [5, 8, 17–19], multimodal generation [23–25, 43], visual enhancement [30, 34, 44], architectural performance optimization [15–18, 31] and human-AI interaction [4, 11, 28, 33, 36, 39].Among these, architectural style analysis remains a critical yet relatively underexplored.

Systematic analysis and quantitative comparison of architectural styles across regions and historical periods are of critical academic and practical importance to architectural design and cultural studies [27]. Architectural styles reflect not only regional culture and social transformation but also spatial logic and aesthetic ideologies shaped by historical contexts [1]. Therefore, such investigations help reveal underlying evolutionary mechanisms and offer new theoretical foundations for architectural discourse and innovation in contemporary design approaches.

Traditional studies of architectural styles have primarily relied on expert interpretation and historical-visual analysis [20]. However, these approaches are inherently constrained by subjectivity and cognitive limitations, often leading to inconsistencies in classification. Moreover, manual analytical methods lack scalability when dealing with large-scale and heterogeneous datasets. In recent years, with the increasing integration of architectural cultures across regions, traditional methods have struggled to provide reproducible and quantitative comparative frameworks, posing significant challenges to the rigor of architectural style research. Advances in computer vision, deep learning, and natural language processing have shifted architectural style research toward automated and data-driven paradigms.Machine learning enables efficient extraction of stylistic features from large-scale datasets, outperforming traditional expert-based methods. Meanwhile, the rise of Vision Large Language Models (VLLMs) further enhances multimodal analysis, supporting automatic classification and descriptive articulation of architectural styles—laying the groundwork for a systematic and integrated analytical framework.

Accordingly, we proposes ArchiLense, a novel framework based on VLLMs to quantitatively analyze architectural style variations. By integrating deep learning and natural language processing, ArchiLense enables automatic recognition and linguistic articulation of stylistic differences in architectural imagers: (1) VLLM extracts visual features and generates candidate style descriptions; (2) A ranking module selects the most distinctive and expressive descriptions; (3) Selected descriptions guide a text-to-image generation model to synthesize architectural images, which are validated via expert evaluation. Through ArchiLense, we are able to enrich academic inquiry and applications in architectural style studies.

2 Related works

2.1 Research on Architectural Style

Architectural style research is a core area in architecture, art history, and heritage studies, centers on identifying design features across historical periods, regions, and movements [9].

Traditional architectural style studies often rely on literature review and expert visual assessment, which, though insightful, are subjective and inefficient at scale. Recent advancements in computer vision have enabled data-driven methods, such as CNN-based models that extract stylistic features and classify large-scale architectural images [27]. However, these approaches are typically confined to classification tasks and lack mechanisms to analyze stylistic evolution or perform cross-style transfer. Additionally, conventional models emphasize low-level features, limiting their ability to capture semantic nuances in architectural styles.

2.2 Application of Visual Large Language Model (VLLM)

In recent years,Vision Large Language Models (VLLMs) represent a major advancement in multimodal AI, combining computer vision and natural language processing for enhanced semantic reasoning.

While VLLMs like CLIP and ViT-GPT perform well in cross-modal tasks, their use in architecture remains limited. Recent work has leveraged CLIP for architectural image-text alignment, such as improved Chinese image labeling via CLIP-enhanced DreamBooth [2], and diffusion models for style transfer and prompt-based generation [10]. However, most studies focus on image synthesis, with limited attention to style analysis and quantitative comparison.

3 Dataset

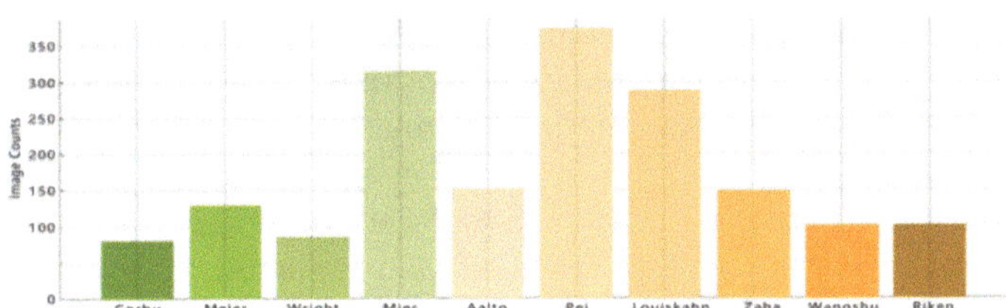

Fig. 2. The Distribution of Each Category of Images In Dataset

To systematically analyze architectural styles, we developed ArchDiffBench, a dataset integrating expert evaluations and textual analysis. The images were sourced from ArchDaily® and gooood®, two major global repositories of architectural design. ArchDiffBench comprises 1,765 images across various periods and regions, encompassing facades, interiors, and exteriors to ensure visual richness. Images are categorized into

10 groups by architect identity. To analyze stylistic relationships, 81 paired subsets were constructed for inter-group comparison. Figure 2 illustrates the category-wise image distribution.

4 Methodology

Inspired by Visdiff [22], we developed the ArchiLense framework, a two-stage analysis system designed for style feature extraction and quantification. As shown in Fig. 3, the framework consists of a Style Extractor, which generates discriminative textual descriptions from randomly sampled images, and a Style Evaluator, which validates and quantitatively assesses these descriptions in style differentiation tasks. This approach addresses the limitations of directly training neural networks on complex architectural imagery, where general-purpose models often struggle to balance encoding efficiency and predictive accuracy.

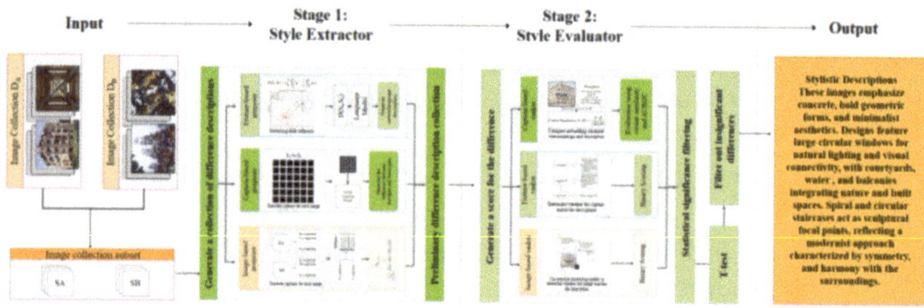

Fig. 3. Archi Lense Framework

4.1 Style Extractor

The Style Extractor aims to identify distinct stylistic features between two architectural groups D_A and D_B while ensuring the extracted traits remain group-specific and not easily generalized. To explore feasible implementations, we tested three VLM-based approaches: (1) Image-based Analysis: Gridded images from both groups are input to VLMs (e.g., GPT-4 V) to identify differences in façade design, geometry, materials, and spatial configuration. (2) Embedding Difference Computation: Style embeddings from each group are averaged and subtracted. The resulting vector is input into a language model (e.g., BLIP-2) for textual style descriptions. (3) Text-based Description Comparison: VLMs generate detailed text for each image, while GPT-4 compares texts between groups to produce stylistic difference descriptions.

Experimental results show that Text-based descriptions more effectively capture architect-specific styles, surpassing visual-only methods. Hence, it is chosen as the primary extraction strategy, with others as baselines..

4.2 Style Evaluator

The Style Extractor's output can be invalid or redundant. Therefore, A style discriminator is used to assess the description's validity and quantify its ability to distinguish architectural styles. It calculates the differentiation fraction S_y of the style description y and sorts it as follows:

$$S_y = \sum_{x \in D_A} \theta(x, y) - \sum_{x \in D_B} \theta(x, y)$$

θ(x,y):The degree of match between image x and style description y.

We explored three computational methods to evaluate $\theta(x,y)$:(1) VQA-based Matching: LLaVA-1.5's strong multimodal reasoning enables us to classify whether an image x aligns with a style description y in a binary manner: "Does this image match the description?" (Yes/No). (2) Caption-to-Description Matching: Each image is captioned by BLIP-2 to obtain *c*, which is evaluated by Vicuna-1.5 through QA(*c*,*y*) for alignment with the style description. The task adopts a (Yes/No) classification format. (3) Embedding Similarity Computation: CLIP ViT-G/14 generates vector representations for both the image *x* and the textual description *y*. The semantic alignment between the two is then quantified using cosine similarity, computed as follows:

$$\theta(x, y) = \frac{e_x \cdot e_y}{\|e_x\| \|e_y\|}$$

As the value is continuous, AUROC is used to assess the discriminative power of style descriptions between D_A and D_B.

The embedding-based evaluator shows superior accuracy and efficiency, and is thus used as the main evaluation method, with others as baselines. Additionally, a t-test (significance level 0.05) is conducted on $\theta(x,y)$ distributions to exclude statistically insignificant descriptions.

5 Experiment Result

This section presents the feature analysis results of ArchiLense, trained on ArchDiff-Bench, focusing on: (1) generation of architect-specific stylistic descriptions; (2) evaluation of their effectiveness in capturing architectural style differences; (3) application of these texts in Text-to-Image Generation to synthesize representative architectural images, validated via expert review.

5.1 Discriminative Effectiveness and Cognitive Consistency of Style Descriptions

To evaluate the effectiveness of ArchiLense-generated style descriptions, we assessed both their discriminative power and alignment with architectural design cognition. Style descriptions generated from image pairs reflect differences in design approach, form, materials, and typology. Evaluated using AUROC, these descriptions are ranked by their effectiveness in distinguishing architectural styles. Despite the absence of architect identities, the outputs align well with established design knowledge. Word clouds of key terms further confirm that ArchiLense results are both discriminative and cognitively consistent with expert understanding (Fig. 4).

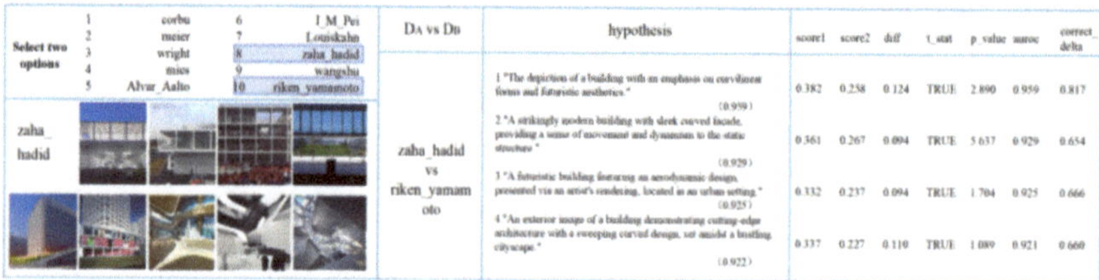

Fig. 4. Using Riken and Zaha Hadid as representative cases, we conducted 81 pairwise comparisons between their respective subgroups in the dataset. Each run involved random sampling of 20 images per group.

Functional: Keyword extraction effectively captures stylistic features, as shown in Fig. 5 word clouds. CLIP and GPT-4 distinguish between Meier's minimalist ("white", "window") and Yamamoto's urban-functional style ("cityscape", "marble").

Accuracy: Model outputs align with architectural cognition, proving its reliability and value. For example, Le Corbusier ("staircase") and Mies van der Rohe ("glass") reflect their design philosophies. Wang Shu ("wooden") captures traditional-modern synthesis, while Aalto ("forest") express organic and monumental themes respectively.

Generalization: The model robustly identifies diverse architectural styles—across regions and periods—without predefined labels, enabling objective, bias-free classification.

Fig. 5. Architectural Projects and Corresponding Word Cloud Map

5.2 Quantitative Validation of Stylistic Difference Descriptions

5.2.1 T-test and Significance Analysis

An experiment using 270 style descriptions from 9 architect groups evaluated stylistic distinctiveness via Score 1 and Score 2 comparisons, followed by t-tests. As shown in Fig. 6, over 80% of the descriptions achieved statistical significance ($p < 0.05$), confirming their effectiveness in differentiating architectural styles.

5.2.2 Data Validation Visualization

To validate the feature extraction and separation capability of ArchiLense, pairwise comparisons were conducted, visualized through Histogram, KDE, and Boxplot. As

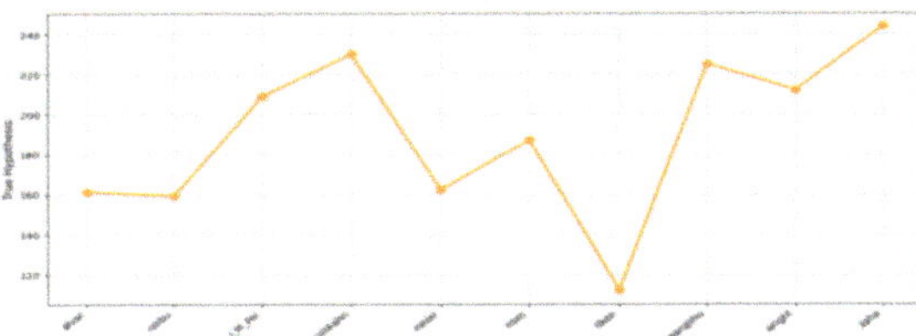

Fig. 6. The number of descriptions T-tested as "true" can reflect the degree to which an architect's style is unique.

shown in Fig. 7, the model's performance is visualized across multiple evaluation metrics. The left histogram compares cosine similarities within Groups A and B: Group A's concentrated distribution indicates robust feature consistency, while Group B's dispersion suggests sensitivity to stylistic diversity. Kernel density curves further highlight ArchiLense's balance between consistency and differentiation. The right-side boxplots show that Group A yields higher medians and narrower IQRs, reflecting stable representations, whereas Group B exhibits greater variability and outliers, demonstrating responsiveness to complex stylistic variations.

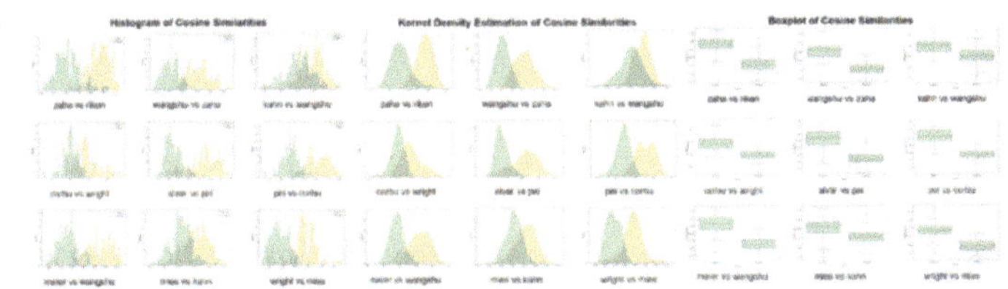

Fig. 7. Cosine Similarity Distribution and Feature Representation Analysis

5.2.3 Architectural Stylistic Difference Attributes

Figure 8 shows textual similarities among architects, with orange indicating high similarity and blue indicating low. Meier, Mies, and Louis Kahn exhibit strong stylistic alignment (0.995), suggesting shared modernist tendencies. Aalto and Wang Shu (0.95) also show notable similarity, possibly indicating stylistic continuity. In contrast, architects with more blue regions tend to express individualistic styles, while orange-dominant patterns may reflect shared academic or traditional influences.

5.3 Feature Style Images Verified by Expert Evaluation

To assess model performance, 20 architecture PhD students and 5 senior architects rated the consistency between generated images and description texts for 10 architects on a 1–5 scale (1 = inconsistent, 5 = fully consistent). Participants matched text-generated images to one of 10 architects. ArchiLense achieved a mean score of 4.62 and an 84.5% matching accuracy, demonstrating strong stylistic reliability and interpretability (Fig. 9).

	alvar	corbu	I_M_Pei	Louiskahn	meier	mies	riken	wangshu
alvar	1	0.98	0.97	0.985	0.99	0.98	0.975	0.99
corbu	0.98	1	0.945	0.96	0.975	0.97	0.96	0.965
I_M_Pei	0.97	0.945	1	0.97	0.965	0.975	0.97	0.975
Louiskahn	0.985	0.96	0.97	1	0.985	0.98	0.97	0.98
meier	0.99	0.975	0.965	0.985	1	0.995	0.985	0.99
mies	0.98	0.97	0.975	0.98	0.995	1	0.98	0.975
riken	0.975	0.96	0.97	0.97	0.985	0.98	1	0.985
wangshu	0.99	0.965	0.975	0.98	0.99	0.975	0.985	1
wright	0.97	0.95	0.965	0.975	0.985	0.97	0.975	0.96
zaha	0.965	0.94	0.97	0.98	0.98	0.975	0.97	0.95

Fig. 8. Architectural Style Similarity and Clustering Visualization

Fig. 9. Architectural Style Cosine Similarity Matrix

6 Discussion

ArchiLense demonstrates the potential of vision-language models for architectural style analysis. However, current limitations include a dataset biased toward modern Western styles and limited analysis of spatial authenticity. Future work will broaden the dataset's diversity and incorporate spatially aware evaluation methods, alongside deeper theoretical engagement, to improve generalizability and interpretability across cultural contexts.

7 Conclusion

Although still preliminary, ArchiLense suggests that vision-language models may help improve the efficiency and objectivity of architectural style analysis. By translating visual features into structured language, it provides a potential basis for quantitative and cross-cultural comparisons. Current limitations in dataset diversity and representativeness point to the need for broader coverage and deeper theoretical engagement. With further refinement and expansion, Archi Lense may serve as a useful tool for supporting future research in architectural historiography, style interpretation, and cross-cultural understanding in heritage studies.

References

1. Canizaro, V.B.: Critical regionalism: Architecture and identity in a globalized world. Princeton Architectural Press (2007)

2. Chen, F., Mai, M., Huang, X., Li, Y.: Enhancing the Sustainability of AI Technology in Architectural Design: Improving the Matching Accuracy of Chinese-Style Buildings. Sustainability. **16**(19), 8414 (2024). https://doi.org/10.3390/su16198414
3. Gao, C., Sun, Y., Yang, K., Xie, Y., Liu, J., & Liu, Y. (2025). Thought-augmented policy optimization: Bridging external guidance and internal capabilities. arXiv preprint https://arxiv.org/abs/2505.15692. https://arxiv.org/abs/2505.15692
4. Gao, W., Lu, S., Zhang, X., He, Q., Huang, W., Lin, B.: Impact of 3D modeling behavior patterns on the creativity of sustainable building design through process mining. Autom. Constr. **150**, 104804 (2023)
5. Huang, Y., Zeng, T., Jia, M., Yang, J., Xu, W., Lu, S.: Fusing Transformer and diffusion for high-resolution prediction of daylight illuminance and glare based on sparse ceiling-mounted input. Build. Environ. **267**, 112163 (2025)
6. He, Y., et al. (2025). Enhancing Low-Cost Video Editing with Lightweight Adaptors and Temporal-Aware Inversion. arXiv preprint https://arxiv.org/abs/2501.04606.
7. He, Y., Wang, J., Wu, J., Zhu, W., & Liu, Z. (2025). Boosting multimodal reasoning with MCTS-automated structured thinking. arXiv preprint https://arxiv.org/abs/2502.02339. https://arxiv.org/abs/2502.02339
8. Jia, Z., Lu, S., & Yin, J. How real-time energy feedback influences energy-efficiency and aesthetics of architecture design and judgment of architects: two design experiments.
9. Lefaivre, L., Tzonis, A.: Architecture of Regionalism in the Age of Globalization: Peaks and Valleys in the Flat World, 2nd edn. Routledge (2020). https://doi.org/10.4324/9780367281182
10. Li, J., Lu, S., Wang, W., Huang, J., Chen, X., Wang, J.: Design and climate-responsiveness performance evaluation of an integrated envelope for modular prefabricated buildings. Adv. Mater. Sci. Eng. **2018**(1), 8082368 (2018)
11. Li, J., Lu, S., Wang, Q., Tian, S., Jin, Y.: Study of passive adjustment performance of tubular space in subway station building complexes. Appl. Sci. **9**(5), 834 (2019)
12. Li, J., Song, Y., Lv, S., Wang, Q.: Impact evaluation of the indoor environmental performance of animate spaces in buildings. Build. Environ. **94**, 353–370 (2015)
13. Li, J., Lu, S., Wang, Q.: Graphical visualisation assist analysis of indoor environmental performance: Impact of atrium spaces on public buildings in cold climates. Indoor Built Environ. **27**(3), 331–347 (2018)
14. Luo, Y., Ma, X., Li, J., Wang, C., de Dear, R., Lu, S.: Outdoor space design and its effect on mental work performance in a subtropical climate. Build. Environ. **270**, 112470 (2025 Feb 15)
15. Lu, S., Yan, X., Li, J., Xu, W.: The influence of shape design on the acoustic performance of concert halls from the viewpoint of acoustic potential of shapes. Acta Acust. united Acust. **102**(6), 1027–1044 (2016)
16. Lu, S., Xu, W., Chen, Y., Yan, X.: An experimental study on the acoustic absorption of sand panels. Appl. Acoust. **116**, 238–248 (2017)
17. Lu, S., Luo, Y., Gao, W., Lin, B.: Supporting early-stage design decisions with building performance optimisation: Findings from a design experiment. J. Build. Eng. **82**, 108298 (2024)
18. Lu, S., Yan, X., Xu, W., Chen, Y., Liu, J.: Improving auditorium designs with rapid feedback by integrating parametric models and acoustic simulation. In: Building Simulation, vol. Vol. 9, pp. 235–250. Tsinghua University Press (2016, June)
19. Ma, X., Zeng, T., Zhang, M., Zeng, P., Lin, B., Shuai, L.: Street microclimate prediction based on Transformer model and street view image in high-density urban areas. Build. Environ. **269**, 112490 (2025)

20. Oommen, T.: The glazed eyes of architectural history: Reflections on the (dis)contents of global history survey courses. Archit. Hist. **10**(1), 1–16 (2022). https://doi.org/10.16995/ah.8280
21. Rousset, T., Kakibuchi, T., Sasaki, Y., & Nomura, Y. (2025). Merging Language and Domain Specific Models: The Impact on Technical Vocabulary Acquisition. arXiv preprint https://arxiv.org/abs/2502.12001.
22. Sukkar, A.W., Fareed, M.W., Yahia, M.W., Abdalla, S.B., Ibrahim, I., Senjab, K.A.K.: Analytical Evaluation of Midjourney Architectural Virtual Lab: Defining Major Current Limits in AI-Generated Representations of Islamic Architectural Heritage. Buildings. **14**(3), 786 (2024). https://doi.org/10.3390/buildings14030786
23. Sun, H., Xia, B., Zhao, Y., Chang, Y., Wang, X.: Identical human preference alignment paradigm for text-to-image models. In: ICASSP 2025–2025 IEEE International Conference on Acoustics, Speech and Signal Processing (ICASSP), pp. 1–5. IEEE (2025, April)
24. Sun, H., Xia, B., Zhao, Y., Chang, Y., Wang, X.: Positive enhanced preference alignment for text-to-image models. In: ICASSP 2025–2025 IEEE International Conference on Acoustics, Speech and Signal Processing (ICASSP), pp. 1–5. IEEE (2025, April)
25. Sun, H., Xia, B., Chang, Y., Wang, X.: Generalizing Alignment Paradigm of Text-to-Image Generation with Preferences Through f-Divergence Minimization. Proc. AAAI Conf. Artif. Intell. **39**(26), 27644–27652 (2025). https://doi.org/10.1609/aaai.v39i26.34978
26. Wang, B., Zhang, S., Zhang, J., Cai, Z.: Architectural style classification based on CNN and channel–spatial attention. SIViP. **17**(1), 99–107 (2023)
27. Wang, Y., He, Y., Wang, J., Li, K., Sun, L., Yin, J., et al.: Enhancing intent understanding for ambiguous prompt: A human-machine co-adaption strategy. Neurocomputing. **130415** (2025). https://doi.org/10.1016/j.neucom.2024.130415
28. Wang, J., Liu, C., Zhu, W., He, Y., & Liu, Z. (2024). Beyond examples: High-level automated reasoning paradigm in in-context learning via MCTS. arXiv preprint https://arxiv.org/abs/2411.18478. https://arxiv.org/abs/2411.18478
29. Wang, J., et al.: MDANet: A multi-stage domain adaptation framework for generalizable low-light image enhancement. Neurocomputing. **627, 129572** (2025)
30. Yan, X., Lu, S., Li, J.: Experimental studies on the rain noise of lightweight roofs: Natural rains vs artificial rains. Appl. Acoust. **106**, 63–76 (2016)
31. Yin, J., Li, A., Xi, W., Yu, W., Zou, D.: Ground-fusion: A low-cost ground slam system robust to corner cases. In: 2024 IEEE International Conference on Robotics and Automation (ICRA), pp. 8603–8609. IEEE (2024, May)
32. Yin, J. et al.: Archidiff: Interactive design of 3d architectural forms generated from a single image. Comput. Ind. **168**, 104275 (2025)
33. Yin, J., , , , , , & (2025). Promptlnet: Region-adaptive aesthetic enhancement via prompt guidance in low-light enhancement net. arXiv preprint https://arxiv.org/abs/2503.08276.
34. Yin, J., Xu, P., Gao, W., Zeng, P., Lu, S.: Drag2Build: Interactive point-based manipulation of 3D architectural point clouds generated from a single image. Front. Archit. Res. **14**, 1602–1620 (2025)
35. Yu, T. et al.: Machine learning prediction on spatial and environmental perception and work efficiency using electroencephalography including cross-subject scenarios. J. Build. Eng. **99**, 111644 (2025)
36. Zeng, T., Ma, X., Luo, Y., Yin, J., Ji, Y., Lu, S.: Improving outdoor thermal environmental quality through kinetic canopy empowered by machine learning and control algorithms. In: Building Simulation, pp. 1–22. Tsinghua University Press, Beijing (2025, March)
37. Zeng, P., Gao, W., Li, J., Yin, J., Chen, J., Lu, S.: Automated residential layout generation and editing using natural language and images. Autom. Constr. **174**, 106133 (2025)

38. Zeng, P., Jiang, M., Wang, Z., Li, J., Yin, J., Lu, S.: CARD: cross-modal agent framework for generative and editable residential design. In: Neur IPS 2024 Workshop on Open-World Agents
39. Zeng, P., Hu, G., Zhou, X., Li, S., Liu, P.: Seformer: a long sequence time-series forecasting model based on binary position encoding and information transfer regularization. Appl. Intell. **53**(12), 15747–15771 (2023)
40. Zeng, P., Hu, G., Zhou, X., Li, S., Liu, P., Liu, S.: Muformer: A long sequence time-series forecasting model based on modified multi-head attention. Knowl.-Based Syst. **254**, 109584 (2022)
41. Zeng, P., Yin, J., Gao, Y., Li, J., Jin, Z., Lu, S.: Comprehensive and Dedicated Metrics for Evaluating AI-Generated Residential Floor Plans. Buildings. **15**(10), 1674 (2025). https://doi.org/10.3390/buildings15101674
42. Zhang, M., Yin, J., Zeng, P., Shen, Y., Lu, S., Wang, X.: TSCnet: A text-driven semantic-level controllable framework for customized low-light image enhancement. Neurocomputing. **625**, 129509 (2025)
43. Zou, H., Ge, J., Liu, R., He, L.: Feature recognition of regional architecture forms based on machine learning: a case study of architecture heritage in Hubei Province, China. Sustainability. **15**(4), 3504 (2023)
44. Zou, Y., Lou, S., Xia, D., Lun, I.Y., Yin, J.: Multi-objective building design optimization considering the effects of long-term climate change. J. Build. Eng. **44**, 102904 (2021)
45. Zhang, Q., Li, Z., Chen, M., Liu, Y., Chen, Q., & Wang, Y. (2024). Pandora's Box or Aladdin's Lamp: A comprehensive analysis revealing the role of RAG noise in large language models. arXiv preprint https://arxiv.org/abs/2408.13533. https://arxiv.org/abs/2408.13533
46. Zhang, M., Shen, Y., Yin, J., Lu, S., Wang, X.: ADAGENT: Anomaly detection agent with multimodal large models in adverse environments. IEEE Access. **12**, 172061 (2024)

Intelligence Assessment Methods for Field-Level Plans Generation of Large and Medium-sized Stadiums Based on Deep Learning

Wang Zhenyu[1(✉)], Meng Yufan[2], Xu Chen[2], and Zheng Fang[1,2(✉)]

[1] School of civil engineering, Beijing Jiaotong University,, ChinaBeijing
wangzhenyulily@126.com, Fang.zheng@me.com

[2] School of Architecture and design, Beijing Jiaotong University,, ChinaBeijing

Abstract. Due to the complexity of functional flow organization, stadium design involves a high workload when using traditional methods. Recently, generative AI has shown significant potential in architectural design. A generative-evaluative collaborative generation method for Large and Medium-sized Stadium field-level plan is proposed, using Stable Diffusion for generation and adapted-Convolutional Neural Networks for assessment. The latter is introduced as an assessment model in Stepwise Generation. By assessing the generation results' functional topology and area ratio, the model sorts out results with higher similarity to outstanding case studies, indicating a greater potential for further design, then feeds them into the subsequent generation step. The combination of two deep learning models establishes an effective human-computer interaction mechanism between 'generative' and 'critical' AI, enhancing the scientificity of design decisions, and offering a technical path for intelligent design of complex public buildings.

Keywords: Deep Learning · Intelligence Assessment Method · Large and Medium-sized Stadiums · Plan Generation · Generation Design

1 Introduction

As a public architecture, the design complexity of a stadium exceeds that of general buildings, particularly the field-level plan, which must consider design specifications, flow and function organization, and post-event usage. These constraints increase the difficulty of the design process, often resulting in significant time and financial costs.

In recent years, the rapid development of generative artificial intelligence models has provided new opportunities to improve the efficiency of human-computer interaction in the generative design of stadium graphics. With its fast, batch, and diverse generative capabilities [1, 2], the Stable Diffusion model can collaboratively process multiple design elements in 2D image generation. However, due to the opacity of the model's internal architecture, generation results can sometimes be overly unstable [3], making it difficult to satisfy the front-end constraints. Therefore, based on the existing generation design experience, this paper proposes a design method for step-by-step generation and step-by-step directional filtering of the field-level plans of large and medium-sized stadiums

Y. Liu et al. (Eds.): CDRF 2025, *Transindividual Intelligence*, pp. 14–26, 2026.
https://doi.org/10.1007/978-981-92-0615-5_2

based on a hybrid of two deep learning models, the Stable Diffusion and the adapted-Convolutional Neural Networks.

2 Artificial Intelligence Methods Apply in Architecture Design

In recent years, scholars worldwide have extensively explored the application of deep learning-based generative AI models in architecture.

2.1 Plan Generation

Since 2022, numerous studies have adopted the Stable Diffusion model as the basis for architectural generation, owing to its efficient generative capabilities. In plan generation, current research predominantly focuses on residential units or settlement layouts. For example, Li and Qiu et al. [4, 5] employed *ControlNet* within *Stable Diffusion*, specifically for the generative design of residential units in China. Shabani [6] and Kuhn [7] respectively employed House Diffusion to generate floor plans for North American and Swiss dwellings. Yet the generative design methods for sports-related public buildings are primarily limited to diagnosing and optimizing existing plan-related issues [8–10]. Algorithms for generating and screening overall layout plans are in urgent need of development.

However, the controllability and accuracy of these generation methods is limited. To address this issue, the cleaning of plan datasets is a critical step. After removing irrelevant information allows the model to focus more effectively on relevant feature. In some studies, plans are simplified by converting them into colored block diagrams representing functional areas and functional topological diagrams (Shabani; Maeng and Hyun) [6, 11]. Other studies combine textual descriptions with topological graphs to construct multimodal floor plan datasets, using text to enhance the model's understanding of images (Qiu et al.) [5]. To further enhance generation performance, a stepwise generation approach integrated with multiple *LoRA* and *ControlNet* modules is introduced (Wang) [12].

In short, improving generation controllability and integrating evaluation systems for automated, targeted result filtering remain key challenges. Subsequent research can integrate additional technical approaches, such as the implementation of filtering mechanisms, and extend the scope of application to large-scale public buildings. This would further enhance the practicality, controllability, detail precision, and *human–machine collaboration* efficiency of plan *Generation.*

2.2 Topology Diagram Generation

Functional topology diagrams, as a kind of diagrams that can visually represent the functional types and adjacency relationships of Functions in architectural plans, are the focus of research in the field of plan generation.

Research on topology diagrams generation methods has gained significant attention in recent years. Shabani et al. [6] and Liu et al. [13] both used automated methods for the extraction and visualization of functional topological relationships. The former used

Python to extract the functional center points and overlapping edges from vector-based residential floor plans to generate topology diagrams. The latter employed Python's OpenCV library to identify functional color blocks and their center points in scalar campus plans, then inferred function adjacency relationships by calculating the distances between the center points. That topology diagram generation method for scalar graphs provides an important technical path reference for this paper.

2.3 Generation Assessment

Recently, different types of deep learning neural network models have started to be introduced as assessment filters into various stages of the generation process to discriminate the scientific validity and controllability of the generated results. The use of discriminators to iteratively filter and optimize generated results has become a prevailing approach in current research. This combination of the deep learning models of generation and assessment has effectively dealt with the stochastic nature of the generated results, and significantly enhanced the controllability of the output results.

Hu et al. [14] used Convolutional Neural Networks in the Graph 2Plan model to supervise the input data, ensuring the generated floor plan layout should align with the given constraints and meets expectations. Fan et al. [15] proposed a multi-objective optimization framework for stadium floor plan layouts, combining a Genetic Algorithm with an RBF neural network prediction model to effectively predict the performance of design results. Yu D, Wan B, and Sheng Q [16] introduced a neural network model (CoAtNet, Convolution + Transformer) for predicting the spatial syntactic parameters of road structures with Stable Diffusion outcomes. This integration between the generated results and the actual urban spatial layout, effectively enhances the scientific accuracy of the generated outcomes. Zhuang W.M et al. [17] also analogised the topology similarity between the residential floor plan case studies and the current design. By means of similarity comparison, it provides suggestions for further design steps (adding, deleting, modifying room types or room morphologies). Hao Z highlighted that some deep learning models having 'critical thinking ability' show promising potential to enhance the outcomes of Stable Diffusion generation in 2024CDRF Conference.

The mechanism of mutual collaboration between the neural network evaluation model and the generative model represents an important step toward higher performance and more customized design solutions, providing a crucial reference for the technical approach in this paper.

3 Methodology

3.1 Stable Diffusion a Creative Model

The U-Net cross-attention mechanism and text encoder in Stable Diffusion enable the model to quickly and efficiently generate diverse, high-quality images under the guidance of conditions such as text, images, and semantic maps [18]. This approach is widely used in image generation, style transfer, image enhancement, etc.

However, Stable Diffusion can sometimes be overly creative [3]. Although LoRA and ControlNet can improve the controllability of generation results to some extent, stadium floor plans differ from other types of architecture. As they require more scientific consideration of functional zoning. Therefore, how can we further establish an automated evaluation and screening mechanism for Stable Diffusion results? Another deep learning model provides a potential solution.

3.2 Convolutional Neural Network a Critical Model

The Convolutional Neural Networks chosen in this paper is a deep learning model which excels at dealing with the classification and regression tasks of images [19]. By emulating the processing mechanism of the human visual cortex [20], these Networks are able to learn the feature patterns of images in the field of view, such as the specific combination of edges and colors and classify images accordingly. (Fig. 1.)

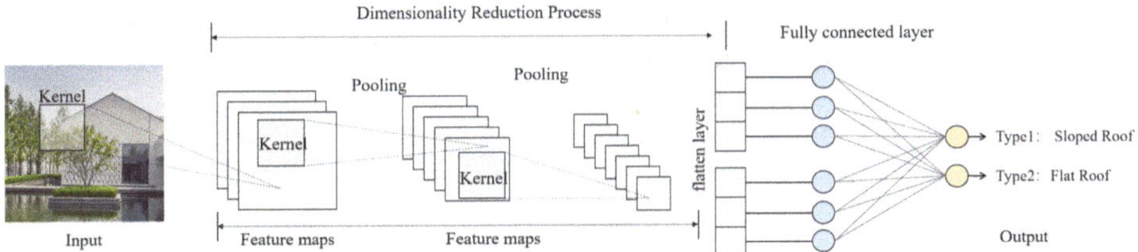

Fig. 1. Convolutional Neural Networks for image recognition (by Wang Zhenyu, Xu Chen)

Among many neural network models, Convolutional Neural Networks are capable of handling multi-dimensional images and are particularly suited for learning local spatial features in images, and are known for their efficiency in training and deployment. The capability of network architecture enables them to act as a discriminator for classifying and filtering the diverse generated results. In addition, they reduce the risk of model overfitting by globally sharing a reduced number of parameters, selectively connecting neurons, and locally sensing each parameter from the data to capture its relevance and learn feature patterns, thus enabling efficient use of limited data. Compared with machine learning and fully connected neural networks, the model significantly reduces the amount of data required for training and presents better training results with small-scale datasets, which effectively copes with the problems of small amount of data and difficult collection and processing of building plans.

Therefore, this paper attempts to integrate the Convolutional Neural Networks into the existing Stable Diffusion workflow, training it with a small-scale dataset to perform classification assessment and gradient screening, thereby regulating the generated results to better align with expectations.

The network architecture of the Convolutional Neural Networks consists of an input layer, network layers (including convolutional, pooling, and fully connected layers), and an output layer, with neurons between layers connected through weights and biases. The combination and number of convolutional or pooling layers within the network layers can be customized [21]. The ResNet50 architecture, widely used in image classification, addresses issues of network degradation and training stability in deep networks through residual learning and skip connections [22], achieving higher computational efficiency and improved generalization ability.

This paper selects ResNet50 as the base network architecture, with adjustments made to the last layers to better suit architectural design tasks. And this improved model is called the adapted- Convolutional Neural Networks.

4 Experiment Framework

In the experiment, the generative model Stable Diffusion and the assessment model CNN collaborated with each other. The specific workflow is as follows: (Fig. 2.)

1. Firstly, the Stable Diffusion model was controlled by LoRA and ControlNet to generate the block plan with functional color.
2. Then, the adapted- Convolutional Neural Networks model was used to evaluate the topological relationship of the generated results from the first step. Accordingly, the model's predicted 'good' and 'bad' labels and their associated scores were used as criteria for filtering the generated results.
3. Next, the better evaluation results were fed into the Stable Diffusion model to generate a detailed plan with room separations. And the optimal solution set was ultimately selected through human-machine collaboration.

Based on the existing Stable Diffusion step-by-step design workflow, the study introduced an adapted-Convolutional Neural Networks as a discriminator between the generation steps to assess the first stage outputs. This aimed to ensure that the functional topologies and area ratios of the previous stage's outputs are as close as possible to those of the outstanding case studies, and thus to improve the scientificity of the generation design.

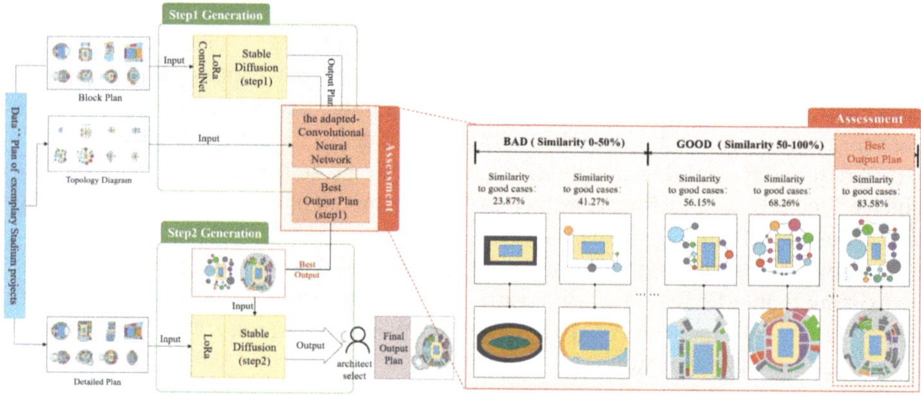

Fig. 2. Workflow (by Wang Zhenyu, Meng Yufan)

During the assessment process, two key aspects of the functional configuration in the first-step block plan generation were examined: 'the topological relationships' and 'area ratios'.

To highlight these two elements, a topology diagram containing those two kinds of information was drawn before being input into the adapted- Convolutional Neural Networks. And the principle of the assessment is to perform feature recognition by comparing those two aspects with the similarity of the outstanding case studies. The higher the feature similarity approaches 100%, the closer the functional area relationship of the generated result is to the outstanding case studies, meaning the layout is relatively scientific and reasonable. The adapted-Convolutional Neural Networks can output two results: 'feature similarity percentage' and 'feature labels' ('Good' or 'Bad'). The labeling rule was based on a 50% feature similarity threshold. In detail, the results higher than 50% are labeled as 'Good', and vice versa are labelled 'Bad'.

5 Data Collection and Processing

The stadium field-level plans dataset used in this research was established through the following three steps: Data Collection and Se-lection, Case Selection, Data processing and Data Enhancement.

5.1 Data Collection and Selection

The study manually collected the data of 120 completed stadiums as outstanding case studies. And the raw field-level plans data included detailed information of various functional rooms, flow lines, areas, etc. In order to ensure that the samples were representative and valid, the collected data were uniformly screened according to the following criteria. And 83 plan data samples were finally selected:

1. With reference to the 'Sports Architecture Design Standards JGJ31–2003'specification, the selected stadiums should be large and medium - sized, and the capacity of their auditoriums should be between 3,000 and 10,000 seats.
2. The selected cases should have a relatively complex functional zoning of the field - level plane, and at least 5 functions are involved.
3. The layouts of the selected plans are complete and centralized rather than discrete.
4. The data samples included diverse outer contour boundaries such as ellipses, rectangles, and sawtooth polygons.
5. Screen out low-quality data with ambiguous functional information that affects functional judgement.

5.2 Data Processing

The raw field-level plans contained information on structural equipment, redundant details, and non-uniform style proportions. Firstly, took the 400 m runway track as the benchmark to align the data proportions. In order to adapt to the learning characteristics of different deep learning models (Stable Diffusion and the adapted-Convolutional Neural Networks), the study carried out different forms of pre-processing on the data in the generation and assessment links respectively.

When processing the data of LoRA training progress, the focus was on retaining the information related to the contour edges and functional partitions of the stadiums. The standard of the International Olympic Committee (IOC) operational design was adopted for labeling the colors in the stadium plan and functional topology diagram to ensure consistency and universality. The color subdivision diagram was then processed into two types: 'Block Plan' and 'Detailed Plan'. The 'Block Plan' data included simple color block diagrams of the functional zones. While the 'Detailed Plan' data further included detailed information about the room divisions. (Table 1.)

Table 1. Functional color labeling rules (by Wang Zhenyu, Meng Yufan)

Color	Hex Color Code	Labeled Function	Labeled Result
	#A3D3DD	Athlete	
	#77B64B	Press	
	#C46494	Commercial	
	#DC4F4B	Security	**Block Plan** with functional color
	#EEAE6E	Venue_Operations	
	#7FB1D3	FOP	
	#AFDAED	Training	
	#D7D9DB	Transportation_space	**Detailed Plan** with room separations
	#F6E5AF	Spectator/Visitor Area	
	#977AB0	VIP Area	
	#72B798	Referees and .Committee Area	
	#939597	auxiliary room	Functional **topology diagram** with node size

The 'topological relationships' and 'area ratios' of function in the floor plan are two important aspects to assessing their reasonableness. Therefore, when dealing with the data of the assessment model, the plans were simplified to a functional topology diagram with node size, which not only showed the proximity of each function, but also reflected the area ratio of each function. The training data of the model contained the training set and testing set which were divided at a ratio of 7:3. The 'Good' label refers to the topology of reasonable generated plans, while the 'Bad' label refers to the topology of unreasonable generated plans.

To enhance the automation efficiency of this process, Python tools (e.g., OpenCV), Multimodal and large language model were used for extracting edge adjacency relationships and calculating node sizes of the topology diagram. The specific steps were as follows. (Fig. 3.)

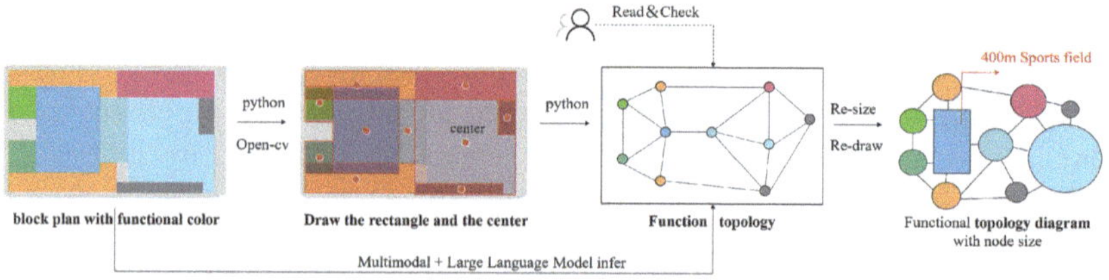

Fig. 3. Topology diagram generation process (by Wang Zhenyu)

5.3 Data Enhancement

The dataset consisted of three forms of data: block plans with functional color, detailed plans with room separations and functional topology diagrams with node size labelled 'Good' or 'Bad'. The 'Bad'-labelled data subset were the generated result with problems like missing key functions and incorrect adjacencies.

Then these data were respectively subjected to data enhancement operations such as mirror flipping and normalization in order to expand the total amount of data in the training set. It is worth mentioning that the special constraints of the stadium's function were also taken into account in the flip processing. For example, the VIP seats of the outdoor stadium should be on the west side, in which case the plane could not be flipped horizontally.

Finally, we obtained three sets of data: 214 block plans, 214 detailed plans, and 214 functional topology diagrams labelled 'Good' or 'Bad'. The first two sets of data were used to train the generative model (LoRA), while the third dataset was used to train the assessment model. (Table 2.).

Table 2. Extracts from datasets (by Wang Zhenyu, Meng Yufan, Xu Chen)

Data for Generation (Stable diffusion-LoRA)		**Data for Assessment** (the adapted- Convolutional Neural Networks)	
STEP1 block plan with functional color	**STEP2** detailed plan with room separations	**GOOD CASE** functional topology with node size	**BAD CASE** functional topology with node size

6 Training and Result

6.1 Training of the Adapted-Convolutional Neural Networks

Then loaded the data into the workflow. Since the training process of Stable Diffusion and LoRA is essentially the same as the existing stepwise generation workflow, this paper will not elaborate on this process. However, study will introduce the training procedure of the adapted-Convolutional Neural Networks, the assessment model.

In model architecture definition, the ResNet50, known for its strong image recognition capabilities and high generalization ability, was employed. It is worth mentioning that the final layer of the networks was modified using the SoftMax function, enabling the model to not only classify labels ('Good' or 'Bad'), but also output their feature probabilities ($\hat{y}c$). This modification enabled the model to adapt to the specific problem in this study. In the context, compared to $\hat{y}_2$, the value of $\hat{y}_1$ is more informative to be followed. In detail, $\hat{y}_1$ was the class probability for the 'Good' tag, which represents the degree of feature similarity (including function topology relationship and area ratio) between the input and outstanding case studies. And vice versa, $\hat{y}_2$ represents 'Bad'

The class probability ($\hat{y}$c) was computed using the SoftMax function (6.1–6.2) [23], which exponentiated the eigenvector (h) and then divided it by the sum of the exponentiated scores of all classes, yielding the probability for class 'c'. The critical parameter in this function was the eigenvector (h), which represented the quantized expression of key features extracted through spatial filtering by convolutional kernels, dimensionality reduction via pooling layers, and feature reorganization through fully connected layers during the processing of input topological graphs. The monotonic increase in the modulus of the eigenvector (|h|) revealed the cumulative effect of discriminative feature acquisition. Specifically in classification with the 'Good' label, larger values of the eigenvector (h_1) indicated that the convolutional neural networks extracted more critical features from the input topology diagram that correspond to outstanding case studies. As a result, the output class probability($\hat{y}_1$), approaches 1. Accordingly, in the context of block plan topological relationship assessment, the value of $\hat{y}_1$ was used as a key indicator for evaluating layout rationality. Its formal definition in this study is presented in Eq. (6.3).

Common formula:

$$\hat{y}_c = softmax(\mathrm{h}_c) = \frac{\exp(\mathrm{h}_c)}{\sum_{j=1}^{c} \exp(\mathrm{h}_c)} \tag{6.1}$$

$$\begin{cases} \hat{y}_1 = \frac{\exp(\mathrm{h}_1)}{\exp(\mathrm{h}_1)+\exp(\mathrm{h}_2)} \\ \hat{y}_2 = \frac{\exp(\mathrm{h}_2)}{\exp(\mathrm{h}_1)+\exp(\mathrm{h}_2)} \end{cases} \tag{6.2}$$

The Topological similarity score formula:

$$\text{Topological similarity to outstanding case} = \hat{y}_1 = \frac{\exp \mathit{of} \text{ eigenvector of the outstanding case}}{\mathit{Sum\ of\ all} \exp \mathit{ofeigenvector}} \tag{6.3}$$

The model employed the cross-entropy loss function (nn.CrossEntropyLoss) (6.4) for optimization and the SGD optimizer to update the parameters. During training, the learning rate was set to 0.001. After 40 epochs of training, the loss curve stabilized and the training loss stabilized between 0.08 and 0.18, while the test accuracy remained consistently above 0.99 approaching 1(6.5). The following figure records the loss and accuracy of the training and test sets for each training epoch (Fig. 4.). To ensure optimal performance, the model's best weights were saved corresponding to the lowest loss and highest accuracy during training. Upon loading the optimal weights, the accuracy of test set achieved up to 100%, which means the model can accurately label the given topology diagram as 'Good' or 'Bad' and display the specific percentage of feature similarity.

$$Loss(y, y\hat{}) = -\left[y \cdot \log(\hat{y}1) + (1-y) \cdot \log \hat{y}2)\right] \tag{6.4}$$

$$\text{Accuracy} = \frac{\text{number of correct predictions}}{\text{total sample size}} \tag{6.5}$$

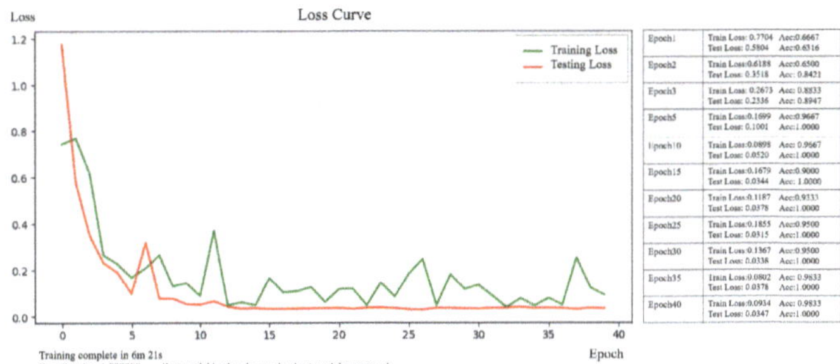

Epoch1	Train Loss: 0.7704 Acc:0.6667 Test Loss: 0.5804 Acc:0.6316
Epoch2	Train Loss:0.6188 Acc:0.6500 Test Loss: 0.3518 Acc: 0.8421
Epoch3	Train Loss: 0.2673 Acc: 0.8833 Test Loss: 0.2336 Acc: 0.8947
Epoch5	Train Loss:0.1699 Acc:0.9667 Test Loss: 0.1001 Acc:1.0000
Epoch10	Train Loss:0.0898 Acc: 0.9667 Test Loss: 0.0520 Acc:1.0000
Epoch15	Train Loss:0.1679 Acc:0.9000 Test Loss: 0.0344 Acc: 1.0000
Epoch20	Train Loss:0.1187 Acc:0.9333 Test Loss: 0.0378 Acc:1.0000
Epoch25	Train Loss:0.1855 Acc:0.9500 Test Loss: 0.0315 Acc:1.0000
Epoch30	Train Loss:0.1367 Acc:0.9500 Test Loss: 0.0338 Acc:1.0000
Epoch35	Train Loss:0.0802 Acc: 0.9833 Test Loss: 0.0378 Acc:1.0000
Epoch40	Train Loss:0.0934 Acc: 0.9833 Test Loss: 0.0347 Acc:1.0000

Fig. 4. Loss and accuracy during training process (by Wang Zhenyu)

6.2 Generation and Assessment Result

In this experiment, the rectangular contour of the outer edges of the site was preset. By controlling the contour through ControlNet and the first-step LoRA, and fine-tuning the LoRA weights, a batch of block plans with functional color was generated.

Then, the pre-trained adapted-Convolutional Neural Networks model was applied to mark and classify the large number of first-step generated results (block plans) into two categories ('Good' or 'Bad'). Accordingly, output the percentage of feature similarity between them and those of the outstanding case studies.

Next, the first-step generation result with the highest score (98.6% feature similarity) was selected as the input for the second-step generation. Combined with the second-step LoRA model, detailed plans with room separations were generated in the second stage. Finally, the ideal second-step generation results (detailed plans) were selected by experienced architects. (Fig. 5.)

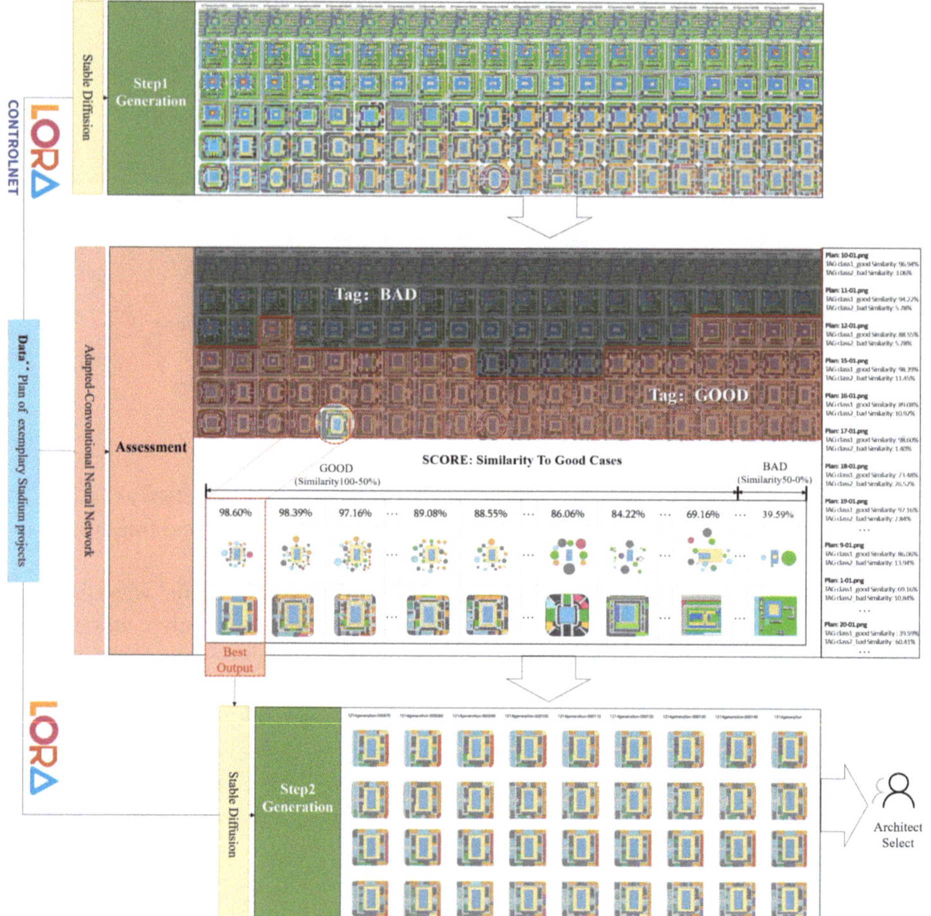

Fig. 5. Stable Diffusion Result and CNN Assessment (by Wang Zhenyu, Meng Yufan, Xu Chen)

7 Discussion and Conclusion

This paper presents a design method combining two deep learning models, Stable Diffusion and the adapted-Convolutional Neural Networks for stepwise generation and generation assessment. Specifically, the latter serves as an assessment model in the former's workflow. It evaluates the similarity of the generated results' functional topology and area ratio to those of outstanding case studies, thus screening the generated results of Stable Diffusion. This effectively improves the scientificity and rationality of the floor plan layout.

An interaction method combining 'generative' and 'assessment' artificial intelligence is proposed in this study by using the adapted-Convolutional Neural Networks to deal with the poor controllability of Stable Diffusion's generation results. It reduces the burden of manual design by improving the efficiency of human-computer collaboration in the designing process, and providing a reference for the intelligent generation of designs for complex public buildings such as stadiums.

However, there are still some limitations in the study. Firstly, the dataset needs further expansion. However, larger datasets also mean that more computing power is required for training, which is also limited by current computer equipment. Secondly, parts of the topology diagram transformation process still require some manual labor, thus future efforts will aim for full automation. Future research could explore additional deep learning models in architectural design, and continue to optimize human-computer interaction efficiency and the scientific basis of design decisions.

Acknowledgments. This work was funded by the National Key Research and Development Program of China (Project No. 2022YFC3801303).

References

1. Rombach, R., Blattmann, A., Lorenz, D., Esser, P., Ommer, B.: High-resolution image synthesis with latent diffusion models. In: Proceedings of the IEEE/CVF Conference on Computer Vision and Pattern Recognition (CVPR), pp. 10674–10685 (2022)
2. Liu, Y., Li, H., Deng, Q., others: Diffusion probabilistic model assisted 3D form finding and design latent space exploration: a case study for Taihu stone spacial transformation. In: Proceedings of the International Conference on Computational Design and Robotic Fabrication, pp. 11–23. Springer Nature Singapore, Singapore (2023)
3. Du, C., Li, Y., Qiu, Z. and others (2024) 'Stable diffusion is unstable', Adv. Neural Inf. Proces. Syst., 36.
4. Li, X., Benjamin, J. and Zhang, X. (2024) 'From text to blueprint: Leveraging text-to-image tools for floor plan creation', *arXiv preprint* https://arxiv.org/abs/2405.17236.
5. Qiu, Z., Liu, J., Xia, Y. et al. (2025) 'Text semantics to flexible design: A residential layout generation method based on Stable Diffusion model', *arXiv preprint* https://arxiv.org/abs/2501.09279.
6. Shabani, M.A., Hosseini, S., Furukawa, Y.: Housediffusion: Vector floorplan generation via a diffusion model with discrete and continuous denoising. In: Proceedings of the IEEE/CVF Conference on Computer Vision and Pattern Recognition, pp. 5466–5475 (2023)
7. Kuhn, E. (2023) 'Adapting housediffusion for conditional floor plan generation on modified Swiss dwellings dataset', *arXiv preprint* https://arxiv.org/abs/2312.03938.
8. Wang, H., He, X., Yan, Z. and others (2024) 'Research on pathology information management of educational architectural heritage based on digital technology: The case of James Jackson Gymnasium', Buildings, 14(4), 1048.
9. Krishnan, S., Yazdanbakhsh, A., Prakash, S., et al.: Archgym: An open-source gymnasium for machine learning assisted architecture design. In: Proceedings of the 50th Annual International Symposium on Computer Architecture, pp. 1–16 (2023)
10. Fan, Z., Liu, M., Tang, S., et al.: Multi-objective optimization for gymnasium layout in early design stage: Based on genetic algorithm and neural network. Build. Environ. **258**, 111577 (2024)
11. Maeng, H., Hyun, K.H.: Data-driven analysis of spatial patterns through large-scale datasets of building floor plan. In: Proceedings of the 26th International Conference of the Association for Computer-Aided Architectural Design Research in Asia, vol. 1. CAADRIA (2021)

12. Wang, J., Li, H., Lin, Y., Zhang, L., Wu, Y., Gao, B.: Incremental control for architectural facade generation: Enhancing design precision through fine-tuning of diffusion models. In: The 30th International Conference of the Association for Computer-Aided Architectural Design Research in Asia (CAADRIA), Tokyo (2025)
13. Liu, Y., Zhang, Z., Hu, K., et al.: Graph constrained multiple schemes generation for campus layout. In: The International Conference on Computational Design and Robotic Fabrication, pp. 125–138. Springer Nature Singapore, Singapore (2023)
14. Hu, R., Huang, Z., Tang, Y., et al.: Graph 2plan: Learning floorplan generation from layout graphs. ACM Trans. Graph. **39**(4), 118:1–118:14 (2020)
15. Yu, D., Wan, B., Sheng, Q.: Automated generation of urban spatial structures based on Stable Diffusion and CoAtNet models. Buildings. **14**(12), 3720 (2024)
16. Su, P., Lu, W., Chen, J. and others (2024) 'Floor plan graph learning for generative design of residential buildings: A discrete denoising diffusion model', Build. Res. Inf., 52(6), pp. 627–643.
17. Guo, S., Zhuang, W.: Analysis, configuration, and optimisation: Prospects of graph theory applied in architectural programming. World Archit. **7**, 76–86 (2024). https://doi.org/10.3969/j.issn.1002-4832.2024.07.010
18. Rombach, R., Blattmann, A., Lorenz, D., Esser, P., Ommer, B.: High-resolution image synthesis with latent diffusion models. In: Proceedings of the IEEE/CVF Conference on Computer Vision and Pattern Recognition (CVPR), pp. 10684–10695 (2022)
19. Haruna, Y., Qin, S., Chukkol, A.H.A., et al.: Exploring the synergies of hybrid convolutional neural network and Vision Transformer architectures for computer vision: A survey. Eng. Appl. Artif. Intell. **144**, 110057 (2025)
20. Mohamed, E., Sirlantzis, K., Howells, G.: A review of visualisation-as-explanation techniques for convolutional neural networks and their evaluation. Displays. **73**, 102239 (2022)
21. Kattenborn, T., Leitloff, J., Schiefer, F., et al.: Review on convolutional neural networks (CNN) in vegetation remote sensing. ISPRS J. Photogramm. Remote Sens. **173**, 24–49 (2021)
22. He, K., Zhang, X., Ren, S., et al.: Deep residual learning for image recognition. In: 2016 (ed.) Proceedings of the IEEE Conference on Computer Vision and Pattern Recognition, pp. 770–778
23. Bengio, Y., Goodfellow, I., Courville, A.: Deep learning, pp. 110–176. MIT Press, Cambridge, MA (2017)

Enhancing Architectural Image Consistency: A Study of Image-to-Image Workflows by Generative AI Models

Yang Yu, Yuqian Liu, and Ding Wen Bao(✉)

RMIT University, Melbourne 3000, VIC, Australia
S4208097@student.rmit.edu.au, nic.bao@rmit.edu.au

Abstract. This study explores the integration of generative AI into architectural design workflows by leveraging the FLUX.1 model and Low-Rank Adaptation (LoRA). It proposes scalable image-to-image (I2I) pipelines that combine semantic image recognition, targeted fine-tuning, and stylistic control using FLUX.1 Redux. Through quantitative evaluation using MSE, SSIM, and PSNR metrics, the approach demonstrates significant improvements in image quality, structural fidelity, and visual coherence. The results highlight the potential of open-source Generative Artificial Intelligence (GAI) technologies to streamline design processes, reduce manual effort, and foster innovative, collaborative design paradigms in architecture.

Keywords: Generative Artificial Intelligence · Image Consistency · Image Evaluation · Diffusion Model

1 Introduction

In architectural workflows, visualizations are central to both iterative development and communication. These images are not static representations; they enable dynamic feedback and refinement across the design process. In leading firms such as Aedas, the alignment between conceptual renderings and built outcomes is critical (Fig. 1), highlighting the importance of preserving visual similarity throughout the design workflow. This study explores how such similarity can be improved through an AI-based image-to-image (I2I) workflow using the FLUX.1 model.

Generative Artificial Intelligence (GAI), including models such as Stable Diffusion [19], MidJourney [9], and DALL·E 2 [1], has transformed visual synthesis across creative domains. Previous works, such as the "Space Opera Theater," highlight the limitations of text-to-image (T2I) models in delivering controlled and repeatable architectural outputs.

Text prompts are inherently ambiguous and prone to interpretation, often resulting in inconsistent or imprecise outputs [18]. In contrast, I2I methods, where image generation is guided by reference visuals, offer more reliability and control, particularly in architectural design where spatial and stylistic precision is critical. Despite their potential, I2I workflows remain underdeveloped, and evaluation practices rely heavily on subjective visual judgment [2].

Y. Liu et al. (Eds.): CDRF 2025, *Transindividual Intelligence*, pp. 27–38, 2026.
https://doi.org/10.1007/978-981-92-0615-5_3

Fig. 1. The comparison of the rendering and reality

To address these issues, objective image similarity metrics such as Mean Squared Error (MSE), Peak Signal-to-Noise Ratio (PSNR), Structural Similarity Index (SSIM), and Histogram Similarity are increasingly employed [10, 20]. These provide quantifiable measures of structural and perceptual fidelity and support more rigorous evaluation of GAI outputs. Accordingly, this study addresses two primary research objectives: (1) To simplify and optimize image-to-image (I2I) workflows tailored for architectural design, ensuring visual coherence with reference inputs; (2) To establish a quantitative evaluation framework using perceptual and statistical metrics to objectively assess the quality and consistency of generated images. The research focuses on conceptual renderings in urban and architectural design, using open-source models, particularly FLUX.1, for accessibility and reproducibility. Closed-source commercial tools are excluded. Key contributions of this study are below:

1. Proposing three open-source I2I workflows;
2. a) Extracting prompts from reference images via Florence-2; b) Fine-tuning generation using LoRA for architectural specificity (e.g., massing, façade, site integration); c) Enabling stylistic variation with structural fidelity using FLUX.1 Redux model.
3. Improving image similarity by over 80%, while enhancing design flexibility and reducing manual input.
4. Introducing a hybrid evaluation method combining visual similarity metrics with practical design criteria.
5. Offering a scalable and practical GAI design workflow applicable to large, stylistically consistent projects.

This research contributes to the foundation for more reliable, automated, and evaluable GAI-assisted design workflows. It also suggests that I2I workflows, compared to text-based approaches, could become central to future GAI model improvements through user-generated performance data.

2 Background

Diffusion-based generative models have significantly advanced the field of image synthesis. Tools such as Stable Diffusion, MidJourney, and DALL·E 2 have gained widespread popularity due to their accessibility and high-quality outputs, facilitating broad adoption in creative and design practices. Among these, ComfyUI [11] stands out as a modular, visual platform for building and customizing diffusion workflows. Its open-source architecture allows seamless integration with various tools and models, enabling both technical experimentation and creative exploration.

ComfyUI supports the incorporation of multimodal models such as Florence 2 [6], which can interpret images and generate captions through cross-modal learning. Other integrated tools focus on increasing output controllability via model fine-tuning or parameter manipulation. Low-Rank Adaptation (LoRA) [12] is one such technique, designed to optimize a small number of trainable parameters while preserving the integrity of pretrained weights.

ControlNet [14] offers another strategy for guided generation by injecting control signals such as edge maps or depth cues into the diffusion process. While it allows for precise conditioning, its creators have acknowledged trade-offs in image fidelity and detectability. For these reasons, ControlNet was not employed in this study.

Despite the capabilities of large pre-trained models, they often lack domain-specific knowledge, resulting in outputs that prioritize general stylistic patterns over fine-grained semantic or contextual accuracy. This gap highlights the importance of fine-tuning for architecture-specific tasks. Traditional approaches such as Adapter Tuning [15] and Prefix Tuning face limitations including inference latency and unstable performance at scale. Prior research [4] further suggests that over-parameterized models often operate effectively within low-dimensional subspaces—an insight that underpins the design of LoRA.

LoRA addresses these challenges by introducing low-rank updates that adapt pre-trained models efficiently, making it particularly suitable for applications requiring both generalization and specificity. In architectural contexts, LoRA has enabled personalized generative outputs—for example, producing interior renderings tailored to user personality traits such as MBTI profiles [13]. When coupled with curated datasets and evaluation modules, LoRA-based workflows can balance logic and aesthetics, enhancing both the creativity and practicality of generative design.

Meanwhile, multimodal pretrained models like Florence 2 have helped bridge the gap between AI research and real-world applications. Leveraging prompt-based interfaces and cross-modal learning, Florence 2 achieves state-of-the-art performance in tasks such as image classification, object detection, and vision-language understanding, further expanding the potential for human-AI collaboration in design workflows.

3 Methodology

High-quality image generation in GAI workflows critically depending on model selection. While numerous advanced models support high-fidelity outputs, including I2I generation, this study focuses on FLUX.1, an open-source diffusion model built on the UNet architecture [8]. Notably, FLUX.1 demonstrates exceptional performance in both speed and quality benchmarks, outperforming several commercial closed-source alternatives [2], making it particularly well-suited to the practical demands of architectural design.

Although FLUX.1 is primarily optimized for Text-to-Image generation, it does not natively support I2I workflows. However, due to its architectural similarity to Stable

Diffusion 3 [18] and its compatibility with CLIP-based models and ComfyUI nodes [17], an I2I pipeline can be adapted by referencing the design logic of SD3. This study builds upon these connections to implement an effective I2I workflow within the FLUX.1 environment.

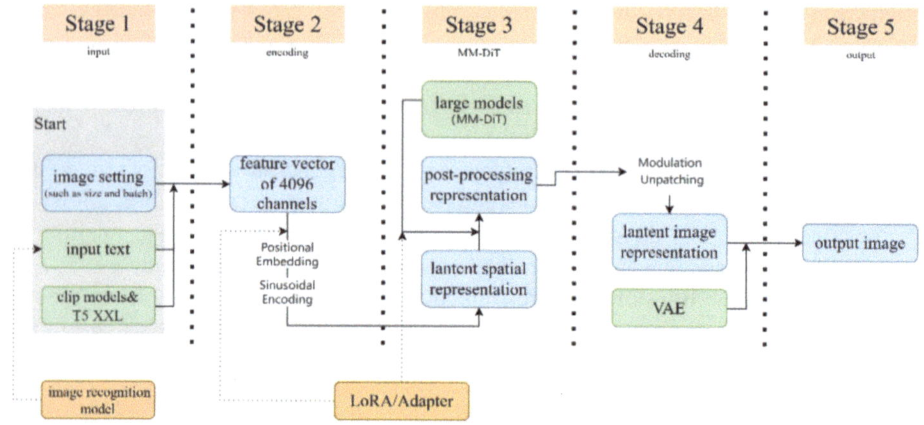

Fig. 2. Working flow of image generation by FLUX.1

The generation process in FLUX.1 comprises the following stages (Fig. 2):

1. Text Input: Input processed via CLIP (e.g., CLIP-G/14, CLIP-L/14) and T5-XXL; image resolution and batch size are predefined.
2. Encoding: Text features are embedded into 4096-dimensional vectors, with added timestep and positional encodings.
3. Diffusion Generation: Feature vectors pass through MM-DiT (Multi-Modal Diffusion Transformer) to produce latent space representations.
4. Decoding: Latent representations are decoded via VAE with patch processing.
5. Image Output: Final high-resolution image is produced.

To extend FLUX.1 for high-fidelity architectural I2I generation, this study implemented three complementary workflows (Fig. 2):

1. Enhancing prompt accuracy via image recognition models.
2. Applying LoRA-based fine-tuning to guide encoding and post-processing.
3. Introducing adapters to regulate the generation process.

Despite FLUX.1's strong overall performance, initial testing revealed that its output quality for aerial architectural renderings was relatively weak. To address this, over 80 high-resolution aerial images ($\geq$2048 $\times$ 1280 pixels) were collected for further experimentation.

3.1 Workflow 1: Image Generation with Image Recognition

To generate images closely aligned with reference images, this workflow used the Florence-2 model to extract detailed captions from input images. These captions were then used as prompts for FLUX.1. Within COMFYUI, Florence-2 analyzes uploaded images and generates descriptive text capturing key attributes such as style, color distribution, and layout (Fig. 3). Prior studies and projects like Space Opera Theater have shown that detailed, structured prompts improve generation quality, some using up to 624 prompts per image. This experiment compared prompts of varying detail levels, confirming that more detailed captions led to images more consistent with the originals. Therefore, all subsequent generations employed Florence-2-generated prompts.

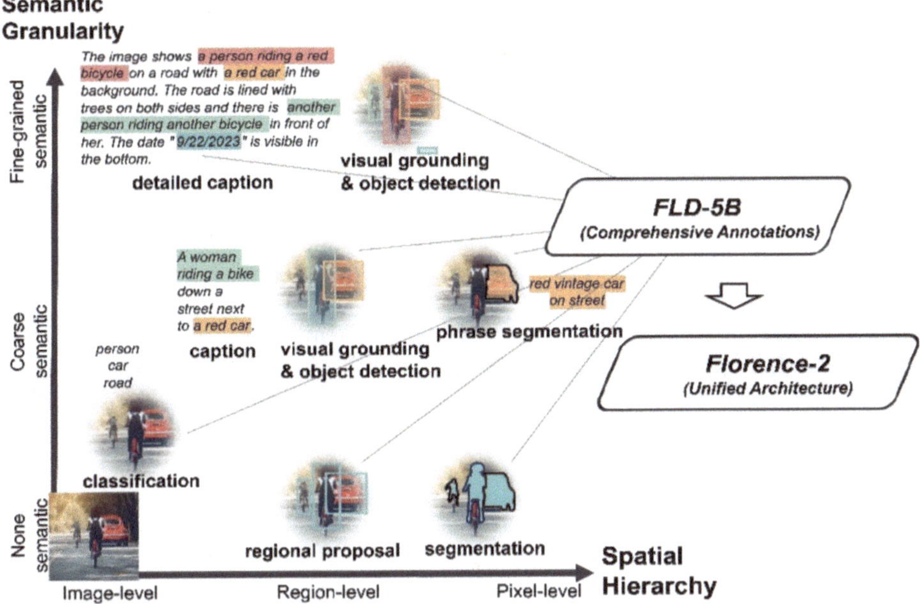

Fig. 3. The development and the function of Florence-2

3.2 Workflow 2: Image Generation with LoRA

Large models like FLUX.1 are typically generalized and lack sufficient specificity for certain tasks. LoRA [12] addresses this by fine-tuning select parameters during the denoising process, allowing more tailored image generation.

However, FLUX.1's large size and unique data structure present technical challenges for fine-tuning. Even its FP8 version requires ~11GB GPU memory; FP16 needs ~22GB, limiting accessibility and dataset size. In practice, LoRA training on FLUX.1 proved minimally effective, likely due to model distillation.

To address this, the study adopted a "de-distilled" version—Flux-dev2pro [5]—and trained LoRA models using an 80-image aerial dataset, annotated by Florence-2. Training was conducted on an NVIDIA 4090 GPU over 75,000 steps, resulting in 10 LoRA models.

Model performance was compared to varying LoRA, and the best-performing model, balancing visual stability and image quality, was selected for integration. Output results from this workflow are shown in Fig. 4.

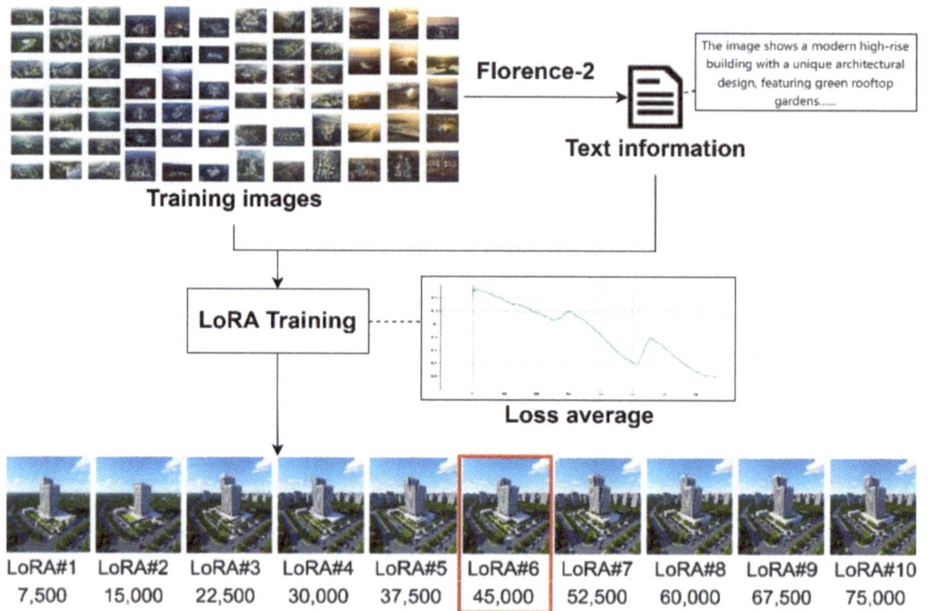

Fig. 4. The development and the function of Florence-2

3.3 Workflow 3: Image Generation with FLUX.1 Redux

FLUX.1 Redux is an official adapter for FLUX.1, designed specifically for image variation generation [7]. It modifies input images via small prompt-based changes, making it ideal for creative iterations.

This method supports stylization and refinement, as demonstrated in previous workflows where Florence-2 generated descriptive prompts to guide Redux-based transformations. Redux proved especially useful for design variation and post-processing optimization.

3.4 Workflow 4: Fine-Tuning the Large Model

Although the above workflows produced high-quality images with notable structural similarity to reference images, aesthetic alignment remained limited. Generated images often resembled reference images in a mechanical way, lacking the nuanced judgment of a human architect. This likely stems from insufficient data precision and scale in the workflows compared to the original FLUX.1 model. To address this, a fine-tuned large model was employed. The model was tested using the three workflows described above, with results significantly improved.

Trained by large architectural datasets, particularly aerial imagery, this model generated outputs with enhanced pixel quality and logical coherence. While compositional alignment was not perfect, the visual and aesthetic quality of the generated images was significantly improved.

4 Result

This study generated images based on 20 reference images from the LoRA training set using six methods, a combination of three workflows and two large models. The detailed results are presented in Fig. 5. For comparison, a manually described text-to-image generation approach was also included in Fig. 6.

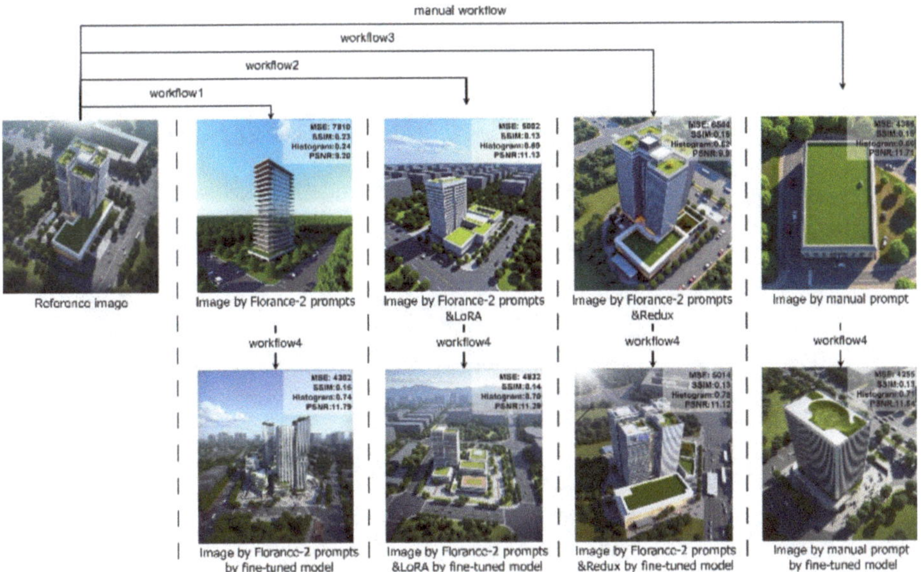

Fig. 5. The result examples of workflows

Image quality and similarity were quantitatively evaluated using OpenCV-supported metrics, including Mean Squared Error (MSE), Peak Signal-to-Noise Ratio (PSNR), Structural Similarity Index (SSIM), and Histogram Similarity.

- MSE calculates the average squared difference in pixel values between two images; lower values indicate higher similarity.
- PSNR, derived from MSE, measures reconstruction quality; higher values suggest better image fidelity.
- SSIM evaluates perceptual similarity by comparing luminance, contrast, and structure. It ranges from 0 to 1, with 1 indicating identical images.

Fig. 6. The results for reference images and output images

- Histogram Similarity reflects the alignment in color and brightness distribution, also ranging from 0 to 1.

The PSNR chart (Fig. 7) shows significant variation across images, suggesting that output quality differs considerably by image. Fine-tuning methods achieved higher PSNR peaks on certain samples, though results were less consistent overall. Despite this, PSNR values remained comparable for most images, with clear divergence only in specific cases.

According to the MSE plot (Fig. 8), fine-tuning methods outperformed baseline approaches, yielding lower and more stable MSE values. Redux and Fine-tuned Redux

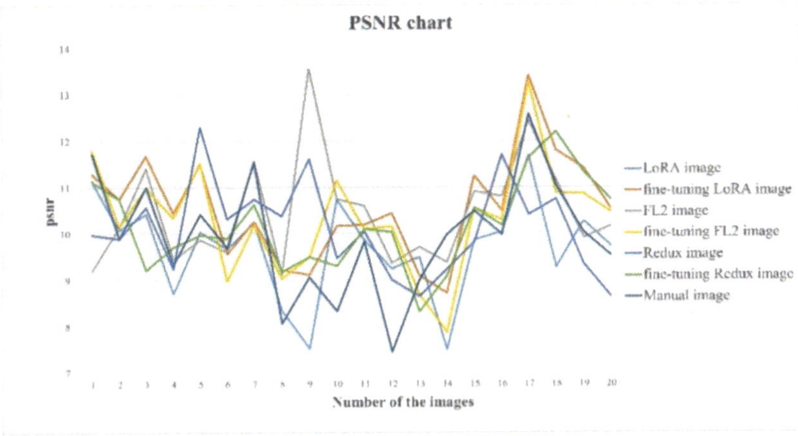

Fig. 7. The PSNR chart

workflows exhibited the best performance, showing minimal error and high consistency. In contrast, LoRA-based results displayed the highest variability, with MSE fluctuations closely tied to image content.

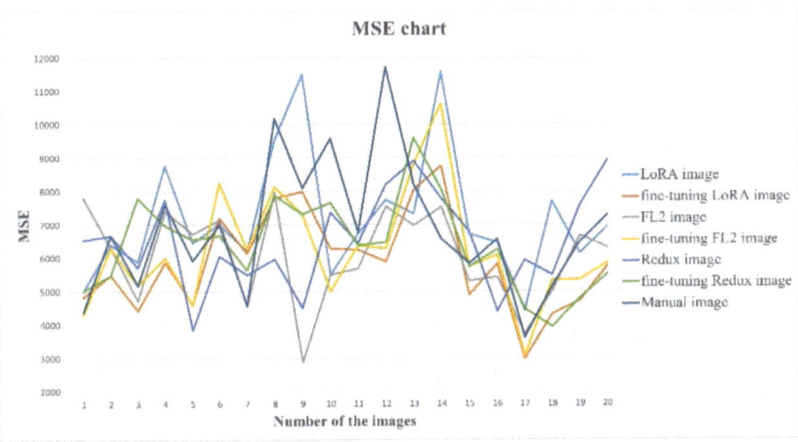

Fig. 8. The MSE chart

The SSIM vs. Histogram Similarity scatter plot reveals a positive correlation between the two metrics (Fig. 9). Fine-tuning consistently enhanced similarity, particularly in Redux and Fine-tuned Redux images, which clustered in the upper-right quadrant of the plot. LoRA and FL2-based images, especially those without fine-tuning, showed lower similarity and more dispersed distributions.

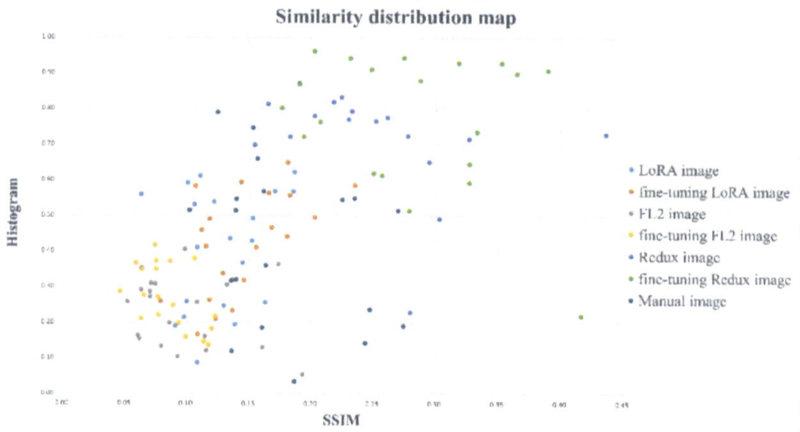

Fig. 9. The similarity distribution map

5 Conclusion and Discussion

The experimental results highlight diverse perspectives on the application of LoRA models and introduce an innovative approach to fine-tuning large models using LoRA. By enabling reference data and images to directly influence the image generation process, this strategy offers valuable insights for advancing generative AI workflows. Beyond enhancing visual consistency, the proposed workflow supports the development of structured data assets, which can facilitate iterative model refinement and systematic data collection. These assets lay the groundwork for improving model performance and delivering more tailored, high-quality services in practical design applications.

This study also demonstrates that combining LoRA fine-tuning with image recognition models can effectively address key limitations of existing GAI-based design workflows. The result is a more streamlined creative process with reduced manual intervention, while promoting greater design diversity and innovation in architecture.

Overall, this research presents a scalable and robust image-to-image (I2I) workflow tailored for architectural applications. It contributes to transforming conventional design practices through generative AI, fostering a more inclusive, creative, and sustainable design paradigm. Moreover, it holds potential to bridge the urban–rural divide by enabling high-quality design outcomes even in low-resource or low-budget contexts.

Acknowledgments. The authors gratefully acknowledge the financial support provided by the Australian Research Council under grant number IH220100016.

References

1. Aditya Ramesh, P. D., Alex Nichol, Casey Chu, Mark Chen. (2022). Hierarchical Text-Conditional Image Generation with CLIP Latents. arxiv.org. Retrieved 04 13., from https://arxiv.org/abs/2204.06125

2. AI, R. (2024). Recraft: Quality, Generation Time & Price Analysis. https://artificialanalysis.ai/text-to-image/model-family/recraft
3. Alves, N.: A New Image-Based Modelling System to Support Architectural Redesign Activities, vol. 1, pp. 322–333 (2022)
4. Armen Aghajanyan, L. Z., Sonal Gupta. (2020). Intrinsic Dimensionality Explains the Effectiveness of Language Model Fine-Tuning. arXiv. Accessed 22 Dec, http://arxiv.org/abs/2012.13255
5. ashen0209. (2024). Flux-Dev2Pro. https://huggingface.co/ashen0209/Flux-Dev2Pro
6. Bin Xiao, H. W. et al.. (2023). Florence-2: Advancing a Unified Representation for a Variety of Vision Tasks. arxiv.org. Accessed 10 Nov, https://arxiv.org/abs/2311.06242
7. Black-Forest-Labs. (2024). FLUX.1-Redux-dev. https://huggingface.co/black-forest-labs/FLUX.1-Redux-dev
8. Black Forest Labs. (2024a). Announcing Black Forest Labs. https://blackforestlabs.ai/announcing-black-forest-labs/
9. Borji, A. (2022). Generated Faces in the Wild: Quantitative Comparison of Stable Diffusion, Midjourney and DALL-E 2. arxiv.org. Accessed 2 Oct, https://arxiv.org/abs/2210.00586
10. Chun yuan Li, H. Farkhoor, Rosanne Liu, Jason Yosinski. (2018). Measuring the Intrinsic Dimension of Objective Landscapes. arXiv. Accessed 24 Apr, http://arxiv.org/abs/1804.08838
11. Comfyanonymous. (2024). Comfy UI. https://github.com/comfyanonymous/ComfyUI?tab=readme-ov-file
12. Edward J., et al., (2021). LORA: Low-Rank Adaptation of Large Language Models. arxiv.org. Accessed 17 Jan, https://arxiv.org/abs/2106.09685
13. Huang, Z.: Adaptive interior design method for different MBTI personality types based on generative artificial intelligence. Archit. Intell., 23–38 (2024)
14. Lvmin Zhang, A. R., Maneesh Agrawala. (2023). Adding Conditional Control to Text-to-Image Diffusion Models. https://arxiv.org/abs/2302.05543
15. Neil Houlsby, A. G. et al., (2019). Parameter-Efficient Transfer Learning for NLP. arXiv. Accessed 13 Jun, http://arxiv.org/abs/1902.00751
16. Neto, P. (2023, 11). About education in architecture: towards an integrative pedagogy in the teaching of communication strategies for architectural design and photography. Towards a New European Bauhaus—Challenges in Design Education & Research, 102–110.
17. Nuñez, M. (2024). Stable Diffusion creators launch Black Forest Labs, Secure $31M for FLUX.1 AI Image Generator. venturebeat.com. Accessed 1 Aug
18. Patrick Esser, S. K., et al., (2024). Scaling Rectified Flow Transformers for High-Resolution Image Synthesis. arxiv.org. Accessed % May, https://arxiv.org/abs/2403.03206
19. Robin Rombach, A. B., Dominik Lorenz, Patrick Esser, Björn Ommer. (2021). High-Resolution Image Synthesis with Latent Diffusion Models. arxiv.org. Accessed 20 Dec, https://arxiv.org/abs/2112.10752
20. Shreyam Gupta, P. Agrawal, Priyam Gupta, 2025. MAUCell: An Adaptive Multi-Attention Framework for Video Frame Prediction. http://arxiv.org/abs/2501.16997

cc
BY NC ND

EditPanorama: An Innovative Approach to Indoor Space Renovation Design Interaction and Presentation Through Diffusion Models

Yang Liu, Xuanming Zhang, Jiawen Yang, and Wen Gao(✉)

College of architecture and urban planning, Beijing University of Technology, Beijing, China
gaow22@bjut.edu.cn

Abstract. With the accelerating pace of urban renewal in China big cities, the demand for renovating existing residential indoor space has surged, while traditional design workflows face bottlenecks in creative expression and client communication efficiency. Generative artificial intelligence (GenAI) technologies, such as Stable Diffusion models, offer novel approaches for design visualization through rapid text-to-image responses. However, their limitations in architectural 3D spatial coherence and real-time interactivity remain unresolved. This paper proposes EditPanorama, an interactive architectural interior design framework that integrates Panoramic Neural Radiance Fields (PERF) with Stable Diffusion model, achieves rapid and high quality generation of spatial renovation proposals. Experimental results demonstrate that EditPanorama preserves architectural structural constraints while enabling real-time adjustment of design elements through natural language instructions and facilitating multi-scheme comparisons within seconds. In case studies of aging residential renovation projects, the framework significantly shortened design feedback cycles and improved client satisfaction compared to traditional workflows.

Keywords: Stable Diffusion · Panoramic Neural Radiance Fields (PERF) · Indoor space renovation · Interactive design · Real-time visualization

1 Introduction

As the pace of urban renewal accelerates in China big cities, the demand for indoor space renovation is increasingly on the rise. In the early design stages of indoor space renovation, designers face the challenge of efficiently conveying their creative concepts to clients. Traditional methods such as hand-drawn renderings and verbal descriptions often fail to fully showcase the intricate details of the design [1]. In recent years, while indoor modeling and diffusion-based rendering methods have been able to express architectural details to some extent, they still suffer from issues such as the need for extensive professional knowledge, lengthy processing times, and less than satisfactory outcomes [2].

With the advent of generative artificial intelligence (GenAI) technologies such as Midjourney, OpenAI's DALL-E, and Stability AI's Stable Diffusion, which can generate images from simple text prompts and image inputs, these tools have begun to

Y. Liu et al. (Eds.): CDRF 2025, *Transindividual Intelligence*, pp. 39–50, 2026.
https://doi.org/10.1007/978-981-92-0615-5_4

be applied in the conceptual design phase of architecture and interior design, including building plans [3–5], 3D massing [6–8], performance analysis [9–12], and human perception [13–15]. Most research use generative adversarial networks (GAN), which produce building facades [16] and layouts [17], to apply machine learning to generative design. Research suggests that AI-generated content has the potential to foster designers' inspiration, creativity, and skill development [18–20]. However, current applications of AI in architecture predominantly focus on building façade design [21] while in interior design, AI-generated results are largely limited to single-view images. This fragmented representation fails to convey the holistic spatial experience of environments defined by multiple walls, ceilings, and floors, ultimately hindering users' ability to fully perceive and engage with the three-dimensional space. This paper proposes an innovative framework for interactive interior design that integrates stable diffusion models with Panoramic Neural Radiance Field (PERF) to streamline design visualization and client communication. We have constructed a comprehensive workflow that allows non-professionals to obtain updated 360° interior previews simply by inputting panoramic images and prompt words.

2 Methodology

We propose a method for generating an editable 3D interior design scene. Our technology combines a panoramic image in-painting model with Panoramic Neural Radiance Field(PERF) techniques, aiming to assist users in completing interior design and scene representation more quickly. Specifically, users only need to capture an indoor panoramic image with their smartphone and use it as main input. Subsequently, users can repaint parts of the panoramic image or adjust the overall style based on personal preferences through simple prompts and partial masking features, allowing for flexible transformation of the imagery. Furthermore, we utilize PERF technology to rapidly construct a view-consistent and completely enclosed indoor 3D scene mesh, providing clients with a stereoscopic and authentic visual experience. (Fig. 1.)

2.1 User Input

2.1.1 Panorama

While current AI-generated images predominantly produce 2D representations or isolated perspective views, these formats struggle to comprehensively convey spatial relationships in interior environments encompassing ceilings, floors, and interconnected walls. In contrast, 360° panoramic imagery serves as an effective visualization tool that holistically captures indoor spatial configurations. Moreover, advancements in mobile imaging technologies (e.g., smartphone-camera systems with gimbals or dedicated panoramic tools) have significantly simplified panoramic data acquisition, creating new opportunities for practical applications. Our proposed workflow enables users to a) capture existing spaces through panoramic photography, b) generate updated interior panoramas via text prompts, and c) leverage VR/3D reconstruction technologies to create immersive spatial experiences based on the AI-enhanced panoramas.

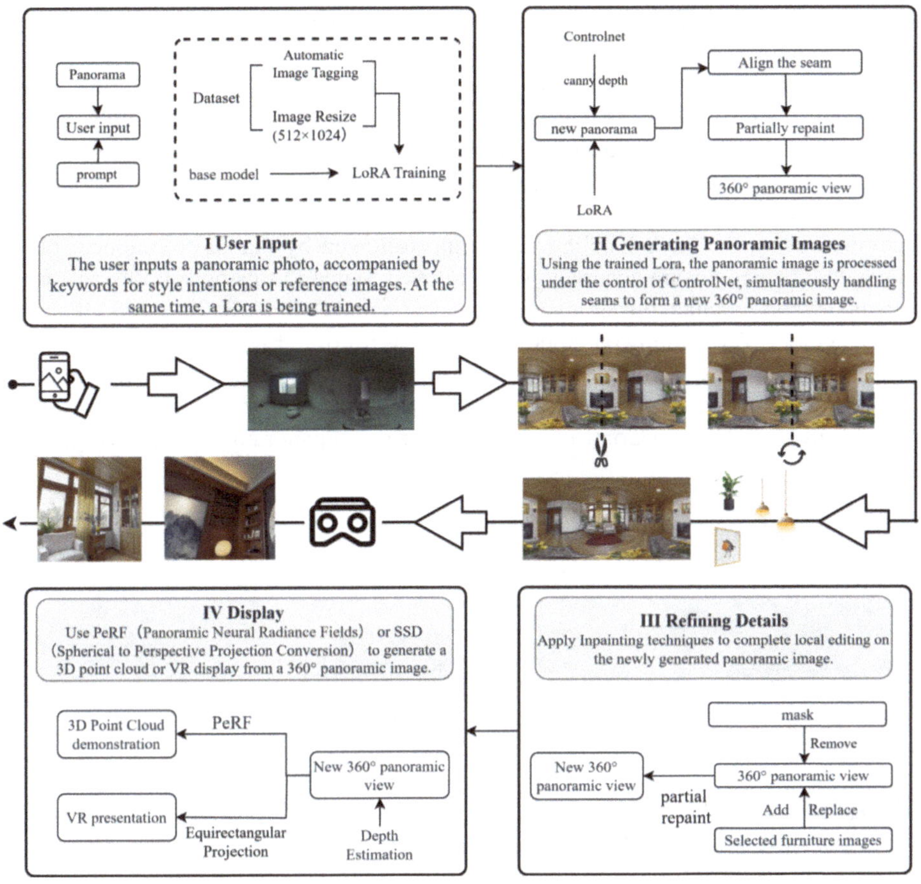

Fig. 1. The workflow of the EditPanorama

2.1.2 LoRA

To address the critical challenges of excessive computational resource consumption and prolonged training cycles in full-parameter fine-tuning of large language models, Microsoft Research introduced Low-Rank Adaptation (LoRA) technology [22], establishing a new paradigm for efficient parameter adaptation. Compared to the substantial computational overhead required for full-parameter fine-tuning of traditional Stable Diffusion models, and the high GPU VRAM demands of lightweight alternatives like Textual Inversion and Dreambooth, LoRA employs a low-rank matrix decomposition strategy. This approach injects trainable bypass rank-decomposition matrices into Transformer modules rather than modifying the original pre-trained weights directly. Through this design, only 0.1%–1% of total parameters require updating during training while keeping the backbone model parameters frozen, thereby reducing VRAM consumption to one-third of conventional fine-tuning and accelerating training speeds by 2–5 times. Experimental results demonstrate that LoRA not only achieves generation quality comparable to full-parameter fine-tuning in Stable Diffusion applications such as style transfer and theme customization, but also exhibits superior parameter efficiency and hardware compatibility.

In this study, we extensively collected indoor panoramic images representing diverse spaces and styles to construct a dataset. From this dataset, we initially randomly selected 100 images (512 × 1024 resolution) as training samples for LoRA fine-tuning. The experimental results demonstrate that this limited sample size achieves excellent performance while significantly reducing computational requirements. The annotation process was performed using the Joy2 labeling algorithm, followed by manual refinement of the generated labels to ensure accuracy. We employed Stable Diffusion F.1_dev-fp8 as the base model and conducted training adjustments on an NVIDIA RTX 4070S GPU with 16 GB memory. The training parameters were set as follows: Epoch = 5, Batch Size = 1, Repeat = 200, with a learning rate of 0.00001. The complete training process required over 2 h, resulting in a final model size of 584 MB. Subsequently, we integrated the trained LoRA model with ControlNet to generate novel panoramic images. (Fig. 2.)

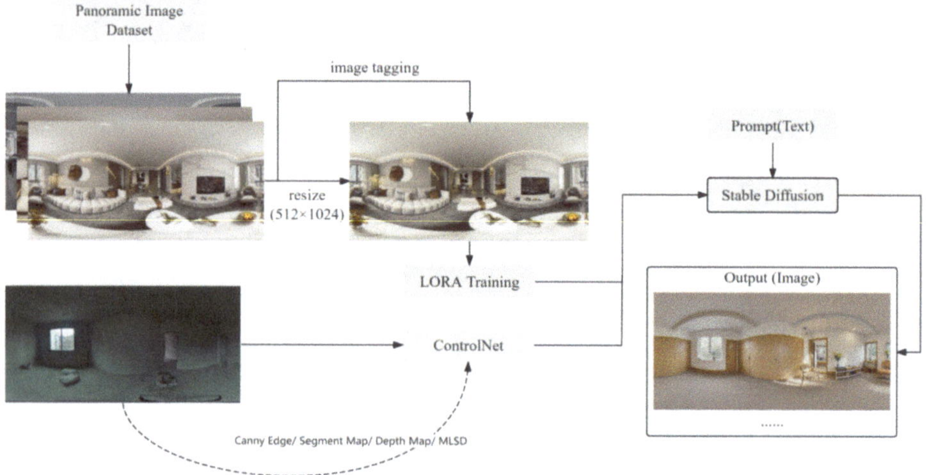

Fig. 2. The Panoramic Image Update Process Flowchart

2.2 Generating Panoramic Images

2.2.1 Stable Diffusion model

Stable Diffusion represents an advanced text-to-image generation framework built upon Latent Diffusion Models (LDMs) [23]. This architecture demonstrates superior performance over traditional GANs across multiple vision tasks including unconditional image synthesis, inpainting, super-resolution, and text-guided generation through its iterative Gaussian noise denoising mechanism. The system architecture integrates three principal components: (1) A dual-purpose Variational Auto-encoder (VAE) that bidirectionally transforms images between pixel space and compressed latent representations, facilitating efficient U-Net processing; (2) A hierarchical U-Net structure employing residual blocks for multi-scale feature extraction and high-fidelity image reconstruction; (3) A text encoding module that generates semantic embeddings to condition the denoising trajectory.

This study employs the ComfyUI implementation of Stable Diffusion as the experimental framework, selected for two principal advantages over conventional Web-UI systems: (i) Enhanced modular flexibility supporting complex operations including panoramic image reprocessing and multi-stage editing workflows; (ii) Parameter persistence functionality enabling reproducible results through workflow template, eliminating redundant parameter re-calibration when processing new inputs.

The methodological pipeline comprises three iterative phases: Initial acquisition of raw panoramic imagery; Progressive refinement through LoRA-optimized SD models under multi-modal ControlNet constraints (depth mapping, semantic segmentation); Targeted image editing operations encompassing object removal, spatial rearrangement, and local inpainting. This cyclical process continues until achieving user-specified visual objectives, ensuring both geometric consistency and semantic coherence in the final panoramic output.

2.2.2 ControlNet

The diffusion model produces images in a highly arbitrary fashion, which complicates the task of managing the produced outcomes. Additionally, it's notably difficult to finely control the end content that is generated when relying on the details supplied in the text. The newly released ControlNet model tackles this difficulty by enhancing the stable diffusion model with additional constraints to govern the image generation process, effectively resolving this challenge [24]. Consequently, manipulating the robustly random outputs of the diffusion model has been simplified. ControlNet incorporates a range of control mechanisms, such as Canny edge detection and segmentation map techniques.

In this work, since panoramic images do not consist of almost straight lines like normal indoor perspectives; instead, they feature more curves, we abandoned the MLSD(Mobile Line Segment Detection), which is most commonly used indoors, and opted to use segment and depth for control instead. At the same time, for unfinished rooms, we hoped to generate more furniture. The use of segment provides the same level of control for components such as ceilings, walls, doors, and windows, which reduces the likelihood of generating furniture in front of walls and lighting fixtures under ceilings. As a result, we processed the images after separating the segment, retaining control only for windows and doors, and then combined this with depth control. This approach allowed for more flexibility in the images based on a certain level of control over doors and windows, enabling better generation of furniture.

The recently emerged FLUX base model has significantly improved the quality of image generation compared to other models, so most of the images in this paper were generated using the FLUX base model. However, since the FLUX basic model currently does not support the Segmentation ControlNet, we had to use depth in combination with Canny Edge Detection to control the images. This improved the quality of the generated images, but without the control of the Segmentation ControlNet, there were instances where the generated images mistook windows for paintings. This issue can be mitigated by Prompt Engineering.

2.3 Panoramic Image Editing

2.3.1 Image Stitching

In the process of AI-generated panoramic imaging, for enhanced visual outcomes, it is common practice to moderately reduce the weight of ControlNet. This adjustment allows for the AI to produce images with superior quality and greater diversity. However, panoramic images generated by AI, such as those produced by Midjouney, often exhibit a seam where the pixels at the left and right extremities of the panorama fail to connect seamlessly.

This issue becomes particularly pronounced when the panorama is viewed through a VR device, leading to a loss of realism for the viewer, which is crucial for the spatial experience. To address this problem, we have developed a workflow within ComfyUI. Initially, we split the panoramic image into two halves from the center(Fig. 3(a)). Subsequently, we exchange the positions of the left and right halves, thereby relocating the seam to the central area of the panorama(Fig. 3(b)). We then apply a mask to the central portion of the panorama and perform a localized redrawing to eliminate the presence of the seam(Fig. 3(c)). This method effectively removes the seam issue without compromising the overall layout of the panoramic image(Fig. 3(d)).

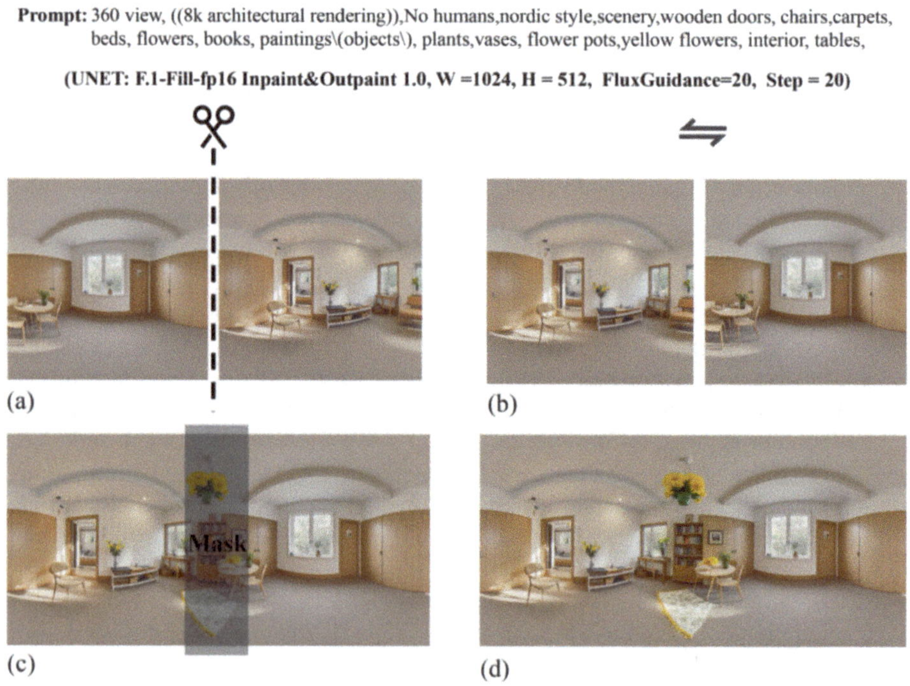

Prompt: 360 view, ((8k architectural rendering)),No humans,nordic style,scenery,wooden doors, chairs,carpets, beds, flowers, books, paintings\(objects\), plants,vases, flower pots,yellow flowers, interior, tables,

(UNET: F.1-Fill-fp16 Inpaint&Outpaint 1.0, W =1024, H = 512, FluxGuidance=20, Step = 20)

(a)

(b)

Mask

(c)

(d)

Fig. 3. The flowchart of the Image stitching

2.3.2 Inpainting

While diffusion models generate globally plausible images, local details often exhibit logical inconsistencies due to "hallucination phenomena" (i.e., the model's over-reliance on prior knowledge), manifesting as structural distortions or textural anomalies. To address this, local editing techniques have become critical for enhancing controllable generation. To this end, Stable Diffusion (SD) proposes an inpainting method based on latent diffusion models [23]. Its core workflow comprises three stages: First, the original image is encoded into a low-dimensional latent space, with user-defined masks identifying target regions for editing. Subsequently, controlled noise (intensity regulated by denoising parameters) is injected into masked regions during the diffusion process, while the U-Net architecture integrates text prompts with contextual information from surrounding pixels to iteratively generate content harmonized with the global structure [25]. Finally, a decoder maps the refined latent representation back to pixel space. By leveraging cross-attention mechanisms and spatial conditional constraints [26], this approach ensures that locally edited content adheres to semantic requirements while seamlessly aligning with the original visual style.This study implements three core functionalities based on image inpainting technology (Fig. 4):

(a) Repainting: Users interactively annotate target regions via masks. Guided by prompt words through the F.1-Fill-fp16 UNet model (Inpaint&Outpaint 1.0), the system generates content that seamlessly integrates with the original image while maintaining semantic consistency (Fig. 4(a)).
(b) Relocation: For background-separated target objects, an image stitching technique embeds them into the scene. The edges are optimized for illumination-geometric consistency via a local inpainting module, ensuring texture harmony and perspective alignment between the transferred object and the scene (Fig. 4(b)).
(c) Removal: Leveraging reverse semantic prompts and structural constraints from ControlNet (mistoLine_rank256), a two-stage optimization strategy is employed:Coarse Repair: Guided by enhanced line art, the LaMa algorithm performs initial reconstruction.Refined Generation: The dpmpp_sde sampler (CFG = 1.4) refines details in the latent space to achieve artifact-free object elimination (Fig. 4(c)).

This workflow significantly enhances editing robustness in complex scenes through multimodal conditional coordination.

2.4 Display the Generated Panoramas

2.4.1 Equirectangular Projection

Equirectangular Projection serves as a standardized methodology for projecting three-dimensional spherical panoramic data onto a two-dimensional plane, fundamentally enabling real-time bidirectional conversion between panoramic imagery and virtual reality (VR) perspectives [27]. This technique unfolds a 360° spherical scene into a rectangular image with 2:1 aspect ratio through linear mapping of longitude ($\lambda \in [-180°, 180°]$) and latitude ($\varphi \in [-90°, 90°]$) to normalized coordinates ($u \in [0,1]$, $v \in [0,1]$) defined

Fig. 4. The result of the Repainting, Relocation and Removal

by:

$$u = \frac{\lambda + 180^\circ}{360^\circ}, \quad v = \frac{90^\circ - \phi}{180^\circ}$$

During VR rendering, real-time headset pose data (yaw, pitch) drives inverse spherical coordinate calculations to dynamically extract perspective-projected view frustums, achieving immersive interaction. Now extensively implemented in platforms like Facebook 360 and Google Street View, this technology allows interactive panoramic exploration via smartphones or computers without specialized VR hardware, offering lightweight solutions that dramatically lower accessibility barriers and empower civilian applications such as virtual property viewing and digital cultural tourism.

2.4.2 Panoramic Neural Radiance Field

Converting a panoramic image into a 3D model can enhance its expressiveness. Using point clouds or NeRF (Neural Radiance Fields) methods, it is possible to generate 3D architectural models from a single image [24].PERF(Panoramic Neural Radiance Field) is a method for training panoramic neural radiance fields from a single panoramic image, overcoming the limitations of traditional NeRF that require multi-view inputs, thereby enabling 3D exploration in complex scenes [28]. The approach first employs a depth estimation model to predict a depth map from the panoramic image and trains a single-view NeRF to reconstruct the visible regions. Subsequently, PERF introduces a collaborative RGBD inpainting framework that integrates a pre-trained Stable Diffusion model with a monocular depth estimator to synthesize RGB images and depth maps from novel viewpoints, thereby generating 3D geometry and appearance beyond the initially visible areas. To address geometric inconsistencies across viewpoints, PERF proposes a progressive inpainting and erasing strategy. By iteratively comparing newly sampled viewpoints with the reference view, conflicting geometric regions are eliminated, ensuring coherent 3D scene completion. Experimental results demonstrate that PERF achieves state-of-the-art performance in single-view panoramic NeRF tasks, with applications spanning panorama-to-3D conversion, text-to-3D generation, and 3D scene stylization. This method leverages PERF technology to further transform generated panoramic images into 3D point clouds, enabling robust 3D scene reconstruction and manipulation(Fig. 5). In this paper, we utilize PERF to generate point clouds, further enhancing the expressiveness.

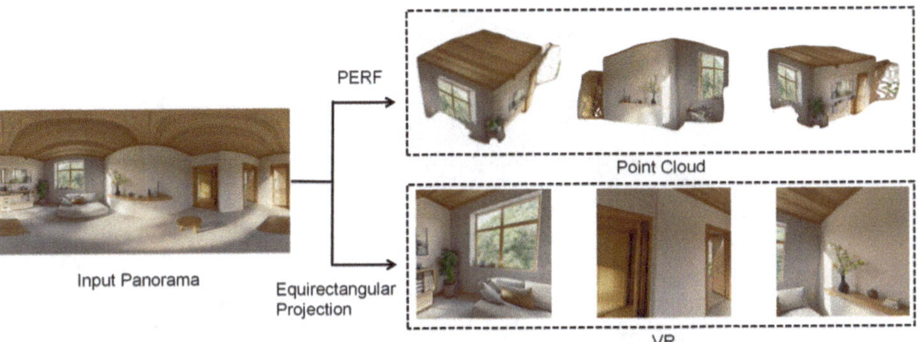

Fig. 5. Generating Point Cloud and VR Display from Panoramic Images

3 Results

By utilizing LoRA fine-tuned Stable Diffusion models, our framework enables the generation of content that significantly diverges from the original training dataset while providing extensive customization options. Through prompt engineering, we synthesize interior panoramic images across diverse architectural styles. In our style transfer experiments, optimal hyperparameters were employed to validate model performance. As shown in Fig. 6, the system successfully generates five distinct interior design styles: Modern minimalist style, New chinese style, European luxurious style, Japanese minimalist style, and Mediterranean style. The LoRA-fine-tuned model demonstrates precise alignment with target stylistic requirements, effectively translating textual prompts into visual outputs. Furthermore, ControlNet ensures geometric consistency between generated panoramas and reference images through perspective constraints derived from edge detection and depth estimation. Integrated with our local inpainting workflow (Sec. 2.2.1), this pipeline enables targeted spatial modifications (e.g., furniture replacement or decorative adjustments) to enhance compositional harmony. Notably, advancements in large language models (LLMs) such as GPT-4 allow automated generation of domain-specific prompts, substantially reducing the technical barriers for panoramic image synthesis. This text-driven paradigm empowers architects to rapidly prototype design concepts through iterative textual inputs, significantly accelerating early-stage conceptual development.

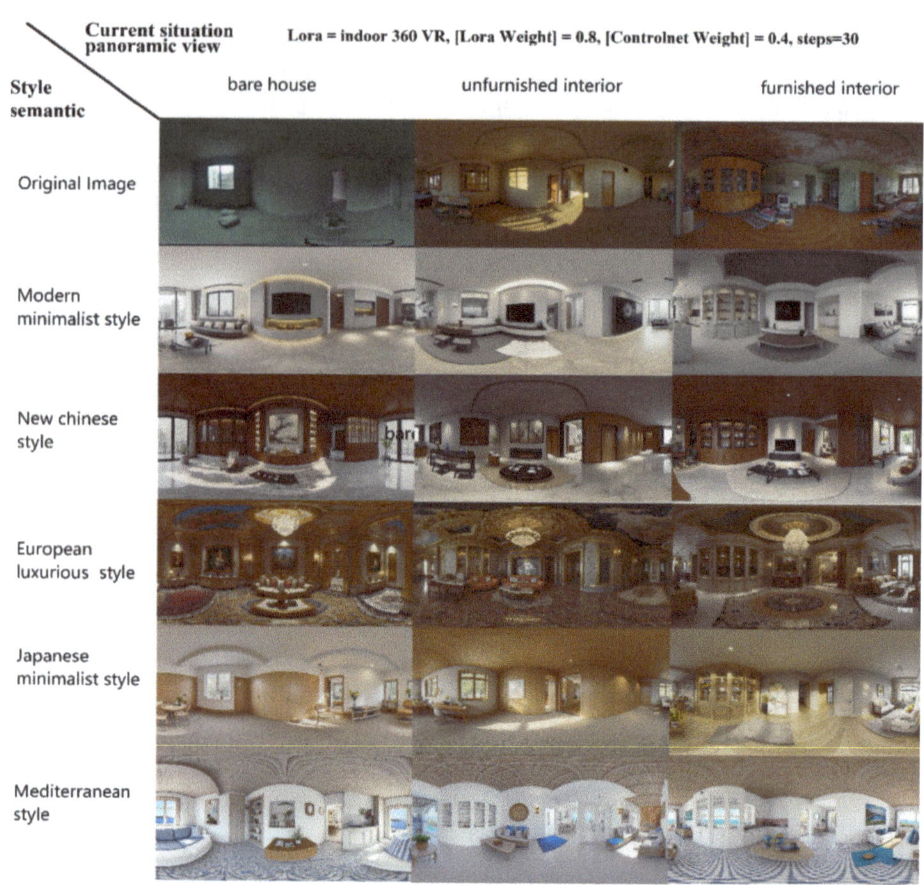

Fig. 6. Different prompt words and panoramic images generate results.

4 Conclusion and Discussion

This paper proposes an innovative interactive interior design framework that integrates Stable Diffusion with Panoramic Neural Radiance Fields (PERF) to streamline design visualization and client communication processes. We developed a comprehensive ComfyUI workflow enabling non-experts to effortlessly obtain updated 360° interior previews through simple inputs of panoramic images and text prompts.

Collectively, our framework enhances both creative expression efficiency and visualization quality in interior design. Through natural language-driven 3D scene generation, designers can rapidly prototype concepts, while clients gain intuitive understanding of design intent, fostering efficient collaboration and accelerating project progression.However, this method is not limited to indoor spaces. It is feasible to apply this approach to outdoor spaces by updating the panoramic images based on the current status of the outdoor environment. Due to time constraints, these explorations are still ongoing, and further research is planned to investigate the application of AI-generated images, VR, and generative models in urban renewal.

Acknowledgments. this work is suppported by the National Natural Science Foundation of China (Grant No. 52308002); the China Postdoctoral Science Foundation(Grant No. 2023 M730148).

References

1. Yee, R.: Architectural Drawing: A Visual Compendium of Types and Methods. John Wiley & Sons (2012)
2. Lu, S., Yan, X., Xu, W., Chen, Y., Liu, J.: Improving auditorium designs with rapid feedback by integrating parametric models and acoustic simulation. Build. Simul. **9**(3), 235–250 (Jun. 2016). https://doi.org/10.1007/s12273-015-0268-x
3. Li, J., Shuai, L., Wang, Q.: Graphical visualisation assist analysis of indoor environmental performance: Impact of atrium spaces on public buildings in cold climates. Indoor Built Environ. **27**(3), 331–347 (2018)
4. Zeng, P., et al.: Residential floor plans: Multi-conditional automatic generation using diffusion models. Autom. Constr. **162**, 105374 (2024)
5. Zeng, P., Gao, W., Li, J., et al.: Automated residential layout generation and editing using natural language and images[J]. Autom. Constr. **174**, 106133 (2025)
6. Yin, J., et al.: ArchiDiff: Interactive design of 3D architectural forms generated from a single image. Comput. Ind. **168**, 104275 (2025)
7. Lu, S., et al.: Supporting early-stage design decisions with building performance optimisation: Findings from a design experiment. J. Build. Eng. **82**, 108298 (2024)
8. Lu, S., et al.: The influence of shape design on the acoustic performance of concert halls from the viewpoint of acoustic potential of shapes. Acta Acust. United Acust. **102**(6), 1027–1044 (2016)
9. Li, J., et al.: Study of passive adjustment performance of tubular space in subway station building complexes. Appl. Sci. **9**(5), 834 (2019)
10. Lu, S., et al.: An experimental study on the acoustic absorption of sand panels. Appl. Acoust. **116**, 238–248 (2017)
11. Huang, Y., et al.: Fusing transformer and diffusion for high-resolution prediction of daylight illuminance and glare based on sparse ceiling-mounted input. Build. Environ. **267**, 112163 (2025)
12. Ma, X., et al.: Street microclimate prediction based on Transformer model and street view image in high-density urban areas. Build. Environ. **269**, 112490 (2025)
13. Zeng, T., et al.: Improving outdoor thermal environmental quality through kinetic canopy empowered by machine learning and control algorithms. In: Building Simulation. Tsinghua University Press, Beijing (2025)
14. Luo, Y., et al.: Outdoor space design and its effect on mental work performance in a subtropical climate. Build. Environ. **270**, 112470 (2025)
15. Yu, T., et al.: Machine learning prediction on spatial and environmental perception and work efficiency using electroencephalography including cross-subject scenarios. J. Build. Eng. **99**, 111644 (2025)
16. P. Isola, J.-Y. Zhu, T. Zhou, and A. A. Efros, "Image-To-Image Translation With Conditional Adversarial Networks".
17. Huang, W., Zheng, H.: Architectural drawings recognition and generation through machine learning. In: presented at the ACADIA 2018: Re/Calibration: On Imprecision and Infidelity, pp. 156–165, Mexico City, Mexico (2018). https://doi.org/10.52842/conf.acadia.2018.156
18. Gao, W., et al.: Impact of 3D modeling behavior patterns on the creativity of sustainable building design through process mining. Autom. Constr. **150**, 104804 (2023)

19. Wang, C., et al.: Effectiveness of one-click feedback of building energy efficiency in supporting early-stage architecture design: An experimental study. Build. Environ. **196**, 107780 (2021)
20. Gao, W., et al.: Quantitative analysis of performance-oriented design efficiency in early divergent parametric design of office building façade. J. Build. Eng. **98**, 111449 (2024)
21. Yan, C., Yuan, P.F.: Phygital intelligence. Archit. Intell. **3**(1), 30 (Aug. 2024). https://doi.org/10.1007/s44223-024-00073-0
22. Hu, E.J., et al.: LoRA: Low-Rank Adaptation of Large Language Models (Oct. 16, 2021, arXiv: https://arxiv.org/abs/2106.09685). https://doi.org/10.48550/arXiv.2106.09685
23. Rombach, R., Blattmann, A., Lorenz, D., Esser, P., Ommer, B.: High-resolution image synthesis with latent diffusion models. In: Presented at the Proceedings of the IEEE/CVF Conference on Computer Vision and Pattern Recognition, pp. 10684–10695. Accessed: 19, Mar. 2025 (2022) https://openaccess.thecvf.com/content/CVPR2022/html/Rombach_High-Resolution_Image_Synthesis_With_Latent_Diffusion_Models_CVPR_2022_paper
24. Zhang, L., Rao, A., Agrawala, M.: Adding conditional control to text-to-image diffusion models. In: 2023 IEEE/CVF International Conference on Computer Vision (ICCV), pp. 3813–3824. IEEE, Paris, France (Oct. 2023). https://doi.org/10.1109/ICCV51070.2023.00355
25. Lugmayr, A., Danelljan, M., Romero, A., Yu, F., Timofte, R., Van Gool, L.: RePaint: Inpainting using denoising diffusion probabilistic models. In: 2022 IEEE/CVF Conference on Computer Vision and Pattern Recognition (CVPR), pp. 11451–11461. IEEE, New Orleans, LA, USA (Jun. 2022). https://doi.org/10.1109/CVPR52688.2022.01117
26. C. Saharia *et al.*, "Photorealistic Text-to-Image Diffusion Models with Deep Language Understanding".
27. L. Barazzetti, A. M. Oteri, R. Pellicano, A. Santoro, and C. Valiante, "Rendering Equirectangular Projections Acquired With Low-Cost 360° CAMERAS," 2022.
28. Wang, G., Wang, P., Chen, Z., Wang, W., Loy, C.C., Liu, Z.: PERF: Panoramic Neural Radiance Field from a Single Panorama (Oct. 28, 2023, arXiv: https://arxiv.org/abs/2310.16831). https://doi.org/10.48550/arXiv.2310.16831

Revealing the Relationship Between Spatial Visibility and Personnel Flow in Hospital Outpatient Departments Using Graph Deep Learning and Agent-Based Modeling

Xinyan Ren, Ying Zhou(✉), and Yangpeng Xin(✉)

School of Architecture, Southeast University, Nanjing, China
{213231775,zhouying}@seu.edu.cn, 2316555903@qq.com

Abstract. This study integrates Graph Deep Learning (GDL) and Agent-Based Modeling (ABM) to analyze the association between spatial visibility and human flow in hospital outpatient departments. Addressing limitations of subjective traditional methods, spatial syntax theory is applied to construct a visibility graph model, quantifying node-level visibility. ABM simulates patient behaviors to capture environment-behavior interactions. Innovatively, Graph Neural Networks (GNNs) are employed to mine spatial network topology, embedding visibility indices and behavioral data into node-edge relationships to reveal underlying spatial-crowd dynamics. Trained on 100 outpatient samples, the model achieves 77.5% prediction accuracy. Results indicate positive correlations between visibility and spatial capacity demands. The research validates GNN's interpretability in spatial analysis, offering a quantitative basis for optimizing hospital layouts and advancing evidence-based healthcare design. This framework bridges spatial metrics with behavioral patterns, enhancing data-driven decision-making in architectural planning.

Keywords: Hospital Outpatient Departments · Graph Neural Networks · Agent-Based Modeling · Spatial Visibility · Evidence-Based Design

1 Introduction

Hospital design critically impacts healthcare efficiency, patient experience, and space utilization. Traditional methods relying on architects' experience and qualitative analysis fail to dynamically assess spatial effects on patient flow. While data-driven approaches are emerging, current research faces two key limitations: first, the absence of quantitative models for dynamic interactions between spatial visibility and crowd distribution; second, the inadequacy of static frameworks to address real-time healthcare demands. This study proposes a closed-loop "spatial-behavioral" analysis paradigm integrating GNNs' topological feature extraction with ABM's dynamic simulation capabilities to overcome these limitations.

Y. Liu et al. (Eds.): CDRF 2025, *Transindividual Intelligence*, pp. 51–61, 2026.
https://doi.org/10.1007/978-981-92-0615-5_5

1.1 Study of Medical Building Plan

The science of hospital spatial layout has a direct impact on healthcare efficiency and patient flow organization. Bayraktar Sari & Jabi (2024) demonstrated the synergy of accessibility, privacy, and natural lighting in spatial optimization. Jiang & Verderber (2017) revealed nonlinear relationships between spatial connectivity and care efficiency through circular corridor studies. Teran-Somohano & Smith (2023) innovatively integrated spatial syntax into the hospital layout framework to verify its effectiveness in reducing the risk of cross-infection. Zhang et al. (2024) further introduced a multi-intelligence deep reinforcement learning approach to achieve intelligent optimization of people flow distribution through dynamic function module combination, marking the paradigm shift of graphic design from static analysis to dynamic regulation and control.

1.2 Pedestrian Simulation and Spatial Visibility Analysis

In quantitative research on spatial visibility, Haq & Luo (2012) linked corridor integration and nurse station visibility to navigation patterns via spatial syntax, setting healthcare spatial assessment benchmarks. Yuan & Zhou (2025) merged multi-agent modeling with spatial syntax, demonstrating circular corridors reduce patient backtracking and boost care efficiency through visibility-driven dynamic visibility-flow feedback. Liu et al. (2024) developed data-driven multi-agent models for precise people flow prediction in complex medical settings. While tools like Massmotion simulate crowd density, their integration with spatial visibility metrics remains limited, highlighting gaps in coupling mechanisms between visibility features and pedestrian dynamics. These studies collectively advance quantitative spatial visibility analysis while underscoring the need for deeper visibility-behavior interaction modeling.

1.3 Graph Neural Networks in Spatial Analysis

GNN provide a new technical path for building spatial analysis, and their development has been able to solve complex spatial analysis problems. Hu (2021) developed a spatio-temporal graph convolutional network, improving traffic flow prediction accuracy through non-Euclidean spatial feature extraction, validating GNNs' efficacy in spatial modeling. Chen & Jiang (2025) proposed NE-Graph-BERT for precise space classification and layout optimization, demonstrating GNNs' superior feature extraction. Yao et al. (2024) merged conditional variational autoencoders with graph structures to generate functionally enhanced layouts, expanding GNNs' role in generative design. These innovations collectively establish a technical foundation for "spatial visibility-people flow distribution" analysis framework.

1.4 Research Objectives

This study proposes an analytical framework that combines GNN with ABM to quantitatively analyze the relationship between spatial visibility and people flow in hospital outpatient departments. The purpose of this study is to provide a data-driven optimization tool for hospital building design through graph neural networks to improve space utilization, optimize patient flow paths, and enhance the science of hospital space management.

2 Methodology

As shown in Fig. 1, the methodology of the study was based on the following steps.

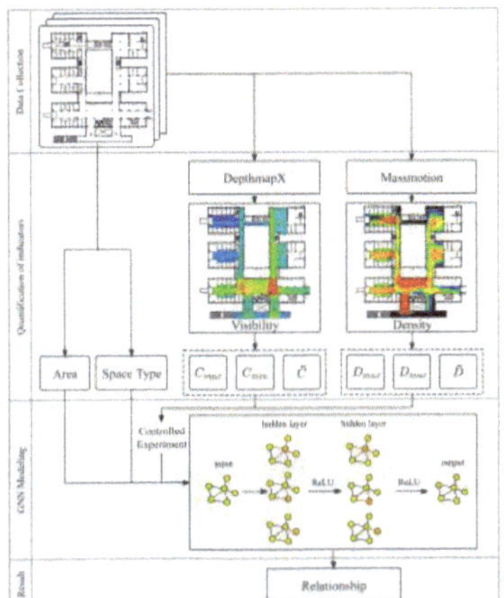

Fig. 1. Technical route

2.1 Data Collection and Pre-processing

The data for this study were obtained from publicly available datasets on the Chinese web, and the floor plan data of outpatient clinics in 100 of these general and specialized hospitals were screened. The building sizes of the sample outpatient departments involved are typical: single-story outpatient departments range from 4,000–25,000 m^2 in size and contain 52–218 functional rooms. This study focuses on the public mobility area of the outpatient department, and the data are limited to the group activity space of the outpatient department, excluding the closed diagnosis units and ultimately forming a standardized graphic dataset containing 2,299 nodes and 2,532 edges.

2.2 Quantification of Indicators

2.2.1 Visibility

This study applies spatial syntax theory (Hillier & Hanson, 1984) using DepthmapX, calculating visibility connectivity by dividing spaces into 1 m × 1 m grids. Based on the analysis of related studies, the line-of-sight analysis sets the sight distance threshold within 20 m. The visible neighboring nodes of each node are determined, and the topological connections are accumulated to form a connectivity index to quantify the spatial visual relationship. The visibility connectivity formula is as follows:

$$C_i = \sum_j V_{ij},\ V_{ij} = \begin{cases} 1, & d_{ij} \le 20 \text{ m and a visual connection exists} \\ 0, & else \end{cases} \tag{1.1}$$

Finally, the standardized visibility connectivity heat maps are generated, extracting the maximum (C_{max}), minimum (C_{min}), and average ($\overline{C}$) values of visibility in each region by room boundary segmentation.

2.2.2 Crowd Density

In this study, Massmotion, a pedestrian dynamics simulation software, is used to quantify the crowd density distribution of a spatial scene through ABM and conflict avoidance algorithms.

Based on field research, a sample of 30 patients with autonomous behaviors was taken, and based on the selection of the sample departments and the length of stay, the general flow of visits to the outpatient department of the hospital was summarized, and a behavioral logic diagram in Fig. 2(a) was drawn based on the flow.

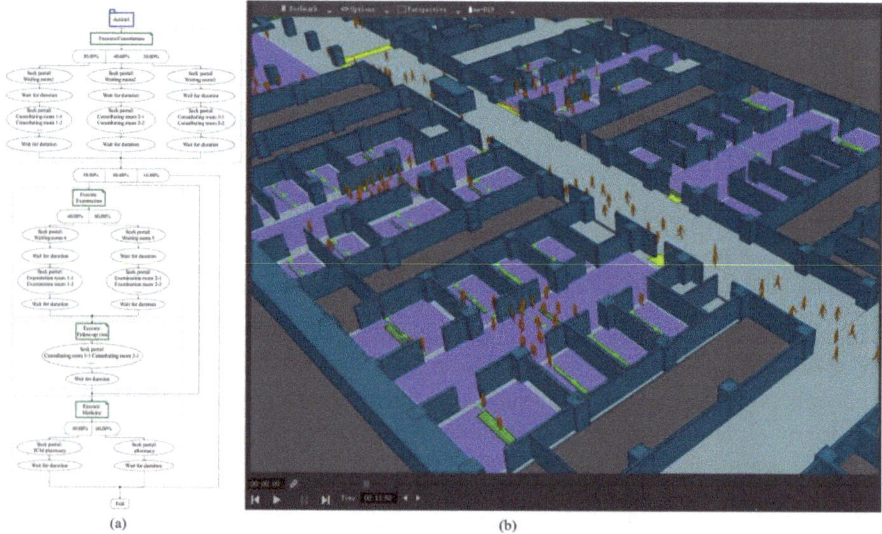

Fig. 2. Behavior logic and pedestrian simulation

The logic of the method is as follows: import the building plan model into the Massmotion platform, define the entrance and exit locations and obstacle constraints, set the crowd flow parameters and behavioral logic as in Fig. 2(a), simulate the pedestrian dynamics flow map as in Fig. 2(b) through parallel computation of multiple intelligences, generate the average heat map of the crowd density over a period of time, and extract the maximum (D_{max}), minimum (D_{min}), and average ($\overline{D}$) values of the crowd density in each region by room boundary segmentation.

2.2.3 Correlation Analysis

Correlation coefficients were used to test the correlation between the visibility ($\overline{C}$) and density ($\overline{D}$) indicators, with the significance threshold set at $p < 0.01$.

2.3 Graph Neural Network Modeling

In this study, an improved GraphSAGE model is constructed based on the GDL framework to realize crowd density prediction through spatial topology and feature fusion (Fig. 3).

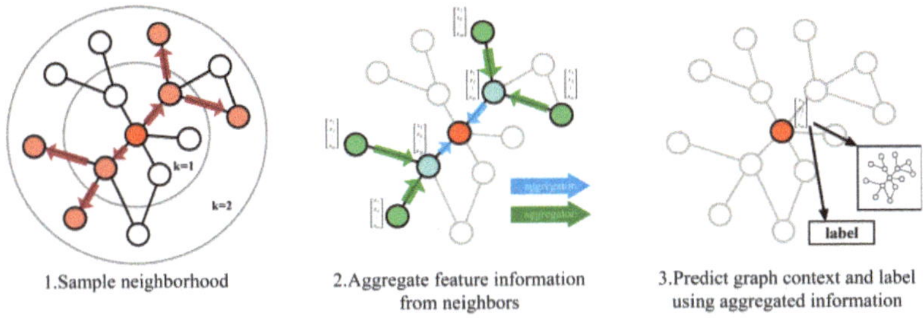

Fig. 3. Visual illustration of the GraphSAGE sample and aggregate approach

2.3.1 Graph Structure Definition

An undirected graph $G = (V, E)$ is constructed using spatial units as graph nodes and physical connections in space as graph edges. The visibility calculated by DepthMapX and crowd density output by Massmotion are mapped to nodes by the Grasshopper plugin. Figure 4 shows the overall flow of the process.

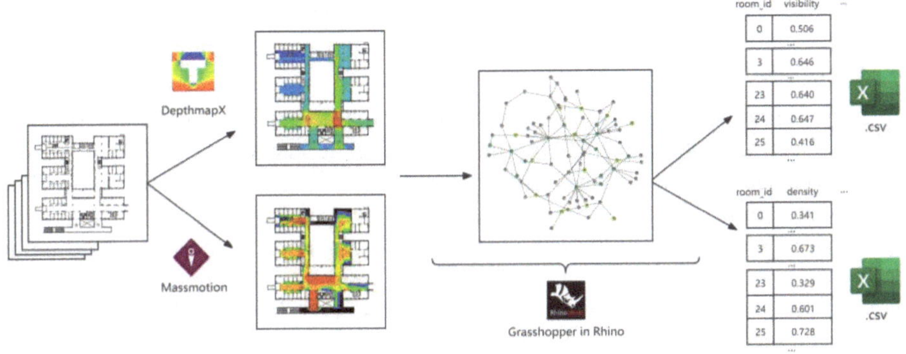

Fig. 4. DepthMap and Massmotion simulation results mapping path to GNN graph structure

In addition to this, the node feature matrix $X \in \mathbb{R}^{|V| \times 3}$ also contains the normalized area, visibility, and spatial type encoding, and the edge features have a default weight of 1. To enhance local feature aggregation, self-loop edges are explicitly added.

2.3.2 Model Architecture

The experiments use the same three-layer GraphSAGE convolution operation (hidden layer 128 dimensions), with each layer performing the following computations:

$$h_{v^{(l+1)}} = \mathrm{ReLU}\Big(\mathrm{W}^{(l)} \cdot \mathrm{MEAN}\Big(\mathrm{h}_{\mathrm{v}^{(l)}} \cup \Big\{\mathrm{h}_{\mathrm{u}^{(l)}}, \forall \mathrm{u} \in \mathrm{d}\,\mathcal{N}(\nu)\Big\}\Big)\Big) \tag{2.1}$$

In this equation, $h_{v^{(l+1)}}$ is the embedding vector of node v in layer l, $W^{(l)}$ is the trainable parameter matrix, and $\mathcal{N}(\nu)$ denotes the set of neighbors of node v.

The input layer is a 3-dimensional node feature and the output layer is a 1-dimensional density prediction. Layer embeddings are batch normalized (BatchNorm) with Dropout (ratio = 0.5) to enhance the model generalization ability.

SmoothL1Loss is chosen as the loss function, and its segmentation property can suppress outlier interference (Setting $\beta = 0.05$):

$$Loss = \frac{1}{n}\sum_{i=1}^{n}\begin{cases} \frac{-x0.5(y_i-\hat{y}_i)^2}{\beta}, & |y_i - \hat{y}_i| \le \beta \\ |y_i - \hat{y}_i| - 0.5\beta, & else \end{cases} \tag{2.2}$$

The optimizer uses AdamW (learning rate 0.001, weight decay 1e-5).

2.3.3 Training Strategy

The optimizer is configured as AdamW (lr = 0.001, weight_decay = 1e-5), 10,000 training rounds are set, the dynamic threshold is set to 10% of the extreme deviation of the normalized crowd density, and the ratio of the training set to the validation set is 9:1.

The absolute error between the predicted crowd density and the true value is calculated every 50 rounds during the training process, and when the error is less than or equal to the threshold it is determined to be a correct prediction, and the final model outputs loss and accuracy metrics and decision coefficients on the validation set.

2.4 Controlled Experiment Design

The graph neural network model in this study was optimized for input feature selection through nine groups of controlled experiments, where the spatial visibility maxima, minima, and averages and the crowd density maxima, minima, and averages were combined with the node base features as independent input data (area + type $+C + D$) for the controlled experiments, and the same GraphSAGE architecture was used for training. Each set of experiments was run for 1000 epochs, and the absolute error between the predicted density and the true value was calculated every 100 rounds during the training process, and the loss and threshold accuracy were recorded.

3 Result and Discussion

3.1 Results of Spatial Visibility and Crowd Density

3.1.1 Visibility

The visibility analysis using DepthmapX demonstrates the spatial connectivity and openness of the outpatient clinic. The analysis generated heat maps as in Fig. 5 (a), with red areas indicating areas of high visibility and blue areas indicating areas of low visibility.

As in reality, the highest visibility values are concentrated in the open halls and main corridors, while enclosed spaces and narrow corridors show significantly lower visibility values.

3.1.2 Crowd Density

The Massmotion simulation results demonstrate the dynamic distribution of people movement within the outpatient clinic. The simulation outputs heat maps as in Fig. 5 (b), with red areas indicating high density areas and blue areas indicating low density areas. Typically, the most crowded areas are located in the registration and charging area, the waiting area and the main intersections, and the results are generally consistent with the actual situation.

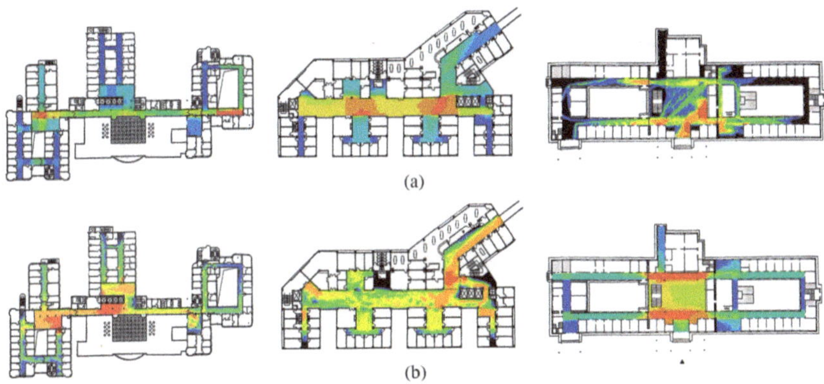

Fig. 5. Visibility and density heat maps

3.1.3 Relationship

According to Fig. 6, the spatial visibility of the lobby ($r = 0.47$, $p < 10^{-13}$), waiting area ($r = 0.35$, $p < 10^{-12}$), and corridors ($r = 0.45$, $p < 10^{-74}$) showed a positive correlation with the spatial carrying capacity, indicating that the open view could lead patients to gather naturally, thus enhancing space utilization. In contrast, the correlation was not significant for the registration and charging area ($r = 0.24$, $p = 0.04$), which may be due to the fact that the visibility of the registration and charging area mainly affects the initial wayfinding efficiency of the patients, but once they arrive at the area, the subsequent behavior are dominated by the service flow, and the moderating effect of spatial visibility on the distribution of density is weakened.

3.2 Comparison Experiment Results

After the comparison experiments with different combinations, the validation set loss values and threshold accuracies for each group with epoch = 1000 are as follows (Table 1):

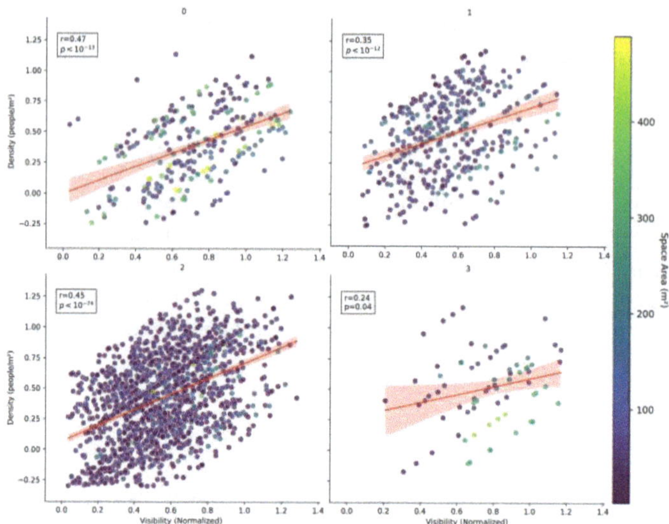

Fig. 6. Scatter plot of visibility and density correlation (0 = hall, 1 = waiting area, 2 = corridor, 3 = registration)

Table 1. Experimental result

area + type +C + D (Loss;Accuracy)	C_{min}	$\overline{C}$	C_{max}
D_{min}	0.2512; 0.3766	0.3101; 0.3117	0.2803; 0.4286
$\overline{D}$	0.3591; 0.3300	0.3257; 0.3854	0.3643; 0.3399
D_{max}	0.3632; 0.6176	0.3565; 0.6127	0.3914; 0.5784

This leads to the optimal combination of experiments as "area + type + $\overline{C}$ + D_{max}".

3.3 Model Test

The trained GraphSAGE model successfully predicts the density of people movement based on spatial visibility and architectural parameters. As shown in Fig. 7, the training process reaches convergence at 9000 rounds, and the loss function stabilizes at about 0.16, with an accuracy of about 77.5% and an R^2 of about 0.69, which indicates that the graph structure effectively captures the spatial-behavioral association.

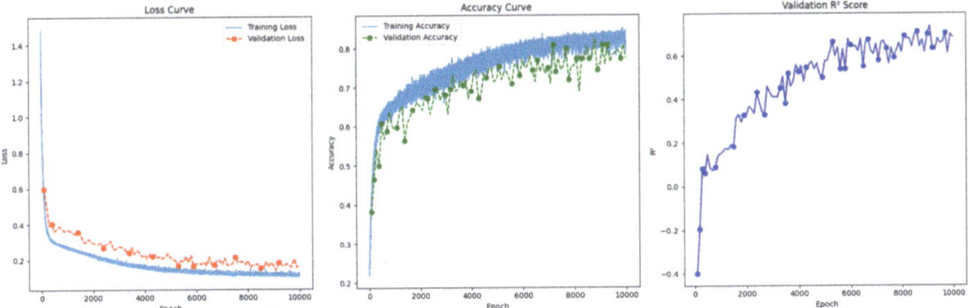

Fig. 7. Model validation visualization

Table 2 shows the prediction results for the first 7 samples of the validation set, the mean absolute error of the model's predicted values versus Massmotion 's simulated values is 0.072 (standard deviation ±0.032), and the maximum error is 0.1168, which shows that the predicted values differ from the true values by a small amount, and are valid.

Table 2. Prediction results of the first 7 validation samples

True Density	Predicted Density	Absolute Error
0.6857	0.7522	0.0665
1.0000	0.8832	0.1168
0.3282	0.4289	0.1007
0.8245	0.8571	0.0326
0.6688	0.7542	0.0854
1.0000	0.9562	0.0438
0.8333	0.8815	0.0482

3.4 Limitations

Although this study was successful in demonstrating a quantitative relationship between spatial visibility and mobility, the following limitations need to be noted:

Effect of Sampling Units: In this study, a 1 m × 1 m spatial syntactic sampling unit was used for visibility analysis, but the effects of different sized sampling units on the results have not yet been explored in depth. 1 m × 1 m grids may ignore microbehavioral differences.

Data Set Bias: The training dataset used in this study only includes the layouts of large general hospitals and specialty hospitals, while other types of healthcare organizations such as small community hospitals are not included in the study. Hospitals of different sizes and types may exhibit different footfall characteristics.

4 Conclusion

This study integrates GNN and ABM to quantify how spatial visibility shapes personnel flow patterns in hospital outpatient departments. Results demonstrate a strong correlation between visibility metrics and flow density, directly influencing patient pathways and healthcare efficiency. By linking spatial structure to behavioral dynamics, the framework offers architects a data-driven tool for optimizing layouts, enhancing space utilization, and improving patient experience through visibility-informed design decisions.

Future directions prioritize addressing current constraints: refining visibility analysis by evaluating sampling unit scale sensitivity, expanding datasets to include diverse healthcare facilities (e.g., community clinics), and incorporating dynamic visibility adjustments for real-time obstacle and movement variations. These improvements aim to advance the model into an adaptive tool for evidence-based architectural planning, capable of addressing institutional diversity and evolving spatial demands while bridging static analysis with real-world operational complexity. The approach marks a paradigm shift toward responsive, human-centric healthcare design grounded in spatial-behavioral interactions.

References

Bayraktar Sari, A.O., Jabi, W.: Architectural spatial layout design for hospitals: a review. J. Build. Eng. **97**, 110835 (2024)

Chen, Y., Jiang, H.: Optimized graph neural networks for spatial recognition to support automatic building information model semantic enrichment - Exploring node features enhancing strategies. Eng. Appl. Artif. Intell. **147** (2025)

Haq, S., Luo, Y.: Space Syntax in health-care facilities research: a review. Health Environ. Res. Des. J. **5**(4), 98–117 (2012)

Hillier, B., Hanson, J.: The Social Logic of Space. Cambridge University Press (1984)

Hu, Y.: Research on city traffic flow forecast based on graph convolutional neural network. In: 2021 IEEE 2nd International Conference on Big Data, Artificial Intelligence and Internet of Things Engineering, ICBAIE 2021, pp. 269–273. Institute of Electrical and Electronics Engineers Inc. (2021)

Jiang, S., Verderber, S.: On the planning and design of hospital circulation zones. HERD: Health Environ. Res. Des. J. **10**(2), 124–146 (2017)

Liu, Y., Zhou, Y., Yang, L., Xin, Y.: Simulating staff activities in healthcare environments: an empirical multi-agent modeling approach. J. Build. Eng. **84**, 108580 (2024)

Teran-Somohano, A., Smith, A.E.: A sequential space syntax approach for healthcare facility layout design. Comput. Ind. Eng. **177**, 109038 (2023)

Yao, Z., Chen, Y., Cui, J., Zhang, S., Li, S., Hao, A.: Conditional room layout generation based on graph neural networks. Comput. Graph. (Pergamon). **122**, 103971 (2024)

Yuan, H., Zhou, Y.: The impact of crowds on visibility in emergency department: integrating agent-based simulation and space syntax analysis. Front. Archit. Res. **14**, 1398–1414 (2025). https://linkinghub.elsevier.com/retrieve/pii/S2095263525000196

Zhang, Z., Guo, Z., Zheng, H., Li, Z., Yuan, P.F.: Automated architectural spatial composition via multi-agent deep reinforcement learning for building renovation. Autom. Constr. **167**(11), 105702 (2024)

Cellular Automata as a Teacher of Machine Learning: A Human-machine Collaboration to Contextualise Environmental Architecture

Provides Ng[1,2(✉)], David Doria[2], Nikoletta Karastathi[2], Carlos Rivera Salaverry[1,2], and Alberto Fernandez[2]

[1] The Chinese University of Hong Kong, Shatin, New Territories, Hong Kong, SAR, China
provides.ism@gmail.com

[2] University College London (UCL), London WC1E 6BT,, United Kingdom

Abstract. The challenge to creating buildings that intelligently respond to local environmental conditions while respecting cultural needs is increasingly dependent on hybrid computational workflows that are flexible and adaptive to designers' creative decisions and needs. We propose a novel approach where Cellular Automata (CA)—simple digital building blocks that evolve via programmed rules—act as teachers to Machine Learning (ML) systems, which learn patterns from data to automate and accelerate tasks. Our framework uses CA states to translate site-specific intelligence (e.g., solar instances or cultural landmarks) into training protocols for ML, enabling rapid generation of context-sensitive solar designs. Through workshops with 20 architects across 12 countries, we tested the framework and results showed how CA successfully encoded environmental constraints into ML and significantly reduced computation time. However, the pipeline required significant human refinement and spontaneity to resolve computational bottlenecks, highlighting human-machine collaboration. Critically, we argue that automation's value lies not in replacing designers but in structuring a pedagogical collaboration: Humans define contextual priorities, CA formalises them into teachable rules, ML accelerates application. Our work ultimately questions whether intelligence can scale without collaboration, offering technology not as a solution but as a framework for negotiating this tension.

Keywords: artificial intelligence · bio-inspired design · solar architecture · discrete computation · human-machine interaction

1 Introduction

Machine Learning (ML) algorithms are the applications through which Artificial Intelligence (AI) is achieved. In other words, if AI describes an intelligent computer program, ML comprehends the algorithms that allow such intelligence to exist. These algorithms usually learn by progressively improving as it iterates over a large dataset, or contexts that provide this data. Through the iterative process, ML algorithms tune itself to achieve the expected behaviour, such as finding deep patterns across data points, or predicting future states of data, or even predicting how it should behave in an unseen scenario. This is what allows it to automate intellectual tasks.

Y. Liu et al. (Eds.): CDRF 2025, *Transindividual Intelligence*, pp. 62–73, 2026.
https://doi.org/10.1007/978-981-92-0615-5_6

As technology quickly progresses, it is not easy to determine what ML algorithms are and will be able to achieve, but one can say that the current application of these is overall limited not simply by hardware or cloud capacity, but creative implementation methods that can overcome such limitations; that said, the development and application of ML algorithms requires cautious reflections. Algorithms will carry over the biases that are present in the data it consumes, or the crafted environments it learns from. Consequently, the process of selecting or creating the data that a ML system will consume is a crucial step in designing its application. A dataset that is not carefully constructed can lead to a training process that outputs an unfit model, or, in social applications, reinforce prejudices and problematic behaviours. This research, however, seeks to use biases as an advantage—it seeks to embed specific bias in the creation of custom datasets as a means of formulating human-machine interaction.

The project's premise is to focus on the creation of the dataset, with the help of other synthetics intelligences, such as Cellular Automata (CA) to determine the ML algorithms behaviour. CA is a discrete computational model that is well-suited to simulating contextual information to a certain extent because they explicitly model local interactions between cells based on their neighborhood states. It is often used in architecture design to simulate complex patterns through controlled input, such as solar radiation. By embedding the characteristics of such rule-based, computationally demanding processes into the dataset that will be used for training, the objective is to capture into a ML model a design process that, originally would take several hours, can be executed relatively instantaneously.

The expected result is an algorithm that is capable of predicting 3-dimensional shapes in a given urban context, that could be used to populate architectural elements in a manner that maximises solar incidence, with the end goal of also maximising the capability of a structure to harvest energy from photovoltaic elements or through passive solar strategies. *How can rule-based systems translate contextual information to teach AI in automating repetitive and computationally heavy tasks, and synergistically enhance human creativity in sustainable architecture?*

2 Literature Review

Generative Adversarial Networks (GANs) is a class of ML models designed for generative tasks, which was quickly adopted by architectural design disciplines, as it presents a novel approach to automate image generation, from building layouts to spatial renderings (Chaillou, 2020). However, as GANs operate on pixel data, questions arise over how it can be embedded with relevant contextual information beyond simple form-finding, such as environmental and cultural considerations, to ensure that the generated designs are contributing to sustainable development goals. Recent advances in bio-inspired computational design highlights the critical tension of how ML algorithms can synergistically enhance human creativity, with three key debates.

First, The Digital Universalism in Computational Design. While Frazer's (2002) evolutionary architecture prioritized top-down optimization, contemporary research leverages bottom-up biological intelligence for sustainability. For instance, Ertan and Adem (2024) showed how discrete aggregation of timber units can encode craftsmanship, presenting data aggregation as a form of cultural expression, responding to Charitonidou's (2022) critique of "digital universalism".

Second, Translating Theoretical Automata to Design Agents While classical Cellular Automata (CA) theory (Wolfram, 2005) emphasized emergent complexity from simplicity, recent work integrates environmental data. Reimagined discrete aggregation discourses (Retsin, 2019; Kohler, 2017) prioritised ecological and topological assembly, but discounted evaluating the knowledge transfer between algorithms.

Third, Data Biases as Design Opportunities. ML's rise intensified debates about agency and contextual sensitivity. Work exploring the relationship between text, images, and form in ML (Koh, 2023; Bolojan et al., 2023) established linguistic and visual principles for stylistic exploration; simultaneously, it highlighted the role of unintentional factors or biases in the learning process. This reframed operational bias as a vehicle for embedding priorities into a dataset.

Across these cases, discrete computation was the trigger for complexity via simplicity. In the case of CA, the outcomes are based on the interplay of basic growth rules in a strict relationship with an environment. The best feasible solution to the interaction of these components in an n-dimensional space is the outcome. As a result, such aggregation logic favours choices above ideal solutions in a dynamic process that is always changing—Dynamism. Therefore, working with aggregated shapes and adding characteristics that define a CA, we may produce a spectrum of outputs that speculatively inform the solution space, with seeds acting as agents and geometrical outputs acting as aggregation rules. This allows for the testing of different data inputs, such as solar radiation to enhance a building's light-harvesting quality and iterate options in an evolutionary manner.

This presents an opportunity to synthesize cross-intelligence systems for tailored cases, responding to specific disciplinary needs. This study uses CA to train ML models for architectural acceleration and contextualisation—a gap that our "CA-as-teacher" pipeline addresses. This positioned CA not as a form-generator but as a bridge between designer intuition (rule definition) and ML efficiency (pattern replication), shifting CA from a computationally-heavy application to an intelligent pedagogical tool for ML.

3 Methods

Addresses the research gap between discrete computation, biomimicry, and ML contextualization in architectural design, this research responds to limitations identified in our prior CA-to-GAN pipeline (Ng et al., 2021, 2022), which encountered three critical challenges:

1. Rigid discretization hindered organic form and cultural adaptation
2. Inadequate control for contextual variables (local resources, aesthetics)
3. Computational bottlenecks in translating CA outputs to design solutions

To resolve these, we developed a novel three-phase methodology.

PHASE 1: CA-ML Pipeline Enhancement with Houdini. Houdini's procedural environment enabled three key innovations, including biologically-grounded growth simulation, adaptive discretisation with dynamic voxel resizing (5–50 cm^3) responding to solar incidence thresholds, and data standardization with lossless translation between CA states to generate adaptive design outputs.

PHASE 2: Cross-context Deployment. A series of workshops were organised where 20 designers from 12 countries worked in teams of 4 to test the pipeline, following a task protocol.

1. Select hometown context with documented climatic/cultural constraints
2. Generate climatic data using CA to train GANs
3. Translate CA states into Houdini to create baseline design solutions
4. Iterate designs with human-GAN feedback to accelerate the process

PHASE 3: Benchmarking Comparing rule-based and ML workflows in terms of time per generation, total computational time, and data volume (Table 1; Figs. 1 and 2).

Table 1. The algorithmic components and specification of the proposed hybrid workflow.

Component	Specification	Role
Houdini 19.5	VEX-based growth algorithms	Bio-inspired form generation
Rhino 3D	Galapagos evolutionary solver	Solar optimization
PyTorch	Custom pix2pix	ML acceleration
RTX 4090	48GB VRAM	Training/inference

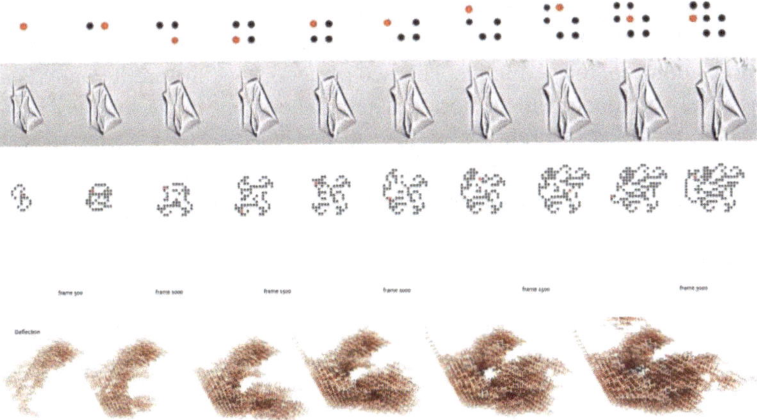

Fig. 1. A CA "seed" may generate a large set of outcomes by activating the discrete grid state from a simple on and off (0 to 1) in its basic form, controlled by predefined rules, such as solar radiation predictions, suitable for context with low data resolutions.

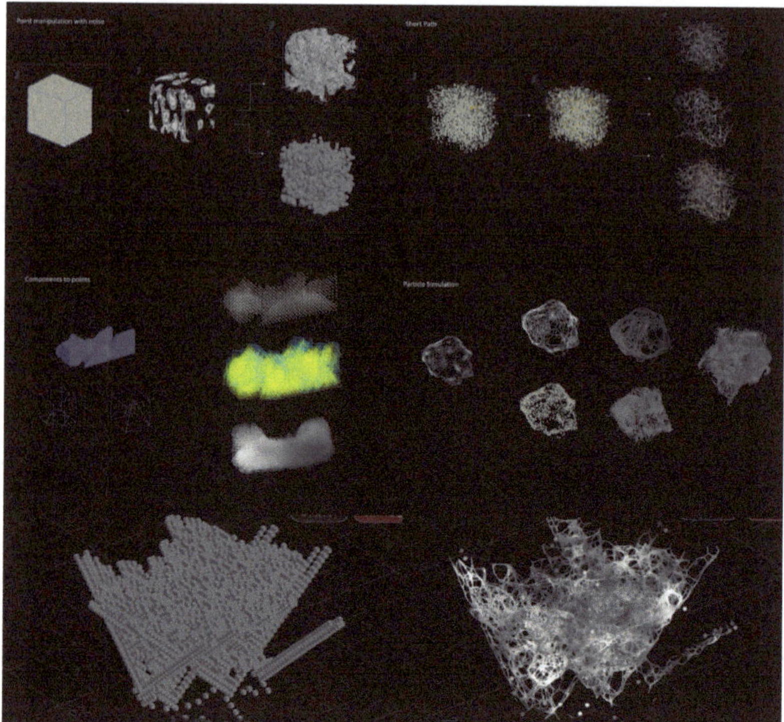

Fig. 2. The 3D data flow creates a feedback loop between Rhino and Houdini using discrete aggregation logic of CA, with voxel as proxy to define the geometry growth.

4 Design Outcomes

These projects hope to demonstrate how the same pipeline can lead to diverse design output with its combinatorial quality. The two projects have different contexts and local organisms as inspiration: one in the northern hemisphere, one in the southern hemisphere. Although both contexts are relatively close to the equator, they suffer from different climatic concerns: desert and highland climates. Both projects tried to combine active and passive solar strategies into one design as the surrounding environments are relatively scarce in resources due to socioeconomic or geographical reasons.

The Sanna Tower is a low-rise refugee shelter design situated in low-income fabrics, next to Alawi Mosque, Sharmah, Yemen. The project feedback to local society by enabling self-sufficient energy structures. As a seaport country, the design took inspiration from commonly found waste materials—discarded shipping containers that can be assembled and disassembled easily as temporary shelter structures—and upcycle them with PV units into energy resilient refugee blocks. The challenge with Modular Integrated Constructions (MiC) is to define a distribution logic that utilizes the site boundary effectively without homogenising the landscape. Participants learnt from the aggregation strategies of Marine Sponge: the organisms stack upon one another without being in direct competition for light. Here, sunlight instances became an instructor of CA distribution.

The coast of the Red Sea (Tihamah) has a desert climate, generally hot but suffers from daily temperature differences that range from 15–35 °C in inland areas. Being relatively close to the equator, daylight often comes from most directions all year round (Fig. 3).

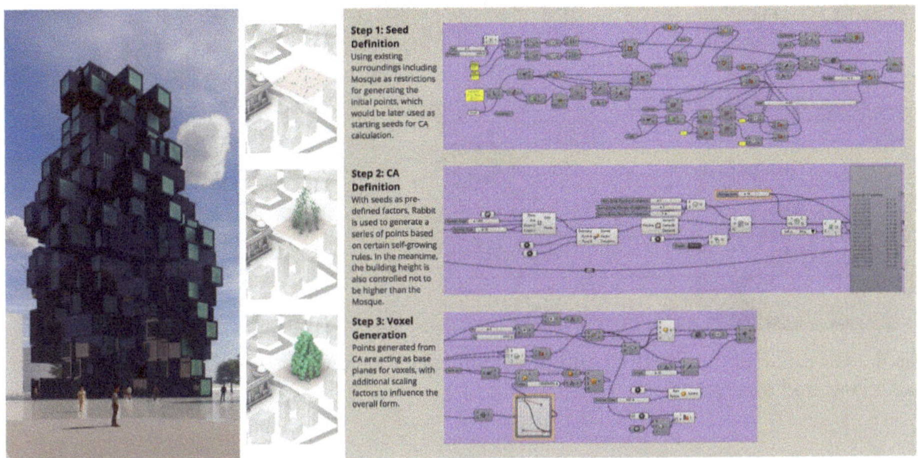

Fig. 3. The Sanna Refugee Tower. Participants credits: Chowon Kang, Yutong Zhang.

The design utilised a combination of active and passive solar strategy by playing with the orientation of units. Participants considered installing windows made with Luminescent Solar Concentrators (LSCs), which are ultra low-cost plastic doped with fluorescent dyes that harvest light while keeping windows semi-transparent.

The design process began with solar simulation to structure a voxel grid with CA, then upsampled the voxel grid with container models. The same process was carried out over a number of sites in adjacent areas to find out which site is the best for light-harvesting. Finally, participants chose this Mosque, which is a cultural site for solidarity, religious services and charity work, and the area has shown to have more Mosques than supermarkets. The flexibility allows for the same workflow to be applied to each of those sites to assess possibilities of temporary towers.

The L1ch3nic Sph3ribl0b5 is a project designed for hikers on the highland of Markawasi Plateau Geopark, Lima, Peru. The hiking tracks are in far locations, the project offers resting spots which would provide water, electricity, and light for the travellers. Participants speculate on micro-spherical solar cells and vapour-harvesting design for a combination of active and passive strategies.

Spherical solar cells are an emerging technique that tries to tackle daylight direction problems. As sunlight is not uniform in the natural environment and the sun is always moving, the typology of the sphere helps to catch light coming from all directions, ensuring at least 50% of the overall surface is in contact with light at all times, even catching those reflected and refracted from clouds and water. The design also includes a structural component and a web that catches and condenses vapour from the temperature change at dawn to provide water (Figs. 4 and 5).

Fig. 4. The LICHENIC SPHERI-BLOBS. Participants credits: Carlos Rivera, Manuel Halim, Gao Xiang, and Mason Mo.

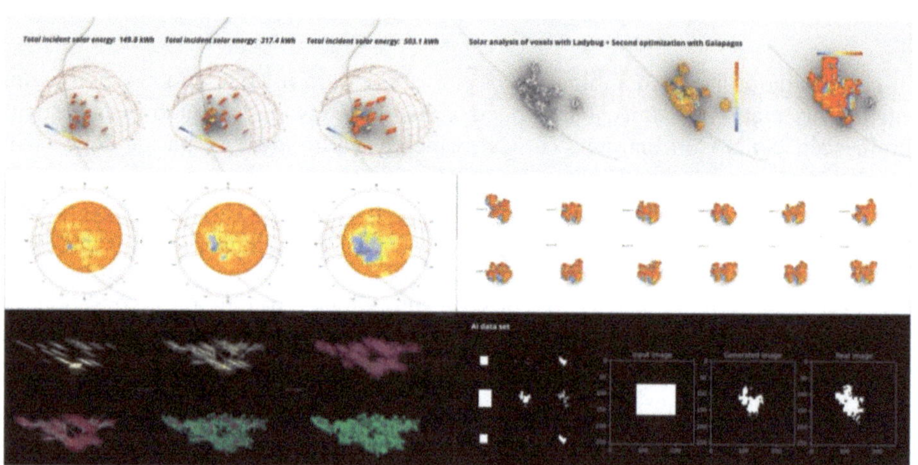

Fig. 5. Learning from Thallus Lichen that is often found in the highlands and able to collect moisture in the air, Participants simulated the growth pattern in houdini, which were then discretized into voxels. Afterwards, they trained the ML pix2pix with solar data, and the output in pixel format was then stacked into voxels. Finally, the second set of voxels output were imported back in houdini to transform the growth pattern.

5 Comparing Computational Performance

Comparing CA as the teacher and GAN as the student, they outperformed each other at different tasks. CA for optimization relied on population-based methods, utilizing selection, mutation, and crossover to refine candidate solutions. These algorithms directly optimize design parameters or structures by evaluating their fitness in Rhino3D. These simulations were computationally expensive, requiring substantial time and resources, with evaluations taking hours to days for a thousand generations. They did not require pre-existing datasets and are suitable for contexts with low-resolution data (e.g. a single screenshot, satellite images); they did, however, need significant data storage for simulation results and population data. Despite the high computational costs, CA maintained a diverse population, allowing for multi-objective optimization and producing high-fidelity solutions based on physics simulations.

GANs operated on the core principle of adversarial training between generator and discriminator networks. They were designed to learn and generate realistic data distributions, which could be adapted to generate optimized designs very quickly when trained (within seconds vs. the hours needed by CA). However, GANs require datasets of images or structures for training, and are often difficult to apply to low-resolution targets, such as those sites chosen by participants of our workshop. The algorithmic quality is highly dependent on the dataset's size and diversity, thus, they make great students for CA, which could evolve solutions from very little data but takes a long time to process. Although GANs training could be computationally intensive, often taking hours on GPUs, once completed, inference was remarkably fast (within seconds). Further, it required much less storage capacity compared to CA.

Overall, CA were best suited for direct optimization problems with explicit fitness functions, while GANs excelled in generative modeling and rapid design synthesis post-training. CA had straightforward evolutionary steps, but the evaluations were costly, whereas GANs involved complex adversarial training. When combined for hybrid approaches, design iterations could leverage the strengths of both methods, balancing quality with time costs (Table 2).

Table 2. Comparing computational performances of CA as the teacher and GAN as the student.

Aspect	CA	GAN
Iterations/ Generations	900–1000 evolutions for convergence.	600 training iterations, 500 samples, 256x256 px each.
Compute Time	6–7 h for 900–1000 generations.	8–10 h on GPU. After training = ~instantaneous.
Data Storage	70 GB for 900–1000 evolutions.	2–4 GB (200 MB checkpoints × 12 saves over 600 iterations).
Intermediate Data	High (simulation outputs, population states).	Moderate (gradient updates, logs).
Hardware	CPU (parallelization limited by simulation bottlenecks).	GPU-accelerated (training); CPU/GPU for inference.

(continued)

Table 2. (*continued*)

Aspect	CA	GAN
Scalability	Time/data grow linearly with generations.	Storage scales with checkpoints; training time depends on the dataset.
Use Case Suitability	Context with little data; generating synthetic data	Data-rich context; almost instantaneous inference.

6 Human-Machine Collaboration

Although the pipeline has been shared in the same manner, each team of participants deployed it with variability. The most common workflow can be summarised as follows.

After preliminary contextual and bio-inspired research for concept framing, designers started the algorithmic process in three ways: (a) a parametrically defined form, (b) a set of manipulatable data points, or (c) solar simulation to structure a voxel grid.

The main challenge was how the complexity of bio-inspired form can be processed by solar analysis and CA with the least amount of computing power and run time so the pipeline can be democratised amongst team members with varying computing capacity.

Two solutions came up. The first was demonstrated by LICHENIC SPHERI-BLOBS: asking the CA to teach ML, which automated tasks in an instantaneous manner when trained. The second was tested by The Sanna Tower: separating the generation of complex geometry and instruction of unit aggregation. The former can be done with (a) & (b), whereas the latter are done with (c) as geometry proxies: the voxels are taken as empty data vessels that can be upsampled with complex geometries in Houdini.

In this way, the computing power and time are largely minimised because CA were only processing a set of boxes, although it could compromise the accuracy on a micro-level, the evaluation on a macro-scale remained relatively the same. The key here is to balance the resolution of the discrete grid over the predefined volume. However, it could raise issues of continuity between discrete units in later design phases, which has to be solved through designer intents. There were generally two approaches to tackle it: discretise a continuous shape or aggregate discrete units with connectors, which the selected projects had illustrated.

The overall design process took 3–5 days (Fig. 6).

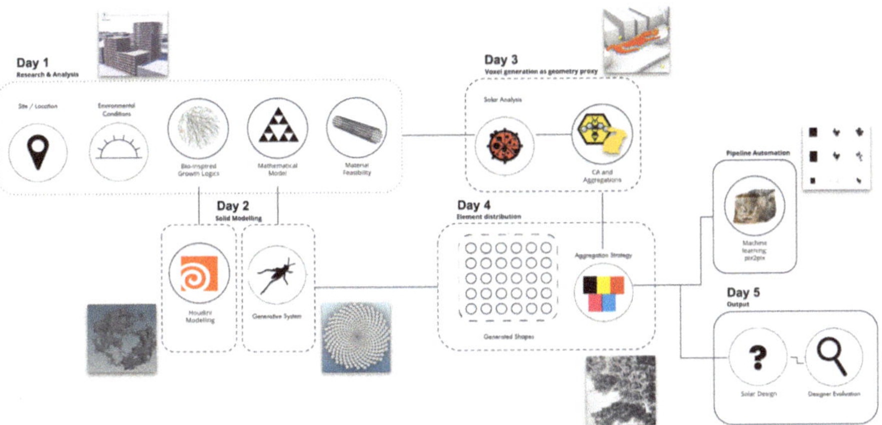

Fig. 6. The proposed hybrid workflow combining CA, ML, and Houdini growth algorithms to balance between computational efficiency and designer needs.

The tensions and opportunities negotiated through this technological framework was critically captured by participants' reflections:

> "Through this iterative design workflow, a proposal is able to adapt to different conditions in the same environment, and opens up possibilities for different alternative designs. We architects base our design skills and criterias on personal learning experiences. AI does pretty much the same, with the difference that it has the capacity of choosing combinatorial paths of data that us, human beings, wouldn't even be capable of thinking of. AI allows us to see things that we are not able to see at first sight"."

> "Through this workshop pipeline, a symbiotic relationship between human and machine has been created. Decisions made by the architect, evaluated by machine, getting a product, and going back again in the loop. A question is posed for the future. What exactly is the architect's role in such a design process?"

> "The future of design relies on self-configuring architecture. With all the design extension tools that architects have got at the moment, we can make such accurate proposals that can adapt to the slightest environmental changes; per month, per day, per hour, per minute. This fact clashes with the reality of today's architectural products; the static, the perpetual."

On this note, the collaboration was not simply human-machine, but also machine-machine and human-human, without whom, the diverse understanding of the experiment would not have been possible (Fig. 7).

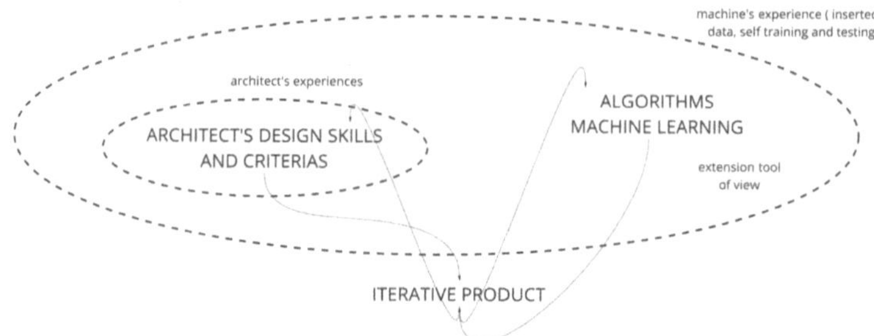

Fig. 7. Participants' diagrammatic reflection on human-machine collaborative workflows.

7 Discussions and Conclusions

By exploring the role of CA as pedagogical agents to translate contextual intelligence into trainable protocols for ML, the study contributed to three critical gaps in architectural computation: 1) the contextual void in AI-driven design, where prior approaches prioritized stylistic generation over environmental-cultural synthesis; 2) the scalability limits of rule-based systems, overcoming compute bottlenecks through ML acceleration; 3) the biomimicry implementation gap, moving beyond aesthetic metaphors to elements distribution logic.

The outcomes highlight a pedagogical paradigm where synthetic intelligence learns from each other into self-advancement, guided by human priorities and needs. The experiemnts tried to navigate the complexity of implementing symbolic automata with connectist ML by treating contextual information as a scaffold for neural networks. By doing so, it also tests the boundaries of translating between discrete computations' rigor with organic design expressions.

However, several limitations must be acknowledged, including the closed-loop feedback between synthetic datasets, which indicated that real-world performance validation remained a future endeavour. Additionally, cultural adaptation relied heavily on designer intervention, indicating a careful design of engagement protocols with decision nodes alongside computational automation procedures. Finally, there is an opportunity to align discrete and voxel based computation with BIM or supplier databases to enhance implementability of the designs.

As sustainable architecture confronts escalating climatic and socioeconomic complexity, this research argues for machines that learn contextually—not autonomously—as students of ecological wisdom and a human-steered interpreter of the world. The findings and limitations pointed to areas for further research in the quest for a more site-sensitive and design-adaptive computational process.

Acknowledgement. We sincerely thank all workshop participants for their contribution, especially Chowon Kang, Yutong Zhang, Carlos Rivera, Manuel Halim, Gao Xiang, and Mason Mo; and Digital Futures for facilitating a workshop platform.

References

Bolojan, D., et al.: Latent design spaces: Interconnected deep learning models for expanding the architectural search space. In: Architecture and Design for Industry 4.0, pp. 201–223. Springer International Publishing, Cham (2023)
Chaillou, S.: Archigan: Artificial intelligence x architecture. In: Architectural intelligence: CDRF 2019, pp. 117–127. Singapore, Springer Nature Singapore (2020)
Charitonidou, M.: Towards a civic approach to urban data: The myths of digital universalism. Computing. **20**(2), 238–253 (2022)
Ertan, İ.B., Adem, P.Ç.: Computational design with Kurtboğaz: The generation of timber structures with an aggregative design algorithm. J. Comput. Des. **5**(2), 235–258 (2024)
Frazer, J.: A natural model for architecture: The nature of the evolutionary model 1995. In: Cyber Reader Critical Writings for the Digital Era, pp. 246–255 (2002)
Koh, I.: Architectural and social 'word forms' slippage with deep learning. In: CAAD F, pp. 112–125. Cham, Springer Nature Switzerland (2023)
Köhler, D.: Large city architecture: the mereological mode of the quantified city. J. Parallel Emerg. Distrib. Syst. **32**(sup1), S163–S172 (2017)
Ng, P., et al. (2021). AI In+ form: Intelligence and Aggregation for Solar Designs.
Ng, P., et al.: An info-biological theory approach to computer-aided architectural design. In: CAADFutures, pp. 436–455. Singapore, Springer Singapore (2022)
Retsin, G. (ed.): Discrete: Reappraising the digital in architecture. Wiley (2019)
Wolfram, S.: Cellular automata. In: Modeling Chemical Systems. Springer (2005)

Optimized Assembly Sequence for Self-Interlocking SL Blocks in Robotic Fabrication

Sama Rajabi[1], Ayda Aliasgarian[1](✉), Reza Taghavifard[2], and Andrei Nejur[2]

[1] Faculty of Architecture and Urban Design, University of Tehran, Tehran, Iran
ayda.aliasgarian@ut.ac.ir

[2] Laboratory of Architecture, Informatics, and Robotics, School of Architecture, Faculty of Environmental Design, University of Montreal, Montreal, Canada

Abstract. Self-interlocking SL blocks offer a modular construction method without adhesives or fasteners, which is suitable for prefabricated building components. Their interlocking geometry ensures stability while enabling rapid assembly and disassembly, making them ideal for robotic fabrication. The robotic assembly of SL blocks reduces labor costs, improves safety, and offers flexibility for various construction needs. The assembly sequence is crucial for robotic fabrication, influencing feasibility and efficiency. Automating this sequence optimizes the process, addressing challenges like component pre-assembly, structural stability, and robot motion planning. This study develops an algorithm for an optimized assembly sequence using a robotic arm. The sequence is derived by reversing a disassembly process. In each loop, a candidate block is selected based on criteria like removable blocks, stability checks, and collision avoidance. Minimizing robotic arm joint rotations and refining motion planning based on block distance from the source ensure structural integrity and operability. The approach narrows possible solutions to determine an optimal assembly sequence.

Keywords: Self-interlocking blocks · SL blocks · Robotic assembly · Assembly sequence planning · Motion planning

1 Introduction

Robotic assembly enables efficient construction through automated processes where robots manipulate parts to form a product, including pick-and-place tasks for object handling [1]. These systems enhance precision, reduce labor needs, and minimize errors [2, 3]. A well-planned assembly sequence ensures structural stability and allows collision-free, feasible placements [4]. This study focuses on the robotic assembly of SL blocks, aiming to optimize assembly sequences for improved robotic performance.

1.1 SL Blocks

SL blocks, developed by Shih [5], advance modular construction by integrating robotics, digital design, and sustainable material reuse. Their octocube shape and self-interlocking

Y. Liu et al. (Eds.): CDRF 2025, *Transindividual Intelligence*, pp. 74–84, 2026.
https://doi.org/10.1007/978-981-92-0615-5_7

capabilities allow for the creation of versatile structures through the concatenation of SL blocks into SL strands. The shape grammar of SL strands relies on six engagement types using specific rotations and translations for stable, continuous, or periodic formations. (Fig. 1). SL blocks can interlock reversibly, enabling the formation of 2D and 3D structures without permanent connections. They can be used for walls, columns, and floor slabs, making them ideal for prefabricated buildings [6].

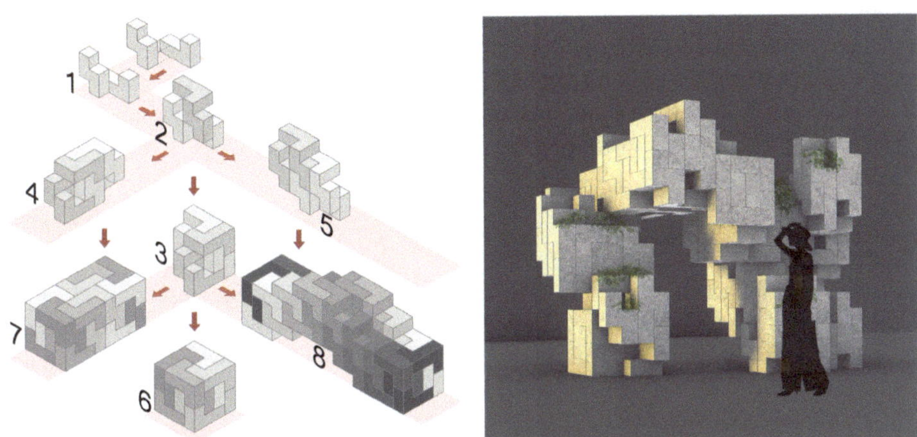

Fig. 1. Left: Formation of SL strands from "a", "h" and "d" engagement types, 1) an SL block, 2) a pair, 3,4,5) engagement "a", "h", and "d", 6,7,8) engagement sequence "a^4", "$(ahha)^2$", and "$(addhddha)^2$", Right: A visualization of a Pavillion formed with SL blocks

1.2 Robotic Assembly

Esmaeili Charkhab et al. [7] explored recursive algorithms to determine effective robotic assembly sequences for SL blocks. They incorporated directional blocking graphs to analyze assembly direction and optimize the positioning of fixation blocks and grasping points. They utilized a voxel-based representation to subdivide SL blocks for detailed spatial analysis and constraint management. Wang [8] developed a mathematical method for automated disassembly sequence planning using an assembly matrix and three auxiliary matrices (contact, space interference, and relation). Belousov et al. [9] explored assembly in 3D with complex blocks, which is non-trivial and more challenging than 2D assemblies. They employed reinforcement learning (TD3 algorithm) for autonomous block placement and linked Rhino/Grasshopper with PyBullet physics engine for real-time stability assessment, enabling robots to refine assembly strategies through simulation feedback. Liu et al. [10] introduced an Overall Assembly Integrity Evaluation for assembly sequence optimization. It considers assembly space operability, assembly stability, fixture changeability, and assembly direction adjustments. Smith et al. [11] improve disassembly efficiency by prioritizing 90° direction changes over 180° rotations to reduce unnecessary reorientations, which enhances speed by reducing large rotations and decreases machinery wear. Murali et al. [12] incorporate directional changes into a fitness function to evaluate assembly sequence quality, with fewer changes indicating a more efficient process.

1.3 Problem Statement

Robotic assembly sequence planning faces several challenges and limitations, primarily due to the complexity and diversity of modern manufacturing tasks. The exponential growth in potential assembly sequences as the number of parts increases is a significant challenge, necessitating advanced computational methods to efficiently determine optimal sequences [13, 14]. This research aims to develop an efficient assembly sequence for SL blocks. The possible solutions are refined through several steps, leading to an optimal sequence and improved motion planning.

2 Methodology

This study develops an algorithm to optimize the assembly sequence of self-interlocking SL blocks for robotic fabrication. The objective is to ensure an efficient and feasible robotic assembly process by minimizing unnecessary robotic movements while maintaining structural stability. The developed algorithm optimizes the assembly sequence of SL blocks by reversing a disassembly sequence. The algorithm consists of five steps, outlined in Fig. 2 and discussed in more detail in the following sections. The process begins by identifying removable blocks within a modular structure. A stability check ensures that removing a block does not compromise integrity. The algorithm then prioritizes blocks based on collision avoidance, minimal joint rotation, and shortest motion distance. This iterative process continues until all blocks are included in the disassembly sequence, which is then reversed to obtain the optimized assembly sequence.

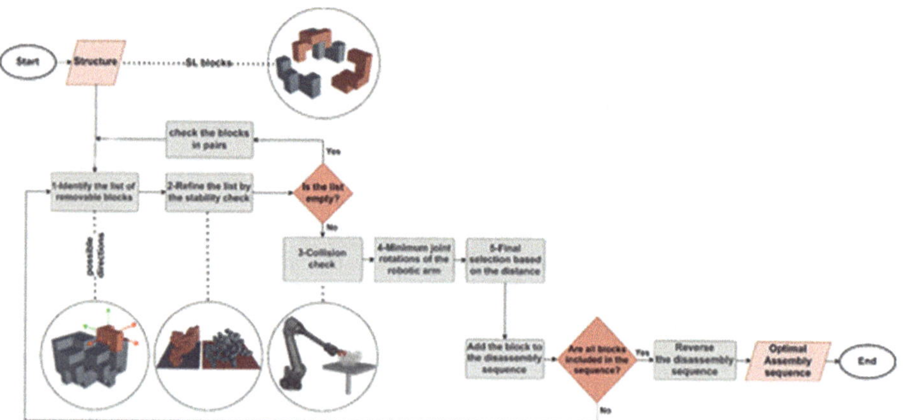

Fig. 2. Overview of the steps for obtaining the assembly sequence of SL blocks

2.1 Identifying Removable Blocks

The first step is to identify removable blocks, which have unblocked, free sides that allow movement in specific directions. Once the list of removable blocks is generated, the algorithm identifies all free sides as possible removal directions for each removable block. The removal directions are determined based on the cubic geometry of the modules as *+z, +x, -x, +y*, and *-y*. The direction *-z* is excluded both because the structure relies on gravity -modules removable downward wouldn't be stable- and due to robotic constraints, as the volume sits on a table, making access from below impossible. Each direction is validated by checking for obstacles along the movement path; if clear, the direction is marked as valid (Fig. 3).

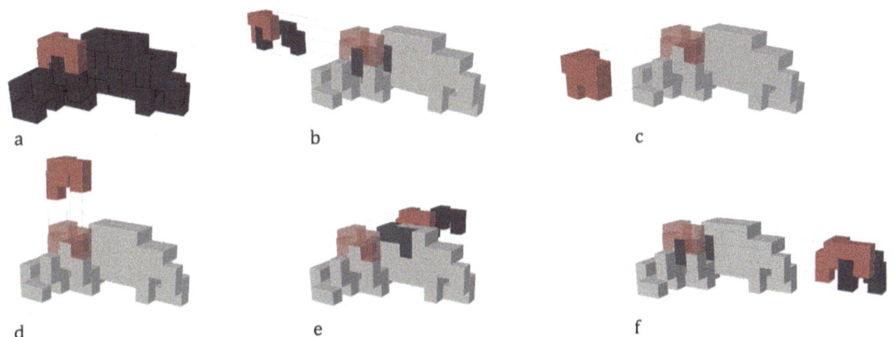

Fig. 3. Illustration of the approach for determining removal directions among five main directions: a) The initial state - the pink block, highlighted for removal direction analysis, b-f) The movements among five main directions. The black blocks in (b), (e), and (f) act as obstacles.

When no single block can be removed due to structural constraints, the algorithm adopts an alternative strategy: identifying pairs of blocks that can be removed simultaneously. The same selection criteria - stability check, collision avoidance, joint rotation minimization, and motion efficiency - are then applied to these pairs.

2.2 Stability Check for Modular Assemblies

The stability check ensures structural integrity at each assembly or disassembly step through real-time feedback between the design environment and the simulation platform. This study uses PyBullet [15], a high-performance physics simulation library, to assess stability, simulating rigid-body dynamics, collision detection, and gravity effects. PyBullet evaluates the stability of SL block assemblies under physical constraints through a bidirectional communication framework linking Grasshopper via a socket-based mechanism (Fig. 4).

Grasshopper extracts and transmits each block's position, orientation, and geometric properties to the simulation platform. PyBullet then initializes the environment, loads blocks as rigid bodies, and applies gravitational forces. Stability is assessed by analyzing displacement and orientation changes under gravity and contact forces, ensuring that

Fig. 4. Stability check process in Pybullet: (a) Initialization of bodies in the PyBullet environment, (b) Simulated bodies undergoing stability analysis

all blocks remain within predefined motion thresholds. The stability status ("stable" or "unstable") is then sent back to Grasshopper, guiding the sequence to maintain structural integrity. This continuous feedback loop ensures efficient modular assembly planning by prioritizing stable configurations.

2.3 Motion Planning and Collision Detection Between Robot and Structure

Efficient robotic assembly requires accurate collision detection to ensure smooth operations and prevent damage. This study employs a computational framework in Rhino and Grasshopper, using the Robot Components library [16] to simulate and evaluate potential collisions. The algorithm detects collisions by simulating the robot's movements and checking for intersections between the robot's body and the placed blocks:

Movement Path Planning and Kinematics The robot's movement is planned using predefined actions in the Robot Components library. Based on the list of removable blocks and their allowed directions, the algorithm defines a start and target plane, applying an offset to improve movement efficiency. The path is generated using the Robot Components library, and forward kinematics (FK) is applied to compute the robot's pose along the path, simulating stepwise motion (Fig. 5(a)).

Collision Detection The collision detection system checks for overlaps between the robot's meshes and the surrounding structure. This study employs a multiple interference detection method [17] to identify collisions. The algorithm samples the robot's trajectories and performs pairwise collision checks between the robot's meshes and the structure's geometry (Fig. 5(b)).

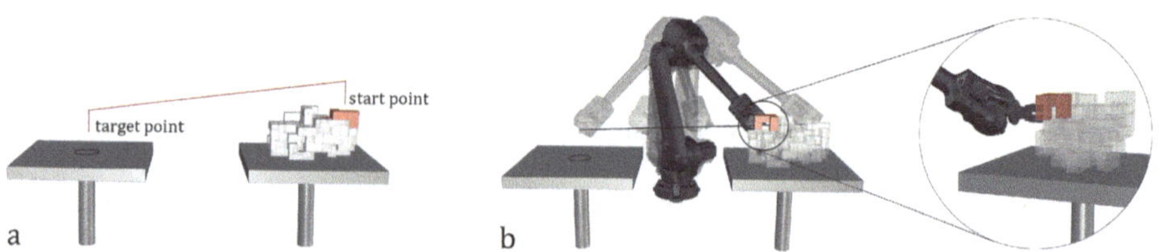

Fig. 5. (a) Path generation for the selected block in the $+z$ direction, (b) Collision detection process: Five samples of the robot's body trajectories. A collision is detected in the second sample

At each step, the Robot Components library evaluates the shortest movement paths for candidate blocks and their removal directions to ensure collision-free assembly. Configurations with collision-free movement are prioritized over those with collisions. Since collision-free motion planning is computationally expensive, configurations that result in collisions are deprioritized. If all possible movements result in collisions, the algorithm selects one based on specific criteria (explained later). Additional motion planning is needed in the future to achieve fully collision-free solutions. This prioritization keeps the robot's movements collision-free, with adjustments made as needed.

2.4 Minimizing Joint Rotations

The algorithm prioritizes minimizing unnecessary joint rotations when selecting the next block for removal, following the directional change minimization principles [11, 12]. Each removable block is evaluated based on the joint rotation it induces in the robotic arm. For each block and its possible removal directions, the algorithm assesses whether the removal direction aligns with the previous movement of the robotic arm. The optimal scenario is when a block can be removed without changing the robot's movement direction. If no such block is available in a given step, the selection prioritizes blocks that require the smallest necessary change in joint angles: **Highest Priority** Blocks with removal directions fully aligned with the previous movement (0° rotation). **Second Priority** Blocks requiring a 90° rotation. **Lowest Priority** Blocks requiring the highest directional change (180° rotation).

Figure 6 illustrates this prioritization. In this example, the previous movement was in the $+y$ direction. The pink arrow represents the highest-priority removal direction (0° rotation). The gray arrows indicate second-priority directions requiring a 90° rotation. The red arrow represents the lowest-priority direction, requiring the largest directional change (180° rotation). By following this prioritization scheme, the algorithm ensures smoother transitions between removal steps, reducing unnecessary joint adjustments and enhancing overall motion efficiency.

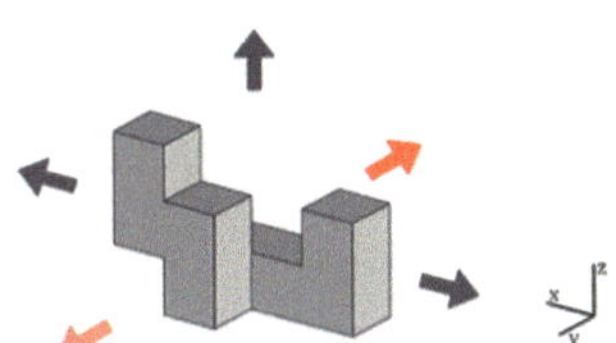

Fig. 6. Prioritization of removal directions based on joint rotation minimization

2.5 Final Selection Based on Distance

In the final stage of the algorithm, one of the remaining candidates is selected based on its distance from the designated placement location. During the process of generating the disassembly sequence, priority is given to blocks that have a shorter distance to their placement position. Figure 7 shows the prioritization of blocks based on distance. This selection strategy ensures that when the disassembly sequence is reversed to generate the assembly sequence, blocks that are farther from the supply point are placed first, while those closer to the supply point are placed later. As a result, the assembly sequence begins with the placement of distant blocks and gradually moves toward the closer ones. This ordering is crucial for preventing potential collisions between the robotic arm and the partially assembled structure. By placing distant blocks first, the need for the robotic arm to navigate around the existing assembly is minimized. This approach reduces unnecessary motion complexity and improves the overall efficiency and safety of the assembly process. The algorithm builds an optimized disassembly sequence by selecting removable blocks based on stability, collision avoidance, minimal joint rotations, and distance. Reversing this sequence yields an efficient and feasible assembly plan for robotic fabrication.

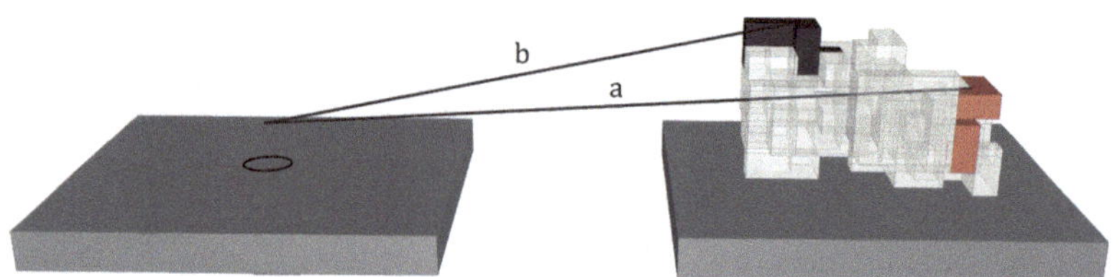

Fig. 7. Prioritization of blocks based on distance. The black block is prioritized over the pink block due to its shorter distance from the designated placement location

3 Results

To evaluate the performance of the proposed algorithm, a case study was conducted using a predefined modular structure composed of SL blocks. The test structure consists of eight SL blocks arranged in an interlocking pattern. The blocks exhibit geometric dependencies, meaning that certain blocks must be placed before others to maintain stability and feasibility during robotic assembly. The initial configuration and the structural dependencies between blocks are illustrated in Fig. 8.

Fig. 8. The initial configuration of the conducted structure

3.1 Determining the Disassembly Sequence

The disassembly sequence was generated by progressively selecting removable blocks based on the stability check, collision avoidance, minimal joint rotations, and motion efficiency criteria. Table 1 presents the block selection process for disassembly in the conducted structure. In each step, a block and an appropriate direction for its removal are selected. In step 4, since no single removable block was available, a pair of blocks (blocks 5 and 6) was selected for removal. In steps 5 and 6, where multiple blocks are eligible for removal, selection is based on the criteria outlined earlier. In step 5, block 7 is removed in the $+z$ direction because this configuration minimizes joint rotation changes and has a shorter distance to the destination than block 3. In step 6, blocks 3 and 4 are considered candidates, but block 3 is chosen, as removing block 4 would compromise the structural stability. The remaining block can be removed in any of the five directions. However, to minimize joint rotations, the $+z$ direction is chosen. The resulting sequence, represented as a list of tuples (block index, removal direction), ensures that each removal step maintains the overall stability of the structure.

Table 1. Block selection process for disassembly

Step	Candidate block	Position of the block in the structure	Stability check	Direction(s)	Detected collision	Joint rotation angle	Distance (cm)	Final selection
1	2		stable	+z	no	0	133	2, +z
2	1		stable	+z	no	0	139	1, +z
3	0		stable	+x	no	90	138	0, +x
4	(5, 6)	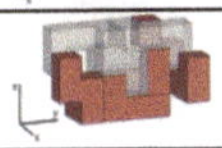	stable	+z	no	90	124	(5,6), +z
5	3		stable	+x	no	90	133	7, +z
				+z	no	0		
	7		stable	-x	no	90	114	
				-y	no	90		
				+z	no	0		
6	3	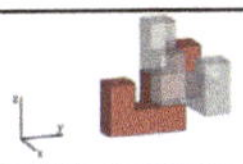	stable	+x	no	90	133	3, +z
				+z	no	0		
	4		not stable	-x	-	-	-	
7	4		stable	+x	no	90	132	4, +z
				-x	no	90		
				+y	no	90		
				-y	no	90		
				+z	no	0		

3.2 Reversing the Sequence to Obtain the Assembly Order

The optimized assembly sequence was obtained by reversing the disassembly sequence, ensuring robotic feasibility. Table 2 outlines the placement order and movement directions. This approach efficiently generates a feasible assembly process, maintaining structural stability while minimizing robotic movement complexity. It offers a structured strategy for modular robotic fabrication in real-world applications.

Table 2. The disassembly and assembly sequence of the conducted structure

	Block number, Direction						
Disassembly sequence	2, $+z$	1, $+z$	0, $+x$	(5, 6), $+z$	7, $+z$	3, $+z$	4, $+z$
Assembly sequence	4, $+z$	3, $+z$	7, $+z$	(5, 6), $+z$	0, $+x$	1, $+z$	2, $+z$

3.3 Algorithm Runtime and Success Rate

Sixty random volumes were generated. The algorithm's runtime for each volume increases as the number of modules increases, and the algorithm's success rate in finding a sequence decreases with a higher number of modules. The volumes were divided into three groups: 5–15, 16–25, and 26–35 modules, each with 20 volumes (Fig. 9).

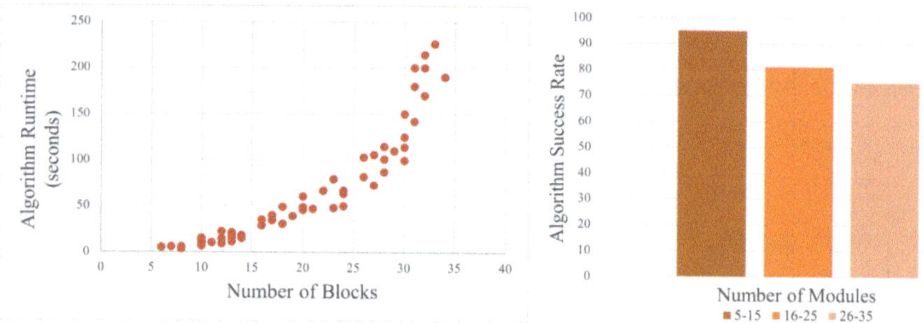

Fig. 9. Left: Algorithm runtime based on the number of modules. Right: Percentage of algorithm success rate based on Number of modules

4 Conclusion

This study introduced a computational approach to optimize the robotic assembly sequence of self-interlocking SL blocks. By applying a reverse disassembly strategy, the algorithm determined an optimal sequence that ensures stability, minimizes robotic movement complexity, and avoids collisions. The methodology incorporated stability validation, collision detection, and motion efficiency through joint rotation minimization and distance prioritization. Reversing the disassembly sequence led to an optimized, assembly process, reducing robotic movements while maintaining structural integrity. These findings underscore the potential of algorithm-driven modular assembly in automated construction. Future research could explore real-world robotic implementations and AI-driven assembly sequence determination.

References

1. Zhu, Z., Hu, H.: Robot learning from demonstration in robotic assembly: A survey. Robotics. **7**, 17 (2018)
2. García, R., Perez, A., Pulido, J.A., Ulloa, P., Forcael, E.: Assembly of stay-in-place concrete blocks using a robot. IOP Conf. Ser. Earth Environ. Sci. **503**, 012077 (2020)
3. Kodama, Y., Yamazaki, Y., Kato, H., Iguchi, Y., Naoi, H.: A Robotized Wall Erection System with Solid Components (1988). https://doi.org/10.22260/ISARC1988/0052
4. Huang, Y., Garrett, C.R., Ting, I., Parascho, S., Mueller, C.T.: Robotic additive construction of bar structures: Unified sequence and motion planning. Constr. Robot. **5**, 115–130 (2021)
5. Shih, S.-G.: The art and mathematics of self-interlocking SL blocks. In: Torrence, E., Torrence, B., Séquin, C., Fenyvesi, K. (eds.) Proc. Bridg. 2018 Math, pp. 107–114. Art Music Archit. Educ. Cult. Tessellations Publishing, Phoenix, Arizona (2018)
6. Liu, Y., et al.: Advancing sustainable construction: discrete modular systems & robotic assembly. Sustainability. **16**, 6678 (2024)
7. Esmaeili Charkhab, M., Liu, Y., Belousov, B., Peters, J., Tessmann, O.: Designing for robotic (Dis-)assembly. In: Dörfler, K., Knippers, J., Menges, A., Parascho, S., Pottmann, H., Wortmann, T. (eds.) Adv. Archit. Geom, pp. 275–288. De Gruyter (2023)
8. Wang, Y., et al.: Interlocking problems in disassembly sequence planning. Int. J. Prod. Res. **59**, 4723–4735 (2021)
9. Belousov, B., et al.: Robotic architectural assembly with tactile skills: Simulation and optimization. Autom. Constr. **133**, 104006 (2022)
10. Liu, P., Wu, L., Wang, Y., Guo, L.: Optimization method for assembly sequence evaluation based on assembly cost and ontology of aviation reducers. Appl. Sci. **14**, 5116 (2024)
11. Smith, S., Hsu, L.-Y., Smith, G.C.: Partial disassembly sequence planning based on cost-benefit analysis. J. Clean. Prod. **139**, 729–739 (2016)
12. Murali, G.B., Deepak, B., Raju, M., Biswal, B.: Optimal robotic assembly sequence planning using stability graph through stable assembly subset identification. Proc. Inst. Mech. Eng. Part C J. Mech. Eng. Sci. **233**, 5410–5430 (2019)
13. Cebulla, A., Asfour, T., Kröger, T.: Beyond feasibility: Efficiently planning robotic assembly sequences that minimize assembly path lengths. In: 2024 IEEERSJ Int. Conf. Intell. Robots Syst. IROS. IEEE, pp. 4076–4083, Abu Dhabi, United Arab Emirates (2024)
14. Shu, C., Kim, A., Park, S.: Subassembly to Full Assembly: Effective Assembly Sequence Planning through Graph-based Reinforcement Learning (2024). https://doi.org/10.48550/ARXIV.2409.13620
15. Coumans, E., Bai, Y.: PyBullet, a Python module for physics simulation for games, robotics and machine learning (2016).
16. Deetman, A., Rumpf, G., Wannemacher, B., Dawod, M., Akbar, Z., Rossi, A.: Robot Components: Intuitive Robot Programming for ABB Robots inside of Rhinoceros Grasshopper (2025). doi:https://doi.org/10.5281/ZENODO.5773814
17. Cameron, S.: A study of the clash detection problem in robotics. In: Proc. 1985 IEEE Int. Conf. Robot. Autom. Institute of Electrical and Electronics Engineers, pp. 488–493, St. Louis, MO, USA (1985)

On Vector-Based Structural Optimization and Design of Wind Bracing of Isostatic Tetragonal Grid Systems for Timber Buildings

Sylvain Rasneur(✉) and Denis Zastavni

SST/LAB, Louvain Research Institute for Landscape, Architecture, Built Environment and Urbanism, 1 Place du Levant, 1348 Louvain-la-Neuve, Belgium
sylvain.rasneur@uclouvain.be

Abstract. This contribution addresses the optimization of bracing systems for innovative and efficient timber structures with the aid of a computational tool based on vector-based graphic statics.

Keywords: Timber construction · Structural design · Digital fabrication · Graphic-statics · VGS · Research-by-design

1 Structural Design: Environment and Innovative Approaches

The design of sustainable and efficient structures demands a multi-disciplinary approach that integrates environmental, geometric, and structural considerations. This article examines the environmental challenges of structural design, emphasizing the role of material optimization and the significance of timber construction as a low-carbon alternative. It explores the principles of graphic statics and the lower bound theorem of plasticity, demonstrating their applicability in analyzing and optimizing force flows within structural systems. Additionally, strut-and-tie models and the integration of elastic-plastic principles in timber structures are presented as tools to enhance structural efficiency. The chapter also highlights the benefits of isostatic design strategies and topology optimization for achieving material efficiency and structural robustness. Finally, the computational tool Vector-based Graphic Statics is introduced, offering interactive capabilities for visualizing and manipulating 3D structural solutions, advancing the conceptual design process towards innovation and sustainability.

1.1 Environment and Structural Design

1.1.1 Environmental Challenges

The design of structures is a multi-factorial process that must remain firmly rooted in the physical context of the project. Beyond ensuring mechanical resistance, it must consider multiple factors such as functionality, geometry, construction, cost and environmental impact, among others. In a world of limited and diminishing resources, it is also crucial to consider the structure's disassembly, recycling, and the reuse of its components [1].

Y. Liu et al. (Eds.): CDRF 2025, *Transindividual Intelligence*, pp. 85–97, 2026.
https://doi.org/10.1007/978-981-92-0615-5_8

A structural solution can only be optimal depending on the importance assigned to each criterion. Therefore, the first challenge lies in defining the comparison parameters and their respective weighting.

1.1.2 Structural Efficiency

The current environmental crisis demands a significant reduction in CO2 emissions to mitigate the effects of global warming. In this context, structural engineers play a crucial role as the construction industry accounts for approximately 37% of greenhouse gas emissions [2]. Load bearing systems contribute the most to embodied carbon emissions and waste generation in construction because of their high material mass and energy intensive production [3].

Given this climate emergency, the article assumes an optimization based on the amount of material used and, consequently, on greenhouse gas emissions.

1.1.3 Timber Construction

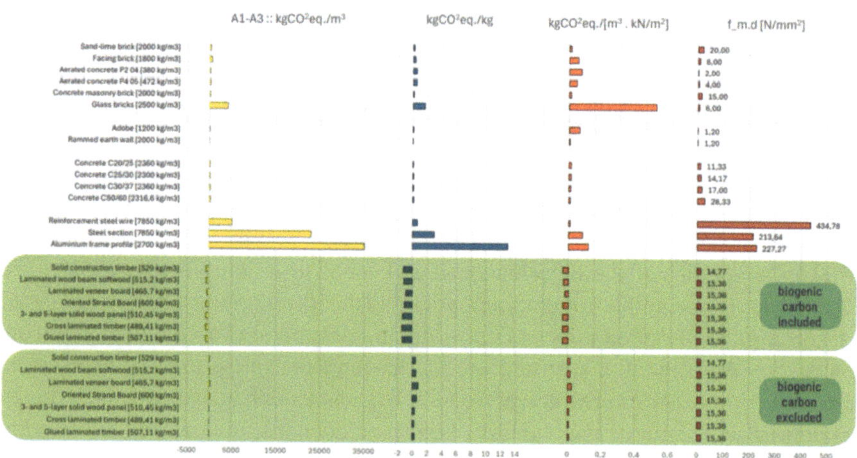

Fig. 1 Comparative diagram of the carbon assessment A1-A3 of construction materials including wood with (highlighted above) or without (below) biogenic carbon (Zastavni & Pollet 2025 - unpublished- dataset Sphera Solution GmbH 2023 and Thünen-Institut für Holzforschung); results are sensitive to hypothesis set.

Timber appears to be a very interesting material because it has the capacity to store CO2 and so is considered as having a positive carbon (negative index) footprint compared to other material.

With comparable performance, the CO_2 footprint of wood in construction is significantly lower than that of steel or concrete, [4] even when calculated without biogenic carbon. Rough numbers show that the GWP impact, excluding the influence of biogenic carbon, is in favour of timber construction even when these values are balanced by their respective stress resisting capacities (see Fig. 1).

Excluding biogenic carbon from the CO2 balance of timber prevent an overestimation of its ecological benefits. The carbon stored in wood during its growth is released at the end of its lifecycle (through decomposition or burning), meaning its long term-impact can be neutral or even negative if forest management is not sustainable. By omitting biogenic carbon from the CO2 balance, the assessment of emissions related to extraction, processing, and end-of-life disposal becomes more realistic, promoting a more rigorous approach to evaluating its environmental impact.

1.2 Graphic Statics and Plasticity Theory in Structural Design

1.2.1 Graphic Statics

Regarding the design and analysis of structures, graphic statics has proven its considerable potential for achieving efficient and elegant structures. This method relies on two interdependent diagrams, namely the form and force diagrams [5]. The first represents the geometry of the structure together with its external force, the latter embeds a synthetic vector representation of the forces applied to each node of the structure, thus representing vectorially the equilibrium of forces acting on the structures.

The interdependency of the two diagrams and their visual convenience provides a visual and intuitive understanding of the relationship between a structural shape and its inner stresses.

1.2.2 Lower Bound Theorem of Plasticity

The theory of plasticity brings an answer to the inconsistencies of elastic theory, considering the ductility of the material and its consequences on the redistribution of stresses [6].

The two theorems of plastic design were theorized in 1936 by Gvosdev and Feinberg, which Greenberg and Prager proved in 1949. The lower bound theorem particularly suited for design purposes establishes the following: for an ideal plastic material, any limiting load obtained from a distribution of internal forces is less than or equal to the actual limiting load. As a result, a graphic static design is one possible solution to the static theorem [7].

1.2.3 Strut and Tie Modelling

Thanks to the lower-bound theorem of plasticity, any continuous structural system made of a plastic material can be modeled as a strut-and-tie network. The strut-and-tie approaches were developed for the analysis of shear walls and structural details in concrete structures based on plastic theorems.

Strut-and-tie models are composed only of rods in compression – struts – or traction – ties – linked together in reticular systems with pin-jointed nodes on which point forces are applied. They can be used as generic abstractions for many types of structures [8, 9].

1.2.4 Plastic Wood

Timber, as a lightweight and sustainable material, presents unique challenges and opportunities in structural design.

Plastic principles are used for structural dimensioning of timber in most structural standards. Stress-strain relationship of timber demonstrates clearly plastic capacities in compression, both parallel and perpendicular to fibers. In usual timber structures, redistributions capacities are reached through plastic behavior of timber in compressed member and steel connections. Even in that case, the application of the plastic theoretical framework requires a special attention that plastic capacities for redistributing forces develop in compression before tensile strength is exceeded.

Therefore, the assumption of an elastic plastic behavior of wood related to its orthotropic characteristics is usual for timber dimensioning and can be considered for a careful application of plastic theorems for structural analysis of timber structures.

1.3 Optimizing Structural Geometry

1.3.1 Isostatic Structure as a Design Strategy

Timber structures represent specific challenges. The optimum use of material can be found in timber structures avoiding bending forces or complex behaviors involving torsion or shear damaging [10]. Only direct transmission of forces with compression or tension members maximizes structural efficiency and reduces the cost of wood joins.

In modelling strategies with FEM, models offer by default embedding characteristic in joints that radically modifies the local behavior, even in trussed systems. Bending-resistant metal connections can no longer be avoided since bending behaviors are activated and hence distorting stress distributions. There are huge consequences on the required degree of complexity of joints. Above this, in the case of complex structures of entire buildings, it is extremely common for the designer to be unable to find how to re-establish the hinging of the nodes of his model while maintaining its global structural equilibrium.

Beyond being less sensitive to settings and dimensional variation, isostatic hinged frames offer several advantages. As far as the structural system is designed isostatic – and is consistent with this from the theoretical point of view – redistributions of forces will not occur under loads that do not change in nature, unlike hyperstatic structures, allowing for direct plastic design. They can be analyzed geometrically only – or with static equilibrium equations – without needing to consider mechanical properties and a preliminary dimensioning. Isostatic structures require careful design of support conditions, non-redundancy and capacities to withstand all load conditions and help to understand and control the structural behavior.

For this, graphic statics – with all its advantages: visual representation, self-checking for structural equilibrium, interactivity – can be used. Isostatic structures require careful design of support conditions, non-redundancy and capacities to withstand all load conditions and help to understand and control the structural behavior. Graphic static designs, as reticular equilibrium that have not reached the level of a complete structural system, can reveal unstable in the face of certain load cases [11].

If their design and construction are simpler, in case of overload or local failure, isostatic structures do not benefit from redundancy as part of a robustness strategy. Among the 16 strategies of robustness proposed by Knoll & Vogel [12, 13], five approaches only are focusing on structural design in itself [14]. Among them strength, stiffness strategies and post-buckling strength fits to isostatic designs, whereas second line of defense and multiple load paths & redundancy are specific to hyperstatic systems. In practice however depending on the structural and nodes' dynamics, bending behavior and alternative load paths can still act as backup mechanisms.

1.3.2 Topology Optimization

Topology refers to the spatial organization and arrangement of a structure's components, regardless of their dimensions or material properties. It focuses on the connections between elements (columns, beams, etc.).

In the specific case of a truss structure, topology corresponds to the layout of nodes and the bars connecting them.

In isostatic systems, each loading case is connected to unique load path that needs to be designed. Topology plays a crucial role in optimizing the bracing of a truss structure made of bars and nodes subjected to horizontal wind loads. It involves identifying the most efficient paths to transmit loads to the supports. By determining critical areas where forces are concentrated, topology guides the optimal placement of bracing bars to minimize deformations and enhance stiffness.

By optimizing the distribution of bracing bars, it is possible to reduce their number or size in less stressed areas.

1.3.3 VGS – A Computational Tool for 3d Vector-Based Graphic Statics

Numerical analysis approaches are not the best suited for the early stage of design since their way of working relies more on an analytical process and structural models requiring several hypotheses to give a result, resulting in a lack of interactivity.

However, the latest theoretical and practical developments of 3d Graphic statics opened new possibilities in terms of complex 3d structural typologies. In this, two main methods were developed, namely the vector-based [15] and the polyhedral-based [16]. Both approaches have their own specificities and benefits.

Vector-based Graphic Statics (VGS) is a plugin in the digital environment of Grasshopper whose main purpose is to automatically generate 3d vector-base interdependent form and force diagrams. The theoretical foundations are based on vector-based graphic statics and its implementation results in a computational tool [17].

The advantage of VGS lies in its geometric nature, as it exclusively manages trusses, reticular and strut-and-tie models without requiring material or section definitions. VGS allows users to manipulate equilibrated models, potentially verify deformations, and evaluate the utilization of structural elements.

Furthermore VGS is naturally suited for analysing and designing both joints and spatial bar networks in static equilibrium, making it particularly well-suited for timber structures that exhibit these characteristics.

2 Structural Explorations

This chapter investigates the optimization of bracing systems for innovative timber structures using a computational tool based on Vector-based Graphic Statics. A research-by-design approach is applied to evaluate bracing strategies for a five-story timber structure exposed to wind loads. The structure is modeled as a tetragonal grid analyzed through various configurations including core, corner and façade bracing systems. Using VGS within Rhino & Grasshopper, the workflow integrates structural performance analysis with carbon footprint assessments, enabling iterative design refinement. The study underscores the potential of computational tools to achieve structurally efficient and environmentally responsible timber designs.

2.1 Research-by-Design

2.1.1 Case Study

The case study involves defining structural bracing strategies for a five-story building to resist wind forces through numerical simulations. It is discretized into a tetragonal grid which represents structural system of post-and-beam truss of equal spans which bracing will be managed with diagonal members or struts as commonly seen in various timber construction systems.

Drawing upon the theories of graphic statics, the plastic lower bound theorem and the strut and tie model, the experiments aim provide insights into the comparative structural performance of the bracing of a timber-based pavilion design.

By employing the computational optional tab Vector-based graphic statics (VGS) and complementary panels in the Rhino & Grasshopper digital environment, a comparative exploration of the bracing system performance is undertaken. The bracing members are assumed to be rigid bars, allowing them to be considered under both tension and compression.

The approach of isostatic models requires a sharp definition of the flow of forces and a clear vision on the structural behavior that both represents a decisive advantage for the structural reliability.

2.1.2 Assumptions

The structure has a rectangular plan consisting of three by five square grids, each six meters on each side. The building rises over five floors, each three meters high. The tetragonal grid system is analyzed under a series of bracing strategies. The bracing elements are composed of timber cores or facade lattice elements such as glulam frames. Key assumptions involve the uniformity of material properties (Glulam GL 28) and the simplification of joints as pinned connections. For environmental assessment, a minimum dimension of 10*10 cm is given to all the bars of the strut-and-tie model.

We take into account the acting forces resulting from a horizontal wind oriented Southwest/Northeast: a positive pressure of 1kN/m2 on the South and West faces and a negative pressure of 1kN/m2 on the 2 other adjacent faces. Neither live nor dead loads are considered.

For the dimensioning of the bracing systems, the following material properties of the glulam GL28 are considered:

- Modulus of Elasticity10 500 000 kN/m2;
- Admissible compressive stress 16900 kN/m2;
- Admissible tensile stress12 400 kN/m2;
- Characteristic resistance to bending 28 000 kN/m2.

2.1.3 Workflow

Although primarily an analysis tool, VGS is combined here with a module for section sizing and carbon footprint calculation. This integration transforms the framework into a design and preliminary sizing tool allowing for iterative exploration of bracing strategies. The capabilities are critical to identify bracing configurations that optimize the structural performance of the tetragonal grid system.

The framework allows the extraction of the characteristics of each variant of the structural bracing system: the material volume and consequently, the amount of CO2 equivalent per square meter.

2.1.4 Pedagocial Framework

The design and analysis exercice for bracing systems presented in this article is part of the "Architectural Structures 3" course (UCLouvain – LOCI faculty). The pedagogical objective of this exercise was to introduce students to the principles of structural design and bracing system analysis through iterative research on form diagram and force evaluations.

These explorations enabled students to develop structural proposals, the results of which are classified and analyzed in the following section. It is important to note that this article does not replicate the exporatory processes involving form diagrams or force diagrams carried out by the students (Figs. 2, 3 and 4) (Table 1).

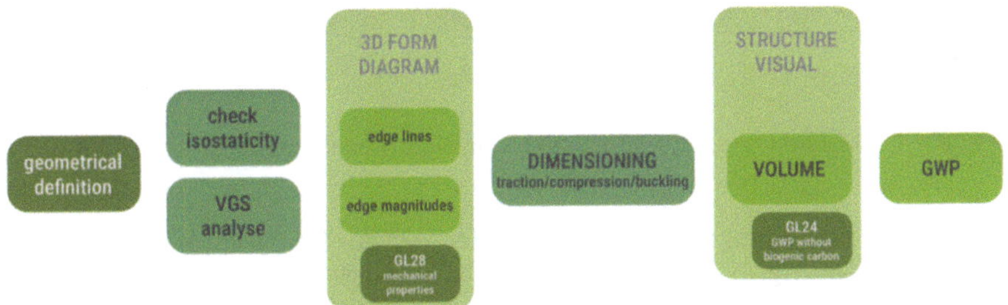

Fig. 2 Flow diagram of the algorithm used for the design of equilibrated and dimensioned structures in the Rhino & Grasshopper digital environment

2.2 Comparative Explorations

2.2.1 Corner Bracing System

The comparison of these two models highlights the advantage of continuous bracing, which more directly balances the stresses. Logically, the sequential repetition of bracing at the corner generates greater horizontal forces, complicating the load path (Fig. 5).

It is worth noting the higher number of horizontal bracings in the M3 model. The more uniform distribution of these bars also contributes to reducing the structural volume required.

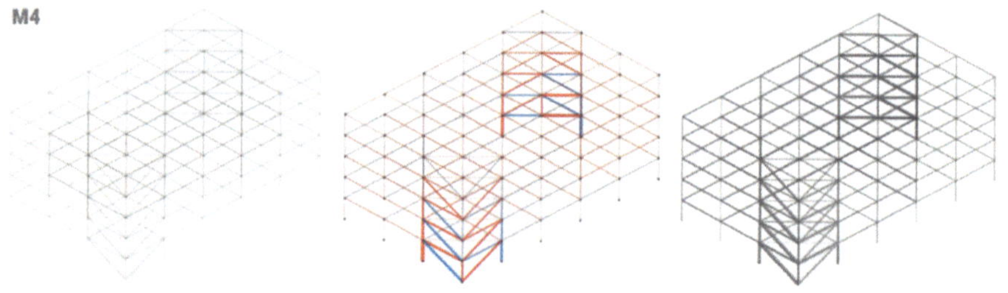

Fig. 3 Isometric views from the southwest of the model n°4: wireframe structure (left), the structure analyzed by VGS (center) and the structure dimensioned for compression, tension and buckling (right)

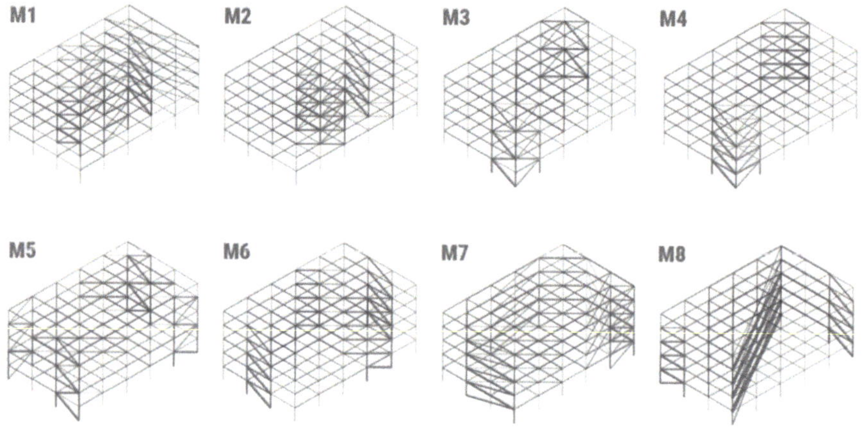

Fig. 4. Isometric views of the eight compared bracing systems

Table 1. Quantitative results from dimensioning. The volume of carbon emissions is calculated based on the carbon weight values of the wood, excluding biogenic carbon.

	BRACING SYSTEM	ANTHROPOGENIC GH. GAS KG CO2 EQ.	WOOD VOLUME M3
M1	OPEN CORE	1163	16,02
M2	CENTRAL CORE	1254	17,268
M3	CONTINUOUS CORNER	1311	18,059
M4	SEQUENTIAL CORNER	1330	18,314
M5	CONTINUOUS FACADE	1410	19,412
M6	SEQUENTIAL FACADE	1547	21,31
M7	SEQUENTIAL FACADE (DOUBLED)	1661	22,87
M8	HYBRID GLOBAL	3789	52,17

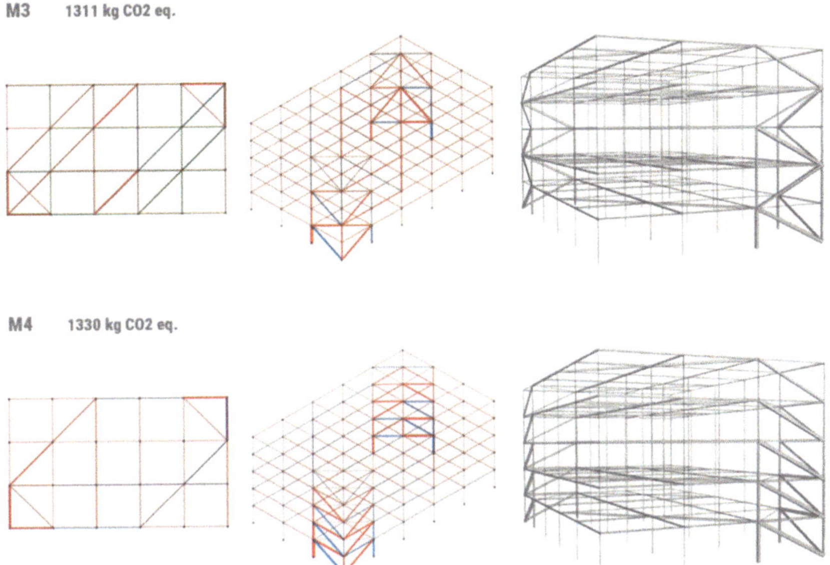

Fig. 5 Plan, isometric and two-point perspective views of the bracing models M3 (continuous corner) and M4 (sequential corner)

Overall, both structures demonstrate remarkable efficiency due to the concentration of forces, which effectively stiffens the facades against the wind, allowing the loads to be transferred more quickly to the supports.

2.2.2 Core Bracing System

Considering supports evenly distributed beneath the 24 columns, the location of bracing at the center of the grid helps to equalize the stresses within the structure. Once again, this approach reduces the material required to stabilize the structure (Fig. 6).

The increased number of loaded supports in the M1 model further optimizes the CO2 emissions generated by the structure by simplifying the load paths.

Similarly, the number of braced grids in the horizontal plane serves as an indicator of the energy efficiency of both structures.

2.2.3 Facade Bracing System

This comparison of façade bracing models highlights the advantages of shorter bars over longer ones, which are sensitive to buckling. This is obvious in the M7 model which suffers from the resizing of its longer bars. On the one hand, the longer bars are resized due to their length and the forces they carry. On the other hand, the length of the bracing reduces the number of loaded supports, which undermines structural efficiency (Fig. 7).

The M5 and M6 models differ in their bracing strategies: continuous bracing (M5) versus sequential bracing (M6). Once again, the continuity of bracing bars around the

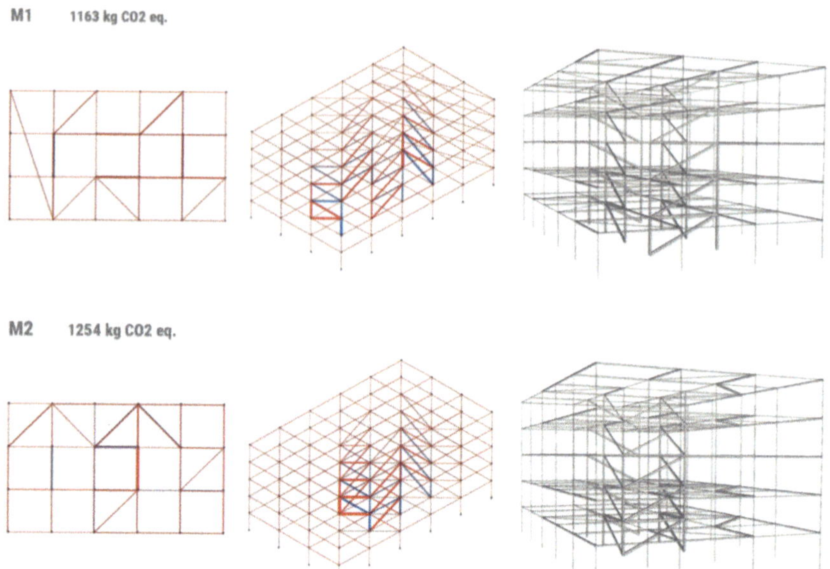

Fig. 6. Plan, isometric and two-point perspective views of the bracing models M1 (open core) and M2 (central core)

nodes ensures greater structural efficiency. This is further enhanced by the number of braced grids in the horizontal plane, helping to balance stresses throughout the structure.

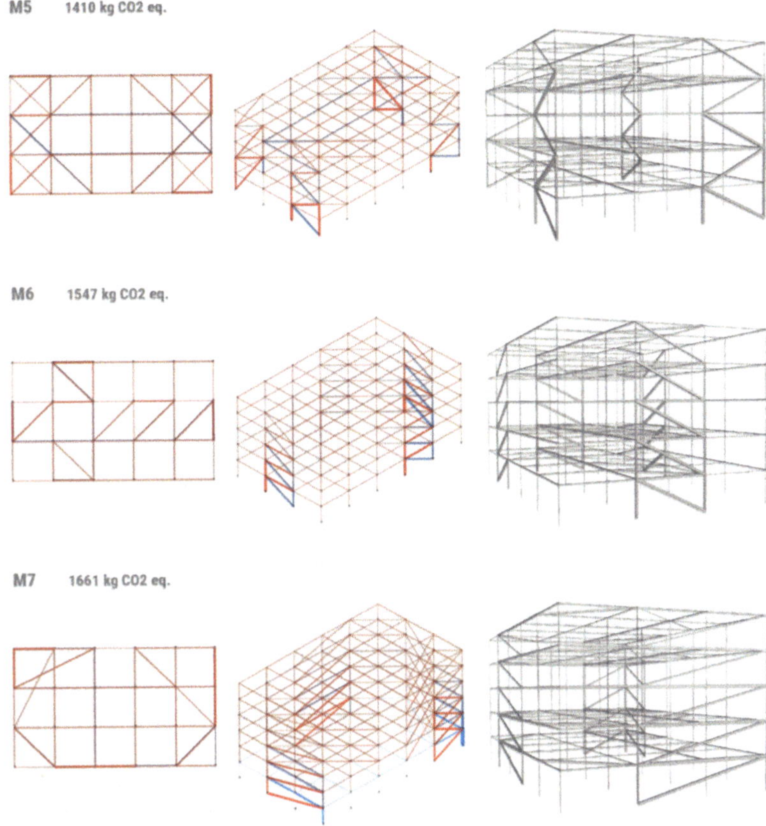

Fig. 7. Plan, isometric and two-point perspective views of the bracing models M5 (continuous façade), M6 (sequential façade) and M7 (sequential facade doubled)

3 Conclusion

This chapter concludes by emphasizing the significant advantages of isostatic systems in timber construction, particularly their predictable load paths and compatibility with digital manufacturing techniques. Key insights into the behavior of isostatic tetragonal grid systems include the efficiency of shorter bars due to reduced buckling risk, the effectiveness of alternating in-plane floor bracing to uniform stress distribution, and the improved stability achieved by aligning bracing elements with corner supports. These findings highlight the potential to optimize wind bracing systems, minimizing material consumption and carbon emissions while enhancing structural efficiency. The integration of Vector-based Graphic Statics with section dimensioning and carbon footprint assessments demonstrates a streamlined and innovative approach for designing efficient, sustainable timber structures.

Isostatic systems present numerous advantages linked to their global behavior. Among them, the certainty of load paths and structural working opens a wide field of application in the direction of contact joints and digital manufacturing with timber.

The above-mentioned exploration reveals several key insights into the behavior of isostatic tetragonal grid systems. First, shorter bars are consistently better than long bars which are more sensitive to buckling. Second, alternating in-plane floor bracing emerges as a highly effective strategy for achieving a uniform stress distribution within the structure. This configuration reduces localized stress concentrations, enhancing the overall resilience of the system. Third, the alignment of bracing elements toward support points, particularly those located at corners significantly improves structural stability. This finding underscores the importance of directing force flows to areas of maximum stiffness and support.

Optimizing wind bracing isostatic tetragonal grid systems for timber buildings presents a valuable opportunity to enhance structural efficiency while minimizing material consumption and carbon emissions. By integrating vector-based graphic statics (VGS) with principles from plasticity theory and topology optimization, this research demonstrates the potential for achieving optimized timber constructions.

This study highlights the crucial role of bracing strategies in governing the overall efficiency and stability of timber structures. Comparative analyses of various bracing configurations highlight the importance of continuous load paths. Configurations that ensure a direct and uniform load transfer demonstrate superior stiffness; reduced material use and lower embodied carbon.

The research-by-design approach supported by computational simulations proves to be an effective methodology for iterating design alternatives and refining bracing configurations.

By integrating VGS with section dimensioning and carbon footprint calculations, this framework extends beyond structural analysis, serving as a preliminary design tool for sustainable timber construction. Ultimately, this research contributes to the development of efficient structural systems, highlighting the importance of digital innovation for sustainability in contemporary architectural practice.

References

1. Rasneur, S.J.-C.M.Z., Denis; Jasienski, Jean-Philippe.: On plastic development of timber structures based on 3D interactive vector-based graphic statics (VGS). In: Architectural Intelligence (2024)
2. United Nations Environment Programme GAfBaC. Global Status Report for Buildings and Construction - Beyond foundations: Mainstreaming sustainable solutions to cut emissions from the buildings sector. 2024.
3. Catherine Elvire, L.D.W.: Low carbon pathways for structural design. In: Proceedings of the IASS Annual Symposium 2018 (2018)
4. Hegeir, O.A., Kvande, T., Stamatopoulos, H., Bohne, R.A.: Comparative life cycle analysis of timber, steel and reinforced concrete portal frames: A theoretical study on a Norwegian industrial building. Buildings. **12**(5), 573 (2022)
5. Maxwell, J.C.: On reciprocal figures and diagrams of forces. Philos. Mag. **27**(4), 250–261 (1864)
6. Baker, J.F.: The Steel Skeleton. Cambridge University Press, New York (1956)
7. Jasienski, J.-P., Zastavni, D., Rasneur, S.: On the development of timber structures based on 3D interactive Vector-Based Graphic Statics (VGS). In: Phygital Intelligence. Springer Nature Singapore, Singapore (2024)

8. Mörsch E. Der Eisenbetonbau - Seine Theorie und Anwendung. 1908.
9. Schlaich Jorg, W.D.: Ein praktisches Verfahren zum methodischen Bemessen und Konstruieren im Stahlbetonbau. In: Comite Euro-International du Beton (CEB). Bulletin d'Information N 150 (1982)
10. Zastavni, D., Rasneur, S., Jasienski, J.-P.: On the application of vector-based graphic statics (VGS) for structural timber optimisation – pavilion example. In: University T (ed.) The 6th International Conference on Computational Design and Robotic Fabrication. College of Architecture and Urban Planning, Shanghai (2024)
11. McRobie Prof, A., Cameron, M.M.S., Denis, Z.P., Baker, P.W.: Graphical stability analysis of Maillart's Roof at Chiasso. Struct. Eng. Int. **32**(4), 538–546 (2022)
12. Vogel, T.: Robustness of structures. In: IABSE Workshop: Safety, Failures and Robustness of Large Structures. IABSE Workshop, Helsinki (2013)
13. Knoll, F., Vogel, T.: Design for Robustness. IABSE, AIPC, IVBH (2009)
14. Zastavni, D., Deschuyteneer, A., Fivet, C.: Admissible geometrical domains and graphic statics to evaluate constitutive elements of structural robustness. Int. J. Space Struct. **31**, 203–214 (2016)
15. D'Acunto, P., Jasienski, J.-P., Ohlbrock, P.O., Fivet, C., Schwartz, J., Zastavni, D.: Vector-Based 3D Graphic Statics: A framework for the design of spatial structures based on the relation between form and forces. Int. J. Solids Struct. **167**, 201–215 (2019)
16. Akbarzadeh, M., Mele, T., Block, P.: Three-dimensional graphic statics: Initial explorations with polyhedral form and force diagrams. Int. J. Space Struct., 31 (2016)
17. Jasienski, J.-P., Yuchi, S., Ohlbrock, P.O., Zastavni, D., D'Acunto, P.: A computational implementation of Vector-based 3D Graphic Statics (VGS) for interactive and real-time structural design. Comput. Aided Des. **103695** (2024)

Architectural Transobjects

The Journey of Multi-Agent Intelligence

Aileen Iverson-Radtke[1(✉)] and Otto Paans[2]

[1] Air-architecture, Buttmanstr 16, Berlin, Germany
aileen@air-architecture.net

[2] BC Consulting, Bulkemstraat 4a, Simpelveld, The Netherlands

Abstract. This text examines architectural objects as porous, open, dynamic entities inseparably entwined with their contexts—hence, *transobjects.* Drawing from contemporary computational design practice and theory (Object-Oriented Ontology, Speculative Realism), we argue that architectural form exists in continuous flux, shaped by a confluence of material, environmental, and cultural forces. Our method, *Cybermodelling*, integrates microsensors into physical "smart" models as analogue-digital interfaces to facilitate *spatiomateriality*—the reciprocal interplay of spatial and material properties—within computational processes. As a hybrid tool, Cybermodelling captures spatiomaterial flux to generate "live" computational transobjects that remain in constant dialogue with context; in doing so, these objects become semi-autonomous and adaptive, capable of responding dynamically to site conditions and performance cues. As a physical (smart) modelling interface, the approach prioritizes craftsmanship—wherein human haptic intelligence, physical making, and tacit knowledge foster an intuitive appreciation of material idiosyncrasies and agency necessary to articulate and individuate transobjects. Beyond morphological transitions, this methodology recognizes the designer as an participant in an active network of human, environmental, and computational agencies. Ultimately, Cybermodelling illustrates how responsive transobjects both shape—and are shaped by—their contextual environments, offering an intuitive, *materially grounded paradigm* for computational design.

Keywords: Transobjects · Cybermodelling · Spatiomateriality · Embodied Computation · Craftsmanship · Human-Computer Interaction

1 Architectural Transobject

This paper takes as a point of departure the near-consensus view of architectural objects as open, porous entities inherently entangled with their context—what we might term transobjects. Here, "trans" indicates that object and context are not separate, but enmeshed within a relational field of historical, cultural, technological, and environmental systems.

Transobjects define architectural objects as a set of interconnected internal and external elements and conditions. Interiority becomes less significant, eclipsed by a recognition that architectural objects are parts of a charged network, wherein every element

Y. Liu et al. (Eds.): CDRF 2025, *Transindividual Intelligence*, pp. 98–106, 2026.
https://doi.org/10.1007/978-981-92-0615-5_9

influences—and is influenced by—many others. No element is entirely autonomous. Hierarchies within this network emerge, dissolve, and re-form, occasionally granting particular elements central positions. Transobjects thus arise as "field solidified"—artfully individuated yet fundamentally linked.

Transobjects require individuation—a cultivated selection among a field of key aspects, gradually coalescing into a concentrated, delimited zone. While intensificating these aspects leads to individualization or objectification—a formal distinction from the broader field—it does not imply total separation. As Simondon (2017) worked out, the object does not detach itself from the field from which it emerges; instead, it retains many features accumulated and assimilated along its developmental path.

Transobjects imply translation—the journey of intelligence and order across various modalities (material, mental, computational, real, or virtual). Within these transactions, there is a process of purification, whereby inherent natures or interpretive frames are guided, transcribed, or shifted.

In practice, designing architectural transobjects involves supporting and sustaining the ongoing dialogue between object and field. This involves preserving the network of connections that flow through the design object. Our approach acknowledges that, although forms (e.g., buildings or neighbourhoods) may appear static, they begin—and remain—dynamic, open entities. Viewing objects as "trans" centres attention on the systems that sustain them, thus framing design as stewardship of "live", continuously evolving entities.

2 Digital (Schematic) Design

In the information age, digital has taken over, transforming media into ubiquitous technological substrate, at once "material" and tool. Engaging media—online or in computational design—we enter a virtual domain composed of myriad actants, some fully specified, and others entirely hidden. Tools such as coding, Sensory Devices, VR headsets, or AI solvers serve to organize this dynamic, ephemeral field along the axes of human habitation and intelligence.

2.1 Digital Transobjects

Within computational design, the concept of the transobject is regularly exemplified. Through digital simulations, we observe virtual buildings as porous entities, absorbing and transforming environmental, phenomenological, and material data flows. Perhaps the starkest illustration arises in parametric modelling, where digital design objects remain umbilically connected to dynamic streams of cultivated data. This image of the artifact as a tangible expression of expansive underlying systems parallels contemporary theories like Morton's *Dark Ecology* (Morton 2018), *Object-Oriented Ontology* (Harman 2018), and Jane Bennett's notions of *vibrant matter* and its associated *actancy* (Bennett 2004, 2010)

2.2 Schematic Transobjects

Within architectural practice traditions, crafting of "trans" design objects—entities thoroughly enmeshed in diverse and extensive contextual parameters—is most visceral in the schematic, or pre-schematic, design phase. As the earliest stage of architectural conception, it is also the broadest and most exploratory, wherein the design "object" is largely a field of competing influences and potential relationships. In this fluid state of variables, transobjects emerge—from what Ingold describes as an 'ephemeral congelate of generative movement' (Ingold 2007).

A key characteristic of this early schematic stage is its "ill-structured" nature, meant'to prevent premature crystallization of ideas and to facilitate the generation and exploration of alternate solutions.' (Goel 1992). It holds latent creativity and underlying order, while eluding rigid models of logic and deliberative reasoning.

This ambiguity finds a corollary in psychology's notion of "transitional objects"—items that allow infants to explore the boundary between "me" and "not-me" (Winnicott 1989). Similarly, pre-schematic transobjects occupy a liminal space between concept and context, bridging mental or virtual constructs and spatiomaterial conditions.

Positioning architectural objects (including outside spaces, such as squares or landscapes) as "trans" closely aligns with these "transitional objects". In both cases, such objects offer a means for testing various forms of engagement with the world. They hover between fiction and reality, encompassing real possibilities alongside imagined scenarios. In doing so, they broaden the scope of how objects interact with our surroundings. Through such objects, we can shift or suspend the typical confines of fiction and reality.

By intentionally expanding the range of possibilities—overlaying the "ideal" upon the "real" (Nelson and Stoltermann 2014: 31)—practitioners engage in a design process marked by methodological porosity:

> '...a specific kind of thinking...that has a complex character...includes the multiplicity of different parameters, which may be conflicting, complementary...an unstable area of exploration as uncharted territory...where...solution is often not the most optimal choice but rather emerges from the synthesis of a multitude of different insights.' (Cross 2008, Quoted in Cahtarevic and Adna 2019: 37)

3 Cybermodelling

It is here, in the domain of the pre-schematic design phase—now almost wholly digital—that our research takes hold; provoked by Frei Otto's observation that it is difficult to conjure the necessary ambiguity and uncharted possibilities using computers alone. Otto notes:

> '... in computers you only find what you are looking for, (whereas) in free experimentation you can find what you have not sought.' (Songel and Otto 2010)

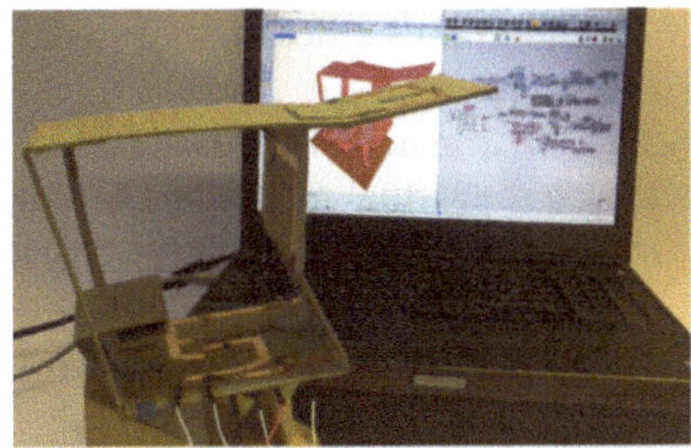

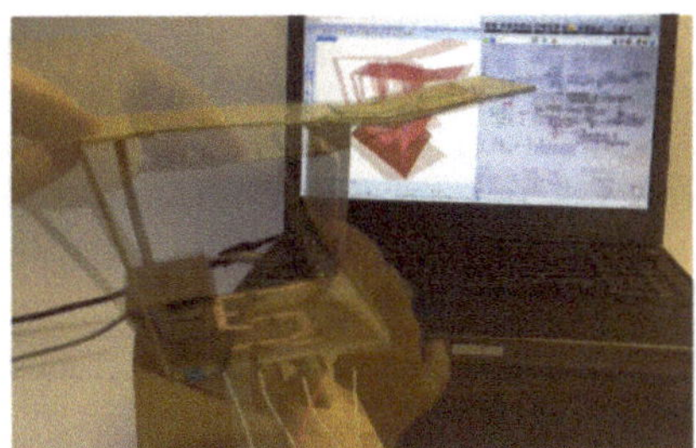

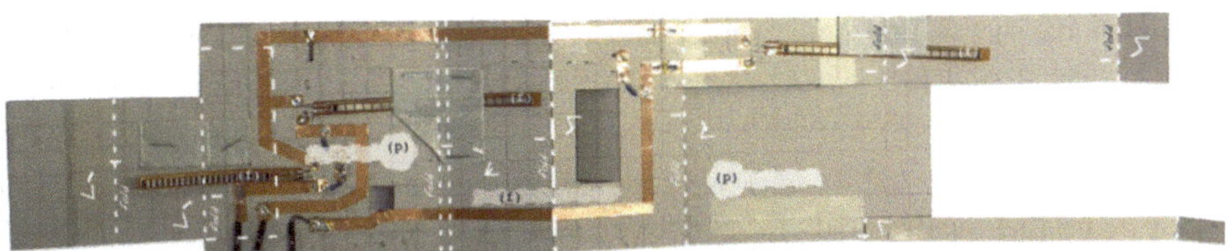

Fig. 1. (top) Case Study 3: "Smart model" (bottom) interface to digital modelling (upper left), interface in action (upper right). [Source: Iveson-Radtke 2022]

Our method, Cybermodelling, seeks to bridge this conceptual divide, making room for the unknown within the otherwise predictable realm of digital precision. The key to achieving this somewhat paradoxical connection lies in *materiality*.

As a hybrid analogue–digital *making* practice, Cybermodelling leverages materiality as a convention shared across physical and virtual spheres—recognized as a principal interface for human corporeal intelligence encountering the external world. Understanding begins at this haptic threshold:

> '…what an encounter between the fingertip and the materiality of the world might have to tell us of a scopic we call place.' (Hetherington 2003, 1938–39, quoted in Ingold 2007)

Cybermodelling integrates microsensors into physical architectural models, replacing conventional input devices (mouse and keyboard) to create a direct analogue interface to digital modelling environments. These physical models, embedded with low-cost sensors (e.g., flex, pressure, sound), are connected via Arduino and Rhino/Grasshopper through the Firefly plugin. The sensors translate physical deformation and environmental data—such as sound waves, bending, or light intensity—into dynamic transformations of the digital model, see Fig. 1.

Cybermodelling highlights *modelling* as a methodological nexus where human, material, and spatial agencies interact and collaborate. While the current workflow uses Firefly, future iterations may incorporate additional data plugins or sensing hardware, such as LiDAR or locally sourced environmental datasets. This work views the analog model as more than a design proposal or prototype; rather as an epistemological interface for engaging *digital materiality*. The goal is to create a data-rich, hybrid making environment as interface to the computational. Such an interface deepens contextual responsiveness in computational making and design education.

In such engagements, manual manipulations and environmental inputs encounter *digital material properties* in feedback loops that shape the virtual model in real time. Rather than serving as an optimization engine, this approach allows technology to function as an expressive medium for human poetic artistic experimentation—cultivating *transindividual* and *transobject* interactions across human, material, and computational systems.

Responding to Otto's concern, Cybermodelling transfers this ethos of open-ended experimentation into the computational realm, where the unpredictability of material conditions—its limitations, propensities, and affordances—re-enters and animates the otherwise abstract space of digital making. Two critical factors underpin this methodology: craftsmanship and spatiomateriality*:*

- **Spatiomateriality**: As hybrid constructs, Cybermodels extend the transobject framework to bridge physical and virtual domains. This fusion reveals a critical condition—spatiomateriality—the profound interdependence of spatial and material properties, wherein materials (e.g., wood swelling or metal rusting in humidity) respond to (and are animated by) environmental context. This relationship becomes starkly evident in the vacuum of pure computational space, where digital materials (e.g., mesh, surface) are often detached from spatial phenomena (e.g., humidity, gravity, sound). By integrating these conditions through hybrid making, Cybermodelling foregrounds spatiomateriality as a conceptual and operative category within computational design (Iverson-Radtke & Paans 2024).
- **Craftsmanship**: Cybermodelling inverts the typical real-versus-virtual relationship. Physical "smart" objects become avatars, transmitting sensor data—derived from manual shaping and environmental inputs—to their virtual counterparts. By positioning physical modelling as human–computer interaction (HCI), the method highlights and accommodates *homo faber* intelligence in the realm of digital design. Tactile engagement with material qualities, nurtures a less cerebral, more ambiguous mode of thinking—recalling an "ill-structured" process. The aim is to establish digital design as a form of hand-tooling computational matter, conferring craftsmanship traditions that honor material logic thus invite external intelligence into digital making, allowing design objects emerge from the idiosyncrasies of their spatiomaterial agency.

3.1 Craftsmanship

Cybermodelling situates computational design within physical making, foregrounding the human–computer interaction (HCI) dimension. By reintroducing the traditional engagement with materials—manual shaping and mastery—this approach aspires to cultivate a sense of craftsmanship within computational practice. More than a set of inherited techniques, craftsmanship here encompasses a perspective wherein the designer becomes a *participant with* rather than an *author of* the emergent object.

Craftsmanship is inherently experiential. It evolves a profound relationship with the particularities of material nature, forged through sustained interaction over time.

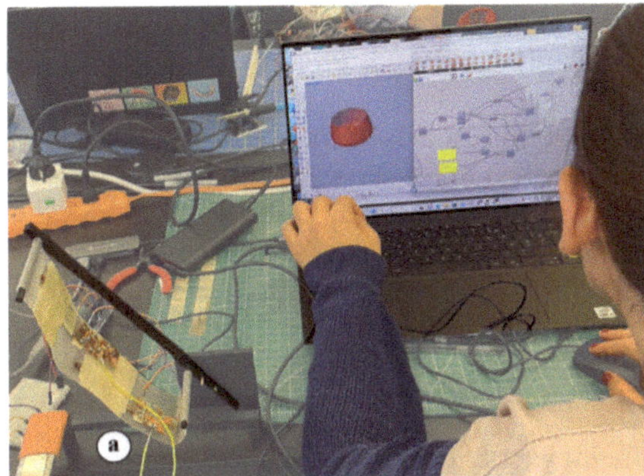

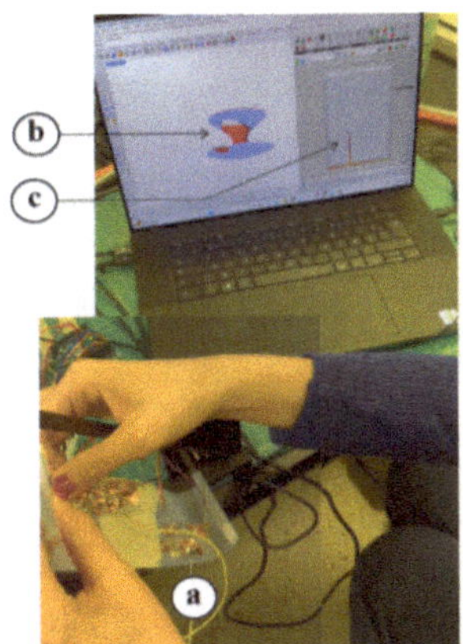

Fig. 2. Smart model interface to digital modelling (left), interface in action (right). Bending the "smart" model, rubber strip with integrated flex sensors (a) alters the section of the digital vessel (b) while whistling creates a hole in the vessel, see sound detection screen monitor (c). [Source: Iveson-Radtke, 2023]

Through this ongoing engagement, human intellect essentially immerses itself in the properties and rules of the material engaged. As Sennett (2008: 210) observes:

> 'How does…using a tool organize…possibilities? The first stage occurs when we break the mold of fit-for-purpose. That break occupies a different part of the imaginative realm…'

By facilitating this break, Cybermodelling reshapes the possibilities inherent in both material and tool, establishing a robust *transindividual ecology* in which human intelligence fuses with data—including phenomena not directly observed but introduced via purely digital sources. Such data represents a form of aggregated experience, observations, and measurements beyond any single lifetime. Within this expanded field, craftsmanship remains pivotal, demanding deep familiarity with, and alignment to, the media at hand. In so doing, it enables the articulation and individuation of the *transobject*, allowing an entity to perceptually emerge from its context while remaining intrinsically linked to it.

3.2 Spatiomateriality

At the molecular level of transobjects we encounter continuous change—spatiomateriality—the ancient, perpetual interplay intrinsic to real spatial and material properties. This interplay animates matter, as materials inherently respond to environmental conditions. In nature, materials are always integrating with their environments—canvas stretches in wind, metal rusts in saline air—yet in virtual contexts, material and spatial properties are seldom or only durationally linked. Even when virtual spatial properties are included, they tend to be simulated, predictable, and homogeneous.

Cybermodelling incorporates microsensors into hybrid analogue–digital models to capture real-time spatial and material data. This live input activates digital material behavior, stimulating digital spatiomateriality—where digital matter registers and reacts to context, yielding "live" design object morphogenesis, see Fig. 2. Contextual conditions begin to shape form directly.

By triggering digital spatiomateriality, Cybermodelling transfers agency to the substance of digital modelling. Freed from purely designer-driven decisions, models adapt based on the internal algorithmic properties of their digital materials. In this way we can, in Kwinter's words:

> '... unlock the door on the universal laws that govern the appearance and destruction of form, and in so doing...free us from the multiple tyranny of determinism and from the poverty of a linear, numerical world.' (Kwinter, 2003:92)

Activating digital spatiomateriality initiates a relational ecology of intelligences — human, material (both physical and digital), and environmental. Cybermodelling design process incorporates material qualities, orientation, and geometry toward integrated functionality across multiple fields of influence, a veritable biotope of responsiveness, see Fig. 3. Harnessing both physical molecular and digital algorithmic structures reflects an interplay of these external intelligence systems, producing transobjects as artifacts of *transintelligent* collaboration.

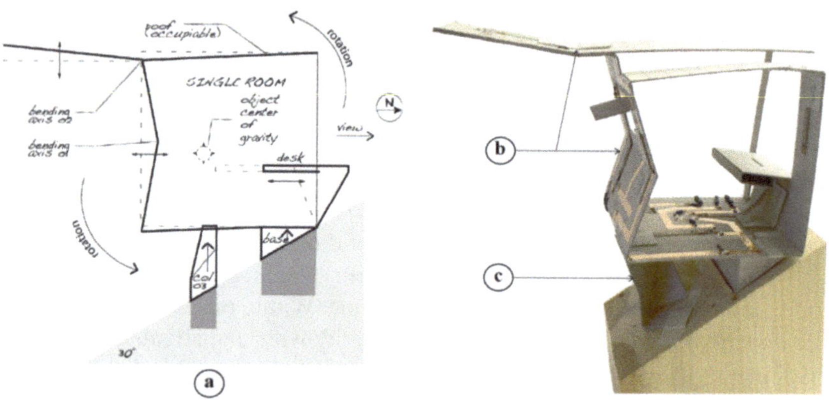

Fig. 3. Diagram of site vector forces and atmospheric conditions (left), smart model HCI (right). Altering (bending, shifting, reshaping) the smart model with integrated sensors adjusts the digital model to balance in its site condition (a), see Fig. 1 (top). [Source: Iveson-Radtke, 2022]

By emphasizing a material-based computational approach, Cybermodelling foregrounds spatiomateriality as essential for enabling objects to adapt semi-autonomously. This approach has precedent in the models of Frei Otto and Antoni Gaudí, who used physical models to compute form. Cybermodelling extends their material experimentation into computational space by incorporating sensor-driven data processing.

From an HCI standpoint, this autonomous object-agency is guided by traditions of craftsmanship and material mastery. Establishing haptic interface with "live" spatiomaterial transobjects—porous and intertwined with fields of internal and external data—preserves human intent and conceptual thinking within the design process. Rather than imposing an aesthetic or formalist schema from the outside, this conceptualization—drawing on DeLanda's (2015) *New Materiality*—enables object-agency, allowing design to unfold from within through the dynamic, emergent qualities of *digital spatiomaterial properties*.

4 Conclusion

Cybermodelling offers a powerful generative framework through which computational design can be reinvigorated by material intelligence, reintroducing haptic exploration and the nuanced qualities of real-world physicality into digital processes. By merging analogue craftsmanship with computational spatiomateriality, this approach generates transobjects—open, porous, and attuned to their broader environmental and cultural contexts. These transobjects, as "embodied computation," (Iverson-Radtke and Paans 2024) reveal how architectural form can be guided not solely by external directives but also by the internal propensities and feedback loops of material systems.

Such a methodology champions human intelligence—defined by creativity, empathy, and aesthetic sensibilities—in concert with the vast processing power and data-handling capabilities of computational systems. In doing so, it broadens the horizon of computational form-making to include the unexpected and the poetic, producing designs that resonate with essential human values while responding dynamically to environmental realities. Moreover, Cybermodelling facilitates a deeper ecological integration, positioning architectural objects in dialogue with seasonal cycles, local ecosystems, and other life forms. By anticipating future data flows—spanning climate models, migration patterns, or traffic streams—the methodology shows potential for progressively more responsive, resilient, and sustainable architectures.

Drawing on Simondon's theory of individuation and Stiegler's concept of transindividuation, Cybermodelling frames digital design not as detached simulation but as active participation in a wider ecology of agency. Ultimately, Cybermodelling serves as both a toolset and a philosophical stance. It demonstrates that embracing material agency and computational synergy can expand technologically-aided design beyond the purely formal, enabling architects and designers to shape "live" objects that negotiate multiple dimensions of context, from structural efficiency to cultural resonance. This hybrid process not only leverages technological innovation but also safeguards human-centric qualities—such as intuition, tactility, and the capacity for wonder—ensuring that future architectural outcomes embody a shared truth recognizable and inspirational across our species.

References

Bennett, J.: The Force of Things – Steps Towards an Ecology of Matter. Political Theory. **32**(3), 347–372 (2004). https://doi.org/10.1177/0090591703260853

Bennett, J.: *Vibrant Matter: A Political Ecology of Things.* Duke University Press, Durham (2010)

Cahtarevic, R., Proho, A.: Geometric Modelling and Complexity – A Conceptual Approach in Architectural Design and Education. *SPATIUM.* No. **42**, 35–40 (2019)

Goel, V.: 'Ill-Structured Representations' for Ill-Structured Problems. In: Proceedings of the Fourteenth Annual Conference the Cognitive Science Society. Lawrence Erlbaum, Hillsdale, NJ (1992)
Harman, G.: Object-Oriented Ontology – A New Theory of Everything. Penguin Books Ltd, London (2018)
Hetherington, K.: Spatial textures: Place, touch, and praesentia. Environ. Plan. A. **35**(11), 1933–1944 (2003). https://doi.org/10.1068/a3583
Ingold, T.: Materials against materiality. Archaeol. Dialogues. **14**(1), 1–16 (2007). https://doi.org/10.1017/S1380203807002127
Iverson-Radtke, A., Paans, O.: Embodied computation and spatiomateriality, exploring complexity through cybermodelling. Agathón J. **16** (2024) Dealing with Complexity
Kwinter, S. (2003) The computational fallacy. Thresholds, 26 (Spring), 90-92.
Morton, T.: Dark Ecology. In: For a Logic of Future Coexistence. Columbia University Press, New York, NY (2018)
Nelson, H., Stoltermann, E.: The Design Way: Intentional Change in an Unpredictable World. The MIT Press, Cambridge, MA (2014)
Sennett, R.: The Craftsman. Yale University Press, New Haven (2008)
Simondon, G. (2017) On the Mode of Existence of Technical Objects. Trans. C. Malaspina and J. Rogove. Univocal Publishing, Minneapolis.
Songel, J.M., Otto, F.: A Conversation with Frei Otto. Princeton Architectural Press, Princeton, NY (2010)
Winnicott, D.W.: Playing and Reality. Routledge, London/New York, NY (1989)

Bridging GANs and Reinforcement Learning: Proximal Policy Optimization for Enhanced Design Outputs

Mertcan Güldilek(✉), Mustafa Emre İlal, and Berk Ekici

Izmir Institute of Technology, Gülbahçe, 35433 Izmir, Turkey
mertcanguldilek@iyte.edu.tr

Abstract. Generative Adversarial Networks (GANs) have demonstrated significant potential in floor plan generation by producing layouts with diverse and structurally coherent understanding. However, the House-Gan++ model often struggles with stability and constraint adherence due to the adversarial nature of their training process. This research introduces a novel hybrid training framework integrating Proximal Policy Optimization (PPO) and GANs to enhance generative performance and stability. The proposed method integrates PPO into the GAN training loop to enhance generator loss optimization through non-adversarial feedback, producing realistic and structurally valid architectural plans. The PPO agent navigates the generator through a policy network that modifies latent abstract features using advantage estimates. The approach enhances the adversarial learning stability by stopping mode collapse and improves how the generative model follows user-defined spatial constraints. Training occurs through the RPLAN dataset because it contains a substantial collection of professional single-story house plans, which ensures great architectural variety and authenticity. Experimental tests between PPO-GAN and House-GAN++ evaluate their performance through four training loss metrics: Generator Loss, Discriminator Loss, PPO Loss, and L1 Loss and two objective quality metrics: Fréchet Inception Distance (FID) and Graph Edit Distance (GED). Including PPO produces layouts that achieve steady convergence, superior spatial constraint compliance, and reduced FID scores, proving their improved authentic quality. The PPO-GAN model demonstrates higher computational expenses and training instability at first but proves reinforcement learning (RL) as an effective GAN-based architectural design enhancement through its superior long-term performance and stability. This study aims to contribute to AI-driven generative design by bridging GANs and RL.

Keywords: Generative Adversarial Networks · Reinforcement Learning · Proximal Policy Optimization · Floor Plan Generation · Hybrid Training Framework

1 Introduction

Floor plan generation is one of the fundamental and challenging fields in architectural design [1]. There are infinite combinations for designers to achieve a desired layout when creating a floor plan. This process takes time and effort to bring multiple architectural

Y. Liu et al. (Eds.): CDRF 2025, *Transindividual Intelligence*, pp. 107–117, 2026.
https://doi.org/10.1007/978-981-92-0615-5_10

spaces with superior understanding; thus, researchers approach this issue differently [2]. In the literature, Weber et al. [3] identified three different methods to automatically create floor plans: top-down, bottom-to-top, and referential. In top-down methods, the designer identifies an envelope of design, and plan layouts are made based on that envelope and some constraints. Bottom-to-top is simply a reversed approach of top-down. The latest floor plan generation method, the referential method, utilizes a collection of dataset(s) and deep learning (DL) models to generate layouts. The referential method is relatively newer than conventional methods [3]. This method has become more popular with the development of DL and artificial intelligence (AI) technologies. Referential methods need a dataset to make a production, so they make their productions according to the data they learn from this dataset.

1.1 Literature Review

1.1.1 Generative Adversarial Networks (GANs)

The DL model structure of GANs contains two components: the generator creates synthetic outcomes, and the discriminator assesses their authenticity [4]. The generator component focuses on producing authentic outputs as the discriminator component analyzes real and fake generator data for distinction purposes. Chaillou's ArchiGAN [5] follows a systematic three-step approach that commences with building massing generation, proceeds to program space distribution, and ends with furniture layout arrangement. The mathematical system takes unprocessed land parcel information to generate architectural drawings while it understands the spatial linkages between original pictures and design outcomes. Pix2Pix HD [6] served Zheng et al. [7] to produce apartment layouts using datasets. Model performance directly correlates to the datasets used according to their research findings. Following more information, such as adjacencies and linkages between rooms in each floor plan layout, Graph neural networks (GNN) are integrated into the DL models. The combination of GNN technology with GAN results in House-GAN [8], which trains on LIFULL HOME data to create architectural drawings using room-type relational graphs. The floor plan diversity improvement through the model occurs by implementing Conv-MPN [9], which operates as a convolutional message-passing algorithm. Adding door-type information through House-GAN++ [10] enhances the circulation route refinement in this approach. GAN training receives enhanced capabilities from the RPLAN dataset [11], which contains 80 K floor plans and their associated wall, door, and room-type information. It allows better-structured design generation through optimized Conv-MPN pooling operations.

1.1.2 Combination of RL and GANs

Using RL provides a systematic structure that improves GANs by overcoming their flaws [12]. Three main strategies exist for merging RL systems with GANs that focus on mere generation, objective maximization, and non-quantifiable feature enhancement [13]. RL is a substitute generative mechanism through the mere generation approach, focusing on problems with non-differentiable loss functions. This study centers on the objective maximization method because it serves as the training technique for this study. GANs use RL to optimize measurable evaluation metrics directly, which they usually assess but do not explicitly optimize during training. GAN training through traditional adversarial loss functions struggles to match domain-specific objectives during the training process. The RL approach adds an organized reward system that lets the generative process achieve specified, measurable targets. Model performance improves when RL joins forces with GANs through the objective maximization framework because it enables better optimization techniques. Through this method, GANs acquire the capability to produce outputs compliant with exact domain requirements, representing a significant advance in generative modeling technologies.

1.2 Significance of This Study

The common shortcoming of the three overstated methods is that they cannot fully protect the user's preferences. It is tough to produce a plan responding to users' preferences. This research proposes a novel methodology integrating PPO to boost the GANs and produce realistic architectural plans that match user preferences. The research method optimizes the GANs' loss function through RL to generate outputs that better match user expectations, rather than solving user preference modeling alone. The model's training process with PPO integration enables GAN to develop better generative abilities, which results in floor plans that maintain architectural consistency.

2 Methodology

This section outlines the methodological framework, including data preprocessing, GANs training, PPO integration, and model evaluation (Fig. 1).

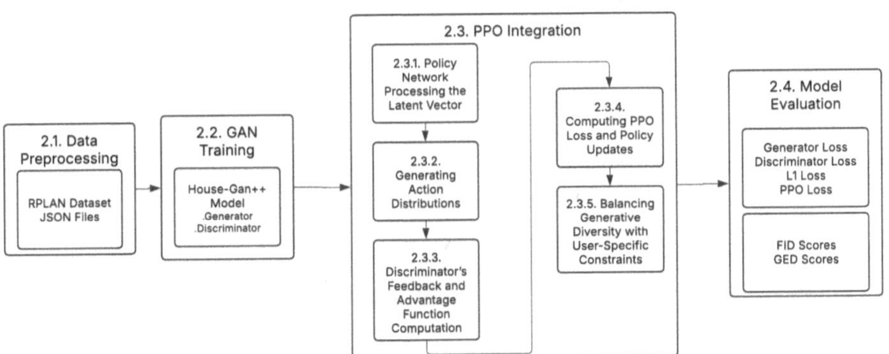

Fig. 1. Methodology Framework

2.1 Data Preprocessing

The RPLAN dataset [11] was selected to train GANs due to the variety of architectural spaces and utilization as research material in the literature. This research utilizes the

House-GAN++ framework in order to preprocess the data according to similar studies. Nauata et al. [10] preprocessed RPLAN into JSON files to minimize the image-based data into numerical values. The GAN model requires a latent vector from a Gaussian distribution as its initial input for generation.

2.2 GAN Training

The GAN training procedure operates through an adversarial training structure. The generator network accepts hidden vector data, node placements edge relations, and masked constraints to produce candidate floor plans. The discriminator network analyzes generated floor plans to determine whether they come from real instances or synthetic sources. The training process stabilizes by combining Wasserstein loss with gradient penalty in the loss function. The adversarial loss optimization of the generator network leads to progressively realistic architectural plan generation. The implementation includes a distance-based loss function for spatial consistency, which controls the generator's position of output masks according to user-defined constraints (Fig. 2).

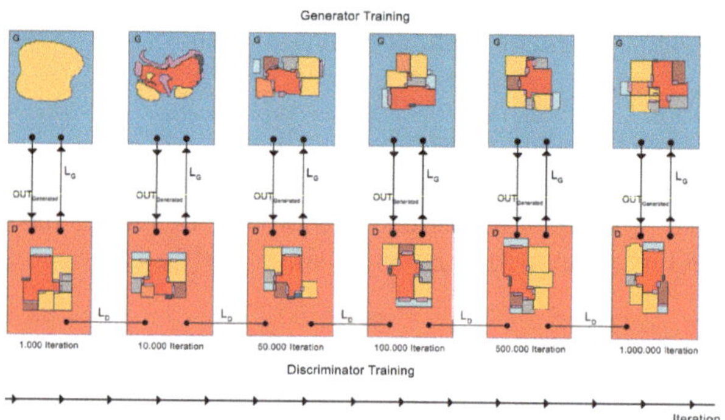

Fig. 2. House-GAN++ Training System

2.3 Proximal Policy Optimization Integration

PPO Policy Mechanism and Generator Interaction

PPO optimization is an important building block toward improving the quality and controllability of the architectural floor plans produced by the GAN model. It runs within the GANs training loop in a way that generation is guided by not just adversarial loss but also RL signals. The latent vector is sampled at the start of every iteration and fed through the PPO policy network to yield a probability distribution over actions [14]. Actions guide the alteration of the latent vector or spatial input arrangement before generator processing. The policy network's fully connected layers calculate these distributions, altering the latent vector in a manner that encapsulates learned design desirability. This policy-aware adjustment allows the generator to incorporate user-informed biases during generation, alleviating a notable limitation of baseline GANs like House-GAN++. The output floor plan generated is then fed through the discriminator, which evaluates realism. The realism evaluation is used to compute the advantage function:

$$A(s, a) = Q(s, a) - V(s) \tag{1}$$

Where Q (s, a) is the expected return for taking action a in state s, V(s) is the predicted value of the current state. If A(s,a) is positive, PPO-GAN modifications improve the floor plan; if negative, they do not contribute positively. This quantifies how beneficial the action taken was in improving the realism of the output. A positive advantage indicates a beneficial policy action, thereby reinforcing that trajectory.

PPO Loss Function and Policy Optimization Strategy

In order to optimize the generator for desirable latent states, PPL employs a clipped surrogate loss function defined as:

$$L(\theta) = E\left[\min(r\theta A, \mathrm{clip}(r\theta, 1 - \varepsilon, 1 + \varepsilon)A)\right] \tag{2}$$

Where rθ is the ratio between the new and old policy probabilities, and ϵ is the clipping parameter, which prevents significant updates. Through clipping operations, the method stops major changes to policy while delivering training stability and protection against overfitting. With this framework, PPO-GAN outperforms traditional GANs by iteratively correcting layout generations and preserving the semantic position of building elements. This marriage guarantees that not only are plans realistic and diverse, but also that they conform to structural codes found in real-world datasets like RPLAN. Figure 3 illustrates the complete PPO-GAN training system, which integrates adversarial learning, graph constraints, and reinforcement-based latent adjustment into a single generation pipeline.

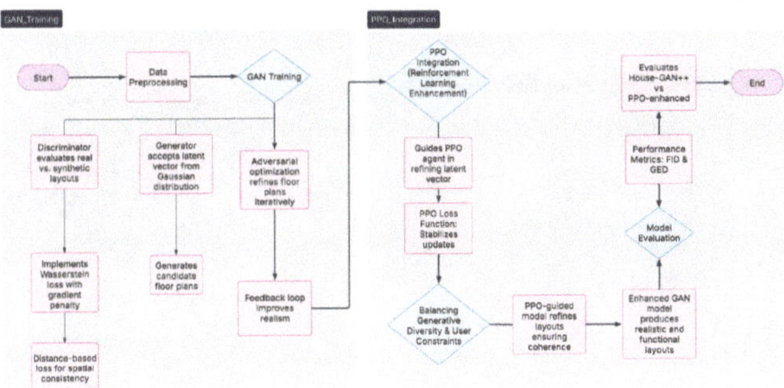

Fig. 3. PPO-GAN Training System

2.4 Model Evaluation

The training process tracks four different loss functions through TensorBoard. Two metrics calculate realism and structural consistency: FID [15] and GED [10]. FID is responsible for the realism score, which is the similarity penalty between the generated output and the real sample in the dataset. GED is responsible for the structural performance of generated inputs compared to the real sample in the dataset. To evaluate the PPO-GAN model and the House-Gan++ model, these metrics compare similarities and performances for both generation processes after the training. Model training was conducted with a complete set of hyperparameters to ensure convergence stability and representational learning adequacy. The generator and discriminator were both trained using the Adam optimizer with learning rates 0.0001 and momentum terms $\beta_1 = 0.5$ and $\beta_2 = 0.999$. Training was performed for 18 epochs with a batch size of 2 (1 million iterations). Latent space dimensionality was set to 128. Gradient penalty of the Wasserstein loss was weighted with $\lambda = 10$ for Lipschitz continuity. PPO-GAN model outputs were stored every 50 training iterations, and training stability was monitored.

The PPO policy network, with two hidden layers of 256 units each, was trained using the Adam optimizer with a learning rate of $1 \text{ x } 10^{-4}$. The PPO update epochs per GAN iteration were set to 10. The clipping threshold of the policy surrogate objective was set to $\epsilon = 0.2$ to prevent overly aggressive updates. The value function was optimized together with mean squared error loss. The above PPO hyperparameters were selected to maintain training stability, convergence speed, and spatial constraints in the resulting architectural designs.

3 Results

3.1 Comparison Results

According to Fig. 4, the comparison results are shown.

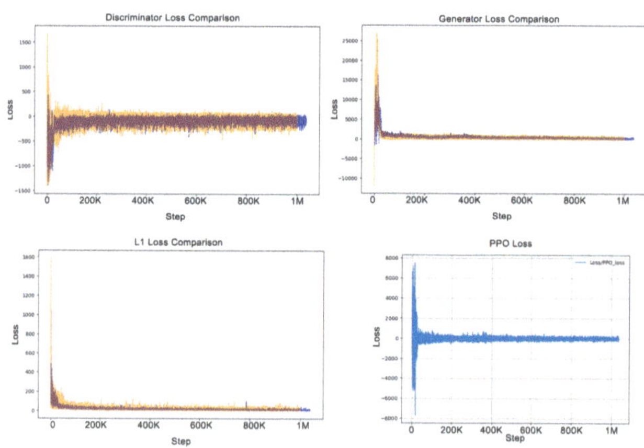

Fig. 4. Loss Comparisons of the PPO-GAN Model (Blue) and House-GAN++(Orange)

Three Losses Comparison.

- During training, HG exhibits significant variations in discriminator loss that continue until the end of the process. The PPO-GAN model tracks loss values in a stable pattern, indicating better-controlled adversarial learning. PPO integration contributes to stable GAN adversarial dynamics, which helps stop the discriminator from taking control of the generator.
- The optimization process of the HG model starts with initial unstable behavior as indicated by increased fluctuation in generator loss during early training stages. The PPO-GAN model shows a faster decrease in loss and sustained low variance from the beginning to the end of training. The PPO-GAN model's generator loss stabilizes better because RL efficiently guides the generator in producing realistic architectural plans during its convergence process.
- The L1 loss in HG maintains higher values for longer, indicating poor base model performance in maintaining fixed constraints. The PPO-GAN model drops its L1 loss rate substantially more quickly until it reaches a reduced level, demonstrating superior spatial constraint and responsiveness to user preference. RL demonstrates improved constraint adherence by optimizing the generator to follow user-defined floor plan elements strictly.

PPO Loss.
From the beginning, the PPO Loss shows erratic behavior until it reaches stable values during training. During the early stages of RL training, we observe high variations because the agent conducts exploration to optimize policy updates. The loss demonstrates stability throughout the training, which signifies that the PPO agent discovered an optimal level between generator enhancement and policy update reliability. The PPO loss convergence demonstrates that the policy system learns to improve generator output quality by reducing mistakes and strengthening architectural quality.

3.2 Model FID & GED Results

According to Fig. 5, FID and GED scores provide insights into the effectiveness of PPO in improving generative performance.

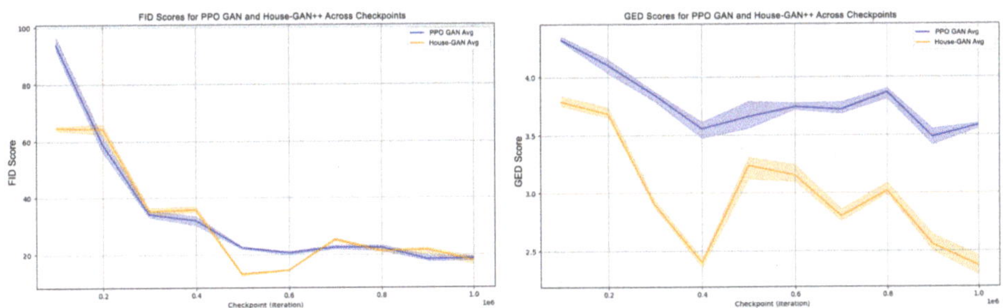

Fig. 5. FID and GED Score Comparison between PPO-GAN and HG model.

FID Scores.
The FID score analysis shows a steady improvement in the training cycles for both PPO-GAN and HG models. Initially, both models exhibit a decrease in FID scores. However, from 400 K iterations onwards, PPO-GAN shows superior stability in score reduction compared to the HG model. RL enables PPO-GAN to produce more detailed architectural plans and reduces the perceptual differences between real and synthetic layouts. Although the HG model experiences a sharp FID reduction of around 500 K iterations, it shows larger fluctuations in the later stages. In contrast, the baseline GAN struggles to maintain stable layout quality, leading to erratic performance. The continuous decrease in PPO-GAN FID scores demonstrates that RL policy feedback capabilities allow generators to independently improve their distribution output quality beyond what adversarial loss optimization provides.

GED Scores.
A lower GED score indicates proper enforcement of space-based geometric constraints and connection requirements in genuine architectural plans. During long-term training, the PPO-GAN and HG models experience varying GED trajectories. The two models show initially decreasing GED scores, but the HG model performs temporarily better than PPO-GAN between 400 K and 500 K iterations. The HG model oscillates with high variability in GED scores, which capture unstable topological integration and room connectivity inconsistency. In contrast, PPO-GAN maintains a more stable GED curve, though with higher values in certain areas. That is due to RL focusing on realistic spatial arrangements and diverse structural features, potentially improving GED as it improves architectural realism. Although the HG model, at times, generates higher GED values, they are not always observed under varying constraints. On the other hand, policy optimization informed by RL has a more balanced spatial realism and structural consistency.

4 Discussion

4.1 Training Discussion

The PPO-GAN model stabilizes adversarial training considerably. Compared to House-GAN++, it offers smoother curves for discriminator loss and mode collapse prevention. RL acts as a balance factor, nudging the generator to produce higher-quality outputs from outside normal adversarial feedback. It promotes faster convergence and improved generator performance via structured optimization signals and accelerates exploration of the design space by combining discriminator feedback with user-defined constraints. Adherence to spatial constraint is measured using the L1 loss metric, and PPO-GAN proves to learn faster in terms of structural coherence, fundamental for sustaining room layout, connectivity, and functional zoning in building design. Yet, earlier PPO instability indicates that learning rates must be adaptive, while reward shaping must be better designed. Moreover, PPO-GAN has a greater computational cost, inviting more effective training schemes. Despite these challenges, PPO integration enhances performance, stability, and constraint satisfaction in architectural design using GANs.

4.2 Model Results Discussion

When the PPO-GAN model is tested against House-GAN++ base models, the analysis proves that RL produces benefits for structured generation tasks. Experimental FID score findings reveal how PPO improves layout visualization realism and creates a steady generation process that maintains realistic architectural designs in the output. Analysis of GED scores indicates that PPO algorithms improve the training system by decreasing undesirable irregularities observed during adversarial learning. The PPO-GAN model performs better than the baseline HG model because it achieves better stability, increased convergence, and enhanced layout quality. Applying PPO inside GAN improves generator performance through RL, which generates improved architectural designs with realistic structures. Future scientific investigations should investigate additional ways to enhance reward shaping and hierarchical RL to improve the balance of realistic building designs with spatial architectural coherence.

4.3 Comparison with Industry Tools and Future Work

Future research can explore using RL to guide GAN architectures toward specific architectural goals. As datasets remain scarce, RL can help GANs generalize better from limited data. In the Architecture industry, the hybrid method promises to automate layout generation, spatial reasoning optimization, and facilitate early-stage design processes. Despite the computational cost and early instability, there is technical robustness. The follow-on work must be benchmarked against leading industry generative tools and investigate computational trade-offs in detail.

5 Conclusion

In this research, the integration of PPO-GAN to enhance architectural floor plan generation quality, stability, and structural coherence is explored. The method combines reinforcement learning and adversarial training to address the weaknesses of simple GAN models. The formal optimization of PPO delivers designs more closely adhering to architecture conventions and user-defined constraints. Assessment metrics for the performance metrics show higher training stability, with PPO-GAN recording better discriminator and generator loss convergence. Reduced L1 loss reflects improved spatial coherence and constraint satisfaction, and lower and stabilized FID scores reflect improved perceptual quality. While early PPO loss instability provided challenges, these were mitigated by employing adaptive learning rates and reward shaping. Effective RL strategies, such as hierarchical RL or meta-learning, must be investigated for future research to further optimize performance and adaptability in real-world design projects to reduce computational demands.

References

1. Baduge, S.K., Thilakarathna, S., Perera, J.S., Arashpour, M., Sharafi, P., Teodosio, B., et al.: Artificial intelligence and smart vision for building and construction 4.0: Machine and deep learning methods and applications. Autom. Constr. **141**, 104440 (2022 Sep)

2. Bahrehmand, A., Batard, T., Marques, R., Evans, A., Blat, J.: Optimizing layout using spatial quality metrics and user preferences. Graph Models. **93**, 25–38 (2017 Sep)
3. Weber, R.E., Mueller, C., Reinhart, C.: Automated floorplan generation in architectural design: A review of methods and applications. Autom. Constr. **140**, 104385 (2022 Aug)
4. Goodfellow, I.J., Pouget-Abadie, J., Mirza, M., Xu, B., Warde-Farley, D., Ozair, S., et al.: Generative Adversarial Networks [Internet]. arXiv; 2014 [cited 2024 Mar 19]. Available from: http://arxiv.org/abs/1406.2661
5. ArchiGAN, C.S.: Artificial Intelligence x Architecture. In: Yuan, P.F., Xie, M., Leach, N., Yao, J., Wang, X. (eds.) Architectural Intelligence [Internet], pp. 117–127. Springer Nature Singapore, Singapore (2020 [cited 2023 Jun 20]) Available from: https://link.springer.com/10.1007/978-981-15-6568-7_8
6. Isola, P., Zhu, J.Y., Zhou, T., Efros, A.A.: Image-to-Image Translation with Conditional Adversarial Networks [Internet]. arXiv; 2018 [cited 2023 Jun 25]. Available from: http://arxiv.org/abs/1611.07004
7. Zheng, H., An, K., Wei, J., Ren, Y.: Apartment floor plans generation via generative adversarial networks. In: Bangkok, pp. 599–608, Thailand (2020 [cited 2024 Aug 24]) Available from: http://papers.cumincad.org/cgi-bin/works/paper/caadria2020_015
8. Nauata, N., Chang, K.H., Cheng, C.Y., Mori, G., Furukawa, Y.: House-GAN: Relational generative adversarial networks for graph-constrained house layout generation. In: Vedaldi A, Bischof H, Brox T, Frahm JM, editors. Computer Vision – ECCV 2020 [Internet]. Cham: Springer International Publishing; 2020 [cited 2023 Jun 20]. p. 162–177. (Lecture Notes in Computer Science; vol. 12346). Available from: https://link.springer.com/10.1007/978-3-030-58452-8_10
9. Zhang, F., Nauata, N., Furukawa, Y.. Conv-MPN: Convolutional Message Passing Neural Network for Structured Outdoor Architecture Reconstruction [Internet]. arXiv; 2021 [cited 2023 Aug 23]. Available from: http://arxiv.org/abs/1912.01756
10. Nauata, N., Hosseini, S., Chang, K.H., Chu, H., Cheng, C.Y., Furukawa, Y.: House-GAN++: Generative adversarial layout refinement network towards intelligent computational agent for professional architects. 2021 IEEE/CVF Conference on Computer Vision and Pattern Recognition (CVPR) [Internet]. Nashville, TN, USA: IEEE; 2021 [cited 2023 Jun 20]. p. 13627–13636. Available from: https://ieeexplore.ieee.org/document/9577959/
11. Wu, W., Fu, X.M., Tang, R., Wang, Y., Qi, Y.H., Liu, L.: Data-driven interior plan generation for residential buildings. ACM Trans. Graph. **38**(6), 1–12 (2019 Dec 31)
12. Rahbar, M., Mahdavinejad, M., Markazi, A.H.D., Bemanian, M.: Architectural layout design through deep learning and agent-based modeling: A hybrid approach. J. Build. Eng. **47**, 103822 (2022 Apr)
13. Franceschelli, G., Musolesi, M.: Reinforcement learning for generative AI: State of the art, opportunities and open research challenges. J. Artif. Intell. Res. **6**(79), 417–446 (2024 Feb)
14. Schulman, J., Wolski, F., Dhariwal, P., Radford, A., Klimov, O.: Proximal Policy Optimization Algorithms [Internet]. arXiv; 2017 [cited 2024 Oct 28]. Available from: http://arxiv.org/abs/1707.06347
15. Heusel, M., Ramsauer, H., Unterthiner, T., Nessler, B., Hochreiter, S.: GANs Trained by a Two Time-Scale Update Rule Converge to a Local Nash Equilibrium [Internet]. arXiv; 2018 [cited 2024 Jun 7]. Available from: http://arxiv.org/abs/1706.08500

BY NC ND

Performance-based Design, Analytics and Optimization

A GNN-Based Surrogate Model for Rapid Energy Consumption Prediction of Residential Floor Plans for the Early Design Stage

Jiaqi Wang[1], Wanzhu Jiang[2], Haotian Li[1], and Llewellyn Tang[1,3](✉)

[1] Department of Real Estate and Construction, The University of Hong Kong, Hong Kong, China
lcmtang@hku.hk
[2] School of Architecture, South China University of Technology, Guangzhou, China
[3] Faculty of Science and Technology, BNU-HKBU United International College, Zhuhai, China

Abstract. Under the global sustainability goal of reducing carbon emissions, it is crucial to consider building energy consumption at the early design stage. However, previous studies mainly focused on physical features like building form and materials, overlooking the impact of spatial layout on building energy performance. This study proposes a graph-based spatial representation method and develops a Graph Fusion Network (GFN) model for rapid room-level energy consumption prediction. Experiment results demonstrate that the proposed model effectively captures the relationship between graph-based spatial features and energy consumption, significantly outperforming traditional machine learning methods in predictive accuracy and generalization capability. This research validates the influence of spatial layout on building energy performance. It provides an efficient and accurate energy assessment tool for early design stages, supporting carbon-conscious spatial decision-making and design optimization.

Keywords: Data-driven · Spatial Graph Representation · Graph Neural Network · Residential Floor Plans · Energy Consumption Prediction

1 Introduction

Reducing carbon emissions has become a pivotal target in pursuing global environmental protection and sustainable development. As a major contributor to carbon emissions, the building industry accounts for nearly 40% of global CO_2 emissions [1]. Throughout the building lifecycle, operational energy consumption typically constitutes the largest share of these emissions at around 75% [2]. To mitigate building-related carbon emissions, decision-making during the early design phase is particularly crucial, as it is widely recognized as the most cost-effective period with the greatest carbon reduction potential, locking in 70–80% of a building's energy performance [3]. Therefore, accurately and efficiently assessing energy consumption during early design stages is crucial for optimizing architectural design and implementing carbon reduction strategies.

Y. Liu et al. (Eds.): CDRF 2025, *Transindividual Intelligence*, pp. 121–132, 2026.
https://doi.org/10.1007/978-981-92-0615-5_11

Simulation-based and data-driven methods are two mainstream approaches for evaluating building energy consumption. The former relies on physical modeling with precise parameterization, resulting in high operational thresholds, equipment demands, and time consumption. It is less suitable for iterative trial-and-errors in early-stage design. The latter, employing machine learning algorithms like ANN [4] and XGBoost [5] to construct surrogate models, has gained widespread popularity due to its efficiency. However, limited by the availability of original data and vector-based building representation, most existing studies focus solely on the impact of form and materials or restrict their analysis to individual rooms, leaving a notable research gap concerning the entire spatial layout.

Therefore, this study proposes a graph-based representation of architectural spatial data and develops a Graph Fusion Network (GFN) to realize the room-level rapid energy consumption prediction for residential building layouts, examining the correlation between spatial configurations and energy demand. This study, leveraging large-scale building data, presents an accurate and efficient method for building energy performance evaluation, supporting the exploration, decision-making, and optimization in early-stage spatial layout design.

2 Literature Review

The rapid advancement of AI technologies has made machine learning-based surrogate models for energy consumption prediction the predominant approach in contemporary Building Energy Simulation research [6]. Based on the different models and data employed, these studies can be categorized into three main types.

(1) Energy Consumption Prediction with Simple Regression Models: The existing literature includes methods such as Multiple Linear Regression (MLR), Support Vector Regression (SVR), and tree models. Catalina et al. [7] applied MLR to predict monthly heating demand in residential buildings, using building form, glass U-value, window-to-floor area ratio, and climate conditions as input factors. Similarly, SVR has been implemented to predict the hourly cooling load of an office building [8]. Yan et al. [5] employed XGBoost to predict energy consumption and daylighting performance across various residential building forms and construction combinations. These models offer the advantage of low training costs and are often used as baseline benchmarks.

(2) Energy Consumption Prediction with Artificial Neural Networks: ANN is the most popular prediction model in current research, accounting for approximately half of the surveyed literature. It has been extensively applied to various prediction targets across multiple spatial scales. At the urban level, Li et al. [4] proposed two ANN-based methods to predict the Energy Use Intensity of complex building clusters, achieving prediction errors within 10%. At the building level, Hossam Wefki and his team [9] established an ANN-based energy prediction method for residential buildings, incorporating factors such as building morphology, orientation, and window placement. At the room level, Zou et al. [10] developed a parametric model for a classroom using ANN to simultaneously predict energy consumption, thermal and visual comfort, replacing simulation tools for subsequent spatial optimization. ANN

demonstrates superior nonlinear fitting capabilities compared to traditional machine learning models.

(3) Energy Consumption Prediction with Convolutional Neural Networks: In contrast to the above models, CNN specializes in processing two-dimensional matrix data (images), leveraging convolution to extract local features between pixels. Yan et al. [11] developed an integrated CNN to predict energy consumption, daylighting, and thermal comfort by incorporating floor plans, construction details, and climate conditions. Singaravel et al. [12] transformed design elements into a two-dimensional matrix for CNN processing. Their results indicated a slight improvement in prediction accuracy over the MLP model.

In summary, these types of surrogate models present limitations. Vector-based models are constrained by data representation, relying on indirect descriptions of building form. As a result, it overlooks many potential spatial features and is primarily suitable for large-scale, geometrically simple building clusters or individual buildings. In contrast, image-based models employ flattened representations to convey spatial forms and relationships, which can introduce irrelevant noise into the prediction process. Furthermore, current studies focus on simple building volumes or single parametric rooms, with limited consideration of spatial layouts, making it challenging to support spatial analysis during early-stage design. Therefore, this paper aims to establish a more practical large-scale dataset and explore a novel graph-based spatial representation and prediction model.

3 Methodology

Focusing on residential floor plans, the research framework includes three stages:

(1) Database Establishment: Based on the 3000 raw floorplan data collected from the open-source dataset "RPlan" [19], a series of automated methods were developed to vectorize image data and convert it into 3D models. Energy simulations were then conducted to get the room-level energy consumption data. Ultimately, a labeled graph dataset was established via graph representation.

(2) Graph Fusion Network Establishment & Training: GFN was developed to work towards energy prediction. A comparative experiment was conducted between three different Graph Neural Network modules (Graph Convolutional Networks, Graph Attention Networks, and GraphSAGE) and the MLP baseline model.

(3) Model Evaluation & Implementation: All models' accuracy and generalization were cross-compared and analyzed. Additionally, some cases were tested to validate this method's advantages in performance prediction (Fig. 1).

3.1 Graph Database Establishment

3.1.1 From 2D Image to 3D Model

The residential plans in this study were extracted from RPlan, a large public dataset [13]. It contains 80,000 annotated floor plans of real Asian residential buildings, with

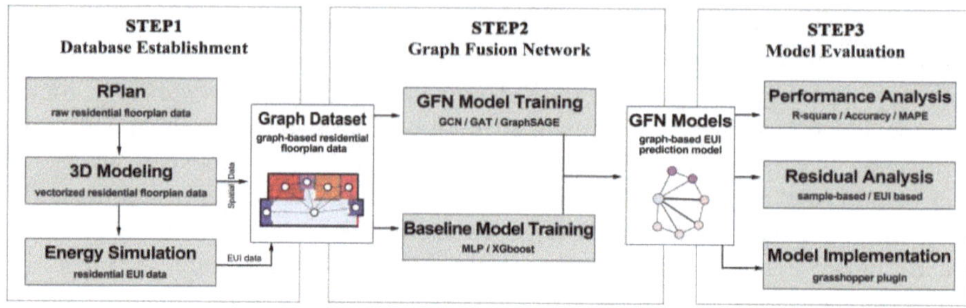

Fig. 1. Research Framework

semantic annotations at the pixel level. As shown in Fig. 2, each data sample includes four layers of digitally labeled semantic information: (1) inside masks, (2) boundary masks, (3) semantics of all rooms along with interior/exterior walls, and (4) room masks. This study selected the first 3,000 samples from the dataset. The unit area ranges from 60 to 120 m^2, with a concentration of around 80 m^2. The room count is predominantly between 6 and 8.

The vectorization of the selected samples is based on the third channel. By detecting the region boundaries with different labels, we extracted the vertex coordinates for each room, unit boundary, and the door/window to generate a JSON file. In Rhino, connecting the vertices sequentially resulted in the raw geometric data. Due to the image-to-data conversion process, alignment issues may occur. Therefore, we developed a wall alignment algorithm that automatically corrects room-to-room and room-to-boundary offsets smaller than 0.3 m, thereby producing refined geometric floor plan data. Then, the 3D residential models were generated by extruding the walls and positioning doors and windows.

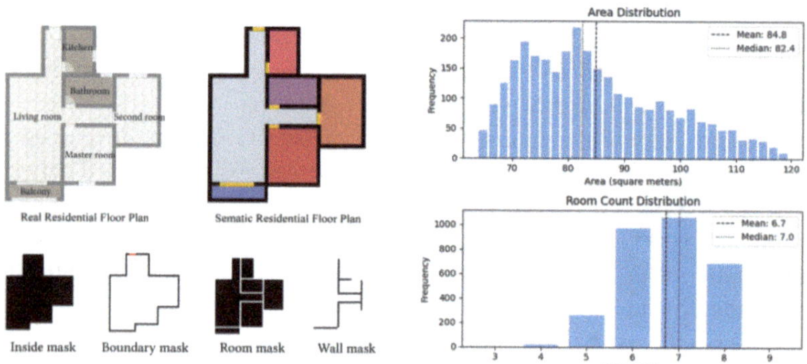

Fig. 2 (a) RPlan Channels, (b) the distribution of Area and Room count in the selected dataset.

3.1.2 Energy Consumption Modeling

In this study, EUI serves as the metric to measure residential energy consumption, calculated as the ratio of annual total energy consumption to gross floor area. Typically, the energy consumption includes cooling supply, heating supply, lighting, electrical equipment, and hot water. This study primarily focuses on the first three factors, which are directly related to the spatial layout. Therefore, the EUI formula used in this study can be expressed as:

$$\text{EUI} = \frac{\text{EC}_\text{C} + \text{EC}_\text{H} + \text{EC}_\text{L}}{\text{Gross Floor Area}} \tag{1}$$

The detailed energy consumption simulation was performed using the Honeybee plugin within the Rhino-Grasshopper platform, which is powered by the EnergyPlus engine. The 3D model generated in the previous step was converted into a corresponding energy model and assigned appropriate construction materials. Table 1 presents the key control parameters configured for the energy simulation setup. In addition to the primary focus on spatial layout, other crucial parameters were organized into three distinct performance sets, which were randomly assigned during the simulations to improve model robustness.

Table 1. The setting of the energy consumption simulation

Category	Variables	Value Range			Units	Variation
Climate	location	Hong Kong[a]			–	constant
	Weather	EPW weather data			–	
Building Layout	Floor Plans 3D models	get form 3D model			–	3000
wall Construction	Thickness	0.2	0.2	0.3	m	3
	Conductivity	1.8	1.0	0.5	W/(m·K)	
	Density	2200	1800	800	kg/m3	
window Construction	U-value	1.2	2.0	2.8	$W/(m^2 \cdot K)$	3
	Solar heat gain coefficient	0.3	0.4	0.5	–	
	Visible transmittance	0.5	0.6	0.7	–	
windows opening	WWR-South	0.4	0.5	0.6	–	3
	WWR-North	0.2	0.3	0.4	–	3
	WWR-East	0.2	0.3	0.4	–	3
	WWR-West	0.2	0.3	0.4	–	3
Equipment	HVAC COP	Cooling:3.5, Heating:2.5			kW/kW	constant
	Light Power	5.0			w/m^2	

[a]Hong Kong was chosen as the background city for this study

3.1.3 Graph Representation and Graph Dataset Establishment

Based on the 3D models and corresponding spatial energy consumption data, the construction of the graph dataset involves two steps: (1) spatial graph representation and (2) feature extraction. As illustrated in Figure 3, the conversion from spatial layouts to graphs follows these principles: (1) Each room with an independent boundary is represented as a node; (2) An edge is established between two nodes if their corresponding rooms share an adjacent wall. Following existing literature, features related to function, form, and components are extracted at each level. At the same time, spatial topological relationships are expressed in the form of an adjacency matrix (Fig. 3). Ultimately, a graph data sample consists of the graph features, edge index, and node features with corresponding EUI data serving as labels. The feature extraction process is automated through a custom algorithm. After data cleaning, this study obtained a total of 2,859 residential unit samples and 19,141 room samples.

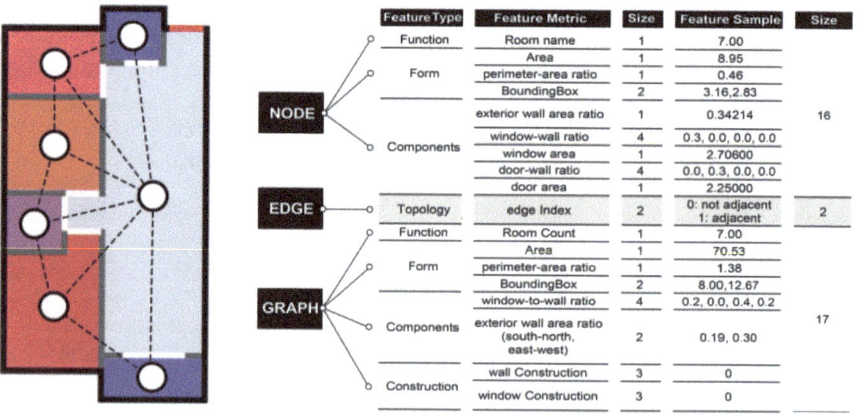

Fig. 3. (a) graph representation, (b) the feature list extracted from the floor plan

3.2 Graph Fusion Network & Model Training

GNNs update node features through a message-passing mechanism [14]. Specifically, at each layer, a node aggregates the features of its adjacent nodes and combines them with its features to derive an updated representation. While GNNs have demonstrated good performance in classification tasks, this study tackles a distinct regression problem. To tailor the model to the scenario of mapping spatial graphs to energy consumption, we propose a novel GNN architecture—Graph Fusion Network (GFN). As shown in Fig. 4, the model accepts three input feature groups (node features, edge index, and graph features) and comprises two core Modules: 1) GNN Module: Each includes a GNN layer followed by BatchNorm, ReLU, and Dropout layer. (2) MLP Module: Each includes the Linear, ReLU and Dropout layer. Specifically, the features extracted from the GNN modules are fused with both the original node features and the graph features before being passed through the final MLP for regression output. This architecture not only preserves the influence of spatial relationships but also mitigates the issue of feature over-smoothing in graph-based learning.

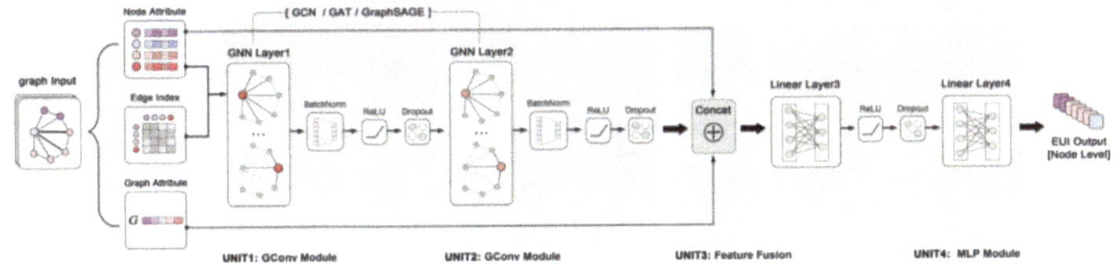

Fig. 4. Architecture of Graph Fusion Network

In terms of training setup, the graph neural network is built using PyTorch with the Adam optimizer and MSE (Mean Squared Error) as the loss function. The graph data set is shuffled and split into training and testing sets with a ratio of 9:1 after normalization. For the experiment design, as shown in Table 3, three parallel experiments were conducted to validate the impact of different graph convolutional algorithms (GCN [14], GAT [15], and GraphSAGE [16]) on the model's performance. Also, this study incorporated two state-of-the-art baseline models (MLP and XGBoost) to assess the predictive accuracy of our approach. The comparison between the experimental and baseline groups, due to the different input feature sets, reflects the influence of spatial layout (especially spatial topological relationships) on energy consumption (Table 2).

Table 2 The parallel experiments setting (▲ means the feature set is included in the training).

Experiment Group	GNN Module	Feature Set		
		node	edge_index	graph_attr
Test model	GCN	▲	▲	▲
	GAT	▲	▲	▲
	GraphSAGE	▲	▲	▲
Baseline model	MLP	▲	X	▲
	XGBOOST	▲	X	▲

4 Results & Discussion

The training was conducted on a laptop equipped with an i7-11370H processor, 16 GB RAM, and an NVIDIA RTX 3070 laptop GPU. The average training time for 3000 epochs was approximately 15 min. Figure 5 illustrates the training curves across five experimental trials. Among them, GCN exhibited the fastest convergence, followed by GraphSAGE, while the final loss values of the three GNN models were relatively close. In contrast, the MLP model demonstrated a higher minimum loss upon convergence compared to the other models.

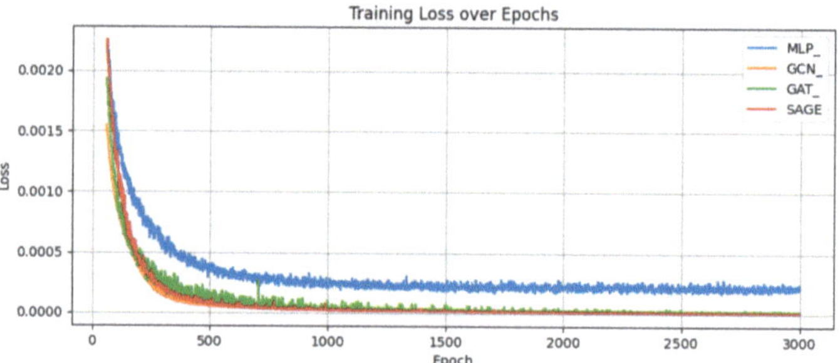

Fig. 5. The Training Curve.

In the model performance evaluation, a diverse set of metrics was employed to assess the prediction accuracy and precision, including R-square, accuracy within 10% and 5% deviation, and MAPE. Table 4 presents the performance of the tested models on both the training and testing datasets. Notably, the GNN models based on the GFN architecture significantly outperformed the baseline models, achieving over a 10% improvement in accuracy within a 10% deviation and approximately a 5% reduction in MAPE compared to the XGBoost model, resulting in a 50% optimization. A comparative analysis of the three GNN variants indicates that GCN and GAT exhibit similar predictive performance, whereas GraphSAGE achieves the best results. This finding suggests that for node-level energy consumption prediction tasks, the locality-focused approach of GraphSAGE is more effective, highlighting the influence of adjacent spatial relationships on energy consumption, which aligns with practical observations. Moreover, the substantial accuracy improvement of the tested models over the baseline models indirectly demonstrates the impact of spatial topology on energy consumption, reinforcing that incorporating spatial layout information can significantly enhance the performance of predictive models.

Table 3. The evaluation results of models on the training set and the testing set.

Experiment Group	GNN Model	Training Set				Optimization Rate
		R^2	Accuracy_ ± 10%	Accuracy_ ± 5%	MAPE	
Test model	GFN-GCN	0.9991	96.74%	87.40%	2.57%	65.50%
	GFN-GAT	0.9978	96.41%	82.76%	3.06%	58.93%
	GFN-GraphSAGE	**0.9994**	**98.05%**	**90.29%**	**2.23%**	**70.07%**
Baseline model	MLP	0.9943	83.69%	69.18%	6.36%	14.63%
	XGBOOST	0.9848	80.29%	68.16%	7.45%	--
Experiment Group	GNN Model	Testing Set				Optimization Rate
		R^2	Accuracy_ ± 10%	Accuracy_ ± 5%	MAPE	
Test model	GFN-GCN	0.9957	94.03%	78.40%	3.64%	58.40%
	GFN-GAT	0.9935	94.08%	77.12%	3.72%	57.49%
	GFN-GraphSAGE	**0.9957**	**96.27%**	**81.44%**	**3.15%**	**64.00%**

(continued)

Table 3. (*continued*)

Experiment Group	GNN Model	Testing Set				Optimization Rate
		R^2	Accuracy_ ± 10%	Accuracy_ ± 5%	MAPE	
Baseline model	MLP	0.9881	82.88%	67.73%	6.36%	27.31%
	XGBOOST	0.9787	74.57%	47.29%	8.75%	--

By comparing the predicted values of the GFN-GraphSAGE model with the actual values, it can be observed that in the training set, all points align with the optimal prediction line. In contrast, in the testing set, the majority of samples fall within the 10% deviation range. To provide a more intuitive representation of the error distribution, the percentage deviation between predicted and actual values was further visualized. Analyzing the deviation percentages across different EUI values reveals that larger deviations are primarily concentrated within the 0–50 kWh/m^2 range, particularly for samples with EUI values below 10 kWh/m^2. This discrepancy is likely due to the inherent skewness of the original dataset, which results in a higher concentration of smaller values within this range. Future improvements could involve mapping adjustments to the original data distribution (Figs. 6 and 7).

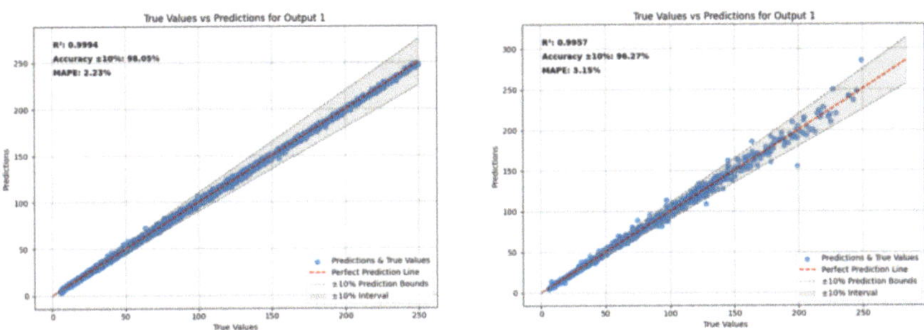

Fig. 6. Parity Plot of GFN-GraphSAGE model on the training and the testing set.

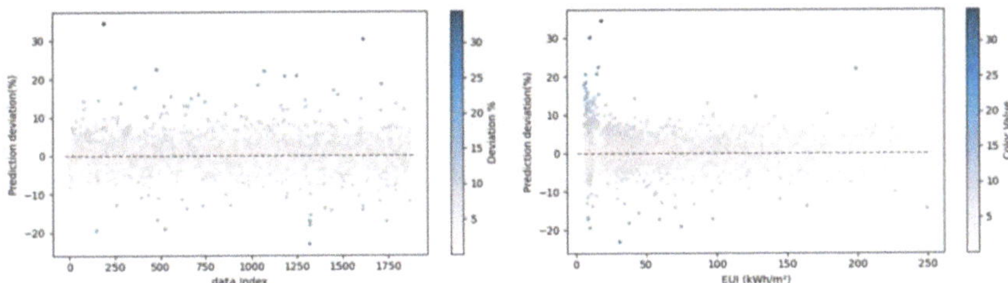

Fig. 7. Residual Plot of GFN-GraphSAGE model based on data index and EUI.

By integrating the trained model into Grasshopper, we evaluated its predictive performance across various cases. As shown in Fig. 8, two cases with regular and complex geometric boundaries were examined: In both cases, the prediction deviations for all rooms remained within 5%, with the overall deviation being even lower for the residence with a complex boundary. Regarding different functional rooms, the prediction error tended to be higher for irregularly shaped spaces such as living rooms. These initial application results demonstrate the practical feasibility of the proposed model.

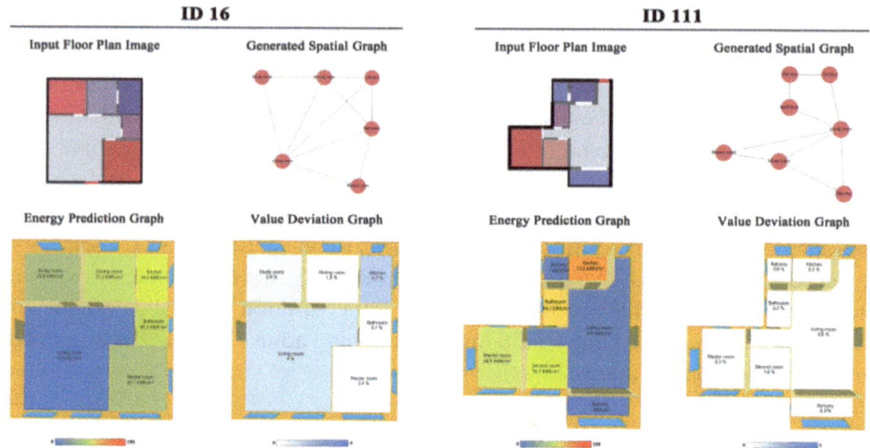

Fig. 8. The case study of the prediction model.

5 Conclusion & Future Work

This study proposes a room-level energy consumption prediction method for residential buildings based on a graph neural network. A large-scale residential floor plan dataset with energy consumption data is constructed, and a Graph Fusion Network specifically for spatial layout energy prediction is designed. By leveraging graph structure to represent spatial layouts, the proposed method can effectively capture spatial morphology and topology, significantly improving prediction accuracy and generalization capability. Experimental results show that GFN models outperform current mainstream models across multiple evaluation metrics, validating the importance of incorporating spatial topological features into energy prediction models. Furthermore, a comparative analysis with traditional representation methods highlights the potential of graph-based spatial encoding and reinforces the influence of spatial layout on energy consumption. As a promising starting point, this study still has certain limitations. Future research directions include expanding the dataset into more climate zones and optimizing data distribution to mitigate small-value prediction biases. Additionally, we are dedicated to developing a user-friendly interactive platform and establishing transformation interfaces for images, 3D models, and BIM models to enable multi-modal energy prediction and decision support based on this method.

References

1. Su, Y., Cheng, H., Wang, Z., et al.: Analysis and prediction of carbon emission in the large green commercial building: A case study in Dalian, China. Journal of Building. Engineering. **68**, 106147 (2023)
2. Zhang, X., Chen, H., Sun, J., et al.: Predictive models of embodied carbon emissions in building design phases: Machine learning approaches based on residential buildings in China. Build. Environ. **258**, 111595 (2024)
3. Schlueter, A., Thesseling, F.: Building information model based energy performance assessment in early design stages. Autom. Constr. **18**(2), 153–163 (2009)
4. Li, Z., Dai, J., Chen, H., et al.: An ANN-based fast building energy consumption prediction method for complex architectural form at the early design stage. Build. Simul. Tsinghua University Press. **12**, 665–681 (2019)
5. Yan, H., Ji, G., Yan, K.: Data-driven prediction and optimization of residential building performance in Singapore considering the impact of climate change. Build. Environ. **226**, 109735 (2022)
6. Amasyali, K., El-Gohary, N.M.: A review of data-driven building energy consumption prediction studies. Renew. Sust. Energ. Rev. **81**, 1192–1205 (2018)
7. Catalina, T., Iordache, V., Caracaleanu, B.: Multiple regression model for fast prediction of the heating energy demand. Energ. Buildings. **57**, 302–312 (2013)
8. Li, Q., Meng, Q., Cai, J., et al.: Applying support; vector machine to predict hourly cooling load in the building. Appl. Energy. **86**(10), 2249–2256 (2009)
9. Elbeltagi, E., Wefki, H.: Predicting energy consumption for residential buildings using ANN through parametric modeling. Energy Rep. **7**, 2534–2545 (2021)
10. Zou, Y., Zhan, Q., Xiang, K.: A comprehensive method for optimizing the design of a regular architectural space to improve building performance. Energy Rep. **7**, 981–996 (2021)
11. Yan, H., Ji, G., Cao, S., et al.: Developing an integrated prediction model for daylighting, thermal comfort, and energy consumption in residential buildings based on the stacking ensemble learning algorithm. Build. Simul. Tsinghua University Press. **17**, 2125–2143 (2024)
12. Singaravel, S., Suykens, J., Geyer, P.: Deep convolutional learning for general early design stage prediction models. Adv. Eng. Inform. **42**, 100982 (2019)
13. Wu, W., Fu, X.M., Tang, R., et al.: Data-driven interior plan generation for residential buildings. ACM Trans. Graph. **38**(6), 1–12 (2019)
14. Kipf T N, Welling M (2016) Semi-Supervised Classification with Graph Convolutional Networks.. arXiv preprint https://arxiv.org/abs/1609.02907.
15. Veličković P, Cucurull G, Casanova A, et al (2017) Graph Attention Networks.. arXiv preprint https://arxiv.org/abs/1710.10903.
16. Ahmed N K, Rossi R A, Zhou R, et al (2017) Inductive Representation Learning in Large Attributed Graphs.. arXiv preprint https://arxiv.org/abs/1710.09471.

Validating the Consistency of Daylight Brightness Perception of Users in Real and Virtual Environments

Wentao Zeng(✉), Alan Lewis, and Benjamin Blackwell

University of Manchester, School of Environment, Education and Development, Oxford Rd, Manchester M13 9PL, UK
wentao.zeng@manchester.ac.uk

Abstract. To explore consistency and differences in daylighting design and research using virtual reality (VR), this study proposes a comparative method of standardised workflows and assessment processes to systematically investigate the effects of real and virtual reality environments with different daylight illumination levels (100Lux, 500Lux, and 1500Lux) on the user's activity performance, perception feedback and physiological responses. The study results showed that participants' reading speed and cognitive abilities in VR environments with different illumination levels were not significantly different from the real environment. In addition, the effect of VR on users' perception of pleasure, calmness, interest and satisfaction at different illumination levels was not significant compared to real environments. The findings provide insights into the validity of VR in simulating daylight environments and its impact on user perception and performance.

Keywords: Daylight Illumination · VR · User Perception and Performance · Comparative Method

1 Introduction

Daylight is essential for energy use, temperature control, and visual comfort in buildings, and it also affects occupants' health and well-being [1–3]. Traditionally, architects have used rules and daylight metric measures, often relying on their training and experience to guide design [4, 5]. Although standards exist for comfort and daylight performance, they don't always reflect how real end-users perceive a space. To bridge that gap, many researchers now use a human-centred approach and turn to virtual reality (VR) because it can create realistic, immersive environments [6, 7].

For example, Moscoso et al. [8] demonstrated that VR gives a better sense of brightness and field of view than traditional stereoscopic images in daylighting studies. Ahmed et al. [9] found that VR helps people perceive space more accurately than 2D and 3D renderings, so they can give clearer feedback on daylighting design. Marzouk et al. [10] used VR to compare how three skylight layouts affect user preferences. In a series of studies, Chamilothori et al. [11, 12] explored how people in different regions perceive

Y. Liu et al. (Eds.): CDRF 2025, *Transindividual Intelligence*, pp. 133–146, 2026.
https://doi.org/10.1007/978-981-92-0615-5_12

daylighting design differently using VR. Altogether, these findings suggest that VR is a powerful tool for users and designers to understand and improve daylighting design.

Research on how consistently and accurately VR can be used for daylighting research and design is still scarce [13]. Architects and researchers worry about VR's limits, like how it handles brightness, clarity, and smooth motion, and how these might affect users and results [14, 15]. This research explores the extent to which a VR-based virtual daylight environment can replicate users' brightness perception experienced in the real environment. Additionally, the study aims to determine through the experiment which VR-based daylighting evaluation methods effectively capture users' perceptions, preferences, and behaviours. Based on these objectives, the study poses the following research questions:

- How to structure a standardised workflow and evaluation process to compare user perceptions, preferences and behaviours in real and VR daylight environments?
- And what aspects of their daylighting perception differ between virtual and real environments? What limitations make the difference?
- Do the different illuminance levels impact the perceptions, preferences, and behaviour of users? What are the main impacts?

In summary, this study aims to explore the consistency and differences in user daylight brightness perception between real and VR daylight environments with different illumination levels to complement the gap in the end-user experience aspect of VR daylighting design and research. This paper will detail the methods, experimental design, and data analysis process in the following section to provide a theoretical basis and practical guidance for exploring the application of VR in daylighting evaluation.

2 Material and Method

To explore the consistency and differences in the perception of daylight illuminance levels among users in real and virtual daylight environments, we proposed a comparative method based on a standardised workflow and assessment process (Fig. 1). A small meeting room at the University of Manchester, equipped with adjustable shading devices, will be selected as the case study. Researchers used these adjustable shading devices to create observation areas with three different illuminance levels (100 lx, 500 lx, and 1500 lx) in the real daylight environment. To compare real and virtual environments, the study utilised Rhinoceros and ClimateStudio (CS) for modelling and daylight simulation to replicate the three daylight illuminance levels as 360° High Dynamic Range Images (HDRIs). These images were integrated with interactive modules in Unity3D to develop a VR-based immersive virtual daylight environment for evaluation. Finally, the consistency and differences between real and VR daylight environments were analysed by comparing users' activity performance, perception evaluation, physiological response and qualitative feedback in the two environments at different illumination levels.

2.1 Experimental Setting

This study was conducted in a group meeting room located on the first floor of the Manchester Business School library. The space measures approximately 5.6 m in length,

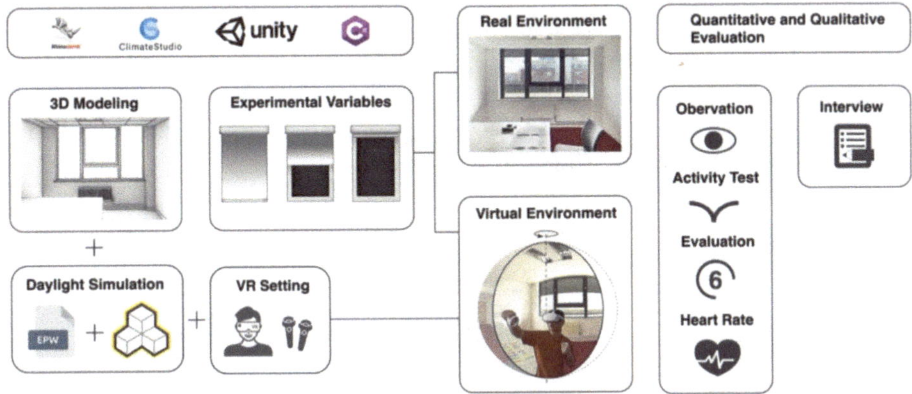

Fig. 1. The comparative method based on a standardised workflow and assessment process.

3.5 m in width, and 3 m in height, covering a total area of about 20 m^2 with a window-to-wall ratio (WWR) of approximately 50%. The room faces southeast and typically receives direct daylight in the afternoon of February. The shading device consists of adjustable vertical curtains made of a low-transmittance white cotton-linen fabric. The observation area for participants was positioned around a desk approximately 2 m from the window (Fig. 2). On sunny and partly cloudy days, the illumination level on this desk area could be adjusted via the shading device to range between 50 and 2000 lx. In the real daylight environment experiments, the researchers controlled the opening and closing of the shading device to set three illumination levels in the observation area—100 lx (±10 lx), 500 lx (±50 lx), and 1500 lx (±100 lx) (Fig. 2). A calibrated LATNEX Light Meter LM-50KL was used to monitor the daylight settings throughout the experiment. The experiments were conducted between early February and early March 2023 during the hours of 12:00–15:00 on days with clear and partly cloudy weather.

2.2 Daylight Simulation and VR Environment Development

Based on the measurement, the space was modelled using Rhinoceros 3D, and daylight simulations were performed using CS. The daylight simulation incorporated EPW weather data of Manchester, with the conditions set to clear weather on February 15, 1 pm. The Konica Minolta CM-600d spectrophotometer was used to record the materials and colours of the space's main surfaces. The materials were selected and replicated in the daylight simulation as accurately as possible according to the material properties available in the CS library (Table 1). The camera position for the simulation was set at the same observation area as in the real environment, approximately 1.2 m above ground. The illumination levels in the daylight simulation were controlled by adjusting the shading devices in the 3D model. The researcher simulated the scene in Falsecolour via CS 's Radiance Render, labelling the daylight illuminance values to ensure that the observation area in VR environments maintained 100 lx, 500 lx, and 1500 lx (Fig. 3). After generating 360° HDRIs for the different illuminance levels, the researchers controlled the images' exposure and gamma values and tone-mapped into low-dynamic

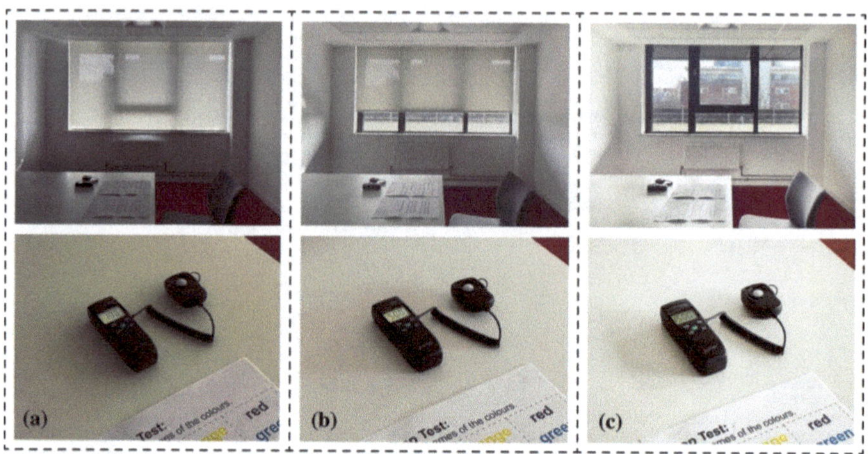

Fig. 2. Setting of observation areas with three different illuminance levels of 100 Lux (a), 500 Lux (b), and 1500 Lux (c) by controlling the opening and closing of the shading device.

range (LDR) images for display on screen. The resulting LDR images were then embedded into VR-based immersive virtual daylight environments developed with Unity3D and synchronised with a VR headset.

Table 1. Radiance material properties for the main surfaces.

Surfaces	Type	Reflectance	R	G	B	Transparency
Ceiling Panel	Plastic	85.67%	0.863	0.856	0.784	
Roof and Walls	Plaster	89.89%	0.908	0.899	0.839	
Floor	Carpet	8.64%	0.231	0.032	0.061	
Door	Wood	31.88%	0.452	0.257	0.105	
Window Frame	Metal	4.74%	0.038	0.039	0.039	
Window	Glass					88%

2.3 Measurement Tools and Metrics

Reading tests were conducted to assess participants' visual capabilities and task performance under different illumination levels. The reading charts, adapted from the MNRead charts and consisting of three short sentences of equivalent length, were printed in A3 with a font size of 8.0 to measure reading time [16]. The recognition test used an adapted version of the Stroop test. The Stroop test is a reliable neuropsychological evaluation tool known for its simplicity, sensitivity to indoor lighting conditions, and close association with visual perception for the user's cognitive analysis [17, 18]. Participants were instructed to verbalise the 27 different ink colours rather than the word quickly. Three

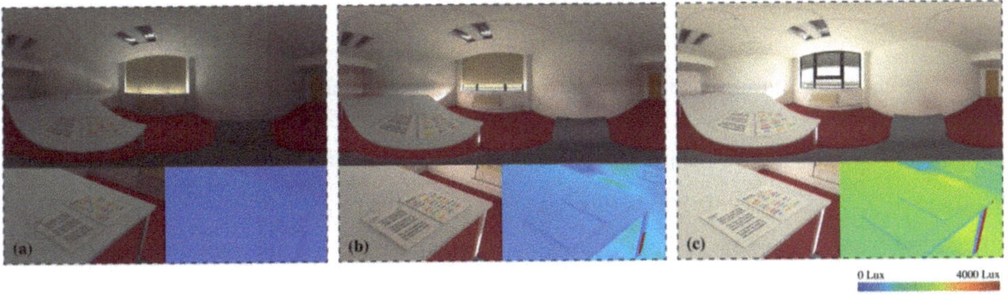

Fig. 3. Using Falsecolor display of the brightness of the marked area to ensure that the illumination level of the observation area in the daylight simulation is 100 Lux (a), 500 Lux (b), and 1500 Lux(c).

key indicators were collected from the recognition test: accuracy (AC), defined as the percentage of correctly identified words out of the total presented; reaction time (RT), defined as the time required to complete the recognition task; and the performance indicator (PI), calculated as AC per unit of RT, as shown in Formulas (1) and (2). In Formula (1), CA represents the count of words with correct colour responses, and TA represents the total number of words answered.

$$AC = \frac{CA}{TA} \times 100 \tag{1}$$

$$PI = \frac{AC}{RT} \tag{2}$$

In addition, a perception evaluation was conducted to understand users' perceptions under different environmental and illuminance conditions. As listed in Table 2, the perception evaluation comprised two components: affective evaluation (assessing pleasantness, calmness, interest, and excitement) and spatial evaluation (assessing brightness, glare, and satisfaction). These evaluative attributes and questions were defined based on previous theoretical research and aligned with the research objectives [12, 19]. All measures were recorded on a seven-point Likert scale. Moreover, participants' heart rate (HR) variation data in each scenario were monitored to compare their physiological feedback in different daylight environments, including baseline HR, average HR, minimum HR, and maximum HR. Semi-structured interviews were also conducted to collect detailed feedback on users' perceptions of the real and VR daylight environments. As shown in Table 3, the interview consisted of three primary questions.

Table 2. Scoring perception of the daylight environment on a 7-point Likert scale questionnaire.

Based on this scene, choose a number for your perception and satisfaction on a scale from 1 to 7, 1 being not at all, 7 being very.	
Pleasantness	How pleasant is this space?
Calmness	How calm is this space?
Interesting	How interesting is this space?

(continued)

Table 2. (*continued*)

Based on this scene, choose a number for your perception and satisfaction on a scale from 1 to 7, 1 being not at all, 7 being very.	
Excitement	How exciting is this space?
Brightness	How bright is this space you perceived?
Glare level	What is the level of glare you can perceive in this space?
Satisfaction	How satisfactory is the space brightness if you're working here?

Table 3. Main questions of the semi-structured interviews.

Answer the following questions based on your first feelings.	
1	Do you think the VR daylight environment felt similar to the real environment? In what aspects was it most or least realistic? (e.g., brightness, colour temperature, and contrast)
2	Did you feel that the illumination of daylight affected your task performance and evaluation? Was this effect similar in both real and VR environments?
3	Did you encounter any difficulties or discomfort during the experiment, such as wearing the VR headset, adjusting to the environment, and dizziness?

2.4 Experimental Procedure

A total of 30 participants (12 males and 18 females) were involved in the project. Participants ranged in age from 21 to 47 years (mean age = 28.1 years, SD = 5.5) and were primarily postgraduate and doctoral students at the University of Manchester. A within-subject experimental design was employed, in which each participant was tested under the same six daylight environments (three real environments and three virtual environments) to minimise inter-individual variability. The headset used for VR window view observation in the study was an Oculus Quest 2 with a field of view of 100°, a resolution of 1832 × 1920 pixels per eye, and a refresh rate of 90 Hz. In addition, the study used the Polar OH1+ wrist HR monitor to collect HR data. The experimental protocol received ethical approval from the School of Environment, Education and Development at the University of Manchester and approval for the equipment risk assessment. In the preparation phase, researchers first informed participants about key aspects of the experiment, reviewed their basic information, and assisted them in completing the consent form. Subsequently, participants received approximately 5 min of experimental and equipment training in a neutral scene to familiarise them with the detailed procedure and VR environments. During the experiment, researchers randomly set one of the three daylight illuminance levels in the real environment. Each participant observed the space's daylight conditions and completed activity tests and perception evaluation for approximately 2 min. Furthermore, researchers recorded participants' HR data during this process. Once all real environments have been tested, participants have a 2-min break before proceeding to the same tests, evaluations, and recordings in VR environments (Fig. 4). Finally, researchers conducted semi-structured interviews to collect qualitative feedback on participants' perceived consistency and differences between the real and VR daylight environments.

Fig. 4. A participant observed a scene with the same illumination level in the real environment (left) and VR environment (right).

3 Results

Before conducting the data analysis, a normality test was performed for all variables using the Shapiro-Wilk test to ensure that the assumptions underlying subsequent statistical methods were met. Based on the data characteristics, descriptive statistics and independent-sample t-tests were employed to compare reading times under different illumination levels between real and VR environments for participants' reading performance. For cognitive and perception evaluation, a two-way analysis of variance (ANOVA) was conducted to investigate the effects of illumination levels, environments, and their interaction on the outcomes of the Stroop test, affective rating, and spatial rating. In addition, a mixed linear model (MLM) was used to analyse participants' HR variation. In this study, a significance level of $p < 0.05$ was set to denote statistical significance, while $p < 0.01$ or $p < 0.001$ indicated high significance.

3.1 Activity Performance Analysis

3.1.1 Reading Test

The study compared reading time between real and VR environments with different illumination levels through descriptive statistical analysis and independent samples t-tests (Table 4). The statistical analysis revealed that in the 100 Lux and 500 Lux conditions, the mean differences in reading time between the Real and VR environments were minimal, with differences of less than 0.1 s, and none of them showed statistically significant difference ($T_{100Lux} = 0.140$, $p = 0.890$; $T_{500Lux} = -0.056$, $p = 0.956$). In the 1500 Lux condition, the mean reading time in the VR environments was slightly higher than in the real environments, although the difference was not statistically significant ($T_{1500Lux} = -0.197$, $p = 0.845$). As shown in Fig. 5, the box plots visualise the distribution of reading time for the two environments under different illumination levels, showing a high degree of consistency of performances between the real and VR environments. Overall, these results show that VR environments can effectively simulate real reading tasks within a suitable range of illumination levels.

Table 4. Descriptive statistics and independent t-test results of participants' reading time in different illumination levels and environments

Illumination Level	Avg Reading Time in the Real Environments	Avg Reading Time in the VR Environments	Reading Time Difference	T	p
100 Lux	14.20 s	14.14 s	+0.06 s	0.140	0.890
500 Lux	14.30 s	14.27 s	+0.03 s	−0.056	0.956
1500 Lux	13.27 s	13.36 s	−0.09 s	−0.197	0.845

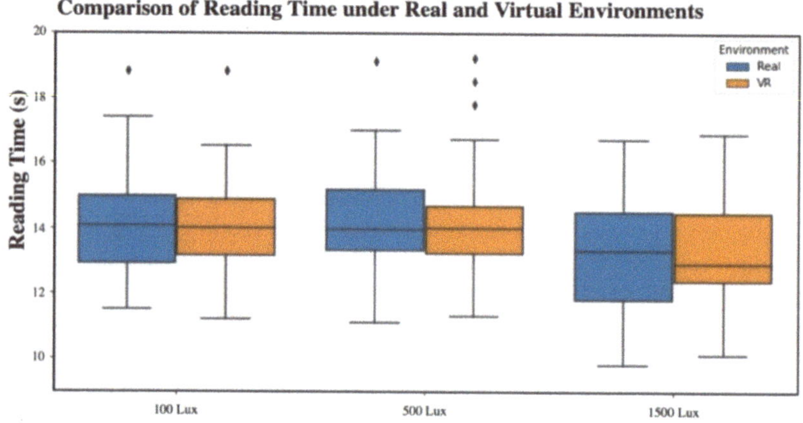

Fig. 5. Boxplots comparing the distribution of participants' reading time in VR and Real environments with different illumination levels.

3.1.2 Recognition Test

A two-way ANOVA was conducted to investigate whether the PI of the Stroop test varied significantly under different illumination levels and environments, as well as to explore any interaction effect between these two factors. As shown in Table 5, there was a significant main effect of illumination on PI ($F_{Illumination} = 4.134$, $p = 0.018$), indicating that changes in illumination levels had a notable impact on cognitive performance. In contrast, no significant main effect was found for the environments ($F_{Environment} = 0.631$, $p = 0.428$), showing that the overall Stroop test performance was comparable in the

VR and real environments. Additionally, the interaction between illumination levels and environments was not significant ($F_{Illumination/Environment} = 0.694$, $p = 0.501$), implying that the influence of illumination levels on PI was consistent across both environments. Figure 6 illustrates the distribution of PI scores under different illumination levels for real and VR environments, demonstrating that higher illumination levels generally increased participants' mean PI scores and further confirming the similarity of performance across environments within each illumination level. Overall, these results underscore the importance of illumination levels in modulating cognitive performance while also showing that the VR environments can closely replicate the recognition performance observed in a real daylight setting.

Table 5. Results of two-way ANOVA to test the effect of different illumination levels and environments on the PI of participants.

Variation	ss	df	F	p
Illumination	4.541	2.000	4.134	0.018
Environment	0.347	1.000	0.631	0.428
Illumination/Environment	0.763	2.000	0.694	0.501

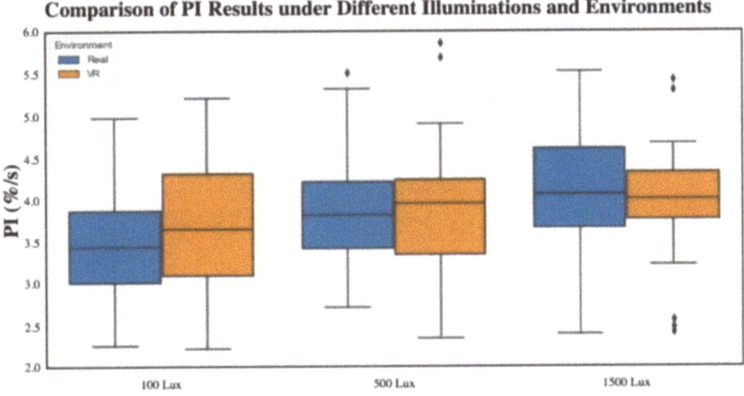

Fig. 6. Boxplots comparing the distribution of participants' PI scores in VR and Real environments with different illumination levels.

3.2 Perception Evaluation Analysis

3.2.1 Affective Evaluation

A two-way ANOVA was conducted to examine whether different illumination levels and environments had significant effects on participants' affective perception and to explore potential interaction effects (Table 6). The results revealed that illumination levels had a significant effect on participants' pleasantness, interesting, and excitement

($F_{Illumination\_Pleasantness} = 93.565$, $p < 0.001$; $F_{Illumination\_Interesting} = 73.187$, $p < 0.001$; $F_{Illumination\_Excitement} = 64.879$, $p < 0.001$). In contrast, Calmness did not differ significantly across illumination levels ($F_{Illumination\_Calmness} = 2.155$, $p = 0.119$). As shown in Fig. 7, higher illumination levels generally yielded more positive emotional evaluations for pleasantness, interesting, and excitement, while Calmness appeared relatively unaffected by changes in illumination. Furthermore, statistical analysis revealed no significant effect of the environments on Pleasantness, Calmness, and Interesting ($F_{Environment\_Pleasantness} = 0.096$, $p = 0.757$; $F_{Environment\_Calmness} = 1.523$, $p = 0.219$; $F_{Environment\_Interesting} = 1.337$, $p = 0.249$), indicating that participants' affective responses in these dimensions were similar across real and VR environments. However, a significant difference was observed in excitement between the two environments ($F_{Environment\_Excitement} = 5.33$, $p = 0.022$), indicating that the VR environments may elicit a different level of excitement than the real environment. Participants excitement was significantly higher in VR environments compared to the real environments (Fig. 7). Furthermore, no significant interaction effects were found for any of the affective factors ($p > 0.05$), implying that the trend of affective perception remained consistent between the real and VR environments across all illumination levels. Overall, these findings suggest that VR can reliably replicate participants' affective responses under controlled illumination levels.

Table 6. Two-way ANOVA results for the effects of illumination levels and environments on affective perception.

	Pleasantness				Calmness				Interesting				Excitement			
Variation	SS	df	F	p	SS	df	F	p	SS	df	F	p	SS	df	F	p
Illumination	173.078	2	93.565	0.000	6.933	2	2.155	0.119	136.811	2	73.187	0.000	165.678	2	64.879	0.000
Environment	0.089	1	0.096	0.757	2.450	1	1.523	0.219	1.25	1	1.337	0.249	6.806	1	5.33	0.022
Illumination/Environment	1.011	2	0.547	0.580	1.200	2	0.373	0.689	2.7	2	1.444	0.239	0.211	2	0.083	0.921

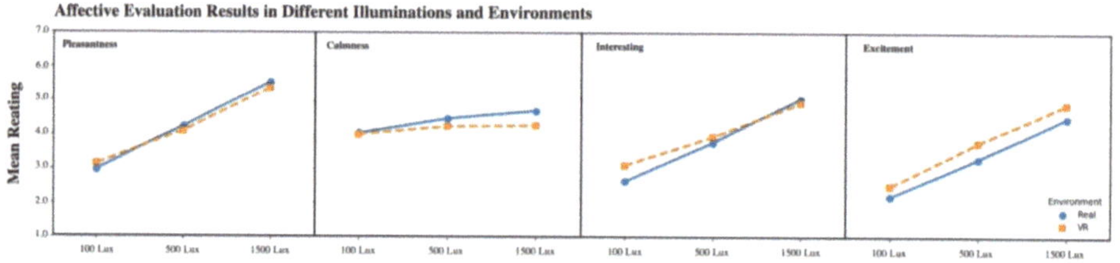

Fig. 7. Interaction plot for average ratings of pleasantness, calmness, interesting, and excitement across different illumination levels in real and VR environments.

3.2.2 Spatial Evaluation

A two-way ANOVA was conducted to determine whether illumination levels and environments significantly influenced participants' spatial perception and satisfaction with daylighting and to examine potential interaction effects (Table 7). Statistical analysis revealed that illumination levels had a significant impact on participants' perception of brightness, glare, and satisfaction ($F_{Illumination\_Brightness} = 326.938$, $p < 0.001$; $F_{Illumination\_Glare} = 43.386$, $p < 0.001$; $F_{Illumination\_Satisfaction} = 160.314$, $p < 0.001$). In contrast, the environment had no significant effect on glare perception or satisfaction with daylight ($F_{Environment\_Glare} = 3.772$, $p = 0.054$; $F_{Environment\_Satisfaction} = 0.152$, $p = 0.697$), suggesting that participants' space perception of glare and satisfaction with daylighting were consistent across Real and VR environments. No significant interaction effects were found for brightness or glare ($p>0.05$), indicating that the influence of illumination intensity on spatial perception probably followed a similar pattern across real and VR environments. As depicted in Fig. 8, participants generally reported higher brightness and glare in the VR environments compared to the real environment. However, at the highest illumination level, satisfaction ratings were higher in the real setting than in the VR environment, suggesting a potential threshold beyond which the VR environment may be perceived as less satisfactory under intense illumination.

Table 7. Results of two-way ANOVA to test the effect of different illumination levels and environments on the space perception of participants.

	Brightness				Glare				Satisfaction			
Variation	ss	df	F	p	ss	df	F	p	ss	df	F	p
Illumination	381.678	2.000	326.938	0.000	147.744	2.000	43.386	0.000	292.311	2.000	160.314	0.000
Environment	4.050	1.000	6.938	0.009	6.422	1.000	3.772	0.054	0.139	1.000	0.152	0.697
Illumination/ Environment	0.100	2.000	0.086	0.918	2.211	2.000	0.649	0.524	8.578	2.000	4.704	0.010

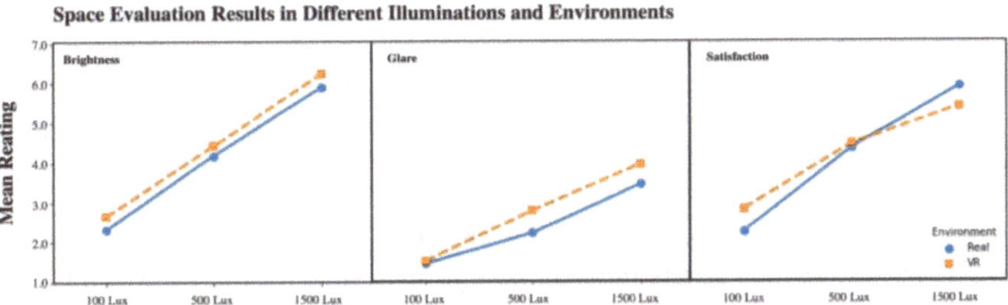

Fig. 8. Interaction plot showing the average ratings of brightness, glare, and satisfaction at different illumination levels in real and VR environments.

3.3 Heart Rate Variation Analysis

An MLM was employed to analyse whether environment and illumination intensity significantly affected participants' heart rate variation. The statistical analysis revealed that heart rate variation in the VR environment was significantly lower than in the Real environment ($\beta = -2.83$, $p = 0.031$), suggesting that participants may have experienced a more stable and focused mental state in VR (Fig. 9). Additionally, illumination levels had no significant effect on heart rate variation ($p > 0.05$), indicating that within the experimental range (100–1500 Lux), illumination intensity did not significantly alter users' heart rate fluctuations (Fig. 9). Furthermore, no significant interaction effect was found between environments and illumination levels ($p > 0.05$), suggesting that the influence of illumination on heart rate variation followed a similar pattern across VR and Real environments. The random effect variance for inter-individual differences (Group Var = 14.18) was relatively low, indicating that individual differences had limited explanatory power for variations in heart rate. Overall, these findings underscore the potential of VR to induce a stable physiological response, irrespective of illumination intensity.

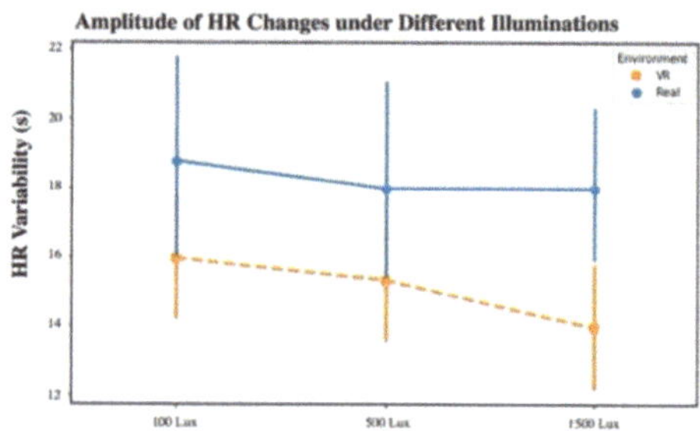

Fig. 9. Interaction plots of HR variation in different illumination levels and environments.

3.4 Feedback Analysis of Semi-structured Interviews

The research first analysed participants' feedback on perceived consistency and differences. Most participants provided feedback that VR reproduced the perception of daylight in real environments very well. "Brightness perception is similar for real and VR scenes." A few participants indicated that the light intensity and colour temperature of real daylight environments were softer and more variable than those shown in the VR, which may be related to the influence of weather conditions during the test. Some participants provided feedback that the design, functions, and external environment of the space may affect their perceptions of the VR environments. In addition, the research analysed participant feedback regarding problems and suggestions encountered with the experimental setup, equipment, and user interference. Some participants indicated confusion about the questions on the evaluation questionnaire. Among them, the participants showed the most difficulty in assessing the ratings of calmness and excitement. Subsequent experiments need to select emotional factors that are less likely to confuse users when assessing their perception based on the research objectives. A few participants provided feedback on how the VR equipment interfered with the experiment, including issues with resolution, the field of view, and the device's weight affecting their observations and assessments. Some participants reported that they felt a giant feeling of objects in the VR scenes. Only one participant expressed mild dizziness during the testing of the VR scenes.

4 Conclusion

This study systematically examined the effects of users' reading speed, cognitive performance, affective perception, spatial perception, and physiological responses by real and VR environments with different illumination levels using the comparative method based on a standardised workflow and assessment process. The findings provide insights into the validity of VR in simulating daylight environments and its impact on human perception and performance. The originality and value of the research are as follows:

- The multidimensional evaluation method based on the reading test, Stroop test, affective and spatial perception ratings, and HR monitoring can be effectively applied to VR daylighting research.
- The study provides empirical evidence that VR can effectively replicate the effects of a real daylight environment on activity performance, perception feedback, and physiological responses under controlled conditions (100–1500 Lux), supporting its application in daylight perception research.
- Higher daylight illumination levels significantly improved participants' activity performance and enhanced their evaluation of pleasure, interest, excitement, and satisfaction value.

The parameter range of the evaluation method proposed in this study limits the illuminance levels (100–1500 Lux) of the test based on hardware attributes, not exploring darker or brighter scenes. In addition, the simulation of the VR environments in the study did not take into account the effects of the external environment and weather changes. Subsequent studies should test brighter illumination levels, adding more weather scenes and external settings. While perceptions of VR are largely consistent with reality, users' excitement levels and HR responses differ, suggesting that the potential for specific psychological effects of VR daylight environment is worth exploring further.

References

1. Galatioto, A., Beccali, M.: Aspects and issues of daylighting assessment: A review study. Renew. Sust. Energ. Rev. **66**, 852–860 (2016)
2. Boubekri, M.: Daylighting, Architecture and Health. Routledge (2008)
3. Lee, E.S., Matusiak, B.S., Geisler-Moroder, D., Selkowitz, S.E., Heschong, L.: Advocating for view and daylight in buildings: Next steps. Energ. Buildings. **265**, 112079 (2022)
4. Lewis, A.: The Mathematisation of daylighting: A history of British architects' use of the daylight factor. J. Archit. **22**, 1155–1177 (2017)
5. Tregenza, P., Wilson, M.: Daylighting: Architecture and Lighting Design. Architecture and Lighting Design, Daylighting (2013). https://doi.org/10.4324/9780203724613
6. Amundadottir, M.L., Rockcastle, S., Sarey Khanie, M., Andersen, M.: A human-centric approach to assess daylight in buildings for non-visual health potential, visual interest and gaze behavior. Build. Environ. **113**, 5–21 (2017)
7. Heydarian, A., Pantazis, E., Wang, A., Gerber, D., Becerik-Gerber, B.: Towards user centered building design: Identifying end-user lighting preferences via immersive virtual environments. Autom. Constr. **81**, 56–66 (2017)
8. Moscoso, C., Nazari, M., Matusiak, B.S.: Stereoscopic Images and Virtual Reality techniques in daylighting research: A method-comparison study. Build. Environ. **214**, 108962 (2022)

9. Ahmed, K.G., Omar, M.M., Megahed, M., Alazeezi, S.A.: Involving young Emirati women in the pre-occupancy evaluation of "Modern" housing designs: Simple versus advanced participatory tools. SAGE Open. (2022). https://doi.org/10.1177/21582440221094613/ASSET/IMAGES/10.1177_21582440221094613-IMG9.PNG
10. Marzouk, M., ElSharkawy, M., Mahmoud, A.: Analysing user daylight preferences in heritage buildings using virtual reality. Build. Simul. **15**, 1561–1576 (2022)
11. Moscoso, C., Chamilothori, K., Wienold, J., Andersen, M., Matusiak, B.: Regional differences in the perception of daylit scenes across Europe using virtual reality. Part I: Effects of window size. LEUKOS. **18**, 294–315 (2022)
12. Chamilothori, K., Wienold, J., Moscoso, C., Matusiak, B., Andersen, M.: Regional differences in the perception of daylit scenes across Europe using virtual reality. Part II: Effects of façade and daylight pattern geometry. LEUKOS. **18**, 316–340 (2022)
13. Bellazzi, A., et al.: Virtual reality for assessing visual quality and lighting perception: A systematic review. Build. Environ. **209**, 108674 (2022)
14. Vince, J.: Introduction to Virtual Reality. Springer Science & Business Media (2004)
15. Fuchs, P.: Virtual Reality Headsets–A Theoretical and Pragmatic Approach. CRC Press (2017)
16. Calabrèse, A., Owsley, C., McGwin, G., Legge, G.E.: Development of a reading accessibility index using the MNREAD acuity chart. JAMA Ophthalmol. **134**, 398–405 (2016)
17. Payedar-Ardakani, P., Gorji-Mahlabani, Y., Ghanbaran, A.H., Ebrahimpour, R.: Daylight illuminance levels, user preferences, and cognitive performance in office environments: Exploring an optimal illuminance range using virtual reality. Build. Environ. **258**, 111638 (2024)
18. Chauhan, H., Jang, Y., Pradhan, S., Moon, H.: Personalized optimal room temperature and illuminance for maximizing occupant's mental task performance using physiological data. J. Build. Eng. **78**, 107757 (2023)
19. Zeng, W., Zhang, H.: A virtual reality window view evaluation tool for shading devices and exterior landscape design. Comput. Des. Robot. Fabric. Part. **F2072**, 163–179 (2024)

An Approach to Predicting Buildings Energy Consumption in Urban Block Based on Multimodal Deep Learning Model

Yuchen Qin[1], Yuchen Xie[2(✉)], Kaifan Chen[3], Zixuan Deng[4], Lei Zhen[5], and Minghao Wang[1]

[1] Huazhong University of Science and Technology, Wuhan, China
[2] The University of Hong Kong, Hong Kong, China
1498596577@qq.com
[3] South China University of Technology, Guangzhou, China
[4] Hubei University of Technology, Wuhan, China
[5] Southeast University, Nanjing, China

Abstract. Traditional methods of predicting building energy consumption often fail to comprehensively capture the complex environmental features of urban blocks. This study proposes a multimodal deep learning framework that integrates microclimatic environments, and building information modelling with demographic attributes for the complete construction of underlying scenarios for building energy prediction in urban blocks. Based on this, multi-model comparison experiments are carried out by adjusting the sample size and dataset weighting ratio, resulting in the multi-modal dataset with the best overall performance and corresponding prediction model. Finally, the interactions between multi-modal factors and their effects on block-scale building energy consumption levels are analysed using ablation experiments, SHAP and Grad-CAM methods. The proposed multimodal model-based approach for modelling building energy consumption outperforms the single modal dataset model in terms of multifactor interpretability and prediction efficiency.

Keywords: Urban building energy modeling (UBEM) · Urban block · Complex urban systems · Multimodal data fusion · Deep learning

1 Introduction

The United Nations Environment Programme (2024) reported thatbuilding-related carbon emissions are expected to have an increasingly significant impact on the global energy mix and climate dynamics with urbanization accelerates and construction expanding. Urban blocks, as fundamental spatial units characterized by uniform urban form, microclimates, and landscaping, play a critical role in the development of low-carbon

Y. Liu et al. (Eds.): CDRF 2025, *Transindividual Intelligence*, pp. 147–156, 2026.
https://doi.org/10.1007/978-981-92-0615-5_13

cities. Intensifying urban microclimatic challenges, such as the heat island effect, substantially affect environmental and building performance at the block scale. The complexity of the built environment and its interaction with microclimatic conditions complicate the accurate assessment of energy consumption for building clusters. This underscores the need for a robust UBEM methodology that comprehensively captures multidimensional environmental features while fully considering the interdependent effects of diverse urban elements to ensure energy consumption prediction accuracy.

UBEM is mainly classified into "top-down" and "bottom-up", according to the differences in construction methods and applicable scales. Which integrate climate data, building geometry, construction standards and usage schedules through different paths and methods. The applicability of bottom-up modelling ranges from building to urban block. Among them, simulation is a mainstream approach, where energy models of individual building units are usually developed based on heat and mass balance equations. For block-scale, climate, building, user behaviors and landscape greenery can be built into scenarios to break down the formation of energy consumption in buildings through the invocation of simulation engines with the geometry of 3D modelling methods in specific simulation software (Reinhart and Davila 2016). The machine learning modelling approach known as "black-box modelling" enriches the research scenario of building energy assessment and improves the ability to deal with complex problems (Chen et al. 2022), such as numerical forecasting of time series (Zhan et al. 2025), multi-performance objective optimisation (Sood et al. 2025) and the analysis of non-linear impact relationships (Yuan et al. 2024). Advances in deep learning have significantly enhanced the ability to process multidimensional data, particularly image data, compared to traditional data-driven models. They are widely used in built environment assessment (Jiang et al. 2025), urban risk and environmental performance prediction, (Wang et al. 2025) etc.

Urban blocks comprise diverse environmental and social elements that must be integrated into UBEM through software collaboration and semantic modeling. Simulation methods facilitate the development of a comprehensive energy calculation model that incorporates building structures, site environments, and climate conditions via multi-software collaboration. However, these methods often face limitations in adjustability and computational efficiency. Algorithmic models, such as machine learning and deep learning, mitigate dependence on specific scenarios and platforms, enabling greater adaptability for various energy prediction tasks. Nevertheless, each approach has its shortcomings-machine learning models frequently overlook environmental features during training, while deep learning models that focus on geometric features struggle to represent non-geometric attributes across different dimensions. As a result, simulation-based scenarios constructed using these methods still diverge significantly from real urban environments. Multimodal fusion techniques construct urban computational fusion models and enable the holistic realisation of urban tasks by integrating structured and unstructured data from multi-source datasets, which enhances the understanding of the description of complex urban scenarios and the associated explanatory factors. It has been used for traffic behaviour prediction (Zhang et al. 2017), urban physical environment prediction (Suel et al. 2021) and social risk assessment (Baek and Lim 2024). A few research has applied this method to urban-scale energy efficiency assessments (Sun et al. 2022). However, addressing the inadequacy of traditional UBEM methods in comprehensively capturing the multidimensional urban environment by integrating multimodal data to optimise complex underlying models of building energy consumption in blocks still needs to be discussed more extensively. This study aims to leverage multi-source datasets and semantic feature fusion to capture multidimensional environmental characteristics, developing a scientific and adaptive predictive model for urban block energy consumption. This study aims to predict the building energy consumption level in urban blocks by integrating multimodal data. The main innovations include: (1) designing a framework that can integrate image and tabular data to comprehensively capture the complex features of microclimate, built environment and social environment in urban blocks for constructing new UBEM base scenarios; (2) jointly explaining the differences in predictive performance and their synergistic relationships across different modal datasets.

2 Research Methods and Datasets

2.1 Research Framework

Firstly, a multimodal framework is proposed, integrating a Residual Neural network (ResNet) model with a fully connected neural network model (Du et al. 2024). Publicly available energy consumption and map data are used to construct multimodal dataset, which undergoes feature extraction, dimensionality reduction, and semantic fusion to encapsulate key urban block environmental features. Then, we conducted model training by dividing different proportions of multimodal datasets to derive the best-performing predictive model to assess the level of energy consumption of neighbourhood buildings; further, we jointly applied SHAP and Grad-CAM methods for the characteristics of different data modalities to provide a comprehensive explanation of the independent and synergistic impact effects of the characteristic factors affecting different dimensions of energy consumption. The research framework is shown in Fig. 1.

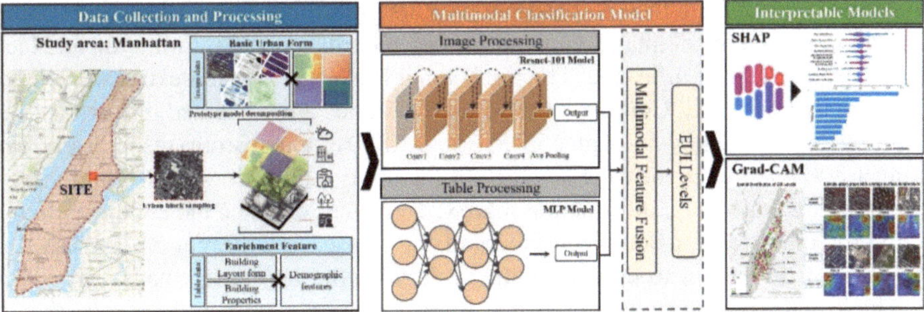

Fig. 1. Research framework

2.2 Multimodal Dataset

To comprehensively capture the complexity of urban block environments, we draw on relevant studies to identify key impact indicators related to microclimatic conditions, built environments, and demographic attributes, forming a multimodal dataset. It creates an urban block prototype that closely mirrors real-world conditions. The microclimate images include monthly average values for solar radiation (SR), land surface temperature (LST), humidity (HD), and wind speed (WS). These data are refined using interpolation methods and data augmentation techniques to align with block scales(resources: https://worldclim.org). The building morphology data consists of both image and tabular formats. The image data depict fundamental urban forms, including aerial views, building footprints, building heights, building functions, and building areas. The tabular data further enrich the morphological description by incorporating the following indicators: building year, building construction, building shape factor (BSF), perimeter-to-area ratio (PAR), sky view factor (SVF), open space ratio (OSR), floor area ratio (FAR), building density (BD), standard deviation of building height (DBH), and standard deviation of building area (DBA). The demographic attributes feature indicators include the number of residents, number of minors, number of elders, gender ratio, and median household income.

The building energy consumption dataset is sourced from the publicly available energy statistics database (resource: http://OpenData.cityofnewyork.us), covering the Manhattan area of New York. To facilitate horizontal comparisons despite significant variations in building sizes across blocks, the study averages energy intensity within block segments at various resolutions and classifies them into five levels to represent energy intensity grades.

2.3 Multi-branch Deep Learning Network Model

The multi-branch DL model constructed contains three parts: feature extraction and fusion, model scaling, and type prediction. In the first stage, built environment morphology and microclimate images are merged, decomposing nine images into 27 single-channel maps, which are then mapped into vectors of the same dimensions and stitched into a composite-channel image, which primarily records merged information and lacks standalone practical significance. The pre-trained ResNet101 model extracts 2048 features from its final layer. Concurrently, 15 text variables are processed through a three-layer fully connected network with 256, 512, and 1,800 output features respectively, generating 1,800 features in the final layer. Both feature types are downscaled and merged to form the final semantic information matrix. To further enhance the understanding of the multimodal dataset, a combined model of six datasets at different ratios was set up at the end of the model to determine the scale with the best performance. This approach highlights the impact of various multimodal dataset combinations, facilitating the selection of the optimal configuration. All six models use a fixed fully connected layer with 1,800 output features but vary in data type proportions. After merging, the features are processed through two fully connected layers with 1,024 and 512 output features respectively, followed by a classification layer that predicts block energy consumption levels. The model structure is illustrated in Fig. 2.

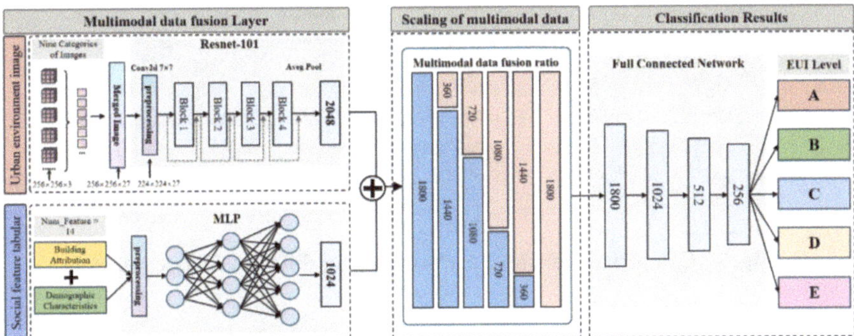

Fig. 2. Model structure

2.4 Interpretable Analysis

To comprehensively explain the impact of multimodal features on building energy consumption, this study jointly interprets the influence of image features and textual indicators on energy consumption using SHAP (SHapley Additive exPlanations), ablation experiments, and Grad-CAM methods (Gradient-weighted Class Activation Mapping), which underscores the advantages of multimodal methods in capturing complex feature. SHAP reinforces the understanding of how enhanced urban form and socio-environmental characteristics non-linearly impact building energy consumption. Grad-CAM extends CAM by integrating gradient information with feature mapping to generate activation images that highlight image feature regions most critical to the model's predictions.

3 Results and Discussions

3.1 Evaluation of the Prediction Model

The performance results of the multi-ratios model indicate: (1) The scaling relationship between the dataset and image size exhibits a synergistic effect on model performance, demonstrating non-linear characteristics. Consequently, the performance of the same dataset varies significantly at different spatial scales; (2) The model performance is most stable in 700 m size, and prediction accuracy remains largely unaffected by changes in the ratio of image to tabular data; (3) Single tabular data models exhibit greater performance fluctuations and generally maintain the lowest accuracy. In contrast, models incorporating image data demonstrate significantly improved accuracy and stability. Moreover, model performance remains consistently high when the proportion of image data exceeds that of tabular data. These findings suggest that the urban environment features captured in 300 m size most accurately reflect building energy consumption levels (Fig. 3).

3.2 Factors Affecting Buildings Energy Consumption in Block

Regarding microclimatic factors, ablation experiments show the model achieves the highest prediction accuracy when LST is the sole climatic input. Solar radiation follows,

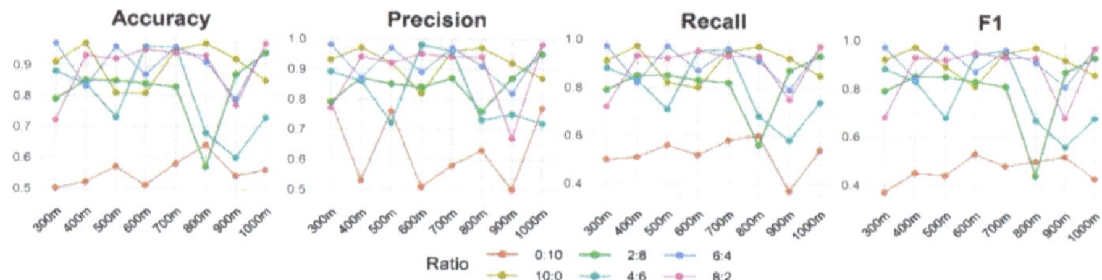

Fig. 3. Comparison of model performance for different dataset scales

while humidity and wind speed exhibit significantly lower impacts. When comparing at the same resolution level, excluding the influence of image resolution, changes in block LST have the greatest impact on building energy consumption. This finding aligns with Mayer's studies(Mayer et al. 2023), which highlight that LST directly influences the thermal exchange processes of buildings and is the most immediately perceptible factor for occupants, shaping their energy use behaviors. Furthermore, aerial imagery outperforms the other groups in terms of accuracy, recall, and F1 scores, suggesting a significant contribution to model prediction performance. Building function follows in importance, while NDVI exhibits the weakest effect. Although aerial images are less precise in describing specific features compared to other types of images, they primarily capture the holistic relationships among different building characteristics, which are reflected through other features (e.g., shaded areas due to shading relationships between buildings). Therefore, aerial images will serve as the foundational input for Grad-CAM in subsequent analyses (Table 1).

The BSF emerges as the most significant indicator, positively affecting building energy consumption. This outcome can be attributed to the expansion of heat dissipation areas on the building's exterior surfaces, as evidenced by the PAR's positive impact effect. Following BSF, the next most influential factors are the SVF, FAR, and BD, while OSR has a comparatively weaker effect. Lower plot ratios, higher sky visibility, greater building densities, and larger proportions of open space are more conducive to reducing building energy consumption. This suggests that an ideal prototype for a low-carbon block is characterized by low to medium building heights combined with high densities, as this configuration effectively minimizes heating energy consumption-a major component of total building energy demand in Manhattan. The Grad-CAM results in further support these findings by identifying point-tall buildings, large-roofed factory buildings, street canyons, building courtyards, and public spaces such as green areas as key areas. Furthermore, variations in building height exert a greater impact on building energy consumption compared to floor area, emphasizing that vertical form features may be more important than horizontal features as they are directly related to shading, ventilation and thermal processes. About building attribution, building age exerts a stronger influence than structural type. Newer buildings demonstrate superior envelope thermal performance, resulting in reduced heat transfer losses and lower energy consumption. This finding may be related to the evolution of residential building construction codes in the United States, which regulate air-conditioning systems in buildings and their energy efficiency.

Table 1. Results of ablation experiments

	Input features	Performance evaluation matrix		
		Precision [%]	Recall [%]	F1 Score [%]
Built environment image	Aerial View+MLP	74.34	70.50	72.37
	Buildings Footprint+MLP	53.90	50.70	52.25
	Buildings Height + MLP	48.20	52.50	50.31
	Buildings Function+MLP	63.50	60.10	61.78
	Buildings area + MLP	43.40	46.20	44.73
	NDVI+MLP	47.12	49.90	48.50
Microclimate image	LST + MLP	58.70	55.80	57.22
	Wind Speed+MLP	45.60	50.10	47.78
	Humidity+MLP	48.50	51.30	49.87
	Solar Radiation+MLP	50.30	47.00	48.62

Both the proportion of elderly residents and household income significantly positively affect building energy consumption, with income showing a greater effect. Previous studies have highlighted the influence of household income, particularly in light of U.S. policies such as the Low Income Home Energy Assistance Program (LIHEAP). Although these programs subsidize rent and energy costs for low-income housing, they may inadvertently increase energy consumption, highlighting the need for energy-efficient retrofits of affordable housing constructed during the implementation of these policies. Moreover, a higher proportion of elderly residents correlates with increased energy use, likely due to heightened temperature sensitivity and greater heating or cooling demands (Figs. 4 and 5).

3.3 Applicability of Multimodal Approaches in Measuring Complex Urban Environments

The complexity of urban energy modeling stems not just from thermodynamic systems, but also from nonlinear interactions among dynamic urban elements including environmental, microclimatic, and socioeconomic factors. Accurate urban environment measurements are consequently crucial for studying urban form and energy demand. Traditional indicator- and typology-based approaches inadequately represent detailed morphological features and environmental contexts, exhibiting weak correlations with energy consumption. The Local Climate Zones (LCZ) framework partially resolves these

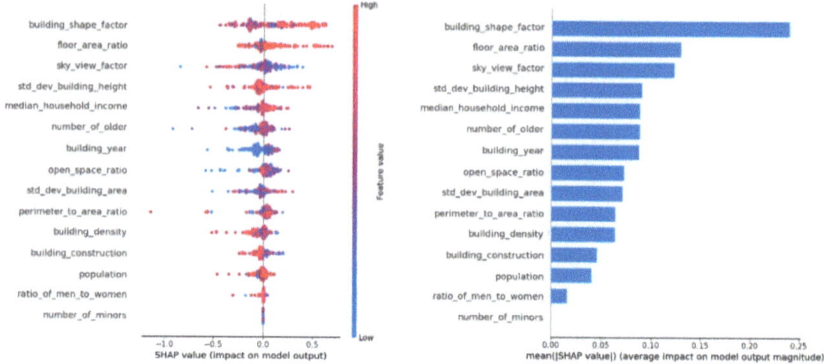

Fig. 4. Results of SHAP model analysis

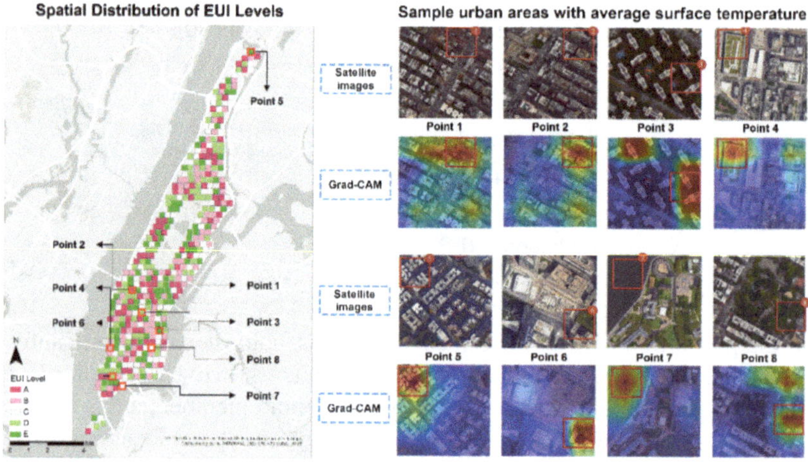

Fig. 5. Results of visual interpretable model Grad-CAM comparison

limitations by integrating built environments, green spaces, and microclimate-related thermal indicators. However, its oversimplification of real-world features and morphological characteristics restricts wider applicability. This study employs a multimodal fusion approach to integrate remote sensing imagery with text-based features, preserving native urban environment characteristics while mitigating modeling uncertainties. This methodology bridges the gap between simulation models and real-world conditions. Nevertheless, further research is needed regarding measurement accuracy, multi-source data granularity and fusion processes, and spatiotemporal dynamics representation in input data.

4 Conclusion

This study proposes a multimodal deep learning framework combining ResNet and MLP models to develop an ideal urban block prototype based on multimodal datasets. The framework outperforms conventional methods by incorporating microclimatic, morphological, building attribute, and demographic features into UBEMs to better approximate real urban environmental features. To evaluate the framework, we conduct a comparative experiment using six differently scaled datasets to identify the optimal multimodal dataset configuration for model performance. Furthermore, we analyze the interactions between cross-modal factors and their influence on block-scale building energy consumption using ablation study, SHAP, and Grad-CAM methods. The framework demonstrates the efficacy of multimodal fusion for urban block building energy consumption prediction, offering superior integration of multidimensional features and enhanced interpretability. It enables rapid energy assessments using publicly available image datasets and flexible integration of textual data while adapting to dynamic urban conditions. This approach facilitates multi-scale urban energy zoning and offers actionable insights for carbon assessment and regeneration strategies, particularly in existing urban blocks.

References

Baek, J., Lim, L.: Unveiling pedestrian injury risk factors through integration of urban contexts using multimodal deep learning. Sustain. Cities Soc. **101**, 105168 (2024). https://doi.org/10.1016/j.scs.2023.105168

Chen, Y., Guo, M., Chen, Z., Chen, Z., Ji, Y.: Physical energy and data-driven models in building energy prediction: a review. Energy Rep. **8**, 2656–2671 (2022). https://doi.org/10.1016/j.egyr.2022.01.162

Du, S., Zhang, Y., Sun, W., Liu, B.: Quantifying heterogeneous impacts of 2D/3D built environment on carbon emissions across urban functional zones: a case study in Beijing, China. Energy Build. **319**, 114513 (2024). https://doi.org/10.1016/j.enbuild.2024.114513

Jiang, Z., Wu, C., Chung, H.: The 15-minute community life circle for older people: walkability measurement based on service accessibility and street-level built environment–a case study of Suzhou, China. Cities. **157**, 105587 (2025). https://doi.org/10.1016/j.cities.2024.105587

Mayer, K., Haas, L., Huang, T., Bernabé-Moreno, J., Rajagopal, R., Fischer, M.: Estimating building energy efficiency from street view imagery, aerial imagery, and land surface temperature data. Appl. Energy. **333**, 120542 (2023). https://doi.org/10.1016/j.apenergy.2022.120542

Reinhart, C.F., Davila, C.C.: Urban building energy modeling–a review of a nascent field. Build. Environ. **97**, 196–202 (2016). https://doi.org/10.1016/j.buildenv.2015.12.001

Sood, D., et al.: Machine learning based multi-objective optimisation of energy consumption, thermal comfort and CO_2 concentration in energy-efficient naturally ventilated residential dwellings. Build. Environ. **267**, 112255 (2025). https://doi.org/10.1016/j.buildenv.2024.112255

Suel, E., Bhatt, S., Brauer, M., Flaxman, S., Ezzati, M.: Multimodal deep learning from satellite and street-level imagery for measuring income, overcrowding, and environmental deprivation in urban areas. Remote Sens. Environ. **257**, 112339 (2021). https://doi.org/10.1016/j.rse.2021.112339

Sun, M., Han, C., Nie, Q., Xu, J., Zhang, F., Zhao, Q.: Understanding building energy efficiency with administrative and emerging urban big data by deep learning in Glasgow. Energ. Buildings. **273**, 112331 (2022). https://doi.org/10.31219/osf.io/g8p4f

Wang, H., Ma, W., Niu, J., You, R.: Evaluating a deep learning-based surrogate model for predicting wind distribution in urban microclimate design. Build. Environ. **269**, 112426 (2025). https://doi.org/10.1016/j.buildenv.2024.112426

Yuan, J., Jiao, Z., Xiao, X., Emura, K., Farnham, C.: Impact of future climate change on energy consumption in residential buildings: A case study for representative cities in Japan. Energy Rep. **11**, 1675–1692 (2024). https://doi.org/10.1016/j.egyr.2024.01.042

Zhan, D., Qin, S., Wang, L.L., Hassan, I.G.: Weather clustering for machine learning-based hourly building energy prediction models at design phase. Energ. Buildings. **329**, 115308 (2025). https://doi.org/10.1016/j.enbuild.2025.115308

Zhang, J., Zheng, Y., Qi, D.: Deep spatio-temporal residual networks for citywide crowd flows prediction. Proc. AAAI Conf. Artif. Intell. **31**(1), 1–7 (2017). https://doi.org/10.48550/arXiv.1610.00081

A Real-Time Control Strategy of Air Conditioning Systems in University Buildings Based on Computer Vision and Deep Learning

Zao Li[1,2(✉)], Rui Shi[1(✉)], Qiang Wang[1,2(✉)], Yulu Chen[1,2], Taoyuan Zhang[3], and Hanyue Tong[1]

[1] School of Architecture and Urban Planning, Anhui Jianzhu University, Hefei, China
lizao@ahjzu.cn, 2227058697@qq.com, wangqiang@ahjzu.edu.cn

[2] Anhui Provincial Engineering Research Center for Regional Environmental Health and Spatial Intelligent Perception, Hefei, China

[3] College of Architecture and Art, Hefei University of Technology, Hefei, China

Abstract. University buildings are characterized by complex, dynamic occupant behavior and high comfort demands, making traditional air conditioning systems inefficient due to slow response times. This paper proposes a pre-control strategy for air conditioning using real-time occupant distribution estimation via computer vision and deep learning. A Multi-Column Convolutional Neural Network (MCNN) model is trained to estimate crowd density and real-time occupant loads from images. Load grading and zoning regulation strategies are developed, and simulation tests assess the impact on temperature, PMV value, and cooling energy consumption. Results show that the proposed strategy enhances indoor comfort and reduces cooling energy consumption by up to 6.5%.

Keywords: Deep learning · Computer vision · Air-conditioning energy consumption · University buildings · Intelligent Control

1 Introduction

Building energy consumption accounts for about 40% of global energy use, with air conditioning being a major contributor. This is particularly significant in public buildings, such as teaching buildings with high occupancy loads, fluctuating occupant flows, and high thermal comfort demands, often leading to excessive air conditioning use.

Effective air conditioning control is essential for energy savings, aiming to meet thermal comfort requirements while minimizing energy consumption. Optimizing control strategies relies on accurately predicting the system's operating load. Ma et al. [1] proposed a method based on a similar day approach with integrated weighting, achieving high accuracy in energy consumption prediction. Li et al. [2] introduced a novel method for optimizing HVAC energy use through load prediction and energy flexibility, demonstrating superior thermal comfort and energy efficiency performance.

Y. Liu et al. (Eds.): CDRF 2025, *Transindividual Intelligence*, pp. 157–169, 2026.
https://doi.org/10.1007/978-981-92-0615-5_14

Previous studies primarily rely on existing energy consumption data and physical environmental factors to predict air-conditioning energy use. However, factors like occupancy, which are uncertain and variable, are often neglected, leading to discrepancies between predicted and actual energy consumption and hindering energy-saving efforts.

Chen et al. [3] highlighted that occupancy rate is a key factor in determining air-conditioning and building energy consumption. They also found that real-time occupancy-based energy optimization is more efficient and cost-effective. However, monitoring occupancy in real-time presents challenges. Zhang et al. [4] used a combination of sensors (PIR, CO^2, sound, temperature, and humidity) to track occupancy in a multifunctional space over two months, using machine learning for better load predictions. Despite improvements, external factors can affect sensor accuracy, and system deployment is costly. Zou et al. [5] proposed a Wi-Fi-based system that adjusts HVAC operation based on occupancy, offering energy savings. However, this system is limited by Wi-Fi signal quality and privacy concerns.

Recent advances in computer vision and deep learning have introduced more direct, realistic, and faster methods for personnel detection in complex environments. This paper presents a real-time occupant load monitoring method for classrooms using computer vision and the MCNN (Multi-Column Convolutional Neural Network) model to inform air conditioning regulation. Surveillance camera images are processed into density maps by the MCNN model to provide real-time occupancy data. This data is then integrated into the air conditioning load regulation system, allowing pre-adjustments based on the current distribution of occupants.

2 Methodology

Convolutional Neural Networks (CNNs) are effective in semantic image segmentation and feature recognition. However, in complex environments like teaching buildings, relying solely on local or global features may hinder accuracy due to diverse scenes and crowd behaviors. Multi-scale CNNs (MCNNs) address this by extracting features at different scales in parallel, integrating them into a comprehensive framework that combines local and global information, improving both accuracy and computational efficiency.

2.1 The MCNN Model Structure

In this paper, we adopt the MCNN architecture proposed by Zhang et al. for crowd density recognition [6]. As depicted in Fig. 1, the architecture comprises three parallel CNNs with filters of varying sizes corresponding to local receptive fields (large, medium, and small). Fewer filters are used for larger receptive fields to improve computational efficiency. Each column follows a conv-pooling-conv-pooling structure, with max pooling applied to 2x2 regions and ReLU as the activation function. The output features from all CNNs are stacked and mapped to the density map. To accommodate arbitrary input image sizes, the fully connected layer is replaced with a 1x1 convolutional layer. The Euclidean distance is used to measure the difference between the estimated density map and the ground truth. The formula for calculating the density map is given by Eq. (1):

$$F(x) = \sum_{i=1}^{N} \delta(x - x_i) * G_{\sigma_i}(x), \ \ with \ \ \sigma_i = \beta \overline{d^i} \tag{1}$$

Where x_i represents the pixel position of the head in the image; $\delta(x - x_i)$ denotes the Dirac delta function at x_i in the image; $G_{\sigma_i}(x)$ represents the Gaussian kernel; $*$ denotes the convolution operation (Gaussian smoothing); $*$ is the total number of heads; and $\overline{d^i}$ denotes the average distance between x_i and other surrounding targets. The loss function used for training is defined as Eq. (2):

$$\mathrm{L}(\Theta) = \frac{1}{2N} \sum_{i=1}^{N} \| \mathrm{F}(\mathrm{Xi}; \Theta) - \mathrm{Fi} \|_2^2 \tag{2}$$

Where Θ is the optimization parameter in the model, N is the number of training images, Xi is the input image, and Fi is the actual density map corresponding to image. F(i; Θ) represents the estimated density map generated by the model. L denotes the loss between the estimated and actual density maps.

This architecture offers several advantages over general CNN models: (1) It adapts to changes in target size due to perspective effects or resolution variations, without needing a perspective view of the image. The density map is computed using a geometrically adaptive kernel. (2) Models trained with this architecture exhibit strong generalization, allowing for easy transfer from the original to the target domain, retaining previously learned features.

This MCNN model performs people counting by summing the pixels of the output density map. At the end of the model, a 1x1 convolutional kernel is used to compress the concatenated feature maps into a single-channel output, where the value of each pixel represents the number of people in that specific area. The total number of people in the image is then calculated by summing all the pixel values, as the Eq. (3) describes:

$$N_{px} = \int F(x_i; \theta) dx \tag{3}$$

Where N_{px} is the number of people within the image, and $F(x_i; \theta)$ is the people density map generated by the MCNN model. Building on this, the real-time indoor crowd load can be estimated using Eq. (4):

$$Q(t) = kN_{px}(t)q \tag{4}$$

Where $Q(t)$ is the indoor crowd load at time t, k is the cluster coefficient, and q is the body load index in watts per person (W/person).

2.2 Model Training

The model training starts with pre-training on a large, authoritative dataset, followed by fine-tuning on a smaller, custom dataset using the optimal model parameters. This flexible approach ensures accurate predictions while minimizing the need for an extensive

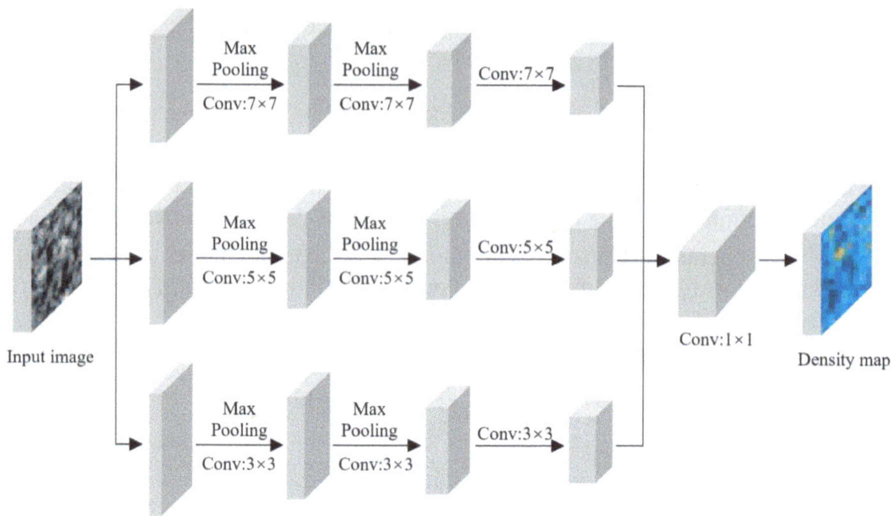

Fig. 1. Schematic of the MCNN structure used

database. Small, scenario-specific datasets can also be used for targeted training based on particular prediction goals.

The model is pre-trained on ShanghaiTech, a widely recognized crowd recognition dataset comprising approximately 1,200 labeled images from diverse real-world scenarios, encompassing varying perspectives and crowd densities. The dataset is divided into Part A (denser crowds) and Part B (sparser crowds). Given that the crowd conditions in Part B better reflect the target environment—educational buildings—716 samples from Part B are selected for pre-training.

Before training, the dataset is preprocessed to generate density maps based on the provided labeling information. A random subset of 400 samples is used to form the training set, with the remaining 316 samples designated as the test set. During each training epoch, forward propagation is conducted for each batch, followed by loss computation, gradient clearing, and backpropagation. At the end of each epoch, model weights are saved, and the cumulative training loss is recorded. The model is then evaluated on the test set, and the version with the lowest test error is retained for subsequent training.

The training loss (Train Loss) is calculated using the mean squared error (MSE), while the test error (Test Error) is calculated using the mean absolute error (MAE). The specific calculations are shown in Eqs. (5) and (6).

$$MSE\ Loss = \frac{1}{N}\sum\nolimits_{i=1}^{N}(y_i - \hat{y}_i)^2 \tag{5}$$

$$MAE = \frac{1}{N}\sum\nolimits_{i=1}^{N}|y_i - \hat{y}_i| \tag{6}$$

Where N represents the sample size, y_i is the actual number of people, and $\hat{y}_i$ is the predicted number of people.

The entire training process was conducted on a single GPU with a constant learning rate of 1.00E-06. Stochastic Gradient Descent (SGD) with momentum was used as the optimizer, and the training progress was visually monitored using the Visdom tool. A total of 2,000 epochs of training were completed. The Train Loss and Test Error trends during training are shown in Fig. 2. It can be observed that the overall trend of Train Loss is decreasing, indicating that the model is learning effectively. At epoch 1910, the test error reaches its lowest value, indicating that the model trained at this epoch exhibits the best performance (Tables 1 and 2).

Table 1. Model training parameters

Norm	Training Cycle		Other Information			
	epoch	Iteration per epoch	learning rate	Parameter update	Momentum	Hardware
	2000	400	1e-06	SGD	0.95	single GPU

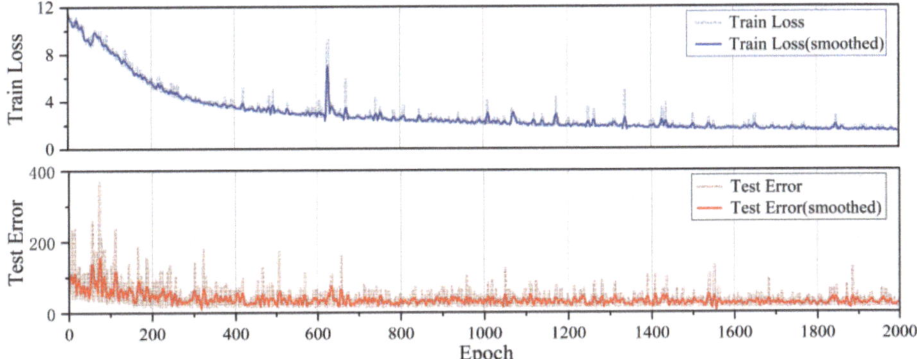

Fig. 2. Training curve

Table 2. Training results

Pre-training database			Localization training database		
Best parameter	Train loss	Test error	Best parameter	Train loss	Test error
Epoch1910	1.46	16.94	Epoch377	2.41	10.00

A localized database was created for retraining using photographs taken by surveillance cameras at different times in the classrooms under study. Each image was accurately labelled with crowd points using LabelMe, resulting in 30 sample sets. Of these, 20 sets were used for the training set and 10 for the test set. The optimal model parameters from the pre-training phase were loaded, and the model was retrained on the newly constructed dataset using the same architecture and training method for a total of 400 epochs. The minimum test error was achieved in the 378th epoch.

2.3 Validation of Model Prediction Accuracy

Sixty sample images were randomly selected from the two datasets to validate the model's performance. Half of the sample images contained an actual number of people ranging from 0 to 100, while the other half contained more than 100 people. Table 3 illustrates some of the specific detection results, and Fig. 3 shows the statistical comparison between the number of people detected by the model in the two types of samples and the actual number of people.

The performance of the MCNN model was evaluated using two metrics: the mean absolute error (MAE) and the root mean square error (RMSE), calculated as shown in Eqs. (6) and (7). The MAE primarily reflects the magnitude of the discrepancy between the predicted and actual values, while the RMSE indicates the precision of the prediction. Based on task requirements and the performance of other personnel recognition models, the accuracy criteria for the model are defined as MAE $\leq$ 10 and RMSE $\leq$15.

$$RMSE = \sqrt{\frac{1}{N}\sum\nolimits_{i=1}^{N}(y_i - \hat{y}_i)^2} \tag{7}$$

presents the evaluation results of the model's performance. When validated on datasets containing fewer than 100 individuals in the image, the model achieves MAE and Root Mean Square Error (RMSE) values below 5, indicating high accuracy in estimating the number of people. However, as the dataset size increases and the number of individuals exceeds 100, both the MAE and RMSE values increase, with the MAE surpassing 10. This suggests that the model's performance diminishes when applied to more crowded scenes. Overall, when evaluated across the entire dataset, the model yields an MAE of 7.2 and an RMSE of 9.61, demonstrating relatively satisfactory performance despite the challenges presented by larger datasets (Table 4).

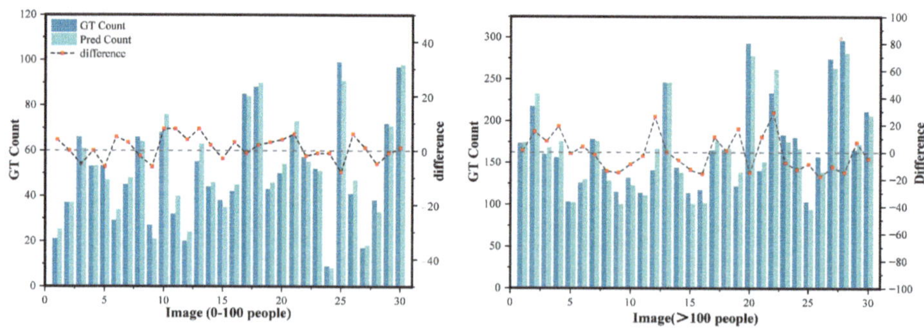

Fig. 3. Comparative results of population density estimates

Table 3. Part of the prediction results show

Localization training database			Pre-training database		
Image	Density Map		Image	Density Map	
Sample	Ground Truth	Estimated	Sample	Ground Truth	Estimated
56	56	55	44	44	46
59	59	62	45	45	48
54	54	55	246	246	246

Table 4. Verification results

	All image	Image (0–100 people)	Image (>100 people)
MAE	7.2	3.67	10.73
RMSE	9.61	4.41	12.88

3 Simulations Design

Conventional air conditioning control systems usually consist of an outer-loop temperature control module and an inner-loop cooling capacity control module. The outer-loop module compares the real-time temperature with the set target temperature to determine the required cooling capacity. The inner-loop module then adjusts the cooling capacity to stabilize the temperature within the designated region.

However, due to limitations in sensor monitoring capacity, it is difficult to provide timely feedback on personnel load changes to the air conditioning system's cooling load. This delay results in energy waste and thermal comfort fluctuations. To address this, this paper proposes an inner-loop air conditioning control strategy based on real-time personnel predictions. The model outlined earlier accurately estimates the real-time occupant heat load in a building space. Building on this, a hierarchical control mechanism is introduced, categorizing crowd density into three levels: low, medium, and high. This classification avoids increased regulation costs associated with too many load categories. The cooling load corresponding to these levels is maintained until the next monitoring cycle. As the load level changes, the cooling load is adjusted promptly, enabling the air conditioning system to respond quickly to shifts in indoor load, thus improving the control of the thermal environment. Figure 4 illustrates the operation of this control mechanism.

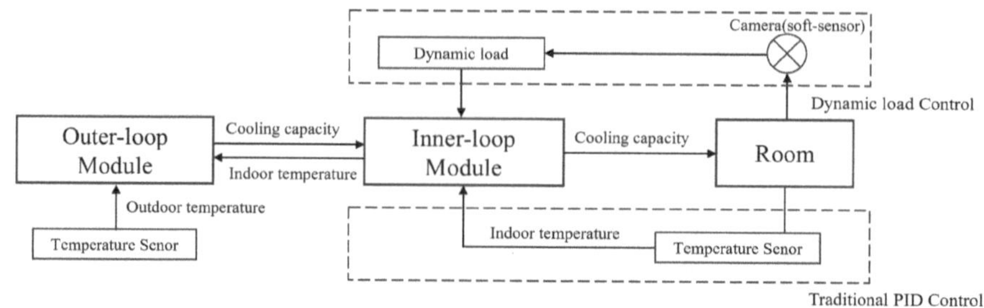

Fig. 4. Schematic diagram of air conditioning control method

The real-time density map generated by the MCNN model provides spatial distribution information of the crowd, enabling zone-based air conditioning control in large spaces. The space is divided into control zones based on air conditioning unit locations and usage conditions. The real-time density map is then used to determine the crowd load in each zone, and the air conditioning terminal load is adjusted according to the load classification. This approach allows for more precise crowd load feedback, resulting in energy savings and improved thermal comfort.

A typical university classroom in Hefei City, Anhui Province, China, is used to design and verify the feasibility of this control strategy through simulation. The target building, constructed in 2002, is a complex structure combining classrooms, offices, and staircase study rooms. The classroom, located on the third floor and facing north, is equipped with four air-conditioning units, two of which ensure full coverage of the space, as shown in Fig. 6.

3.1 Simulation Input Information

The classrooms were equitably categorized into two control zones, A and B, based on the distribution of air-conditioning units and the usage patterns of the occupants. A population study of the target classrooms was conducted through field observations and a review of real surveillance video to determine the specific load rating. The number of people in the classroom and the two control areas were recorded at 15 min intervals from 7:00 a.m. to 6:00 p.m. on a typical weekday. The results are shown in Fig. 5.

Considering the classroom size, the maximum capacity is approximately 75 people, with a maximum density of 0.69 m^2/person. Based on this information, Indicators in Table 5 are classified

Table 5. Personnel density, load class and cooling load

Personnel density	density class	Load class	Cooling load ratio
>0.5 people/m^2	High	III	100%
0.5–0.3 people/m^2	Medium	II	75%
<0.3 people/m^2	Low	I	50%

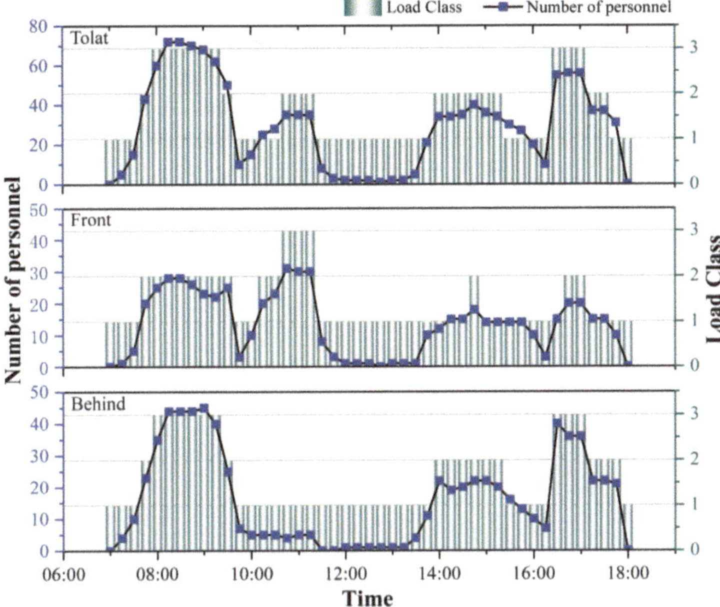

Fig. 5. Changes in classroom crowding loads

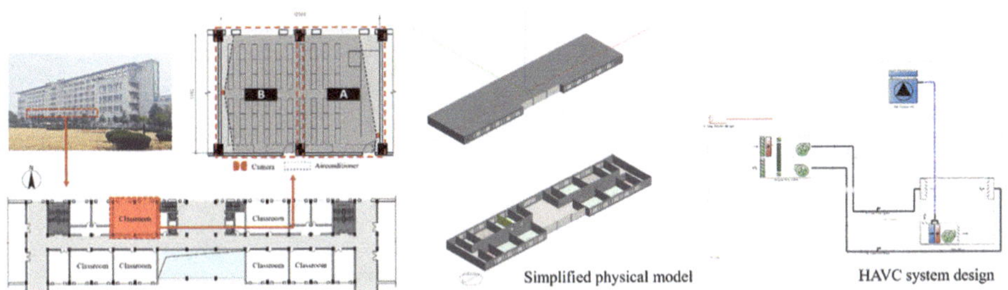

Fig. 6. Simulation objects and simulation models

Simulation validation was carried out using DesignBuilder as the simulation software. DesignBuilder is a building energy efficiency and environmental simulation tool based on the EnergyPlus simulation engine, which is widely used in building design and performance evaluation. The software's Detailed HVAC Design feature enables users to define, configure, and optimize the HVAC system within the model.

A simplified physical model of the classroom and its floor was created based on realistic conditions, with the detailed parameters shown in Table 6. Using the detailed HVAC settings in the software, a thorough simulation setup of the VRF air conditioning system applied to the classroom was conducted.

Three groups of simulations are planned under different control schemes. All simulations will use the same people flow change scheme and the building's physical environment, with only the control of the air-conditioning system being altered. The first group uses the conventional schedule control scheme, where the cooling load operates constantly during working hours. The second group employs a hierarchical control scheme based on the dynamic load of the personnel. In contrast, the third group introduces zone control on top of the hierarchical control in the second group. A building dynamics simulation for a typical summer week (June 17th–June 21st) is performed for each group. The standard annual weather data for Hefei City, provided by the EnergyPlus website, is used for meteorological data. Analysis of simulation results.

Table 6. Input Parameters for Simulations

Items	Contents
Location	Hefei, China
Latitude	31.861 N
Room type	A north-facing university classroom
Room size	W × H:12 m × 9 m; H: 4.2 m; area: 108 m^2
Air infiltration	1 ACH (air change per hours)
Window area	17.072 m^2
Window type	Clear double pane window solar heat gain coefficient = 0.691; U value = 1.96 W/m^2-K
External wall	R = 3.444 m^2·K/W
Internal partition	R = 1.173 m^2·K/W
Roof and ground floor	R = 4.001 m^2·K/W
Air conditioning system	VRF (Cooling capacity: 4KW × 4); No mechanical ventilation
Types of air conditioner inner units	Wall-mounted air conditioner
Number of air-conditioning units	4
Rate air flow rate	1200 m^2/h
Cooling capacity	4 KW
Cooling setpoint	26 °C
Max personnel density	0.69 people/m^2
Light loads	15 W/m^2
Miscellaneous load	1.5 W/m^2
Weather file	Typical meteorological year (TMY) data
Time step of per hour	10
simulation period	June 17th–June 21st

3.2 Simulation Results and Discussion

To demonstrate the effectiveness of the control strategy, the data obtained from three sets of different simulation methods are compared and analyzed. Figure 7 illustrates the detailed simulated data changes for June 19 and 20. The changes in the average indoor air temperature under the three control modes reveal that, except during periods of high outside temperatures, the room temperature remains between 24 °C and 25 °C across all modes. Conventional regulation results in more stable temperature fluctuations due to the constant cooling load. Graded regulation, however, reduces refrigeration efficiency when the personnel load is low, leading to slightly higher room temperatures compared to conventional methods. Conversely, when the occupant load is high, graded regulation brings the room temperature closer to the target of 24 °C. Introducing the zonal regulation mechanism results in smoother temperature fluctuations compared to the simple hierarchical regulation strategy. This demonstrates that zonal regulation is more adaptable to changes in crowd load in this space. Regarding indoor comfort, the predicted mean vote (PMV) in the classroom under all three regulation modes generally remained between 0 and 0.5, indicating that all modes effectively maintain a high level of comfort, even under high temperatures.

By comparing the total cooling rate curves of the cooling coil under the three schemes, it is observed that the load fluctuation in the hierarchical zonal regulation has a more pronounced antecedent, meaning that the load decreases and increases more rapidly compared to conventional regulation. This indicates that this approach better aligns with the real-time demand situation, allowing for more timely adjustments to the cooling load. As a result, it helps prevent energy waste and insufficient cooling capacity caused by fluctuations in personnel load. Table 7 presents the total cooling energy consumption for the three scenarios during the simulation period. Compared to conventional air conditioning regulation, dynamic grading can save approximately 2.27% of the total cooling energy consumption, while zoning grading can achieve a savings of about 6.5% in cooling energy consumption.

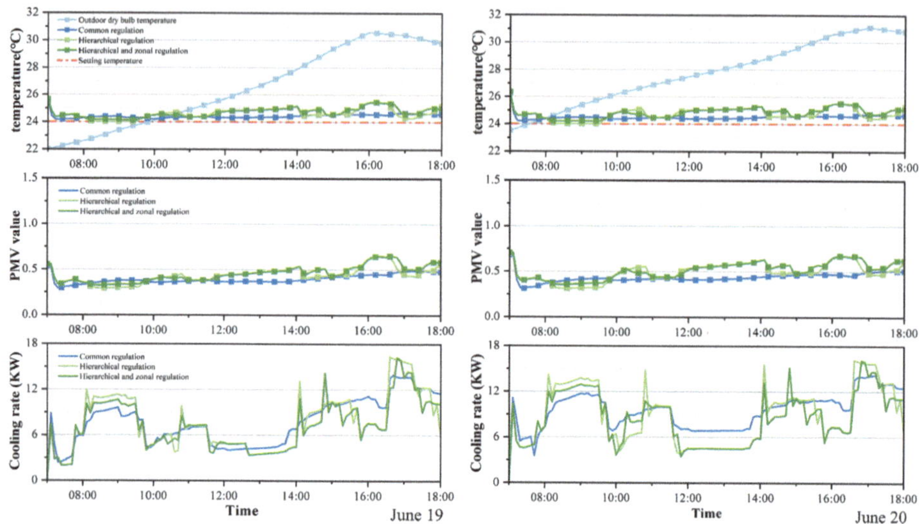

Fig. 7. Average temperature, PMV, and total cooling for June 19 and 20 in the simulations

Table 7. Simulated cycle cooling energy consumption

Simulation scheme	Common regulation	Hierarchical regulation	Hierarchical and zonal regulation
Consumption	136.71 kW·h	133.60 kW·h	127.81 kW·h

In summary, the hierarchical zoning dynamic air conditioning control strategy can maintain a similar indoor comfort level as the conventional control method while saving more cooling energy.

4 Conclusions

This paper proposes a real-time personnel occupancy load estimation method and an air-conditioning predictive control strategy based on computer vision and deep learning technologies. The aim is to improve indoor environmental quality in public buildings, reduce energy consumption, and promote the integration of control systems with artificial intelligence during the operational phase of buildings.

The study results indicate that: (1) The MCNN-based occupancy load estimation method achieves high accuracy, particularly when the number of occupants is below 100, providing a simple and effective approach for estimating indoor occupant density and load; (2) The hierarchical predictive control strategy for air-conditioning zoning, based on real-time occupancy load variations, enhances environmental quality and reduces cooling energy consumption by up to 6.5%, demonstrating strong potential for energy savings in other public buildings; (3) Due to limitations in data acquisition and experimental conditions, the research primarily focuses on the design and simulation of the control scheme for a single-behavior classroom. Future work will explore intelligent air-conditioning control in university buildings with more diverse occupant behaviors.

References

1. Ma, Z., Song, J., Zhang, J.: Energy consumption prediction of air-conditioning systems in buildings by selecting similar days based on combined weights. Energy Build. **151**, 157–166 (2017)
2. Li, W., et al.: A method for energy consumption optimization of air conditioning systems based on load prediction and energy flexibility. Energy. **243**, 123111 (2022)
3. Chen, S., Zhang, G., Xia, X., Chen, Y., Setunge, S., Shi, L.: The impacts of occupant behavior on building energy consumption: a review. Sustain. Energy Technol. Assess. **45**, 101212 (2021)
4. Zhang, X., Zhou, T., Kokogiannakis, G., Xia, L., Wang, C.: Estimating the number of occupants and activity intensity in large spaces with environmental sensors. Build. Environ. **243**, 110714 (2023)
5. Zou, H., Zhou, Y., Yang, J., Spanos, C.J.: Device-free occupancy detection and crowd counting in smart buildings with WiFi-enabled IoT. Energ. Buildings. **174**, 309–322 (2018)
6. Zhang, Y., Zhou, D., Chen, S., Gao, S., Ma, Y.: Single-image crowd counting via multi-column convolutional neural network. In: 2016 IEEE Conference on Computer Vision and Pattern Recognition (CVPR), pp. 589–597. IEEE, Las Vegas, NV, USA (2016)

Evaluation of the Visual Comfort and Energy Performance of Shading Façades with Control Strategy in Response to Evaluation Indices

Chuanrong Cui[1,2,3,4], Yilin You[1,2,3,4], Qiao Zheng[1,2,3,4], Feng Shi[1,2,3,4(✉)], and Xiaoqiang Hong[2,3,4(✉)]

[1] Shenzhen Research Institute of Xiamen University, Shenzhen, China
shifengx@126.com

[2] Fujian Province University Key Laboratory of Intelligent and Low-Carbon Building Technology, School of Architecture and Civil Engineering, Xiamen University, Xiamen 361005, PR China
hongxq@xmu.edu.cn

[3] Fujian Key Laboratory of Digital Simulations for Coastal Civil Engineering, Xiamen University, Xiamen 361005, China

[4] Xiamen Key Laboratory of Integrated Application of Intelligent Technology for Architectural Heritage Protection, Xiamen, China

Abstract. Effective shading control strategies are crucial for optimizing the performance of shading devices. However, the control strategies that optimize the performance of shading façade require further investigation. A response to evaluation index (REI) shading control strategy was proposed, and the performance advantages of three shading façades under this strategy were analysed. This study developed a model for assessing the visual comfort and energy performance of buildings by integrating Radiance, EnergyPlus, and Python. The accuracy of this model was calibrated and validated through a continuous ten-day experimental test that measured indoor temperature and illuminance distribution in a test room. The findings indicated that: (1) In terms of *DGP*, multi-sectional shading façade (MSF) and roller shading façade influenced by REI completely eliminated glare. This finding indicated that the selected indices were critical factor in glare prevention. (2) $\text{sUDI}_{300\text{–}3000\text{lx}, 50\%}$ demonstrated superior performance across all shading façades when influenced by REI. The performance of $\text{sUDI}_{300\text{–}3000\text{lx}, 50\%}$ under MSF was comparable to levels with no shading façade (NSF). (3) MSF exhibited the lowest lighting energy consumption, achieving a 15.04% reduction in total energy consumption relative to NSF.

Keywords: Shading façades · Control strategies · Visual comfort · Energy performance

1 Introduction

Daylighting provides numerous advantages for both buildings and their occupants. It decreases the energy demand for artificial lighting while promoting a comfortable indoor visual and thermal environment for users [1, 2]. However, excessive daylighting can

Y. Liu et al. (Eds.): CDRF 2025, *Transindividual Intelligence*, pp. 170–182, 2026.
https://doi.org/10.1007/978-981-92-0615-5_15

result in significant heat loads on the building, thereby increasing cooling energy consumption. Furthermore, an overabundance of daylight may disrupt the typical learning environment for users, leading to visual discomfort [3]. Therefore, shading façades, which effectively regulate the distribution of indoor lighting and thermal comfort while conserving energy, have become a crucial element of architectural design.

Effective shading control strategies are crucial for optimizing the performance of shading devices. Current research has primarily focused on control strategies for single blinds and single roller shades. Tzempelikos et al. [4] previously conducted a comparative investigation of roller shade control strategies, examining their effects on daylighting and energy performance under varying thresholds of solar radiation in-tensity. In a related study, Kunwar et al. [5] employed experimental tests to explore continuous shading control strategies for roller shades, which were based on the in-tensity of external solar radiation and the positioning of work surface illuminance sensors. In recent years, there has been increasing attention on shading control strategies aimed at enhancing visual comfort, thermal comfort and managing energy de-mand. Chan et al. [6] employed a comparative analysis of two glare protection control algorithms for Venetian blinds, highlighting their benefits in improving daylight autonomy and mitigating glare. Concurrently, Alkhatib et al. [7] identified the most effective control parameters for adjusting roller blinds that consider both energy demand and occupant comfort. Multi-objective control strategies can more effectively address the diverse needs of users. Lee et al. [8] identified three key objectives—anti-glare, energy savings, and illuminance enhancement—that constitute a control strategy for adjusting shading states of an adaptive façade. For MSF that combine roller shades and Venetian blinds, the control strategy has become increasingly significant. Tzempelikos et al. [9] introduced an early concept combining automatically controlled Venetian blinds with manually operated roller shades. Chan et al. [10] proposed a system with independent control for the upper blinds or light shelves and the lower roller shades; however, they did not develop a comprehensive control system. Do et al. [11] advanced the evaluation of a shading control strategy for MSF that incorporated a combination of roller shades and Venetian blinds. This shading control strategy was capable of back-checking based on two reference points; however, it only addressed glare and limited its investigation to daylighting performance. Therefore, this study aims to investigate the daylighting performance, thermal comfort, and energy-saving capabilities of MSF integrating a combination of roller shades and Venetian blinds, considering the combined influence of multiple evaluation indices. The study was conducted in an existing office environment using a building physics simulation model that has been calibrated with experimental data.

This study proposed a shading control strategy based on REI, utilizing four control indices: Daylight Glare Probability (*DGP*), illuminance at the far-window point (IFWP), Predicted Mean Vote (*PMV*), and Energy Use Intensity (*EUI*). The performance of MSF with REI was compared with no shading façade and single shading façades with REI. The significance of this study lay in its provision of a selectively designed shading control strategy that had the potential to enhance both energy efficiency and occupant comfort in shading façades.

2 Methodology

2.1 Experiment Set Up

As shown in Fig. 1(a), in this study, a test room equipped with MSF was established in Xiamen, Fujian. The test room, oriented southward, had no windows on the other three sides. The floor area measures 5.3 m^2 (2.45 m × 2.15 m), with a ceiling height of 2.50 m. Two identical windows with clear glazing are installed. The windows are positioned 0.50 m above the ground, each measuring 0.94 m wide and 1.93 m high. The clear glazing accounts for 39% of the total area of the south wall, measuring 0.72 m wide and 1.45 m high. The material characteristics of the clear glazing, Venetian blinds, and roller shades are shown in Table 1. In the experiment, the upper section of MSF was equipped with Venetian blinds set at an angle of 0°, while the lower section was fitted with roller shades lowered to a height of 0.2 m. During the experiment, air conditioning and electric lighting equipment remained turned off and there were no personnel activities in the rooms.

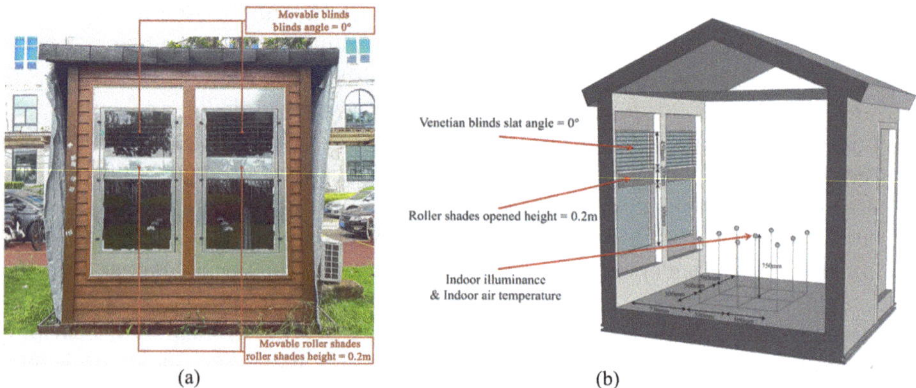

Fig. 1. (a) The test room equipped with multi-sectional shading façades and (b) the specific measurement locations in test room.

The Agilent data acquisition system was utilized for data collection at 1 min intervals. The measurement instruments used in the experiment are listed in Table 2. As shown in Fig. 1(b), daylight illuminance was measured by using 9 illuminance meters, all of which were uniformly placed at a height and distance of 0.75 m from ground, and the spacing between the sensors was set to 0.5 m.

2.2 Shading Façade

As shown in Fig. 2, four different configurations of shading façades were examined in this study. There was no shading façade (NSF) in the baseline case (Case A). Case B and C were installed with a Venetian blind façade (VBF) and a roller shade façade (RSF), respectively, serving as comparison cases with individual shading systems. Case D was

Table 1. Material properties of clear glazing, Venetian blinds and roller shades.

Property	Unit	Clear glazing	Venetian blinds	Roller shades
Thickness	m	0.006	0.001	0.003
Conductivity	W/m/K	1.4	203	0.1
Width	m	–	0.05	–
Separation	m	–	0.05	–
To glass distance	m	–	0.05	0.05
Solar transmittance	–	0.78	0	0.05
Solar reflectance	–	0.08	0.7	0.5
Visible transmittance	–	0.88	0	0.05
Visible reflectance	–	0.0611	0.7	0.5
Infrared hemispherical emissivity	–	0.84	0.9	0.9

Table 2. List of experimental testing and monitoring devices.

Measured parameters	Measurement equipment	Unit	Accuracy
Indoor and outdoor air temperature	T-type thermocouple	°C	±0.6
Horizontal solar radiation	Pyranometer	W/m^2	<5%
Vertical solar radiation	Pyranometer	W/m^2	<5%
Diffuse solar radiation	Shaded Pyranometer	W/m^2	<5%
Wind speed	Anemometer	m/s	±0.05
Daylighting illuminance	Illuminance meter	lx	±7%

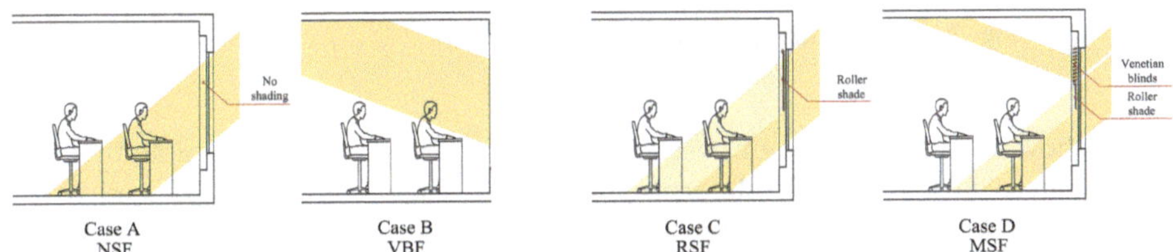

Fig. 2. Side view of studied configurations for four different building shading façades.

equipped with multi-sectional façade (MSF) that integrated Venetian blinds in the upper section and a roller shade in the lower section.

NSF had only a shading state. Each 30° change in the angle of the Venetian blinds meant a new shading state, resulting in a total of seven adjustable shading states for VBF. Similarly, each 0.2 m change in the height of roller shades meant a new shading state, yielding a total of eight adjustable shading states for RSF. MSF integrated the adjustability of both Venetian blinds and roller shades, featuring seven adjustable angles for the Venetian blinds and six adjustable heights for the roller shades, culminating in a total of 42 adjustable shading states.

2.3 Control Strategy

This study presented a REI that was designed to select different evaluation indices based on the specific characteristics of shading façades, the performance requirements of the building, and the diverse expectations of users. Consequently, it could develop unique shading control strategies for different shading façades. As shown in Fig. 3, based on the different characteristics of Venetian blinds and roller shades, IFWP, *PMV*, and *EUI* were established as control indices for VBF. *DGP*, *PMV*, and *EUI* were designated for RSF. For MSF, the four control indices of *DGP*, IFWP, *PMV*, and *EUI* were sequentially established.

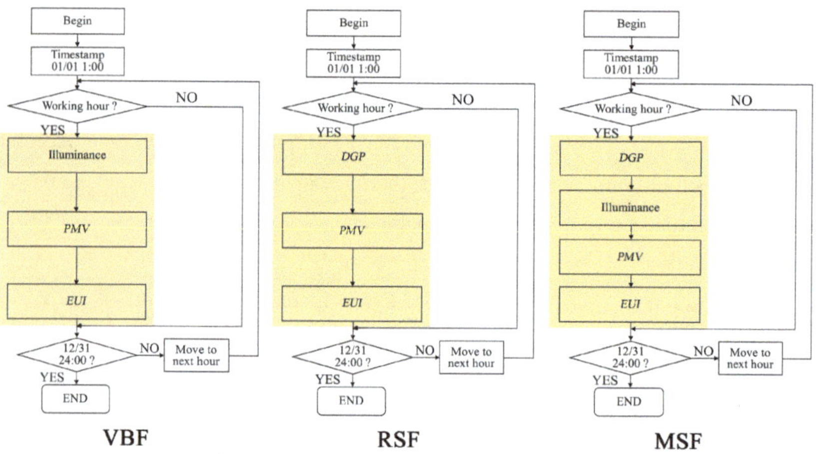

Fig. 3. REI shading control strategy of VBF(a), RSF(b) and MSF(c).

As shown in Fig. 4, the room was divided into the near-window area and the far-window area, each with two reference points were set to calculate *DGP* and IFWP respectively. The glare reference point was positioned at a height of 1.2 m, corresponding to the eye level of a user seated facing the window, while the illuminance reference point was placed at a height of 0.75 m, representing the height of the user's work surface when seated.

Since the selection logic for the shading adjustment program was consistent across the three shading façades when utilizing REI, this study provides a detailed description of the most complex MSF. The first step involved evaluating *DGP*, which must ensure that two glare reference points exhibit imperceptible glare ($DGP < 0.35$). If the shading state met the requirements of the scheme, then it was recorded in set N_{DGP}; otherwise, it was discarded. In the second step, the shading states corresponding to illuminance levels of 300 ~ 3000 lx, <300 lx, and >3000 lx within the set N_{DGP} were identified, to

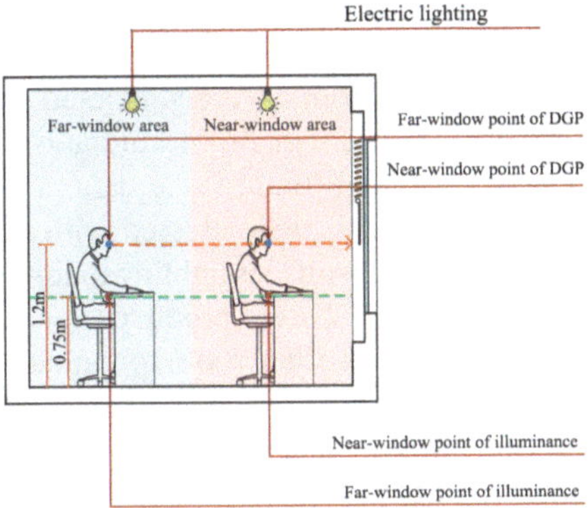

Fig. 4. The arrangement of the reference points and electric lighting.

derive set N_{Illu}. A daylighting environment of 300~3000 lx adequately meets the illumination requirements for work surfaces in the far-window area. This range minimized dependence on supplementary artificial lighting and helped to avoid potential discomfort caused by glare resulting from excessive light. In contrast, a shading state of <300 lx was effective only in preventing uncomfortable glare, while a shading state of >3000 lx proved to be the least effective. In the third step, *PMV* was utilized to select shading states from set N_{Illu}. The shading states were categorized based on their *PMV* values as follows: $-0.5 \le PMV \le 0.5$, $-1 \le PMV < -0.5$ or $0.5 < PMV \le 1$, and $-3 \le PMV < -1$ or $1 < PMV \le 3$, in sequence, resulting in the formation of set N_{PMV}. Finally, *EUI* was chosen as the control index. The unique shading state that demonstrated the lowest total energy consumption within set N_{PMV} was selected. In cases where multiple shading states existed, the state with the highest illuminance at the far-window point was designated as the final shading adjustment state determined by MSF at that specific moment.

2.4 Model Validation

This study utilized EnergyPlus and Radiance to develop building performance simulation models for energy efficiency and daylighting performance, respectively, in the test room. The meteorological parameters during the experimental period were used as weather data within the simulated calculation time period. The measured indoor temperature and illuminance calculated by the model will be compared with the experimental results. The Mean Relative Error (*MRE*) was also used to evaluate the ac-curacy of the model [12].

$$MRE = \frac{1}{N}\sum_{i=1}^{n}\left|\frac{y_i - y_i^{'}}{y_i}\right| \tag{1}$$

where y_i and y_i' are the measured and simulated values of the indoor air temperature and daylighting illuminance at each time step, n is the number of these two parameters. The *MRE*'s best value is 0.

2.5 Simulation Set Up

Three-dimensional models and sensor points were generated using Rhinoceros plugins (Grasshopper, Ladybug, and Honeybee) for daylight simulations conducted in Radiance, which provided annual hourly *DGP* and illuminance data under various shading conditions. Concurrently, EnergyPlus models were created with adaptive lighting controls that were adjusted based on calculations of daylight availability. Synchronized hourly outputs for shading performance and *EUI* were produced.

The test room was used as a reference office room for annual simulation. The simulation set up including external wall specifications, design inner loads, operating schedules of various loads and schedules of heating and cooling set-point are reference to the standards for daylighting design of buildings in China (GB 50033-2013) and our previous study [13]. The simulated weather data is obtained from EnergyPlus. The annual working time used to evaluate building performance is 2920 h (8 h/day × 365 days). To enhance the representativeness of the simulation results, the dimensions of the room, as well as the parameters for glazing, Venetian blinds, and roller shades, were consistent with those of the reference office room.

2.6 Evaluation Index

2.6.1 Discomfort Glare Probability

This study calculated *DGP* values using high dynamic range imaging (HDRI) rendered with Radiance. Its formulation is:

$$DGP = 5.87 \times 10^{-5} E_V + 9.18 \times 10^{-2} \log\left(1 + \sum_i \frac{L_{s,i}^2 \omega_{s,i}}{E_V^{1.87} P_i^2}\right) + 0.16 \quad (2)$$

where E_V is the vertical eye illuminance (lux), L_s is the source luminance(cd/m^2), ω_s is the solid angle of the source, and P is the position index.

2.6.2 Spatial Useful Daylight Illuminance

Spatial useful daylight illuminance (sUDI) represents a ratio between floor area or work plane that has useful daylight autonomy within defined range of illuminance [14, 15]. In this study, the metrics $\text{sUDI}_{<100\text{lx}, 50\%}$, $\text{sUDI}_{100\text{-}300\text{lx}, 50\%}$, $\text{sUDI}_{300\text{-}3000\text{lx}, 50\%}$, and $\text{sUDI}_{>3000\text{lx}, 50\%}$, which describe the proportion of areas achieving these four illuminance ranges for over 50% of the daytime hours, were used to evaluate the illuminance distribution of daylight in the reference office. The annual daytime used to calculate sUDI is 2920 h (8 h/day × 365 days). It can be expressed that:

$$\text{sUDI}_{<100\text{lx},50\%} = \frac{A_{UDI_{<100lx,\geq 50\%}}}{A} \times 100\% \approx \frac{n_{UDI_{<100lx,\geq 50\%}}}{N} \times 100\% \quad (3)$$

$$\text{sUDI}_{100\sim 300\text{lx},50\%} = \frac{A_{UDI_{100\sim 300lx,\geq 50\%}}}{A} \times 100\% \approx \frac{n_{UDI_{100\sim 300lx,\geq 50\%}}}{N} \times 100\% \quad (4)$$

$$\text{sUDI}_{300\sim 3000\text{lx},50\%} = \frac{A_{UDI_{300\sim 3000lx,\geq 50\%}}}{A} \times 100\% \approx \frac{n_{UDI_{300\sim 3000lx,\geq 50\%}}}{N} \times 100\% \quad (5)$$

$$\mathrm{sUDI}_{>3000\mathrm{lx},50\%} = \frac{A_{UDI_{>3000lx,\geq 50\%}}}{A} \times 100\% \approx \frac{n_{UDI_{>3000lx,\geq 50\%}}}{N} \times 100\% \quad (6)$$

where n is the number of calculation points respectively having $UDI_{<100lx,\ \geq 50\%}$, $UDI_{100\sim 300lx,\ \geq 50\%}$, $UDI_{300\sim 3000lx,\ \geq 50\%}$ and $UDI_{>3000lx,\ \geq 50\%}$, and N is the total number of calculation points.

2.6.3 Energy Use Intensity

The total electric power consumption per unit of indoor area, energy use intensity (EUI), was chosen as the indicator of building energy consumption, which is calculated as follows:

$$EUI = \left(E_{lighting} + E_{cooling} + E_{heating}\right)/A_{indoor} \quad (7)$$

where $E_{lighting}$ represents the electric lighting power, $E_{cooling}$ and $E_{heating}$ represent the cooling and heating power consumption respectively, and A_{indoor} represents the indoor area of the office room.

3 Results and Discussion

3.1 Model Validation

The experiment was conducted from 18:00 on June 20, 2024 to18:00 on June 30, 2024, then the indoor air temperature and daylighting illuminance were calculated for validation.

The simulated versus experimental indoor temperature and daylight illuminance of the test room with MSF are shown in Fig. 5, respectively. The MREs for indoor temperature and daylight illuminance were 2.11% and 6.41%–12.80%, respectively. Therefore, these results indicated that the model effectively represented the experimental conditions, thereby validating the accuracy of the simulation and offering a foundation for future research.

3.2 Daylighting Performance

3.2.1 Discomfort glare probability

As shown in Figs. 6(a) and (b), it is evident that the room with VBF influenced by REI experienced discomfort glare for a longer duration at two points compared to other shading façades. As shown in Figs. 6(c) and (d), $DGP_{<0.35}$ in the rooms with shading façades were higher than that with NSF. It is evident that all shading façades could improve daylighting comfort. Under the influence of REI, the proportion of discomfort glare of the office room with VBF were the highest among all the rooms with shading façades, with 38.70% at the near-window point and 10.07% at the far-window point, respectively. In summary, these results indicated that while REI could mitigate glare, it was essential to select DGP as the control index.

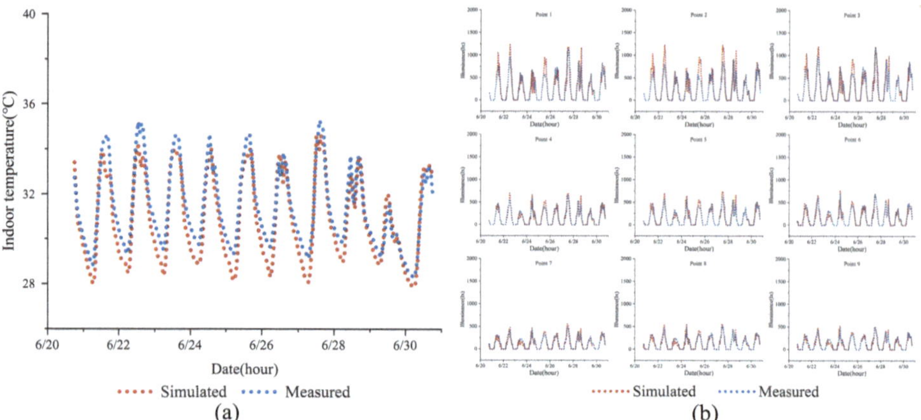

Fig. 5. (a) Comparisons between simulated results and measured data of the indoor temperature and (b) daylight illuminance.

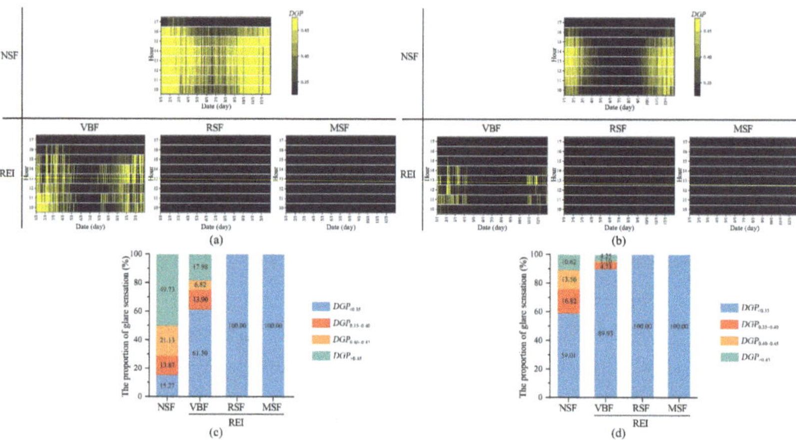

Fig. 6. (a) and (b) Annual *DGP* distribution, and (c) and (d) the proportion of glare sensation for different shading façades at two reference points of *DGP*.

3.2.2 Spatial Useful Daylight Illuminance

As shown in Fig. 7, $\text{sUDI}_{>3000\text{lx}, 50\%}$ of the rooms influenced by REI arrived zero. It is evident that the use of the shading façade influenced by REI has the potential in improving visual comfort and preventing too much undesired sunlight. $\text{sUDI}_{300\text{–}3000\text{lx}, 50\%}$ in the office room with VBF influenced by REI reached 100%. It is evident that daylighting comfort could be enhanced when REI was utilized in the shading façade. Under the influence of REI, although the desired $\text{sUDI}_{300\text{–}3000\text{lx}, 50\%}$ of the office room with MSF was lower than that with VBF and higher than that with RSF, it close to that with NSF (with a decrease of only 1.56%). The results demonstrated that REI was more effective in enhancing the daylighting performance of rooms with MSF in Xiamen.

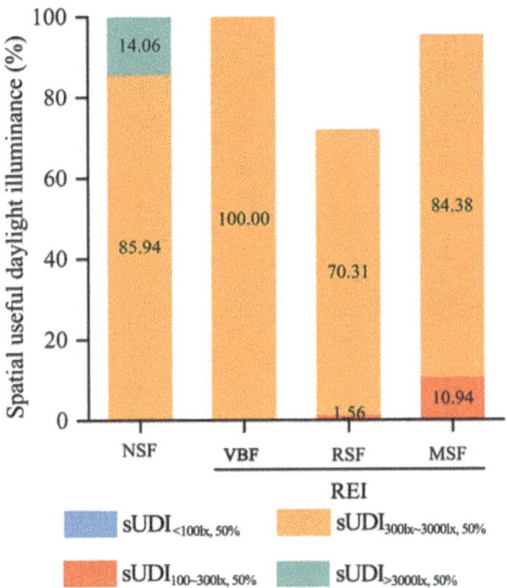

Fig. 7. $sUDI_{<100lx,\ 50\%}$, $sUDI_{100\sim300lx,\ 50\%}$, $sUDI_{300\sim3000lx,\ 50\%}$, and $sUDI_{>3000lx,\ 50\%}$ for different shading façades.

3.3 Energy Performance

As shown in Fig. 8, NSF had worst performance in *EUI*. When the office room was influenced by REI, the annual *EUI* and cooling *EUI* of the office room with VBF were the lowest, and those of the room with RSF were the highest. This result suggests that VBF was more effective than RSF in reducing cooling *EUI* and annual *EUI*. Although the office room with MSF had a slightly higher annual *EUI* of 2.1% than that with VBF, it had a better performance in terms of lighting *EUI*, which was only 41.78% of that with VBF. This is mostly due to the quantity difference of adjustable shading states and it provided advantages for satisfying daylighting requirements. On the whole, the office room with the MSF influenced by REI could more effectively reduce annual energy consumption when applied in Xiamen.

4 Conclusions

In this study, REI shading control strategy was proposed and applied to MSF that combined Venetian blinds and roller shades. Simultaneously, REI was established based on different control indices for VBF, which featured single Venetian blinds and RSF, which featured single roller shades. The visual comfort and energy performance of these shading façades were comparatively analyzed using building performance simulation. The model was calibrated and validated through a continuous ten-day experimental measurement. The results demonstrated the excellent performance of MSF influenced by REI in terms of glare prevention, enhanced indoor illuminance, as well as reduced building energy consumption. The following conclusions were drawn:

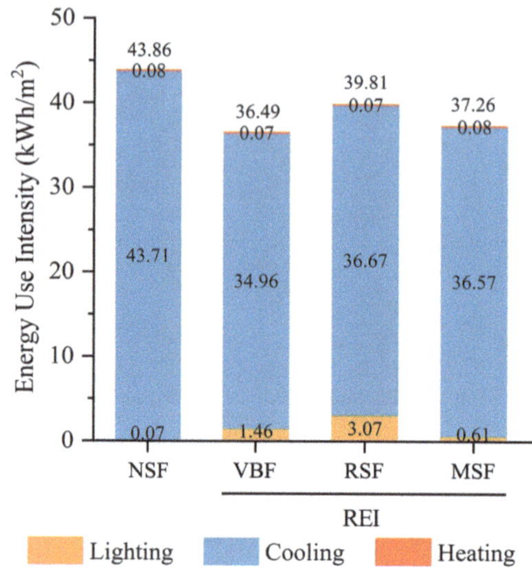

Fig. 8. Annual energy used intensity for ten different shading façades.

(1) In terms of *DGP*, MSF and RSF influenced by REI completely eliminated glare. This finding in-dicated that the selected indices were critical factor in glare prevention.
(2) Regarding sDUI, the $sUDI_{300\text{-}3000lx,\ 50\%}$ demonstrated superior performance across all shading façades when influenced by REI. The performance of $sUDI_{300\text{-}3000lx,\ 50\%}$ under MSF was comparable to levels with NSF.
(3) Regarding *EUI*, MSF exhibited the lowest lighting energy consumption, achieving a 15.04% re-duction in total energy consumption relative to NSF.

Acknowledgments. This work was supported by the National Natural Science Foundation of China (No. 52078443), the Fundamental Research Funds for the Central Universities (No. 20720240072), the Fujian Provincial Natural Science Foundation of China (No. 2024 J01004), Shenzhen Science and Technology Program (JCYJ20240813145509013), and the Construction Science and Technology Project of Xiamen Municipal Housing and Construction Bureau (No. XJK2024-1-07).

References

1. Leslie, R.P.: Capturing the daylight dividend in buildings: why and how? Build. Environ. **38**, 381–385 (2003). https://doi.org/10.1016/S0360-1323(02)00118-X
2. Al Horr, Y., Arif, M., Kaushik, A., Mazroei, A., Katafygiotou, M., Elsarrag, E.: Occupant productivity and office indoor environment quality: a review of the literature. Build. Environ. **105**, 369–389 (2016). https://doi.org/10.1016/j.buildenv.2016.06.001
3. Xie, J., Sawyer, A.O.: Simulation-assisted data-driven method for glare control with automated shading systems in office buildings. Build. Environ. **196**, 107801 (2021). https://doi.org/10.1016/j.buildenv.2021.107801

4. Tzempelikos, A., Shen, H.: Comparative control strategies for roller shades with respect to daylighting and energy performance. Build. Environ. **67**, 179–192 (2013). https://doi.org/10.1016/j.buildenv.2013.05.016
5. Kunwar, N., Cetin, K.S., Passe, U., Zhou, X., Li, Y.: Full-scale experimental testing of integrated dynamically-operated roller shades and lighting in perimeter office spaces. Sol. Energy. **186**, 17–28 (2019). https://doi.org/10.1016/j.solener.2019.04.069
6. Chan, Y.-C., Tzempelikos, A.: Efficient venetian blind control strategies considering daylight utilization and glare protection. Sol. Energy. **98**, 241–254 (2013). https://doi.org/10.1016/j.solener.2013.10.005
7. Alkhatib, H., Lemarchand, P., Norton, B., O'Sullivan, D.T.J.: Comparison of control parameters for roller blinds. Sol. Energy. **256**, 110–126 (2023). https://doi.org/10.1016/j.solener.2023.03.042
8. Lee, D., Cho, Y.-H., Jo, J.-H.: Assessment of control strategy of adaptive façades for heating, cooling, lighting energy conservation and glare prevention. Energy Build. **235**, 110739 (2021). https://doi.org/10.1016/j.enbuild.2021.110739
9. Tzempelikos, A., Athienitis, A.K.: The impact of shading design and control on building cooling and lighting demand. Sol. Energy. **81**, 369–382 (2007). https://doi.org/10.1016/j.solener.2006.06.015
10. Chan, Y.-C., Tzempelikos, A.: Daylighting and energy analysis of multi-sectional facades. Energy Procedia. **78**, 189–194 (2015). https://doi.org/10.1016/j.egypro.2015.11.138
11. Do, C.T., Chan, Y.-C.: Evaluation of the effectiveness of a multi-sectional facade with Venetian blinds and roller shades with automated shading control strategies. Sol. Energy. **212**, 241–257 (2020). https://doi.org/10.1016/j.solener.2020.11.003
12. Shi, F., You, Y., Yang, X., Hong, X.: Annual evaluation of the visual-thermal comfort and energy performance of thermotropic glazing in a reference office room of China. Build. Environ. **254**, 111378 (2024). https://doi.org/10.1016/j.buildenv.2024.111378
13. Hong, X., Shi, F., Wang, S., Yang, X., Yang, Y.: Multi-objective optimization of thermochromic glazing based on daylight and energy performance evaluation. Build. Simul. **14**, 1685–1695 (2021). https://doi.org/10.1007/s12273-021-0778-7
14. Mangkuto, R.A., Siregar, M.A.A., Handina, A., Faridah: Determination of appropriate metrics for indicating indoor daylight availability and lighting energy demand using genetic algorithm. Sol. Energy. **170**, 1074–1086 (2018). https://doi.org/10.1016/j.solener.2018.06.025
15. Fela, R.F.: The effects of orientation, window size, and lighting control to climate-based daylight performance and lighting energy demand on buildings in tropical area. In: Building Simulation 2019 - The 16th IBPSA International Conference At, pp. 1075–1082, Rome (2020). https://doi.org/10.26868/25222708.2019.210677

Physics-Informed Neural Networks with SIMP Hot-Starting for Nonlinear Structural Analysis in Architectural Geometry Optimization

Liang Yuan[1,2(✉)], Yang Taohan[1,2], Fukuda Hiroatsu[1,2], and Gao Weijun[1,2]

[1] The University of Kitakyushu, Kitakyushu, Japan
yuanliang.damlab@gmail.com
[2] Triton Pumps, Melbourne, Australia

Abstract. The intersection of computational design and structural engineering has significantly advanced architectural capabilities, enabling the creation of geometrically complex, high-performance structures. However, nonlinear structural analysis, essential for real-world optimization, remains computationally intensive, with traditional finite element methods (FEM) struggling to balance accuracy and efficiency. This research introduces a novel Physics-Informed Neural Networks with SIMP hot-starting Topology Optimization (PINNSTO) framework that addresses these challenges while achieving superior optimization outcomes.

PINNSTO integrate physical laws directly into neural network loss functions, enabling accurate solutions without extensive labeled data. Our approach employs sinusoidal activation functions that significantly enhance boundary definition and structural representation capabilities, creating designs that are both mathematically optimal and more manufacturable for architectural applications. With SIMP hot-starting strategy, this framework effectively captures nonlinear behaviors crucial for evaluating complex architectural forms.

Benchmark validations demonstrate 14–23% compliance reduction compared to traditional methods, with consistently smoother boundary definitions. Notably, the PINNSTO approach successfully converges for nonlinear problems, while maintaining comparable computational efficiency. It eliminates meshing errors and handles irregular geometries seamlessly—critical advantages for architectural applications. The implementation efficiently balances performance gains with computational costs, requiring only marginally more processing time.

This research establishes PINNSTO as a robust foundation for architectural structural analysis, enabling simultaneous consideration of performance, material efficiency, and aesthetic intent in contemporary design challenges, particularly for applications involving additive manufacturing and complex material behaviors. By combining enhanced optimization capabilities with physical fidelity, PINNSTO offer a transformative approach to performance-driven design methodologies for sustainable, efficient architectural forms.

Keywords: Physics-Informed Neural Networks · Nonlinear Structural Analysis · Architectural Geometry Optimization

Y. Liu et al. (Eds.): CDRF 2025, *Transindividual Intelligence*, pp. 183–194, 2026.
https://doi.org/10.1007/978-981-92-0615-5_16

1 Introduction

Topology Optimization (TO) is an advanced structural optimization method which can achieve optimal structure configuration and maintain minimum material consumption. It received a significant attention from multiple research fields including architecture, civil engineering, aerospace, automotive, etc. [1]. Since the initial paper by Bendsoe and Kikuchi developed the optimal topology generating method in structural design [2], various topology optimization methods have been introduced such as Solid Isotropic Material with Penalization (SIMP) [3], Bidirectional Evolutionary Structural Optimization (BESO) [4] and Level Set Method (LSM) [5]. Topology optimization is now a well-established field and has seen multiple applications in high-tech architectural area, especially in additive manufacturing buildings, high performance and light weight structures, providing different solutions of optimal designs effectively for architects.

In recent years, neural networks (NN) is evolving rapidly, an algorithm considering physical constraints embedded in loss functions has been introduced as Physical Informed Neural Networks (PINN) by Raissi in 2019 [6]. PINN focuses on solving nonlinear partial differential equations and has been proved a solid method in mathematical level [7].

Building upon these foundations, Chandrasekhar [8] integrated NN and TO by using the NN's activation function to represent the SIMP density field. In a similar vein, Zhang [9] combined NN with TO replacing the conventional sensitivity analysis with automatic differentiation technology. These works open a new realm for both NN and TO, expanding the possibilities of TO can achieve in structural designs.

The wide application of digital design, additive manufacturing, and digital fabrication based on robotic processes have facilitated the realization of complex geometry in architecture nowadays [10]. However, the structural analysis and optimization methods still rely on finite element analysis (FEA), the design of material distribution with minimum compliance is very challenging due to the high dimensionality of the design space, and the computational cost becomes very high due to the similar reason [11]. In addition, the FEA based topology optimization method has difficulties of solving structure with strong nonlinearity and complex materials as well as significant challenges under dynamic loading and subsequent fracture [12].

Motivated by the aforementioned considerations, this paper presents a novel PINNSTO framework for nonlinear structural analysis in architectural geometry optimization based on the minimum energy principle introduced by Jeong [13]. Due to the inherence of NN, other architectural constraints such as buckling issue, stress constraints, or structure distribution are easy to integrated to the framework. The adopting of sinusoidal activation enables the method to explore different patterns of an architecture structure by adjusting the frequency. To validate the proposed approach, we conducted comprehensive benchmark case studies and performed comparative computational analyses. Furthermore, we extended our investigation to nonlinear structural behavior utilizing SIMP hot-starting strategy and sinusoidal activation, complemented by detailed displacement field visualization and convergence analysis.

2 Methodology

Energy-based PINN(DEM-PINN) is chosen for this paper since it is computationally efficient and does not rely on FEA while maintain the accuracy and performance. The physics equations are embedded in the loss function so that no large dataset is needed. The proposed method can be concluded overall as:

$$\min : \mathrm{C}(\rho) = \mathbf{U}^{\mathrm{T}}\mathbf{K}\mathbf{U}$$

$$\text{s.t.} : \mathbf{KU} = \mathbf{F},$$

$$\mathrm{V}(\rho)/\mathrm{V}_0 = \mathrm{f},$$

$$0 \leq \rho_{\mathrm{i}} \leq 1, \mathrm{i} = 1 \ldots \mathrm{N} \tag{1}$$

- *C(ρ)* denotes the structural compliance (a measure of flexibility to be minimized, equivalent to twice the strain energy).
- ***U*** and ***F*** are the global displacement and force vectors, respectively.
- ***K*** is the stiffness matrix.
- *f* is the prescribed volume ratio.

In the proposed PINNTO framework, the displacement field ***U*** is approximated by a physics-informed neural network (PINN) rather than solving the equilibrium equation ($\boldsymbol{KU} = \boldsymbol{F}$) via traditional finite element methods. The compliance *C(ρ)* is evaluated through the network's output, while the density field ρ is iteratively optimized to satisfy both the volume constraint and boundary conditions.

The displacement field is used to calculate the strain and stress as:

$$\varepsilon_{\alpha\beta} = \frac{1}{2}\left(\frac{\partial u_\alpha}{\partial x_\beta} + \frac{\partial u_\beta}{\partial x_\alpha}\right) \tag{2}$$

$$\sigma_{\alpha\beta} = \lambda\delta_{\alpha\beta}\varepsilon_{\gamma\gamma} + 2\mu\varepsilon_{\alpha\beta} \tag{3}$$

$\delta_{\alpha\beta}$ is the Kronecker delta function.

- λ and μ are the Lame constants.

As Fig. 1 illustrates, the neural network utilizes x_α, x_β coordinates as input and displacement u_α, u_β as outputs with total potential energy as loss function in terms of strain and stress calculated by *u*.

2.1 Improvements in adapting the PINNTO framework to building geometry

This section introduces a hybrid framework that seamlessly integrates PINN with SIMP methodology. The core innovation resides in the replacement of traditional Finite Element Analysis (FEA) with a computationally efficient, mesh-free PINN approach

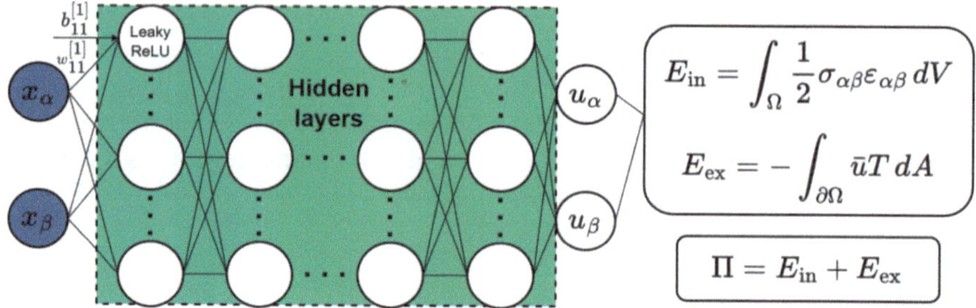

Fig. 1. PINNTO Diagram

for predicting strain and stress fields, while strategically preserving the SIMP-based optimization loop for material distribution updates.

Unlike traditional nonlinear FEA approaches that suffer from element distortion and require iterative nonlinear equation solving, PINNTO provides continuous field representation for exact sensitivity analysis, ensuring robust convergence and smooth boundaries.

For expedited validation of standard benchmark cases, we employ ReLU activation functions, whereas for complex nonlinear problems requiring enhanced representational capacity, we utilize sinusoidal activation functions. The inherent periodicity and precisely controllable frequency characteristics of these sinusoidal functions facilitate meticulous regulation of structural complexity throughout the optimization process. This represents a significant innovation in geometric morphology control, allowing unprecedented precision in manipulating structural features across multiple scales. These sophisticated mathematical properties generate well-defined boundaries without the problematic checkerboard patterns that frequently manifest with conventional activation functions and discretization methods, thereby enabling more nuanced control over the final architectural form.

This methodological adaptation yields three significant advantages in architectural geometry optimization:

- seamless handling of irregular geometries common in architectural design
- elimination of meshing errors
- automatic gradient computation via PINN's differentiation capabilities. The workflow, as illustrated in Fig. 2, iteratively couples PINN-driven physics prediction with SIMP-driven density updates, achieving geometry-adaptive optimization without labelled data or manual sensitivity analysis.

  The proposed PINNSTO framework as Fig. 2 illustrates operates through following phases:

1. Design Domain Initialization: Define the geometric boundaries, material properties, and loading conditions of the target structure.
2. PINN-based Physics Prediction: A physics-informed neural network takes the current material density distribution as input and outputs strain/stress fields by solving governing equilibrium equations through automatic differentiation, avoiding traditional meshing process.

3. SIMP-driven Optimization Loop:
   a. Sensitivity Analysis: Calculate the impact of density changes on compliance using PINN-predicted strain energy.
   b. Sensitivity Filtering: Smooth sensitivity values to suppress numerical instabilities (e.g., checkerboard patterns).
   c. Optimality Criteria: Update material densities via the SIMP heuristic $p_{new} = p_{old} \bullet (D/D_0)^{\eta}$, where D is filtered sensitivity.
4. Convergence Check: Terminate the process if compliance stabilizes (<1% change in 5 iterations); otherwise, repeat Steps 2–3.

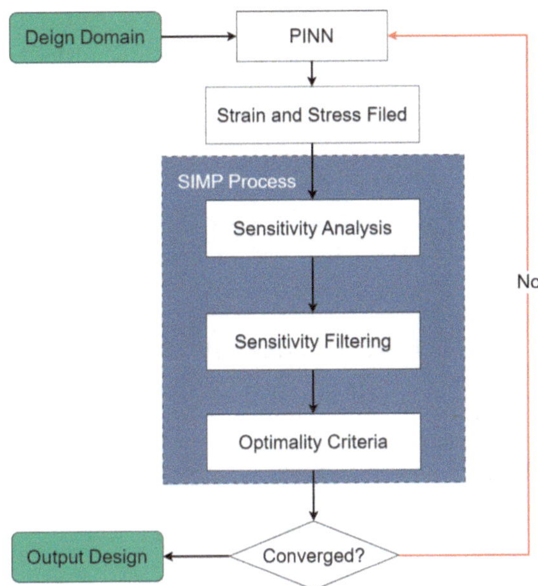

Fig. 2. PINNTO flowchart

2.2 Nonlinear Analysis Module

While topology optimization methods based on linear elastic constitutive models have sufficiently addressed the vast majority of theoretical research challenges, the architectural industry faces emerging complexities with the proliferation of additive manufacturing technologies. These advancements have facilitated the integration of highly elastic materials that exhibit significant mechanical behavior beyond the scope of linear approximations. Such materials undergo substantial displacements and strains under operational conditions, rendering conventional linear models inadequate for accurate structural analysis and optimization. To overcome these limitations, this study extends the PINNSTO framework to effectively solve geometrically nonlinear problems, thereby enhancing its applicability in contemporary architectural design scenarios where material nonlinearity represents a critical consideration.

2.2.1 Geometric Nonlinearity

In geometric nonlinear analysis, the deformation gradient tensor plays a crucial role. Consider that a material point x^0 within the initial configuration deforms to x^t in the current configuration. The mapping relationship is given by:

$$x_\alpha^t = x_\alpha^0 + u_\alpha \tag{4}$$

The deformation gradient tensor can be defined as:

$$x_{\alpha,\beta}^{t0} = \frac{\partial x_\alpha^t}{\partial x_\beta^0} = \delta_{\alpha\beta} + u_{\alpha,\beta}^{t0} \tag{5}$$

The Green-Lagrange strain tensor, which accounts for large deformations, is defined as:

$$\varepsilon_{\alpha\beta}^{t0} = \frac{1}{2}\left(u_{\alpha,\beta}^{t0} + u_{\beta,\alpha}^{t0} + u_{\alpha,\gamma}^{t0} \cdot u_{\beta,\gamma}^{t0}\right) \tag{6}$$

This research employs Neo-Hookean hyperelastic constitutive model to accurately capture nonlinear behaviors. The strain energy density function for this model is expressed as:

$$\Psi(I_1, J) = \frac{1}{2}\lambda\left[\log(J)\right]^2 - \mu\log(J) + \frac{1}{2}\mu(I_1 - 3) \tag{7}$$

Where:

- $I_1 = \mathrm{trace}(C)$ is the first invariant, where C is the right Cauchy-Green deformation tensor.
- $J = \det(F)$ is the determinant of the deformation gradient.

2.2.2 PINNSTO Framework for Nonlinear Analysis

For geometric nonlinear optimization, we propose a novel hot-starting strategy that leverages the efficiency of the SIMP method to provide a favorable initial solution for Physics-Informed Neural Networks. This approach combines two neural networks:

- DEM-PINN (Deep Energy Method): An energy-based physics-informed neural network for predicting displacement fields under given density distributions
- S-PINN (Sensitivity PINN): A sensitivity physics-informed neural network for generating optimized density fields.

The overall optimization process consists of three phases:

- Phase 1: SIMP Hot-Starting
  - Rapidly solve a linear problem using traditional SIMP methodology
  - Obtain initial density distribution and displacement field
- Phase 2: Network Initialization

- Initialize DEM-PINN using SIMP displacement results
- Initialize S-PINN using SIMP density field

- Phase 3: Nonlinear CPINNSTO Iterative Optimization
  - DEM-PINN predicts nonlinear displacement fields
  - S-PINN updates density field based on strain energy
  - Iterative optimization until convergence

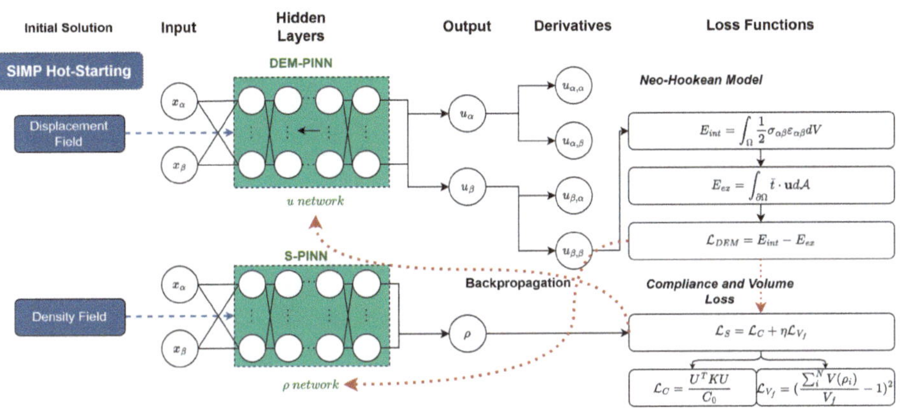

Fig. 3. Hot-starting PINNSTO framework for nonlinear analysis

This approach is particularly effective for addressing geometric nonlinearity problems as it avoids the challenges of directly solving highly nonlinear equation systems while preserving design freedom.

3 Results

3.1 Benchmark Validation

To validate the proposed PINNSTO method, four fundamental cases are examined in the following configurations:

- Material properties: the young's modulus $E = 1$, $E_{min} = 1e^{-6}$ and Poisson's ratio ν = 0.3.
- The design domain is discretized into 60 × 30 mesh grids.
- The optimization parameters include the volume fraction $v_f^* = 0.5$ for the first 3 cases and 0.3 for the Michell Beam, the filter radius $rmin = 3$ and the penalty is set to 3 for all cases.
- The depth and width of NN are 5 × 20 neurons. The learning rate is set to 0.001. For benchmark validation, we utilize Leaky ReLU activation function, for nonlinear module, the sinusoidal activation function is applied.

As shown in Fig. 4, comparisons between PINNSTO and SIMP method reveal distinct characteristics in both the resulting topologies and performance metrics. Despite the theoretically similar optimization objective, the PINNSTO framework achieves compliance values that are consistently lower than those obtained via SIMP implementation, with reductions of approximately 23.14%, 23.25%, 22.85%, and 14.93% for the tip cantilever, mid cantilever, MBB beam, and Michell beam cases, respectively. In addition, PINNSTO generates smoother boundaries due to its mesh-free representation of the design field. This analytic representation eliminates the jagged edges typically associated with element-based discretization in SIMP, potentially offering advantages for manufacturing applications. However, it takes more iterations for PINNSTO to achieve convergence. The overall time consumptions for 4 cases are 25.33 s, 37.97 s, 19.75 s and 15.83 s, it is a little higher than that of SIMP but is still efficient considering its better results.

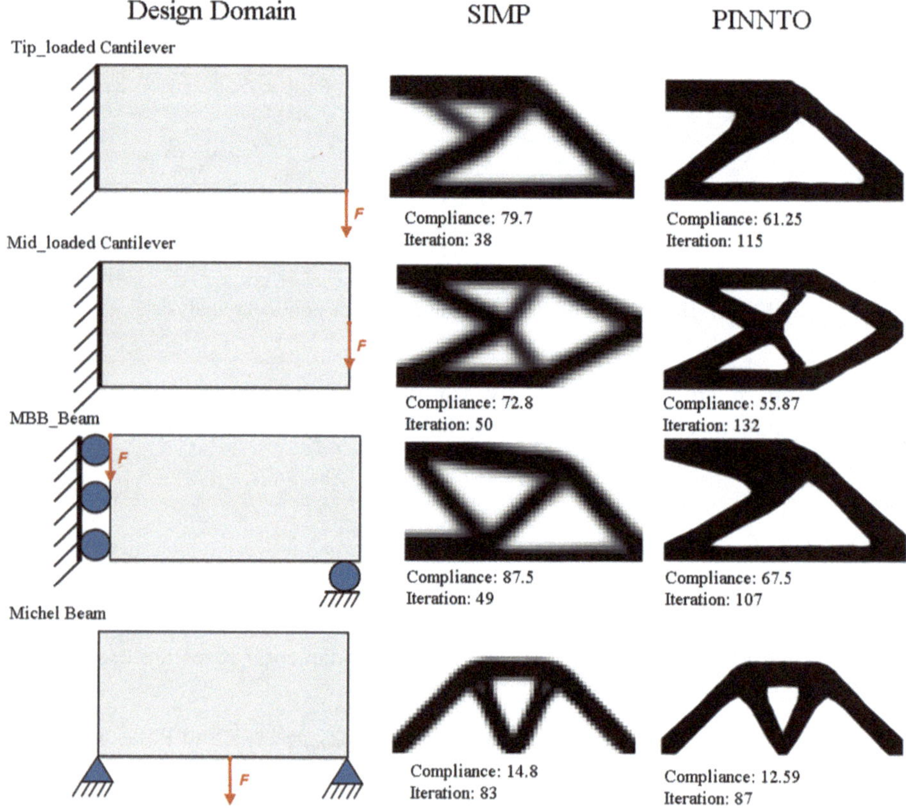

Fig. 4. Comparison with SIMP using four cases

Figure 5 presents the convergence histories for four benchmark cases, illustrating the evolution of relative compliance (blue dotted lines) and volume fraction (red dashed lines) throughout the optimization process. Quantitative analysis reveals that compliance values exhibit rapid initial descent followed by relative stability after approximately 30–40 iterations, demonstrating the framework's efficiency in identifying near-optimal structural configurations. However, the volume fraction constraint displays a noticeably slower convergence rate across all examples, particularly in the tip and mid loaded cantilever cases where subtle oscillations persist even in later iterations. This disparity in convergence rates between objective function and constraint satisfaction results in delayed fulfillment of the global optimality criterion. To address this phenomenon, increasing the penalty factor represents one potential solution, though such modification may adversely affect structural detail resolution by penalizing intermediate densities more aggressively.

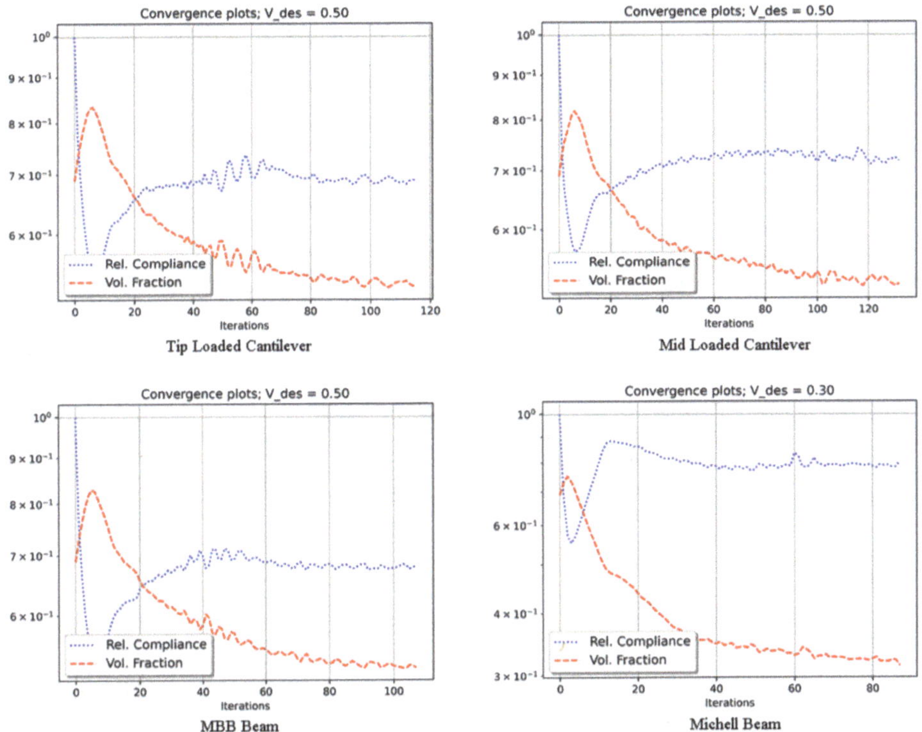

Fig. 5. Compliance and volume fraction convergence history

3.2 Nonlinear Structural Response Analysis

To validate the nonlinear analysis module proposed in Sect. 2.2, we conducted experiments using a mid-loaded cantilever case with a design domain of 4 m × 1 m discretized into a 120 × 30 mesh grid. Due to the inherent complexity of nonlinear optimization problems, we enhanced the neural network architecture by increasing both its depth and width to 8 × 80 neurons. Additionally, the conventional ReLU activation function was replaced with sinusoidal activation functions to better accommodate the geometric and structural complexities involved in nonlinear analysis. Other configuration adjustment can be seen in sect. 2.2.

As illustrated in Fig. 6, the proposed method successfully converged to an optimal solution after 353 iterations. The convergence plot clearly demonstrates the optimization process, with the grey element ratio (green dotted line) progressively decreasing to below 1% at termination. The final topology exhibits well-defined boundaries and a structurally coherent organization. Quantitatively, the optimized design achieved a compliance value of 337.86, demonstrating the effectiveness of our approach for nonlinear structural topology optimization. Notably, the displacement field demonstrates a well-distributed deformation pattern without excessive localized displacements. This stability is particularly significant in nonlinear analysis, where geometric nonlinearities could potentially lead to instability or buckling phenomena if not properly addressed during the optimization process.

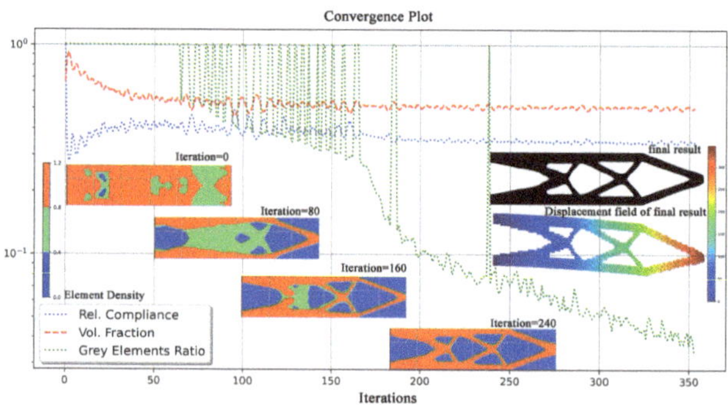

Fig. 6. Convergence plot for nonlinear problem analysis

4 Conclusion

This research has demonstrated the efficacy of Physics-Informed Neural Networks for Nonlinear Structural Analysis (PINNSTO) as a viable alternative to traditional finite element methods in architectural geometry optimization. Through benchmark validation and nonlinear case studies, the proposed framework consistently outperformed conventional SIMP methods and addresses fundamental limitations of traditional nonlinear FEA-based TO approaches including mesh distortion, convergence instabilities, and boundary quality issues.

The novel hot-starting strategy for nonlinear analysis has proven particularly effective in addressing geometric nonlinearity challenges, leveraging the efficiency of SIMP for initialization while preserving design freedom during optimization. Despite requiring slightly more computational iterations than traditional methods, PINNSTO delivers superior results that better accommodate the complex geometries and material behaviors increasingly prevalent in contemporary architectural design.

The mesh-free representation inherent to the PINNSTO approach eliminates discretization artifacts and enhances manufacturability, offering significant advantages for applications in tensile membranes, responsive building skins, natural material grid shells, and seismic-resilient structures where geometric nonlinearity governs performance. As architectural design continues to embrace computational methods and complex geometries, the PINNSTO framework provides a robust foundation for advanced structural optimization, enabling designers to navigate the balance between structural performance, material efficiency, and aesthetic considerations.

References

1. Jihong, Z.H., Han, Z.H., Chuang, W.A., Lu, Z.H., Shangqin, Y.U., Zhang, W.: A review of topology optimization for additive manufacturing: Status and challenges. Chin. J. Aeronaut. **34**(1), 91–110 (2021)
2. Bendsøe, M.P., Kikuchi, N.: Generating optimal topologies in structural design using a homogenization method. Comput. Methods Appl. Mech. Eng. **71**(2), 197–224 (1988)
3. Bendsøe, M.P.: Optimal shape design as a material distribution problem. Structural optimization. **1**, 193–202 (1989)
4. Querin, O.M., Steven, G.P., Xie, Y.M.: Evolutionary structural optimisation (ESO) using a bidirectional algorithm. Eng. Comput. **15**(8), 1031–1048 (1998)
5. Wang, M.Y., Wang, X., Guo, D.: A level set method for structural topology optimization. Comput. Methods Appl. Mech. Eng. **192**(1–2), 227–246 (2003)
6. Raissi, M., Perdikaris, P., Karniadakis, G.E.: Physics-informed neural networks: A deep learning framework for solving forward and inverse problems involving nonlinear partial differential equations. J. Comput. Phys. **378**, 686–707 (2019)
7. Cuomo, S., Di Cola, V.S., Giampaolo, F., Rozza, G., Raissi, M., Piccialli, F.: Scientific machine learning through physics–informed neural networks: Where we are and what's next. J. Sci. Comput. **92**(3), 88 (2022)
8. Chandrasekhar, A., Suresh, K.: TOuNN: Topology optimization using neural networks. Struct. Multidiscip. Optim. **63**(3), 1135–1149 (2021)
9. Zhang, Z., Li, Y., Zhou, W., Chen, X., Yao, W., Zhao, Y.: TONR: An exploration for a novel way combining neural network with topology optimization. Comput. Methods Appl. Mech. Eng. **386**, 114083 (2021)
10. Pantazis, E., Gerber, D.J.: Beyond geometric complexity: a critical review of complexity theory and how it relates to architecture engineering and construction. Archit. Sci. Rev. **62**(5), 371–388 (2019)
11. Lu, L., Pestourie, R., Yao, W., Wang, Z., Verdugo, F., Johnson, S.G.: Physics-informed neural networks with hard constraints for inverse design. SIAM J. Sci. Comput. **43**(6), B1105–B1132 (2021)
12. Kumar, P., Schmidleithner, C., Larsen, N.B., Sigmund, O.: Topology optimization and 3D printing of large deformation compliant mechanisms for straining biological tissues. Struct. Multidiscip. Optim. **63**, 1351–1366 (2021)
13. Jeong, H., Bai, J., Batuwatta-Gamage, C.P., Rathnayaka, C., Zhou, Y., Gu, Y.: A physics-informed neural network-based topology optimization (PINNTO) framework for structural optimization. Eng. Struct. **1**(278), 115484 (2023 Mar)

Advanced Materials and Digital Fabrication

A Digital Framework for Fragmented Modern Tile Wall Assembly System

Xiheng Yan(✉) and Yangzhi Li

The Chinese University of Hong Kong, Shatin, Hong Kong, SAR, China
xihengyan54@gmail.com

Abstract. This study proposes an innovative system for the automated assembly of Fragmented modern tiles (FMT) walls, which integrates intelligent recycling techniques with robotic technologies. This research creates a comprehensive dataset to extract geometric information from broken tiles and applies a multi-objective optimization algorithm (NSGA-II) to maximize waste utilization and optimize tile arrangements. Two UR10E robot arms collaborate to assemble the tiles and extrude cement, which works simultaneously without colliding with each other. This method improves the speed of construction and prevents cement from drying out due to long exposure to air during each layer assembly of the FMT wall. Experimental results demonstrate significant improvements in utilization rates and reducing gaps between tiles with irregular shapes. Utilization rate achieves 100%. The successful construction of the FMT wall consisting of six layers of tiles and cement confirms the feasibility of the workflow from algorithm-based tile matching to dual-robot collaborative construction for sustainable construction practices. This research contributes to advancing waste material utilization in building practices, offering a scientifically grounded approach to reduce environmental impact and improve the sustainability of construction methodologies.

Keywords: Tile Fragments · Sustainable Construction · Robotic Fabrication · Multi-robot Collaboration · Computational Design

1 Introduction

Fragmented modern tiles represent significant waste from construction and demolition. The main method for recycling these tiles is grinding them into powder and using them as aggregates in concrete production (Ray et al., 2021). This grinding process is energy-intensive. Direct use of untreated waste avoids this environmental impact. The Church of S. Maria in Portuno, Italy, employs a construction technique that alternates fragment tiles with concrete layers. This method reduces the energy consumption associated with conventional waste treatment processes (Quagliarini et al., 2014).

In recent research on S. Maria walls, modern tiles have been applied to this wall's assembly method (FMT Wall), which has satisfactory compressive strength (Quagliarini et al., 2023). However, there is a research gap on how to apply waste tiles to this FMT assembly technology and how to integrate digital technologies and automated assembly

Y. Liu et al. (Eds.): CDRF 2025, *Transindividual Intelligence*, pp. 197–205, 2026.
https://doi.org/10.1007/978-981-92-0615-5_17

to enhance the utilization of waste tiles. To address this gap, this study proposes a system for the arrangement optimization and automatic assembly of waste fragment tiles in FMT walls.

This workflow includes four steps: 1) Creation of Broken Tiles dataset. 2) Optimizing the arrangement of each layer of tiles to maximize utilization. 3) Implementing robot collaborative construction systems to assemble the FMT wall. This research integrates intelligent analysis and automated construction techniques to enhance the accuracy and sustainability of FMT wall construction. It will contribute to the broader field of waste material utilization in building practices.

2 Background

In ancient Rome's masonry technology, many waste tiles were reused as materials for structural components. The 11th-century structural porch of Holy Trinity Church in Colchester directly used Roman bricks and tiles from demolished buildings as raw materials (Fleming 2016). Portuno Church (Corinaldo, Marche, Italy) walls are made of tile and brick fragments, and use cement as a bonding agent between the tiles (Quagliarini et al., 2014). Church wall was built using this construction method (Lancioni et al., 2013).

Based on the construction technology of the Portuno Church, a study proposed a sustainable technical structural system (FMT Wall) made of modern tile fragments. The study summarized the arrangements of each layer and verified that the FMT Wall has good mechanical properties (Quagliarini et al., 2023). However, manual assembly is relatively random, and it is impossible to accurately record the tile arrangement, utilization rate, and construction process.

Existing research on digital technology to assist tile assembly includes tile arrangement and assembly systems (Wu et al., 2025). In terms of tiles arrangement, the Evolutionary algorithm (Wu et al., 2022a) and multi-objective optimization algorithm can be used to optimize the arrangement of rectangular tiles, which allows users to select the most material-saving solution (Wu et al., 2022b); BIM software can build semantic relationships of tiles to optimize tiles arrangement (Ergen & Bettemir, 2025). For tile assembly technology, current research is about designing robots for tile laying (Liu et al., 2018) and developing hardware for the robot system (Samarakoon et al., 2024). These studies are about floor laying using rectangular tiles; there is no research on irregular shape tiles arrangement and the assembly technology of tiles in structure components.

3 Method

This study introduces an intelligent recycling and automatic assembly workflow. First, a tile dataset is created. Second, the multi-objective optimization algorithm is applied to optimize the arrangement of broken tiles to maximize the utilization rate and the best alignment of each tile boundary. Finally, two robotic arms are collaboratively assembled into the FMT Wall by inputting the built action code and communication code into the robotic arm's control center (Fig. 1).

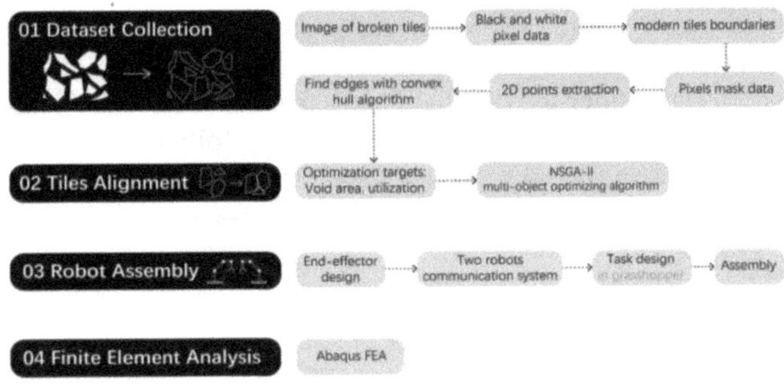

Fig. 1. Workflow

3.1 Dataset Collection

Extract pixels from the tile image for color analysis, filtering out pixel points that contain color information about the tiles. Identify the pixel points at the edges and connect them with polylines to form the tile boundary lines.

3.2 Arrangement and Combination

Among multi-objective evolutionary algorithms (MOEA), the NSGA-II algorithm is a MOEA based on non-dominated sorting. A selection operator is presented that creates a mating pool by combining the parent and offspring populations and selecting the best N solutions (with respect to fitness and spread). Simulation results show that for most problems, compared with the Pareto-archived evolution strategy and the strength-Pareto evolutionary algorithm, NSGA-II is able to find better solution distribution and better convergence near the true Pareto optimal front (Deb et al., 2002). Therefore, this paper used the NSGA-II algorithm to achieve the highest utilization and the most compact arrangement by inputting three optimization objectives.

Randomly rotate the tiles and place them in random positions within the assembly area while ensuring no collisions occur. Calculate the Waste rate (R) (see Eq. 1), gap value (G) (see Eq. 2), and void area (V) (see Eq. 3):

$$R = Ru/Ra \tag{1}$$

$$G = Ct - Cb \tag{2}$$

$$V = S - St \tag{3}$$

Where R_u is the number of tiles that is not used, R_a is the number of tiles, C_t is the perimeter of the pixel boundary occupied by the tile arrangement, and C_b is the perimeter of the outermost pixels of the tile arrangement; S is the area of the assembly region, and S_t is the area covered by the tiles. R, G, and V are input into the NSGA-II algorithm, and their minimum value is the optimization target.

Since multi-objective optimization algorithms produce a set of solutions rather than a single solution (Cui et al., 2017). Therefore, this study introduces weighted ranking within the solution set to filter the optimal Pareto front solution (Hosseini et al., 2015) as the best tile arrangement. The weighting is applied as follows:

$minR : minG : minV = Wr : Wg : Wv(4)$

W_r, W_g, and W_v represent the SV weights for the waste rate, gap value, and void area.

3.3 Automated Assembly

The task sequence of the robot tile assembly and cement printing is planned to avoid collision between the two robot arms. One robot extrudes cement after the other robot assembled a portion of tiles. This collaborative approach can increase assembly speed and prevent concrete from drying during tile assembly. The robot task and communication code are converted to URScript, and the URScript is input into the robot control system to perform the assembly of the FMT wall.

3.4 Finite Element Analysis

A finite element analysis was conducted using Abaqus to evaluate the structural performance of FMT Wall under normal service conditions. The fragmented tiles were ideal elastic materials with a Young's modulus of 200 GPa and a compressive strength of 30 MPa. The wall system was constructed with appropriate geometry to represent the real-world application of the tiles accurately. A bonded interaction was established between the mortar and tiles, neglecting any sliding effects. Boundary conditions were applied to simulate typical support and loading scenarios, including fixed supports at the base and vertical loads. A finite element mesh was generated with sufficient refinement to capture stress distribution accurately, and mesh quality was verified for convergence.

4 Result

4.1 Dataset Collection

Convert the tile image into 1,508 × 1,127 pixels. Use B&W value comparison to filter out points that match the tiles' color characteristics. Employ the Convex Hull algorithm to extract the boundary of tile pixel points. Scale the boundary back to their actual dimensions based on the reference points at the image (Fig. 2).

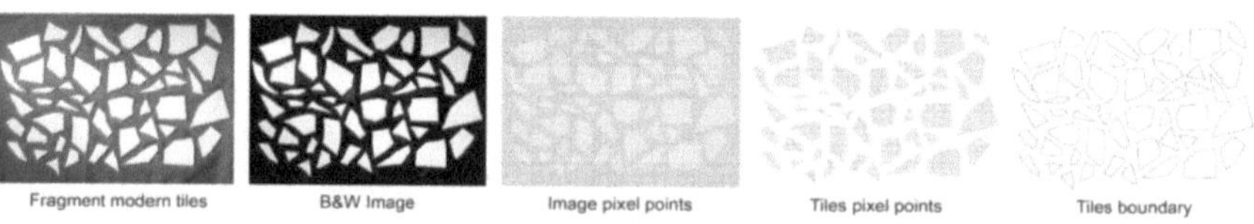

Fig. 2. The process from image to extracting tiles boundary polylines

4.2 Arrangement and Combination

A 200 mm x 250 mm area on the FMT wall was selected as the assembly area. While avoiding collisions, the tiles were moved to random positions within the assembly area to generate a tile arrangement plan. The geometric boundaries of the tiles were converted into 4 mm*4 mm pixel units, where the area not covered by the tile pixels was the gap value (G), and the area ofthe island pixels not covered inside the tile pixels was the void ratio (V). The waste rate (R), G, and V of the plan were calculated and set as the optimization target (Fig. 3).

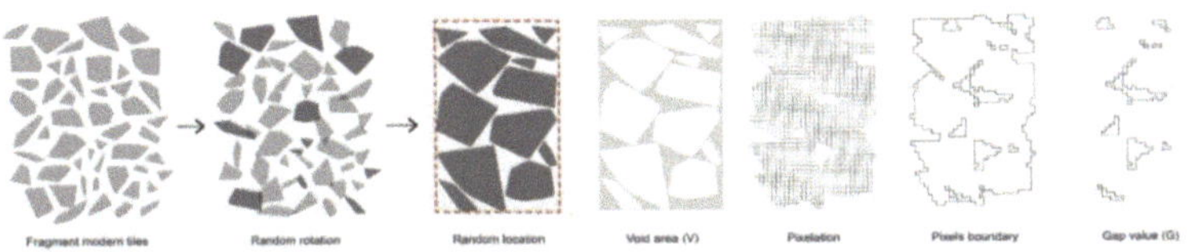

Fig. 3. The process from random arrangement to extracting optimized target values

Perform 200 iterations of calculations on the input data, selecting 70% of the previous generation as elite values for the next iteration. All three optimization objectives showed convergence during the calculations. Throughout 168 generations, the gap value, waste rate, and void area stabilized, ranging between 1792 mm to 2004 mm, 0% to 2%, and 94950mm^2 to 103460mm^2. To enhance the wall's stability, gaps within the tiles should be minimized. With weights of Wr, Wg, and Wv set at 0%, 1.85%, and 98.14%, the gaps in tiles are the lowest. In this configuration, the utilization rate reached 100%. Compared with the unoptimized solution, the utilization rate increased by 14% and the gap rate of each layer of tiles was reduced by about 7%. (Fig. 4).

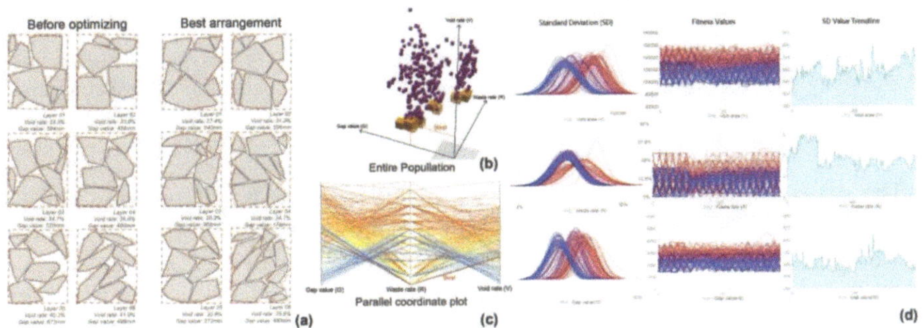

Fig. 4. (a) Best laying after optimization. (b) Distribution of entire population values in a three-dimensional chart. (c) Parallel coordination of three optimization objectives. (d) Distribution and trend of values of each optimization objective during the optimization process

4.3 Robot System Design and Automated Assembly

Two UR10e robotic arms assemble the FMT wall, with digital signals facilitating communication between the two arms and controlling the end-effectors (Fig. 5). Once the tile assembly reaches 80% completion, the other robot begins to print cement, thereby enhancing assembly efficiency while preventing moisture in the cement from evaporating due to prolonged exposure to air. After each layer of tile assembly is completed, the robot responsible for tile assembly returns to its initial position and waits for the cement printing to finish before commencing the assembly of a new layer of tiles.

Interspersed between the task statements for each assembly layer are codes for reading and output signals, facilitating communication between the robotic arms. The tasks for both robotic arms are converted into URScript and input into the robotic control system. The digital output and input pins of the control boxes for the two robotic arms are interconnected to enable information exchange between the arms. Additionally, the output pins of the control boxes are connected to the cement printing end-effector and the vacuum end-effector to control cement print and tiles picking. Both robotic arms are activated simultaneously for assembly.

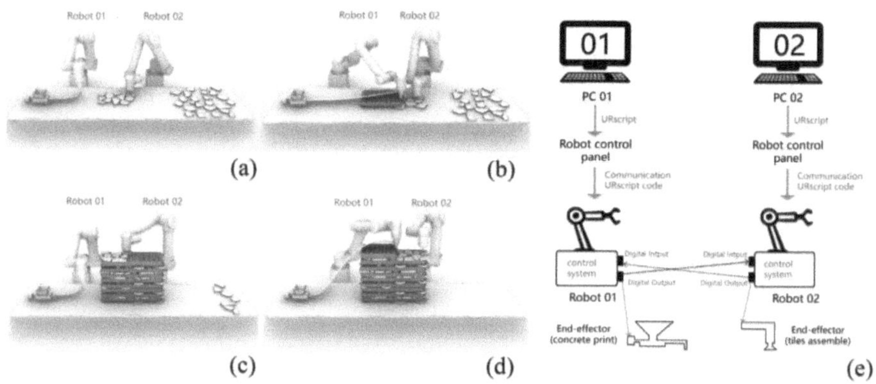

Fig. 5. (a) First layer's tiles assembly. (b) Pouring concrete on the first layer. (c) Eighth layer's tiles assembly. (d) Pouring concrete on the eighth layer. (e) Multi-robot collaborative construction system

The experimental robotic assembly test includes six layers of tile assembly and cement printing. The seamless execution of the programmed construction sequence was verified, with no observed collisions between tile units or violations of the angle and position constraints in the assembly instructions. Following these preliminary tests, the feasibility and accuracy of the integrated digital workflow have been successfully validated, allowing for the subsequent automated assembly testing of the complete FMT wall (Fig. 6).

4.4 Finite Element Analysis

The analysis modeled the fractured tiles as a perfect elastic material with a Young's modulus of 200 GPa and a compressive strength of 30 MPa. The interaction between

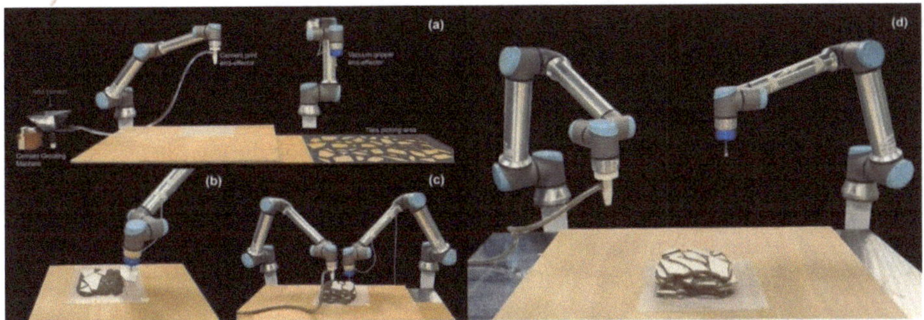

Fig. 6. (a) Assembly settings. (b) Tiles picking and placing in the second layer. (c) Robot collaboration. (d) Assembled six layers of Fragment Tiles Wall.

the mortar and the tiles was defined as a bond condition, neglecting any sliding effects. The wall system is 3 m high and 300 mm thick and is in regular service conditions. The maximum stress obtained from the analysis is 8.885 MPa (Fig. 7), below the maximum compressive strength and within the safe range. These results show that the design meets the necessary safety and performance criteria.

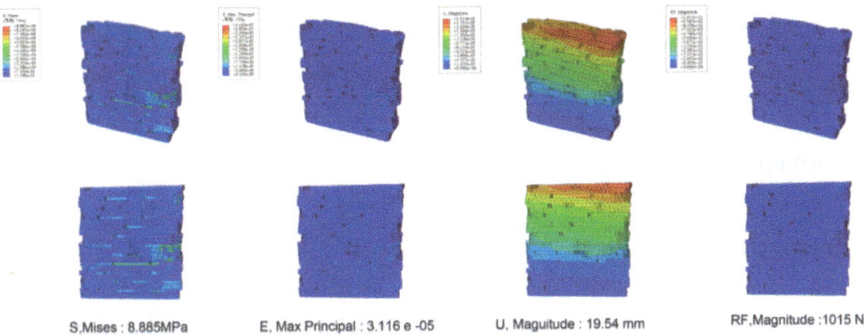

Fig. 7. Fragmented Tile Wall Finite Element Analysis

5 Discussion

There are many environmental benefits of recycling FMT. By directly utilizing broken tiles, we significantly lower carbon emissions associated with traditional tile disposal methods. The finite element analysis conducted in this study further emphasizes the structural viability of FMT walls, demonstrating their capacity to meet or exceed the performance metrics of conventional materials. Applying the NSGA-II multi-objective optimization algorithm significantly advances the arrangement and utilization of waste-broken tiles. The introduction of weighted rankings within the optimization process enhances the applicability of our methodology across diverse construction scenarios.

However, the optimization process may not achieve the desired precision in arrangement. Factors such as real-world imperfections can affect the optimization outcomes, necessitating further refinement of the algorithms used. The deployment of collaborative robotic systems to assemble FMT walls marks a pivotal shift toward automation in tile construction. Proactively managing cement application during tile assembly mitigates potential moisture loss and curing issues, improving overall structural integrity. The methodologies developed in this research have primarily been tested in controlled experimental settings. Translating these processes to larger-scale construction projects may present practical challenges that require further investigation.

6 Conclusion

This study introduces a comprehensive approach to the intelligent recycling and automated assembly of fragmented modern tiles (FMT). It addresses pressing challenges related to waste management and sustainability in the construction industry. By integrating optimization algorithms with robotic assembly techniques, we have demonstrated a significant improvement in using construction waste. Additionally, finite element analysis has validated the structural integrity and mechanical performance of FMT walls, reinforcing the viability of using recycled materials in modern building practices.

The innovative workflow outlined in this research minimizes environmental impact and enhances economic efficiency by reducing labor costs and energy consumption associated with traditional construction methods. Future research will focus on the potential for real-time adaptive assembly processes. By advancing the field of waste material utilization, this study establishes a foundation for a more sustainable and resource-efficient built environment.

References

Cui, Y., Geng, Z., Zhu, Q., Han, Y.: Multi-objective optimization methods and application in energy saving. Energy. **125**, 681–704 (2017)

Deb, K., Pratap, A., Agarwal, S., Meyarivan, T.A.M.T.: A fast and elitist multiobjective genetic algorithm: NSGA-II. IEEE Trans. Evol. Comput. **6**(2), 182–197 (2002)

Ergen, F., Bettemir, Ö.H.: BIM-driven software and algorithm for optimal floor tile layout minimizing material waste. Autom. Constr. **173**, 106115 (2025)

Fleming, R.: The ritual recycling of Roman building material in late 4th-and early 5th-century Britain. Post-Class. Archaeol. **6**, 147–170 (2016)

Hosseini, S.S., Hamidi, S.A., Mansuri, M., Ghoddosian, A.: Multi objective particle swarm optimization (MOPSO) for size and shape optimization of 2D truss structures. Period. Polytech. Civ. Eng. **59**(1), 9–14 (2015)

Lancioni, G., Lenci, S., Piattoni, Q., Quagliarini, E.: Dynamics and failure mechanisms of ancient masonry churches subjected to seismic actions by using the NSCD method: The case of the medieval church of S. Maria in Portuno. Eng. Struct. **56**, 1527–1546 (2013)

Liu, T., Zhou, H., Du, Y., Zhang, J., Zhao, J., Li, Y.: A brief review on robotic floor-tiling. In: IECON 2018-44th Annual Conference of the IEEE Industrial Electronics Society, pp. 5583–5588. IEEE (2018, October)

Lv, H., Shi, B., Li, N., Kang, R.: Intelligent manufacturing and carbon emissions reduction: evidence from the use of industrial robots in China. Int. J. Environ. Res. Public Health. **19**(23), 15538 (2022)
Quagliarini, E., Carosi, M., Lenci, S.: Novel sustainable masonry from ancient construction techniques by reusing waste modern tiles. Sustainability. **15**(6), 5385 (2023)
Quagliarini, E., et al.: Experimental analysis of Romanesque masonries made by tile and brick fragments found at the archaeological site of S. Maria in Portuno. Int. J. Archit. Herit. **8**(2), 161–184 (2014)
Ray, S., Haque, M., Sakib, M.N., Mita, A.F., Rahman, M.M., Tanmoy, B.B.: Use of ceramic wastes as aggregates in concrete production: A review. J. Build. Eng. **43**, 102567 (2021)
Samarakoon, S.B.P., Muthugala, M.V.J., Elara, M.R.: Tiling robotics: A new paradigm of shape-morphing reconfigurable robots. Adv. Intell. Syst. **7**, 2400417 (2024)
Wu, K., Zhang, Y., Kong, X., Zhang, S., Gao, L.: Quality control of robotic floor-tiling by the modifications on technology parameters and adhesive properties. J. Field Robot. **42**(1), 356–372 (2025)
Wu, S., Zhang, N., Luo, X., Lu, W.Z.: Multi-objective optimization in floor tile planning: Coupling BIM and parametric design. Autom. Constr. **140**, 104384 (2022)
Wu, S., et al.: Automated layout design approach of floor tiles: based on building information modeling (BIM) via parametric design (PD) platform. Buildings. **12**(2), 250 (2022)

Repurposing Raw Edge Wood Offcuts Waste for Facade Covering with Minimal Cutting

Zeyin Song(✉), Maria Luiza Gomes Torres, and Oswaldo Veliz

Institute for Advanced Architecture of Catalonia, Catalonia, Spain
song.zeyin@students.iaac.net

Abstract. Manufacturing standardized wood components typically generates 20–30% waste, predominantly as irregular raw edge offcuts. These offcuts are frequently discarded or incinerated, increasing carbon emissions. While the wood industry seeks recycling solutions, the highly variable dimensions and irregular bark edges of these offcuts challenge conventional manufacturing, often requiring secondary cutting that increases embodied carbon. To address this, we propose an integrated computational design and robotic fabrication methodology to upcycle raw edge offcuts into architectural facade coverings with minimal processing. First, a photogrammetry-based scanning method extracts physical dimensions to establish a 1:1 scale digital twin library. Subsequently, a bespoke packing algorithm evaluates library data to optimally position offcuts within a facade design, minimizing waste. Finally, coordinates are sent to a robotic cell that automates material picking, placement, and nailing onto a backing frame. Experimental testing demonstrates that 77% of the wood elements match the design without any post-processing after the initial calculation. For the remaining elements, a secondary algorithmic pass achieves a 100% success rate with minimal cutting. This approach offers a sustainable, automated workflow for timber waste mitigation in construction.

Keywords: Wood Waste · Circularity · Facade · Robotic Fabrication · Computational Design

1 Introduction

Wood processing generates substantial waste, with standardized methods causing 20% to 30% material loss, including cutoffs, sawdust, and other residues [1]. Among these, raw edge cutoffs—irregular pieces discarded due to non-standard dimensions—are particularly challenging to repurpose within traditional computer-aided design (CAD) and manufacturing workflows [2]. Traditional recycling often involves chipping or combustion, contributing to carbon emissions and underutilizing the material's potential, whereas direct closed-loop upcycling into high-value architectural components presents a critical opportunity for carbon mitigation [3]. The woodworking industry has long struggled with standardized sizing, which excludes irregular timber portions from mainstream manufacturing due to the strict reliance of industrial machinery on geometric

Y. Liu et al. (Eds.): CDRF 2025, *Transindividual Intelligence*, pp. 206–217, 2026.
https://doi.org/10.1007/978-981-92-0615-5_18

uniformity [4]. Research shows that digital cataloging and algorithmic material matching can enhance utilization while avoiding extensive subtractive processing, thereby significantly reducing the final product's embodied carbon [5].

Advancements in computational design and robotic fabrication offer a viable solution by integrating raw edge cutoffs into architectural applications, particularly facade construction, without forcing natural materials to conform to standardized forms [6]. Capturing the unique physical traits of irregular wood into digital twins allows generative algorithms to guide material placement, while robotic automation bridges the gap between complex digital variants and physical assembly without expanding labor complexity [7]. To address existing technology gaps in end-to-end timber upcycling, this study introduces a framework combining photographic scanning (OpenCV), computational packing algorithms (Grasshopper), and robotic assembly to efficiently incorporate raw edge cutoffs into prefabricated facade panels [8]. Aligning with circular economy and sustainable design principles, this research demonstrates how robotic fabrication enables high-precision assembly while minimizing waste, advancing sustainable construction, and offering a scalable solution for repurposing irregular wood waste.

2 Methods

This method comprises three main stages: (1) constructing a digital twin library of raw wood cutoffs via robotic scanning and point cloud extraction; (2) applying a computational packing algorithm to optimally match cutoffs to façade panel designs while minimizing material waste; and (3) automating robotic assembly to position and secure cutoffs onto the structural framework. Initially, raw materials were identified, labeled, and imaged to generate precise digital representations. Each cutoff was measured and classified by its dimensions and geometry. Open CV detected individual wood pieces, removed extraneous background, and extracted outer and inner edge data. These edge profiles were imported into Rhino, where Grasshopper enabled visualization, numbering, and dimension assignment in digital modeling. The computational packing algorithm analyzed façade design constraints, optimizing cutoff placement from the digital library to ensure minimal material waste. Through iterative refinement, it significantly improved material utilization, reducing unnecessary cutting. Finally, optimized digital positioning data was transmitted to a robotic system, automating the picking, placing, and securing of cutoffs onto the framework. Experimental results showed that 77% of cutoffs were directly used without modification, and after a second iteration, 100% of material was incorporated with minimal cutting. This approach significantly enhanced efficiency and sustainability in façade panel production, highlighting the potential of computational design and robotic fabrication for construction (Fig. 1).

3 Result

The Results section outlines the development and application of a Digital Library for wooden planks, aimed at optimizing material use through computational tools. Starting from woodworking waste, the study records plank dimensions using OpenCV edge detection and Grasshopper-based analysis. A packing algorithm matches planks to design

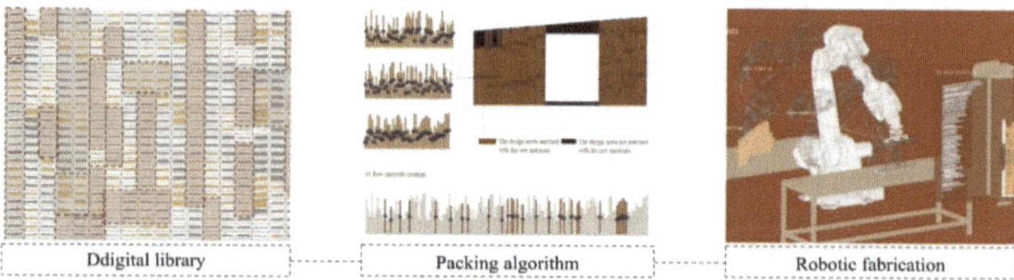

Fig. 1. Workflow

needs, reducing waste in facade cladding and modular construction. The research also explores algorithmic facade design with vertical and 3D panel arrangements, optimized through computation. Robotic fabrication is then demonstrated using automated handling and CNC machining, streamlining the assembly process. This workflow promotes sustainable reuse and highlights the role of computational design in architectural and robotic applications.

3.1 Creating Digital Library

This study was conducted in Barcelona, Spain, in collaboration with a mountain woodworking studio to source discarded wooden planks. These planks, rejected due to inconsistent color or minimal width, retained their natural bark and were uniformly sliced to 2 cm thickness—ideal for constructing sandwich-like panels. The planks were trimmed to manageable sizes and centrally cut to obtain one raw edge, reducing facade load by 50% while preserving the visible area. To ensure secure robotic handling, only planks wider than 100 mm were retained, matching the suction cup diameter of the robotic arm.

The processed planks were cataloged into a Digital Library that integrates computational tools for material reuse. Using OpenCV-based edge detection and Grasshopper algorithms, each plank's geometry was recorded and indexed. A custom packing algorithm matched planks to specific design needs, improving material utilization in facade cladding, interior paneling, and modular construction. As the database grows, refined selection algorithms will enable automated recommendations based on project constraints, enhancing both efficiency and sustainability (Fig. 2).

Fig. 2. Raw material collection

3.1.1 Edge Detection using Open CV

Each wooden plank image was processed using Open CV to extract edge information. First, the input image was converted to grayscale, followed by adaptive shareholding to generate a clear binary image. A contour detection algorithm was then applied to extract external edges, which were visualized both on the original image and against a black background. Finally, the processed contours were converted into an SVG file format, serving as input for Rhino to extract physical dimensions (Fig. 3).

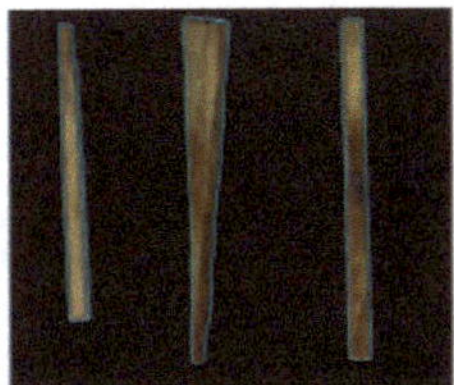
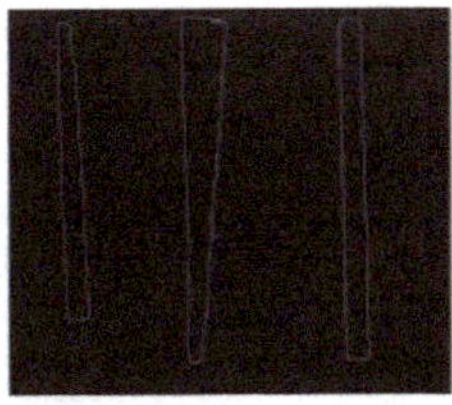

Fig. 3. Generated using Open CV Results

3.1.2 Physical Data Annotation

The generated SVG files were imported into Rhino and processed in Grasshopper using the Squid plugin, which converted line data into curves. Curve lengths, widths, and heights were extracted using standard components like Curve Length and Bounding Box. Alternatively, Python scripts were used to directly analyze SVG data. The resulting curves were converted into surfaces and extruded to match the plank thickness, creating accurate 3D models. Each piece was numbered for tracking and integration into the facade system.

All processed planks were cataloged into a Digital Library, where a Packing Algorithm matched them to design requirements. This system improves material efficiency and supports flexible applications in facade cladding, interior panels, and modular construction. In addition to reducing waste, it offers a structured method for selecting wood based on size and appearance. As the database grows, refined algorithms will enable automated, constraint-based recommendations (Fig. 4).

3.2 Packing Algorithm

The Packing Algorithm plays a crucial role in optimizing material utilization by systematically arranging wood cutoffs to fit predefined design templates with minimal cutting. The algorithm first sorts the available wood pieces based on length, width, and edge quality. It then employs an iterative computation process to match these pieces with design layouts, ensuring that the maximum number of cutoffs are used without modification. By leveraging computational geometry techniques, the algorithm minimizes gaps and material waste, allowing for efficient palatalization. Future improvements include incorporating machine learning for enhanced prediction of material fit and further reducing cutting requirements (Fig. 5)

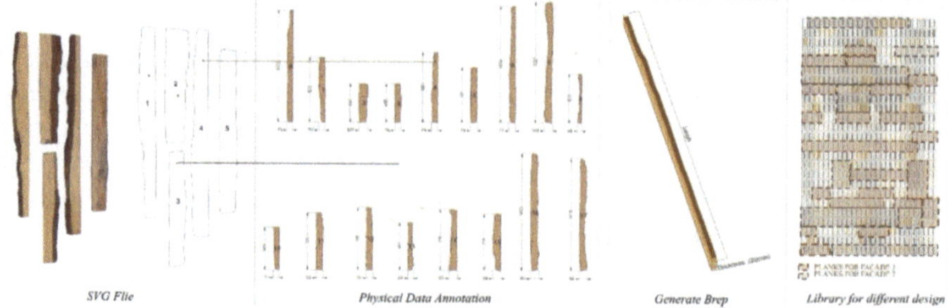

Fig. 4. SVG files converted to virtual models

Fig. 5. Design application results

The compact cabin design features strategically placed doors and windows to meet essential living needs. Vertical wooden elements enhance structural stability and serve as a framework for installing façade panels, aligning with both aesthetic and sustainability goals. Tiny homes have gained popularity in the hospitality industry for their affordability, environmental benefits, and minimalist appeal. In 2022, the global tiny home market was valued at $6.49 billion, projected to reach $7.41 billion by 2028 with a CAGR of 8.07%. Many companies focus on custom designs that integrate natural materials, blurring the boundary between indoor and outdoor spaces for an immersive experience. The façade pattern is structured based on the thickness of wooden panels, guiding baseline alignment. It incorporates vertical distribution and Z-axis rotation, applying distinct computational strategies in both 2D and 3D domains. While the vertical arrangement ensures enclosure principles, the 3D configuration enhances algorithmic complexity and introduces additional control over wood veneer coloration (Fig. 6).

3.2.1 First Operation of Vertical Pattern

This study involved two procedural steps. First, all raw materials were matched with the design patterns. During this process, no materials were cut. The results indicate that some raw materials were successfully integrated into the design, while the remaining materials remained in the catalog. Construction lines are used as the central axis to generate basic geometric shapes. These geometric forms serve as the foundation for visual representation and the subsequent matching of raw materials. This approach ensures a systematic

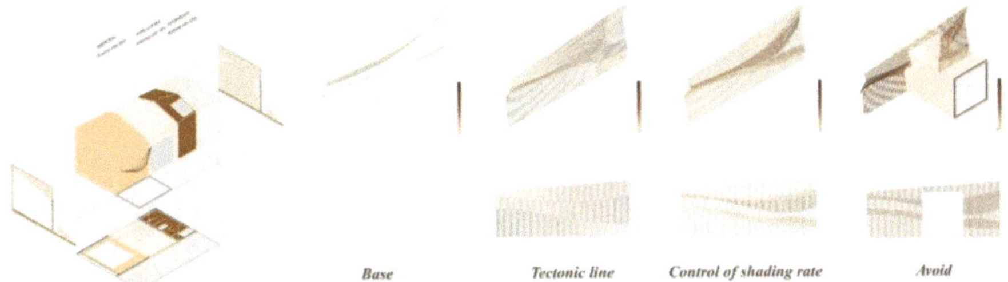

Fig. 6. Designed to set up the cabin

and efficient method for color categorization and material allocation in the design process. This study involved two main procedural steps. First, 100% of the raw materials were digitally matched against predefined design patterns without any physical cutting. The matching process successfully integrated approximately 77% of the raw materials into the design, while the remaining 23% were retained in the material catalog for future use. Construction lines served as central axes to generate fundamental geometric shapes, which formed the basis for both visual representation and precise material allocation. This systematic approach enabled efficient color categorization and optimized material distribution throughout the design process, ultimately promoting resource efficiency and minimizing waste (Fig. 7)

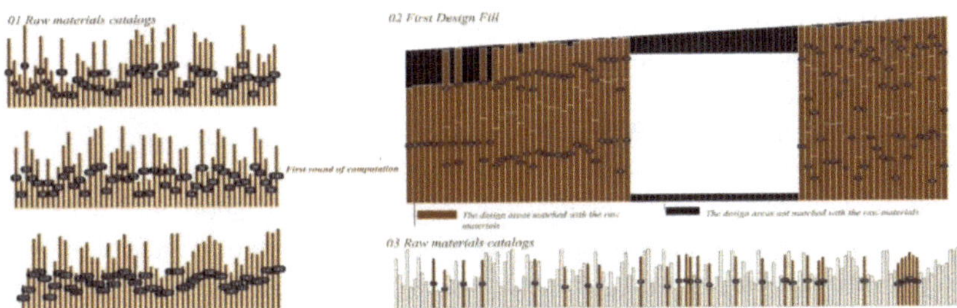

Fig. 7. First packing algorithm operation

3.2.2 Second Operation of Vertical Pattern

The second computation will reassess the areas that were not initially matched with raw materials and reallocate them using the remaining materials. A precise identification of cutting lines and their corresponding indices will be conducted. Subsequently, this portion of the data will be reorganized and renumbered before being retrospectively matched with the remaining raw material inventory. The primary distinction between these two steps lies in their approach: the first step involves matching raw materials to the design, whereas the second step focuses on aligning the unassigned design areas with the remaining materials. Ultimately, this process ensures optimal material utilization, minimizing material cutting while achieving the intended design, with waste generation being nearly negligible (Fig. 8).

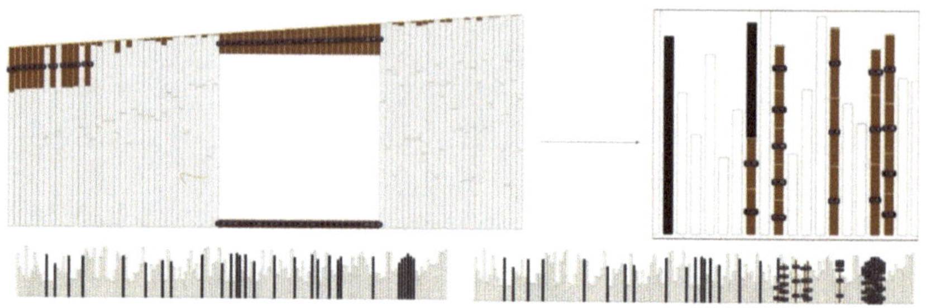

Fig. 8. Second packing algorithm operation

3.2.3 Façade generation with Z-axis rotation

Several spatial connections were identified within the building due to its three distinct functional zones. Consequently, it is necessary to define varying shading requirements for different areas. Additionally, an outdoor seating area is incorporated adjacent to the cabin, necessitating a facade design that ensures adequate spatial allocation. Integrating shading elements into the facade would further enhance environmental comfort and functionality. The color identification of wooden sticks in the digital library involves classifying them into three distinct colors, labeled as 0, 1, and 2. Intervals are created based on the distance to a reference plane. When the endpoints of each wooden stick fall within a specific interval, there is a 70% probability of matching it to the corresponding color. Construction lines are used as the central axis to generate basic geometric shapes. These geometric forms serve as the foundation for visual representation and the subsequent matching of raw materials. This approach ensures a systematic and efficient method for color categorization and material allocation in the design process (Fig. 9).

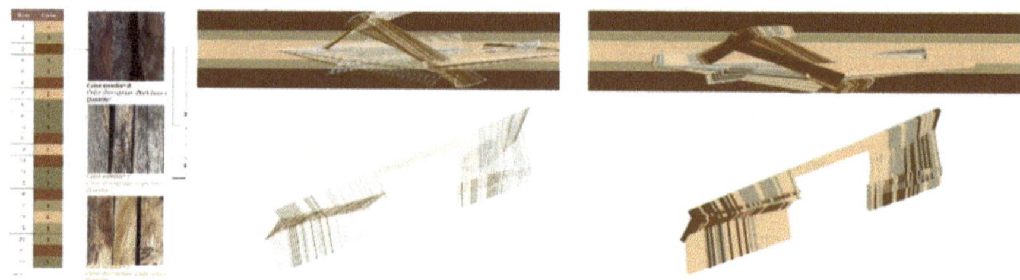

Fig. 9. Color level control and distribution logic

3.2.4 Packing algorithm three-dimensional facade

From a typological standpoint, the wood veneer on the roof is inherently planar, whereas the wood veneer on the façades adopts a three-dimensional configuration. As evidenced by the connection nodes, the material length required for the three-dimensional pattern must exceed the dimensions of the basic geometric shapes defined in the design. In contrast, such a length constraint is not necessary for the planar pattern. Consequently, each wooden stick will be prioritized for matching within the three-dimensional pattern, ensuring that material lengths align with the design's spatial and structural needs. The remaining cut portions of the material, once the three-dimensional pattern is fully allocated, can be efficiently matched with the planar pattern using a packing algorithm. This algorithm ensures optimal material distribution, enabling the remaining pieces to seamlessly fit within the planar design. This method not only minimizes material waste but also maximizes the overall efficiency of the material usage across different design patterns. This approach highlights the importance of context-specific material matching, where different patterns—three-dimensional and planar—require distinct handling strategies. By applying a strategic packing algorithm, it becomes possible to achieve both functional and aesthetic objectives in the design while adhering to sustainable practices that minimize resource wastage. This strategy exemplifies how computational tools can be leveraged to address complex design challenges, facilitating both material optimization and the realization of intricate design patterns (Fig. 10).

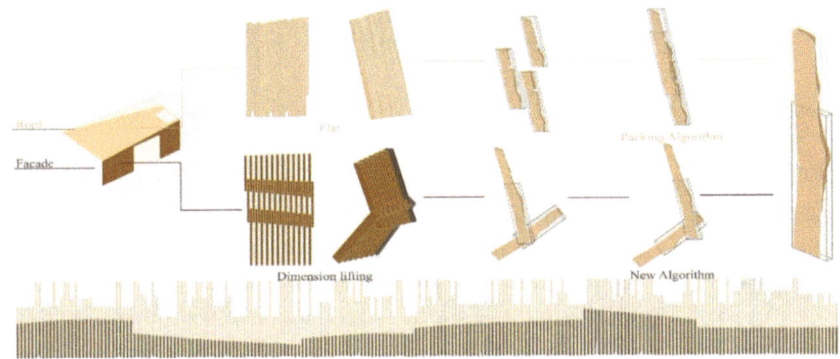

Fig. 10. Packing algorithm for three-dimensional Facade

3.3 Robotic Fabrication

This study explores an automated robotic fabrication process that optimizes wood assembly through design and advanced manufacturing. The workflow ensures precise handling, safety, and accurate placement, enhancing efficiency, reducing waste, and supporting sustainable construction with irregular components.

This end effector is primarily used for the automated handling task of moving wooden boards from the feeder to the assembly position, suitable for wood handling requirements in industrial automation production lines. The actuator features a three-suction cup structure, supported by a metal frame to ensure stability and an even distribution of suction force. The suction cups include two small cups with a diameter of 2.5 cm and one large cup with a diameter of 4.5 cm. The reasonable arrangement of the three cups ensures adaptability to wooden boards of different sizes. The spacing between the suction cups is 32.13 cm, enabling uniform force distribution when grabbing long rectangular wooden boards, thus improving handling efficiency. The end effector utilizes a vacuum system to provide suction force, allow instable gripping without rigid clamping, there by reducing damage to the wooden board surface. Structurally, the device is connected to a six-degree-of-freedom robotic arm and can adjust the effector's angle through rotational joints to meet different pick-and-place requirements. This type of suction-cup end effector effectively enhances the automation level of wood handling, reduces manual intervention, increases production efficiency, and ensures the positional accuracy and stability of the wooden boards during handling (Fig. 11).

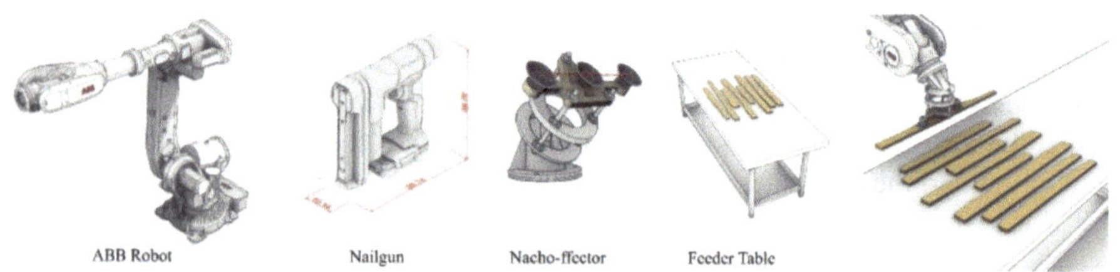

Fig. 11. End-effector design and application simulation

3.3.1 Prototype Construction

Based on the workflow, a representative portion of the complete design was selected for prototype construction by matching the digital model with the available raw materials. To facilitate display, a supporting structure was added at the base, consisting of a wooden board equal in length to the prototype's width and a horizontal wooden rod with cross-sectional dimensions of 100 mm × 60 mm. No additional supports were employed, as the design aimed to achieve balanced weight distribution across the facade. The wooden board attached to the bottom support panel was CNC-machined to create joints, a joint design applicable to large-scale constructions. Although the machining process results in minor material loss—less than 1% of the total raw material area—more precise joint designs could further minimize this loss. Theoretically, alternative construction methods, such as externally mounted curtain wall systems, could eliminate this waste entirely (Fig. 12)

3.3.2 Robotic Path and Prototype

The robotic workflow consists of three main actions. First, the robot grasps the wooden board from the feeder, where boards are numbered and transported via a conveyor to a calibration point. The end effector's suction cup aligns with the board's center fold,

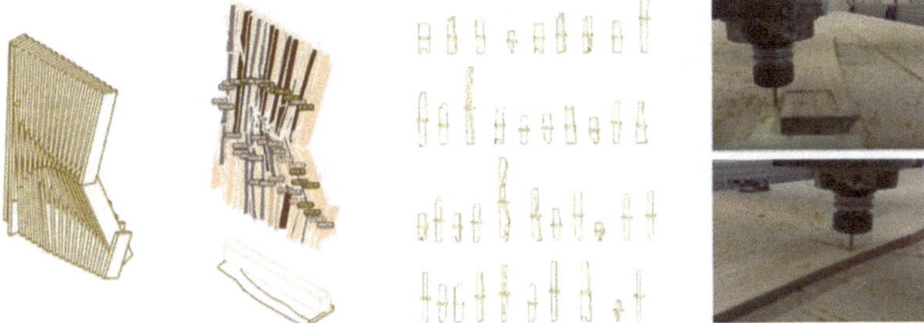

Fig. 12. CNC Node Machining

ensuring the TCP x direction matches the board's long edge. Second, safety points guide the robot's motion, avoiding collisions and singularities. Finally, the board is placed at the assembly position, aligned with the initial reference. Each board, after being grasped, follows a path with a mandatory intermediate point 500 mm vertically upward from both the grasping and placement locations. This vertical path is LIN, ensuring smooth movement. The remaining sections are PTP, involving direct movements. The LIN path ensures accurate alignment, while PTP optimizes speed and efficiency during horizontal movements, creating a well-coordinated robotic system (Fig. 13).

Fig. 13. Robot construction process

4 Discussion

Computational design and robotic fabrication present a viable alternative to traditional wood processing, offering significant advantages in adaptability, cost efficiency, and sustainability. The research project demonstrates how an algorithmic approach can optimize panel sizes to fit architectural facades, increasing design flexibility beyond conventional prefabricated methods. Additionally, the use of raw sawn-edge sticks and robotic assembly reduces costs, embodied carbon, and manufacturing waste while leveraging locally sourced materials. Despite the higher initial setup costs for robotic systems, the long-term benefits in sustainability and cost savings outweigh these concerns. However, challenges remain, particularly in material variability and the need for adaptive fabrication strategies. Although the project has achieved zero waste in material usage through optimized packing logic, the approach still relies on relatively uniform plank dimensions and tight tolerances in thickness. Expanding the system to process a broader range of wood waste, including irregularly shaped offcuts or varying thicknesses, could further enhance sustainability. Advances in scanning and sorting technologies will be critical in improving the efficiency and feasibility of such adaptations. Moreover, while the proposed method has proven effective for small-scale applications, scaling it to larger architectural structures, such as multi-story buildings, introduces new challenges. Structural analysis and the development of robust connection details are essential to ensuring load-bearing capacity and long-term durability. Computational methods could play a key role in optimizing facade anchoring systems and integrating prefabricated structural components, facilitating the adoption of this technology in high-rise construction. Addressing these considerations will be crucial in unlocking the full potential of raw edge wood cutoffs in sustainable architecture, paving the way for a more adaptive, cost-effective, and environmentally friendly construction approach

5 Conclusion

This research explored robotic fabrication and material processing, focusing on utilizing raw wood waste in design and manufacturing. We demonstrated how integrating advanced technologies improve precision and efficiency while minimizing waste. Key findings were made in four areas: data extraction, sorting, packing algorithms, and fabrication. We implemented robotic scanning with photogrammetry to extract point cloud data, from which we derived dimensional parameters. We simplified data extraction using only two curves. The scanning method could be further refined through automation, improving accuracy with future integration of ROS and other hardware. Sorting is essential in managing raw material waste. We identified the need for automated systems capable of sorting materials based on design parameters. Currently, sorting is largely human-driven or parameter-based, but computer vision could enhance decision-making and allocation. Sorting helps customize designs and ensures optimal material allocation for robotic manufacturing. We developed a packing algorithm to optimize material usage. By matching the design catalog with the raw material library, we minimized cutting and successfully matched 70% of materials without cuts. Further iterations reduced remaining waste with minimal cutting. The algorithm reduced waste, improved performance, and contributed to sustainable manufacturing by minimizing embodied carbon. Robotic systems enhanced precision and reduced production time in the AEC industry. However, working with irregular materials posed challenges, requiring human intervention to manage deviations and tolerances. Issues like misalignment and unexpected material behavior highlighted the need for oversight when working with unpredictable materials. In conclusion, the study shows the potential of robotic systems in enhancing design and fabrication but also emphasizes the challenges of irregular materials. Future research should refine automated sorting, packing algorithms, and scanning techniques to improve material utilization, time, and cost efficiency, while considering human expertise in managing material unpredictability.

References

1. Bertoni, M., Ros, J.: Robotic fabrication for sustainable architecture. J Robot Fabr. **15**(2), 23–41 (2020)
2. Bologna, M., Montagna, G.: Optimizing the use of offcuts in manufacturing processes. J Sustain Manuf. **18**(6), 1751–1764 (2023)
3. Clarke, L.: The principles of circular economy in construction. Environ. Sci. Technol. **29**(5), 123–137 (2015)
4. Eriksson, K., Arvidsson, M.: Automation and robotics in the wood industry. Wood Sci. Technol. **51**(4), 785–803 (2022)
5. Gamarra, W., Pirotti, A.: Parametric design in architecture: Use of waste material. J Archit Des. **22**(5), 118–132 (2021)
6. Raviv, L., Reuven, O.: Digital fabrication in sustainable construction. Build. Environ. **163**, 128–145 (2021)
7. Salvatori, P., David, G.: Circular economy and architecture: Exploring the relationship between construction and waste. Environ Des J. **32**(3), 44–59 (2022)
8. Vesartas, P.: Design-to-Fabrication Workflow for Raw-Sawn-Timber using Joinery Solver. Doctoral thesis, École polytechnique fédérale de Lausanne (EPFL), Lausanne, Switzerland (2021). https://doi.org/10.5075/epfl-thesis-8928

From Global to Detail: The Research on Enhancing Construction Accuracy with Cross-scale Multi-order Norm Localization Technology

Quan Zhou[1](✉), Shitong Wang[2], and Jingxuan Li[2]

[1] Xi'an University of Technology, Xi'an, China
334652434@qq.com

[2] College of Architecture and Urban Planning, Tongji University, Shanghai, China
2330113@tongji.edu.cn

Abstract. In the context of high-quality development in the construction industry, bridging the gap between design and construction through industrial and information technologies has become a key challenge. This study proposes a cross-scale, multi-order norm-based positioning method to integrate heterogeneous sensing data from various devices and scales. By fusing multi-source data, the method enables precise control from first-order (global) to third-order (detailed) positioning, significantly enhancing construction accuracy and efficiency. While multi-scale technologies offer rich information, data compatibility issues hinder integration. The proposed method addresses this bottleneck, improving information flow and data utilization. Applications in complex engineering projects demonstrate their feasibility and practical value, contributing to the digital transformation and high-quality advancement of the construction sector.

Keywords: Multi-source heterogeneous data fusion, · Cross-scale positioning technology · Norm-based error optimization · Construction accuracy control

1 Introduction

The multi-order positioning in architecture can be understood as a positioning strategy that organizes and allocates spatial relationships of different building components through spatial hierarchical levels. Specifically, this study employs norm theory as the core method and integrates data from different devices and scales using cross-scale multi-order norm-based positioning technology. Through multi-source data fusion, construction accuracy is enhanced, and precise control is achieved from first-order (global positioning) to third-order (detailed positioning). This strategy is divided into three parts. On a macro-scale, UAVs and satellite technologies provide efficient support for large-area environmental mapping (Steenbeek and Nex 2022). The integrated application of multi-order positioning technologies not only resolves data fragmentation in conventional construction but also establishes a robust technical foundation for high-quality industry development.

Y. Liu et al. (Eds.): CDRF 2025, *Transindividual Intelligence*, pp. 218–227, 2026.
https://doi.org/10.1007/978-981-92-0615-5_19

In large-scale architectural surveying, UAVs and satellite remote sensing enhance mapping accuracy and efficiency. UAVs excel in flexibility, efficiency, and cost-effective data collection. Key advances include: Li (2024) developed a UAV multi-sensor system combining cameras and LiDAR for terrain capture to 3D modeling; Maciąg et al. (2024) created an automated UAV method achieving centimeter precision with faster data capture, including GIS tools for cadastral mapping; Calisi et al. (2023) integrated LiDAR with UAVs for millimeter accuracy through point cloud fusion, improving building documentation; and Wang et al. (2022) used UAV-camera 3D reconstruction for millimeter-level construction monitoring via wireless techniques.

For building-scale sensing and positioning, instruments like total stations and laser scanners precisely capture structural geometry, enabling data fusion and supporting construction optimization. Kang et al. (2012) created a robotic total station (RTS) integrated with BIM for millimeter-level real-time measurement. Their method reduces surveying time through automatic target recognition and reflectorless measurements, proving effective for monitoring concrete and steel components. Zhou et al. (2020) enhanced this with a multi-station RTS system, achieving sub-millimeter accuracy in complex sites through dynamic data fusion. Liu et al. (2021) further combined BIM with laser scanning for structural health monitoring, detecting minute deviations and providing retrofit data across the building lifecycle.

In building component installation, indoor GPS (iGPS) enables architectural industrialization with wide coverage, real-time monitoring, and millimeter-level accuracy. Chen and Du (2017) showed iGPS creates measurement fields through distributed base stations, with increasing precision when adding more stations. Xiong et al. (2011) developed the 4D-PosCon system for millimeter-level installation monitoring through automated measurement and 4D modeling. Combined with laser scanners and rangefinders, iGPS forms a "global-local" system (Zeng et al. 2020), where iGPS handles coarse positioning while local instruments achieve submillimeter interface measurements. When integrated with robot-mounted rangefinders (Zeng et al. 2020), in the context of high-quality construction development, multi-scale perception and positioning technologies have enhanced building precision but still encounter major challenges in integrating multi-source heterogeneous data. Data from satellite remote sensing, UAVs, and other sources differ in formats (e.g., RPC vs. FBX), spatial references (geodetic vs. local construction coordinate systems), and accuracy levels (Chen and Du 2017). Equipment limitations also impact effectiveness—iGPS experiences data interruptions from dynamic occlusion, while laser scanners have limited range coverage in large-scale environments (Zeng et al. 2020). Cross-scale data fusion is notably challenged by spatiotemporal reference misalignment.

In summary, future research needs to focus on breakthroughs in spatiotemporal alignment algorithms for multimodal data and robust dynamic fusion frameworks to enable efficient information flow across the building lifecycle.

This study develops a cross-scale multi-order norm localization (CMNL) framework to integrate multi-source sensing data (satellite/UAV/iGPS/laser scanning) for improved building component installation. The framework resolves data heterogeneity (formats, coordinate systems, accuracy levels), enhances dynamic adaptability (occlusion/noise mitigation), and establishes hierarchical control (global/local/detail positioning) through norm-based algorithms. It enables end-to-end accuracy from geodetic (±5 cm) to component-level (±0.5 mm) while addressing software fragmentation in construction workflows (Fig. 1).

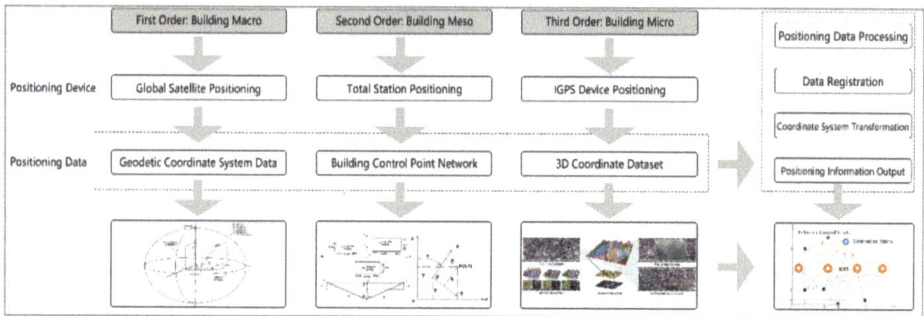

Fig. 1 Optimized flowchart

2 Methodologies

The multi-order positioning in architecture can be understood as a positioning strategy that organizes and allocates spatial relationships of different building components through spatial hierarchical levels. Specifically, this study employs norm theory as the core method and integrates data from different devices and scales using cross-scale multi-order norm-based positioning technology. Through multi-source data fusion, construction accuracy is enhanced, and precise control is achieved from first-order (global positioning) to third-order (detailed positioning). This strategy is divided into three parts:

1. Acquisition of Multi-Source Data through Various Devices: Spatial data is collected across multiple scales—from global to detailed—using devices such as GPS, LiDAR, and drones. By leveraging the strengths of different equipment, precise positioning information is provided to meet accuracy requirements at each construction stage.
2. Integration and Processing of Data Using Norm-Based Methods: Norm theory is applied to standardize and perform weighted fusion on data from various sources, ensuring consistency among datasets with different precision levels. This optimizes the integration outcome and enhances positioning accuracy.
3. Practical Application and Validation of Coordinate Data: The integrated data is applied on-site for positioning, with its accuracy validated by comparing it to the actual building structure. This ensures effective implementation, allows for error correction, and improves construction precision (Fig. 2).

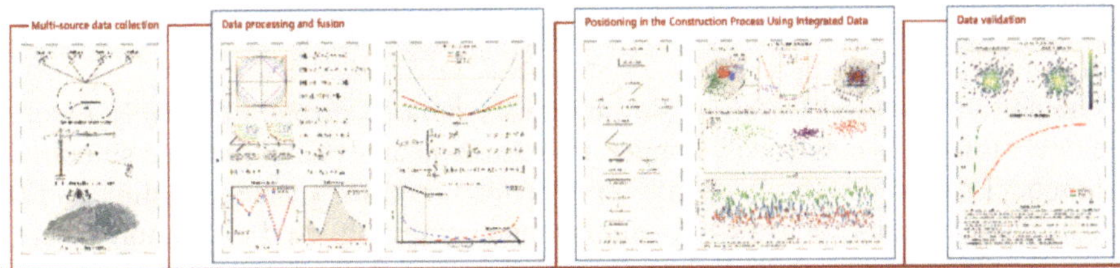

Fig. 2 Methodology flow diagram

2.1 Acquisition and Processing of Multi-Source Data

In the process of acquiring and processing multi-source data, it is essential to collect spatial information from various dimensions using a range of devices and technologies. These sources include satellite positioning systems, total stations, and intelligent positioning systems such as IGPS. Each device plays a distinct role in positioning and measurement across different spatial scales.

During the planning stage, Global Navigation Satellite Systems (GNSS) are employed for large-scale regional positioning. By establishing precise coordinate points based on the geodetic coordinate system, GNSS provides a foundational spatial reference framework for the entire construction project.

In the construction phase, more precise local positioning methods are required. Total stations are widely used to build localized coordinate systems by setting up traverse control networks, serving as secondary reference systems to guide specific tasks such as construction layout and measurement operations.

During component installation, IGPS systems are employed to assist positioning by providing real-time feedback on the spatial location of components, enabling construction personnel to perform assembly tasks quickly and accurately.

Due to the differences between the geodetic coordinate system, the building's local coordinate system, and the coordinate system used by IGPS, data from these sources must undergo coordinate transformation and alignment to achieve unified management and collaborative application (Table 1).

Table 1 Multi-source data comparative analysis

Device Name	Operational Range	Exported File Format	Software	Remarks
SRC	Very large	RPC	ERDAS	
Total Station	Mediun	CSV	Leica office	
IGPS	Small	.csd, .cal	Post Process	

2.2 Methods for Reducing Norm-Based Errors

Norms are mathematical functions used to measure the "length" or "magnitude" of vectors and are widely applied in linear algebra, optimization theory, and machine learning.

In error control and model optimization, norms serve as the foundational form of loss functions, playing a crucial role.

The commonly used L2 norm (1) minimizes squared errors, effectively reducing large deviations. It is extensively applied in regression analysis to produce smooth and stable model solutions. The L1 norm (2) is better suited for sparse modeling and feature selection, offering robustness to outliers, and commonly used for compression and dimensionality reduction in high-dimensional data.

$$\text{Loss} = \frac{1}{N}\sum_{i=1}^{N}\left(y_i - \hat{y}_i\right)^2 \tag{1}$$

$$\|\mathbf{x} + \mathbf{y}\|_1 \le \|\mathbf{x}\|_1 + \|\mathbf{y}\|_1 \tag{2}$$

The L∞ norm, which focuses on the maximum error, helps control worst-case deviations and is often used in robust optimization and safety-critical system design. The Huber loss (3), a compromise between L1 and L2 norms, balances robustness and continuity, making it well-suited for regression problems with noise in non-ideal environments.

$$L_\delta\left(y, \hat{y}\right) = \begin{cases} \frac{1}{2}\left(y - \hat{y}\right)^2, & \text{if } | y - \hat{y} | \le \delta \\ \delta \cdot \left(\left|y - \hat{y}\right| - \frac{1}{2}\delta\right), & \text{if } | y - \hat{y} | > \delta \end{cases} \tag{3}$$

For classification tasks, especially in probabilistic models like logistic regression, the log loss (4) measures the difference between predicted probabilities and true labels, guiding the model to learn more accurate probability outputs. By selecting appropriate norm functions based on specific task requirements, one can effectively control errors and continuously optimize model performance.

$$\text{Log Loss} = -\frac{1}{N}\sum_{i=1}^{N}\left[y_i \log(p_i) + (1 - y_i)\log(1 - p_i)\right] \tag{4}$$

2.3 Validation of Optimized Data

Following the optimization based on norm error reduction, it is necessary to further verify the validity and accuracy of the data to ensure that the results are reliable and practical for real-world applications. One common validation method is to compare the optimized data with actual measurements and theoretical design values, thereby assessing whether the errors remain within an acceptable range.

In this study, the closure error method (5) is employed as the primary means of verifying positioning coordinate accuracy. This traditional yet effective error control technique is widely used in the measurement of closed geometric figures or control networks to determine whether the coordinate deviations between measurement points meet specified accuracy standards.

$$\Delta X = X_{\text{end}} - X_{\text{start}} \tag{5}$$

Specifically, the method calculates the difference between the actual measured path and the theoretical closed path—the closure error—to evaluate the geometric consistency of the data and the degree of accumulated error. When the closure error is small and within the defined threshold, it indicates good internal consistency and external accuracy of the optimized data, signifying a successful optimization process. If the closure error exceeds the tolerance range, it is necessary to re-evaluate the optimization strategy or identify potential sources of measurement error.

3 Case Study

The cross-scale, multi-order norm-based positioning method enables highly precise localization across scales—from first-order (global positioning) to third-order (fine-grained detail)—by integrating multi-source data. This advanced technique was implemented in the installation of 3D-printed panels for the *Xiong'an Wings* hyperbolic curtain wall project. The main structure features a curtain wall that extends 91.2 m in length and spans a surface area of 4,609 square meters, making it the world's largest intelligent construction project utilizing 3D printing technology.

Characterized by a complex double-curved surface and an extensive number of components, the curtain wall posed significant challenges in terms of spatial precision and structural integrity. High-precision installation was essential to ensure both architectural quality and the faithful realization of the design intent. The 3D-printed panels are mainly distributed in two key areas: the curtain wall beneath the twin wings of the main building and the rooftop open-air theater (Fig. 3).

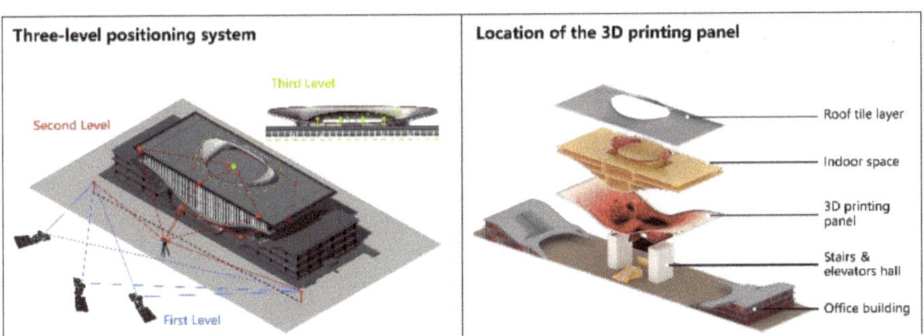

Fig. 3 Location of the 3D printing panel and three-level positioning system

The establishment of a cross-scale positioning network involves four key steps:

① Classify various sensing devices into three hierarchical levels—global, local, and detailed—for scene positioning.

② Conduct batch verification of data from different sources to validate the accuracy and consistency at each level.

③ Perform cross-scale multi-order norm-based data fitting.

④ Carry out error analysis.

Control Point Setup: Control points are marked using ground survey nails. Holes are drilled at designated ground locations to install the nails, which are then marked with red paint for easy identification.

In the global coordinate system, satellite positioning is used to locate the project site within the overall master plan. High-precision satellite positioning systems are employed to obtain accurate site coordinates, ensuring alignment with the planning and design of the Xiong'an New Area. The construction origin point and traverse direction within the site are also defined. The geodetic coordinate data for the site is provided by the general contractor, with point locations shown in the figure. The western point marks the construction origin of the project, based on the geodetic coordinate system. High-precision total stations are used to re-survey the control points. The measured baseline length is 139.700 m, which is then input into the data fitting algorithm to update the coordinates of the eastern point.

In the local building coordinate system, total stations are used to establish survey stations based on satellite-derived coordinate points. The first step involves converting the geodetic coordinate system into the construction grid coordinate system. Six control points are laid out around the main structure.

The overall measurement error is first evaluated using the closure error method. Once the total error is quantified, the L2 norm-based least squares optimization is applied to refine the dataset. This curve-fitting process produces optimized coordinate values and an updated closure error, thereby enhancing positional accuracy within the local coordinate system. As shown in Fig. 4, the optimized coordinates ensure that the error at each point remains within 2 mm.

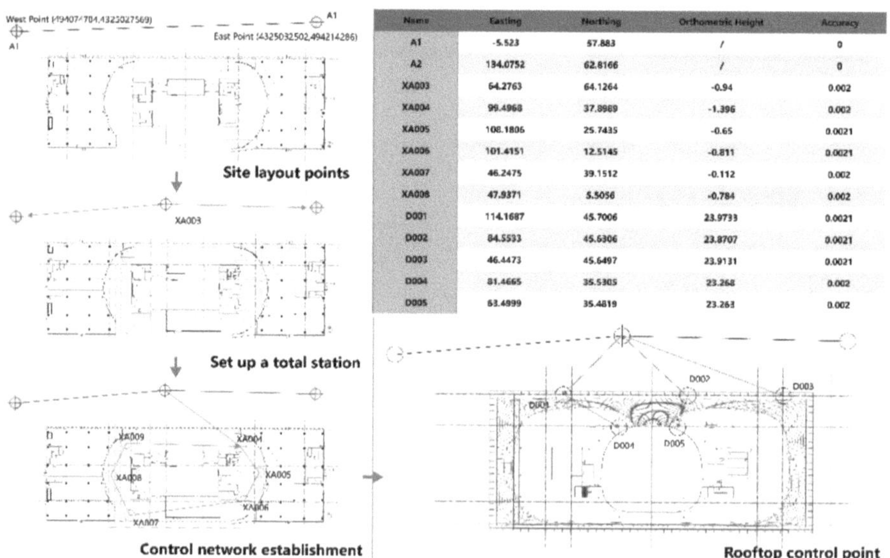

Name	Easting	Northing	Orthometric Height	Accuracy
A1	-5.523	57.883	/	0
A2	134.0752	62.8166	/	0
XA003	64.2763	64.1264	-0.94	0.002
XA004	99.4968	37.8989	-1.396	0.002
XA005	108.1806	25.7435	-0.65	0.0021
XA006	101.4151	12.5145	-0.811	0.0021
XA007	46.2475	39.1512	-0.112	0.002
XA008	47.9371	5.5056	-0.784	0.002
D001	114.1687	45.7006	23.9733	0.0021
D002	84.5533	45.6306	23.8707	0.0021
D003	46.4473	45.6497	23.9131	0.0021
D004	81.4665	35.5305	23.268	0.002
D005	63.4999	35.4819	23.263	0.002

Fig. 4 The total station is used to establish the local coordinate system based on the satellite coordinate point construction

For IGPS data, a feature-point stationing method is adopted. Based on the established control network, three-dimensional coordinate data for detailed building components is constructed. IGPS is capable of simultaneously calculating a large number of point coordinates within its coverage area. However, due to issues such as observation occlusion, extreme errors can occur. To address this, an observation matrix is introduced, and robust regression techniques are applied to complete and stabilize the coordinate system, effectively mitigating the impact of outliers and ensuring the reliability of detailed spatial data (Fig. 5).

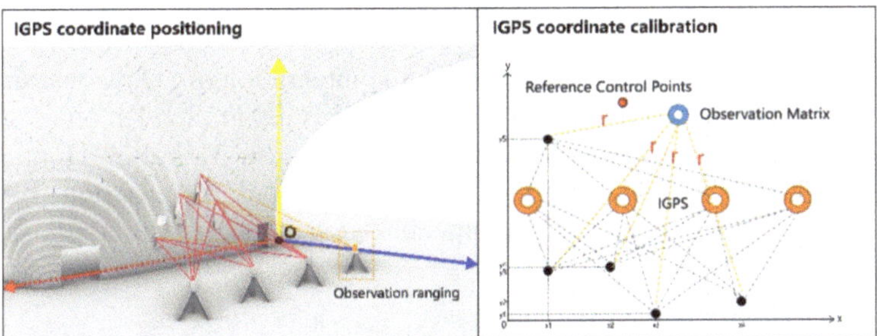

Fig. 5 IGPS is adopted, relying on the established control network to construct the three-dimensional coordinate data for the detailed components of the building

As illustrated in Fig. 6, a portion of the 3D-printed panels has been installed on the rooftop based on the methodology described above. A comparative analysis was conducted between the as-built (scanned) model and the original design model. Despite deviations of up to 5 cm observed in the steel structure installation, the positioning accuracy of the 3D-printed panels was maintained within 10 mm, demonstrating a high level of installation precision and adaptability to structural discrepancies.

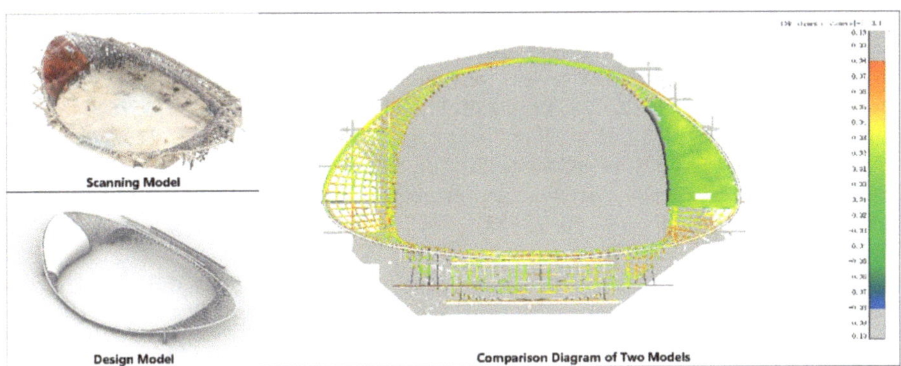

Fig. 6 Comparison diagram of scanning model and design model

4 Discussion

The cross-scale multi-order norm-based positioning technology proposed in this study significantly improves precision and efficiency in construction processes, effectively addressing the issue of poor data integration in current construction practices. By fusing multi-source heterogeneous data across different spatial scales and employing norm-based optimization methods, the approach enables precise positioning from global planning to fine-grained component installation. The study also validates the application of this method in complex engineering scenarios, such as the installation of the hyperbolic curtain wall in the Xiong'an Wings project, demonstrating its feasibility and practical value in real-world engineering projects.

To further enhance construction accuracy, the integration of 3D reconstruction into the workflow is essential. On-site, high-precision 3D point cloud models should be dynamically generated using real-time sensing technologies. This allows for continuous monitoring and real-time updates of the construction environment, ensuring a high degree of consistency between prefabricated components and the digital design model.

References

Brown, B., Aaron, M.: The politics of nature. In: Smith, J. (ed.) The Rise of Modern Genomics, 3rd edn. Wiley, New York (2001)

Calisi, D., Botta, S., Cannata, A.: Integrated surveying, from laser scanning to UAV systems, for detailed documentation of architectural and archeological heritage. Drones. **7**(9), 568 (2023). https://doi.org/10.3390/drones7090568

Chen, Z., Du, F.: Measuring principle and uncertainty analysis of a large volume measurement network based on the combination of iGPS and portable scanner. Measurement. **104**, 263–277 (2017). https://doi.org/10.1016/j.measurement.2017.03.037

Dod, J.: Effective substances. In: The Dictionary of Substances and Their Effects. Royal Society of Chemistry. Available via DIALOG. http://www.rsc.org/dose/titleofsubordinatedocument (1999). Accessed 15 Jan 1999

Fazel, A., Izadi, A.: An interactive augmented reality tool for constructing free-form modular surfaces. Autom. Constr. **85**, 135–145 (2018). https://doi.org/10.1016/j.autcon.2017.10.015

Gudnason, G., Scherer, R. (eds.): eWork and eBusiness in Architecture, Engineering and Construction: ECPPM 2012. CRC Press (2012a). https://doi.org/10.1201/b12516

Gudnason, G., Scherer, R. (eds.): eWork and eBusiness in Architecture, Engineering and Construction: ECPPM 2012. CRC Press (2012b). https://doi.org/10.1201/b12516

Li, H.: Geographic information surveying and mapping methods based on UAV remote sensing technology. China High-Tech. **16**, 33–35 (2024). https://doi.org/10.13535/j.cnki.10-1507/n.2024.16.06

Liu, J., Xu, D., Hyyppa, J., Liang, Y.: A survey of applications with combined BIM and 3D laser scanning in the life cycle of buildings. IEEE J. Sel. Top. Appl. Earth Obs. Remote Sens. **14**, 5627–5637 (2021). https://doi.org/10.1109/JSTARS.2021.3068796

Maciąg, K., Maciąg, M., Leń, P.: Implementation of unmanned aerial vehicles in the automated assessment of geodetic database validity. Adv. Sci. Technol. Res. J. **18**(7), 379–395 (2024). https://doi.org/10.12913/22998624/192264

Slifka, M.K., Whitton, J.L.: Clinical implications of dysregulated cytokine production. J. Mol. Med. **78**(2), 74–80 (2000). https://doi.org/10.1007/s001090000086

Smith, J., et al.: Future of health insurance. N. Engl. J. Med. **965**, 325–329 (1999)

South, J., Blass, B.: The Future of Modern Genomics. Blackwell, London (2001)

Steenbeek, A., Nex, F.: CNN-based dense monocular visual SLAM for real-time UAV exploration in emergency conditions. Drones. **6**(3), 79 (2022). https://doi.org/10.3390/drones6030079

Wang, P., Wang, Y., Wang, M.: Example analysis of digital wireless mapping applied to construction engineering measurement. J Sens. **2022**, 1–10 (2022). https://doi.org/10.1155/2022/6599720

Xiong, L., Lu, M., Zhang, J.-P.: On-site visualization of building component erection is enabled by the integration of four-dimensional modeling and automated surveying. Autom. Constr. **20**(3), 236–246 (2011). https://doi.org/10.1016/j.autcon.2010.10.002

Zeng, Q., Huang, X., Li, S., Deng, Z.: High-efficiency posture prealignment method for large component assembly via iGPS and laser ranging. IEEE Trans. Instrum. Meas. **69**(8), 5497–5510 (2020). https://doi.org/10.1109/TIM.2019.2958579

Zhou, J., Xiao, H., Jiang, W., Bai, W., Liu, G.: Automatic subway tunnel displacement monitoring using robotic total station. Measurement. **151**, 107251 (2020). https://doi.org/10.1016/j.measurement.2019.107251

Inventory Informed Computational Design Method for Upcycling Cross-Laminated Timber Leftovers into Floor Slabs

Qiming Sun(✉), Mathias Bernhard, and Benjamin Dillenburger

Department of Architecture, ETH Zurich, Digital Building Technologies, Institute of Technology in Architecture, Zurich, Switzerland
{sun,bernhard,dillenburger}@arch.ethz.ch

Abstract. Advances in digital methods enable automated design processes with non-uniform materials, promoting an inventory-informed workflow that prevents reclaimed material from being downcycled. This research explores computational design methods for upcycling rectangular cross-laminated timber cutoffs into free-form floor slabs, accounting for varying dimensions and thicknesses. It examines how altering object combinations affects the target design's visual appeal and seam patterns using inventory-informed slicing and adaptive nesting. The slicing process transforms a 3D target design into 2D nesting boundaries with varied layer heights determined by stock availability. The metaheuristic-aided nesting then optimizes leftover combinations to maximize material efficiency, minimize cutting length, and object count. Key challenges include optimizing upcycle logistics across different inventory sizes, finding optimal nesting patterns, and balancing environmental and economic factors. As a proof of concept, case studies configured three 3-m by 6-m slabs from inventories with 50, 100, and 150 leftovers of varying thicknesses, achieving over 60% material efficiency and 30% cutting length reduction compared to new materials. This method can optimize material usage among various target designs and inventory sizes in contrast to heuristic methods. Although focused on cross-laminated timber, this computational framework applies to various planar materials, potentially enhancing circular design principles.

Keywords: Inventory-informed design · Upcycle · Evolutionary algorithm · Floor slab · Circular design

1 Introduction

Standardization allows architects to design with an almost unlimited material supply. However, reusing or upcycling reclaimed materials, such as by-products and demolition waste, often involves dealing with arbitrary dimensions and quality limitations, which adds complexity and uncertainty to the design process. Computational design methods can transform these complexities into a material assignment process, automate matching, and explore optimal design solutions (Cousin et al. 2023).

Y. Liu et al. (Eds.): CDRF 2025, *Transindividual Intelligence*, pp. 228–237, 2026.
https://doi.org/10.1007/978-981-92-0615-5_20

Upcycle is creative usage that offers added benefits compared to traditional scenarios (Oyenuga et al. 2017). High-value wood products like cross-laminated timber (CLT) offer design flexibility, construction efficiency, and environmental benefits (Younis and Dodoo 2022). However, around 5%–22% of CLT panels become leftovers during prefabrication despite their structural equivalence to the master panels (Dupas and Hudert 2024). These leftovers arise when master panels are cut into prefabricated elements, resulting in rectangular window and door cutoffs and irregular scrap pieces. Driven by sustainability and economic incentives, the leftovers are often downcycled into pellets for energy or particleboard production. The main challenge for upcycling CLT leftovers is creating efficient workflows and logistics that enhance material efficiency and reduce cutting costs with limited quantities and varying thicknesses. Upcycling these leftovers based on daily or project-based quantities is more practical when considering transportation, storage, geometric information gathering, and additional cutting requirements.

2 State of the Art

Researchers have explored various methods for reintroducing CLT leftovers into construction. Some studies proposed regularizing cutoffs into uniform pieces and finger-jointing them into master panels (Vamza et al. 2021; Resnais et al. 2021). Vessby et al. presented slicing and laminating leftovers as layers for new CLT or glulam beams fabrication (Vessby et al. 2023). These studies assemble CLT leftovers into new mass timber products, aiming for equivalent structural performance. Robeller and Von Haaren customized CLT leftovers into hexagonal segmented shell structures, using computational tools to define target design and wood connections (Robeller and Von Haaren 2020). Poteschkin et al. standardize CLT cutoffs into uniform widths, enabling the reconfiguration of frame walls and beamed ceilings (Poteschkin et al. 2019). These approaches cut out the desired shape from leftovers, ignoring the original dimensions and thickness variations.

The potential of using computational design tools to optimize material usage in converting CLT leftovers into building components has not been fully explored. Evolutionary algorithms (EAs), which simulate natural selection processes, are widely used in architecture and engineering to explore complex search spaces and optimize design outcomes. They are well-suited for guiding optimal trade-offs, especially in multi-objective optimization (Parigi 2021). To ensure efficient computing, EAs should be carefully structured with a well-reduced solution space. Sequential assignment and inventory sorting are typical strategies when facing an inventory-informed design (Mollica and Self 2016). EAs have been applied to transform rectangular CLT offcuts into cubic wall elements with notched connections (Mangliár and Hudert 2022) or flat reciprocal frame structures (Dupas and Hudert 2024). However, current research using EAs often involves small or undefined inventory sizes, with material assignment and optimization principles frequently lacking detailed information.

This research employs EA to formulate the wood stacking process, optimizing the allocation of CLT leftovers based on their original geometry and thicknesses. It aims to achieve free-form floor slabs while maximizing material efficiency and minimizing cutting effort with fewer objects. The study highlights the solution quality over the heuristic method and tests it with various inventory sizes.

3 Methods

This section outlines an inventory-informed workflow for design with a given stock. The workflow inputs are an object list and a 3D target design. Outputs are the optimal target-object transformation pairs and cutting lines. The computational design methods include: 1) Inventory-informed slicing: slicing the target design into varying layer heights. 2) Adaptive nesting: defining form adaptable nesting rules with dynamically adjusted object preference and seam patterns. 3) EA optimization: balancing the material efficiency, cutting length, and object count. The design process is implemented in the 3D-modelling environment Rhino 8 with Grasshopper CPython. In the following, the target design is a predefined 3D model. The nesting boundaries refer to the 2D cross sections sliced from the 3D model. Objects refer to leftovers represented by their width, height, and thickness. An object can be reused multiple times if usable parts remain after nesting.

3.1 Inventory-Informed Slicing

Traditional slicing methods, which use uniform layer thicknesses, limit customization when dealing with leftover materials of varying thicknesses. The proposed slicing method for 3D models incorporates inventory constraints with varying layer heights. The proportion of different layer heights is determined by stock availability in terms of quantity and area. This approach strategically places thicker slices in regions with larger cross-sectional areas, tailoring the slicing process to the target shape. For each slice, it ensures the area of the stock objects is more than twice that of the current and previous slices, maintaining sufficient stock for nesting. The method iteratively continues until the stock is exhausted or the desired layer count is achieved.

3.2 Adaptive Nesting

The adaptive nesting process strategically allocates objects to a defined boundary, considering the boundary's form properties and nested pattern. The boundaries of interest in this study are characterized by one straight edge and an opposing wavy edge, which contributes to the target design's free-form nature. The goal is to efficiently cover the area using a subset of stock objects, guided by a two-phase heuristic: define sequential placement, and select the suitable object. This nesting focuses on covering the area with an optimal subset rather than fitting all objects, which differs from traditional bin packing. In this research, depthwise and first-fit heuristic rules are employed to automate the adaptive nesting process. These methods effectively maintain pattern coherence and reduce seam complexity.

3.2.1 Positioning Rules for Sequential Nesting

The positioning rules aim to create an optimal seam pattern that can adapt to various form properties. This research compares two strategies, the edgewise approach and the depthwise approach. The edgewise approach places objects sequentially from left to right, aligning edges (Fig. 1 left). It is efficient but nests different shapes using the same pattern, often leaving small pieces at the end (Fig. 2 left). This approach is less robust when handling a limited number of large objects (Fig. 3 left). The depthwise method prioritizes nesting from the greatest amplitude areas of a given boundary to the least (Fig. 1 right). It places pieces sequentially at the deepest pivot, continuously evaluating the nesting boundary in each step. With a given object, it evaluates potential alignments based on a distance parameter, d_{min}. When d_{min} is set to 1, it corresponds to 0.01 m in the real world. If an object can cover one of the nesting edges and the pivot within this distance, it is aligned with the edge; otherwise, it is centered with the pivots. Regardless of its position, the gaps between the object's and target edges must exceed this minimum distance. If not, the next object is considered. This iterative process continues until the entire boundary is covered.

The depthwise approach enhances symmetry around extruded regions while adapting to different nesting boundaries (Fig. 2 right). It prioritizes placing larger pieces in demanding areas, optimizing their use, especially when object availability is limited (Fig. 3 right). However, it can lead to overlapping areas, which might reduce material efficiency or require large materials to fill small gaps, as shown in the third and fourth nesting solutions in Fig. 2 (right).

3.2.2 Object Selection Rules for Efficient Nesting

Given the known geometric characteristics and limited quantity of available objects, this research compared two screening strategies: the first-fit and best-fit methods. The first-fit method involves selecting the first object that fits into a designated location, with a presorted stock by increasing height and then by decreasing width. As illustrated in Fig. 4 (left), the first-fit approach preserves intact extrusion areas, neat seams, and minimal object count. This method is particularly advantageous when the design objective prioritizes computational efficiency and aesthetic continuity. The best-fit method evaluates all available objects to select the one that maximizes material efficiency at each step. While it can achieve higher material efficiency, as shown in Fig. 4 (right), it often increases seam complexity, especially with abundant material or smaller pieces. Over time, configurations can become less optimal, with more objects or incomplete coverage, as demonstrated in the third or fourth nesting solution in Fig. 4 (right).

3.3 The EA Optimization

Unlike a single-boundary focus, 3D target design involves various nesting boundaries and a changing stock. With a 30-layer target design and 100 inventory size, the solution space is over 10^{240}, considering each layer consists of around four objects. To improve the computation efficiency, the EA manipulates d_{min} factors to impact each layer's object selection and alignment preference. As introduced in the depthwise sector, different d_{min} values can lead to varied nesting patterns using the same stock. With d_{min} set to 5, the first selected object aligns differently than setting it to 10 (Fig. 5 left, Fig. 5 middle). It also changes object selection preferences; for example, with a d_{min} of 10, the third nesting step shifts the first-fit object to the next viable object (Fig. 5 middle). An improper value, such as 20 as shown in Fig. 5 (right), may prevent the completion of the nesting process. In a multi-layer setup, the d_{min} factor for one layer can affect material availability for

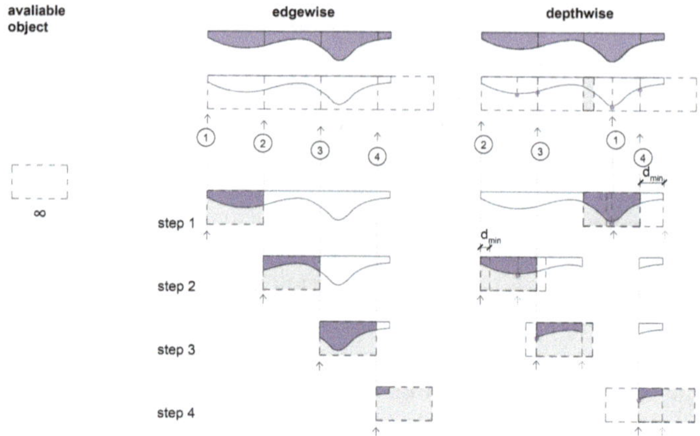

Fig. 1 Comparison of two nesting strategies for object placement

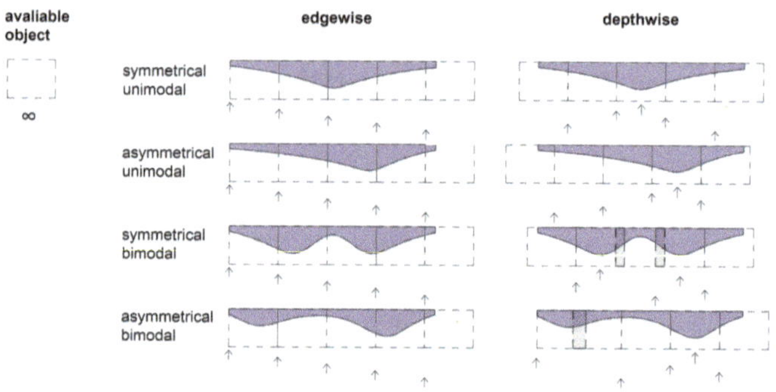

Fig. 2 Comparison of two nesting strategies for form adaptability

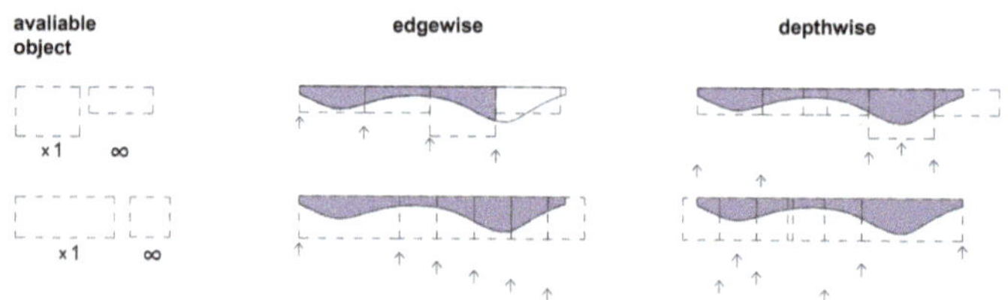

Fig. 3 Comparison of two nesting strategies with limited large objects

subsequent layers. Selecting an appropriate d_{min} for each layer can optimize its seam pattern and object occupations.

Thus, the EA determines an optimal d_{min} value for each layer to influence the nesting configuration, considering each d_{min} as one gene. Each gene can be any even integer from 0 to 20. A genome contains as many genes as there are nesting boundaries. During the evolution process, one generation comprises 50 individuals, evaluated by a fitness

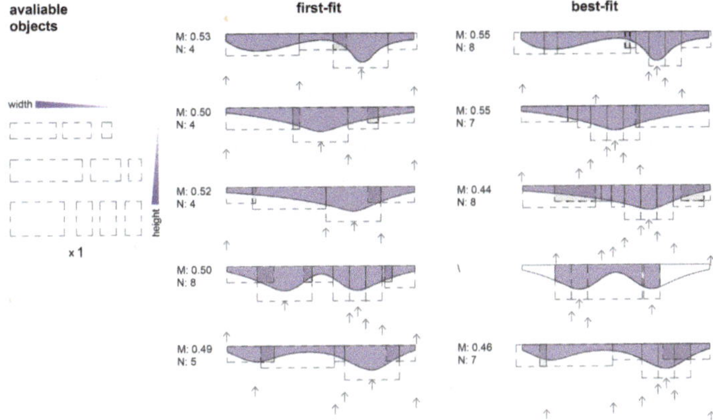

Fig. 4 Comparison of two nesting strategies for object selection

function (Eq. 1). The top two individuals are retained, while others breed based on similarity. Along with individuals from the previous generation, 25 of the lowest-performing individuals are eliminated. The remaining individuals, combined with 24 newly generated individuals, constitute the next generation. This approach reduces the solution space and adapts nesting strategies to stock and boundary geometry changes.

The fitness function (Eq. 1) uses normalized and weighted values to identify the optimal subset of leftover stock *t*, maximizing material efficiency while minimizing cutting length and using fewer objects. Material efficiency *M(t)* is defined as the area of subset *t* divided by the area of its integrated parts, with higher ratios indicating more effective material usage. Cutting length *C(t)* represents the edges of the overlap area between a target and an object in the subset *t*, excluding the object's original contours. The number of objects *N(t)* is the size of the subset *t*. Reducing *C(t)* and *N(t)* decreases the time and labor required for upcycling. Note that, within the scope of this method, a single object may be assigned to multiple target-object transformation pairs, as cutoffs can be returned to the material stock for future nesting operations, enhancing material efficiency.

$$Fitness\ Function = (M(t) \times w_1 - C(t) \times w_2 - N(t) \times w_3) \quad (1)$$

w_1, w_2, and w_3 are the weights for optimization

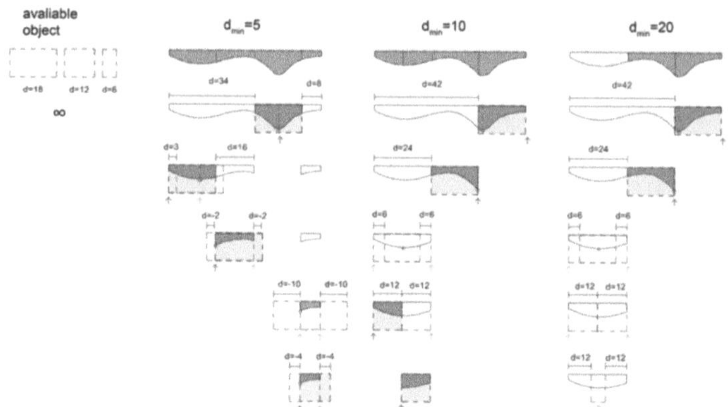

Fig. 5 The impact of d_{min} values on object selection with three objects in an unlimited amount

4 Case Study

Three target slab units were tested, each featuring a span length of 6 m, a width of 3 m, and a maximum thickness of 0.5 m (see Fig. 6). These units differ in shape and have increasing volumes. The slabs are supported by load-bearing walls where they conclude their span, with slicing direction perpendicular to these walls. The material inventory consists of 100 rectangular CLT pieces, randomly generated based on the quantities and dimensions from a daily cutting file sourced from the CLT manufacturing company. Their heights range from 300 mm to 1000 mm, and their lengths range from 800 mm to 2700 mm. The ratio for different thicknesses is set to 4:1. For instance, an inventory of 100 pieces includes 80 pieces at 100 mm thickness and 20 pieces at 160 mm thickness. The fitness function's weights are set at 2:1:1 for material efficiency, cutting length, and object count.

The slicing strategy follows a 4:1 inventory thickness ratio, dividing each design into distinct layers based on shape properties, as shown in Fig. 6. For slab a, six thicker layers are in the bimodal area, resulting in 26 total layers. Slab b features a unimodal with five thicker layers in the middle to maintain symmetry, resulting in 27 layers. Slab c, characterized by a uniform height across layers, strategically places thicker layers in larger cross-section areas on both sides, totaling 26 layers. These variations in slicing patterns reflect a certain adaptability in inventory compositions and target forms, preserving structural and aesthetic balance.

Using an inventory size of 100 objects as an example, each slab's nesting solution is unique, as depicted in Fig. 6. Slab a configures 50 objects with 61% material efficiency and a cutting length of 225.98 m. Slab b achieves 68% material efficiency using 63 objects, with a 199.68-m cutting length. Slab c involves 81 objects, resulting in 64% material efficiency and a 177.28-m cutting length. Leftovers can be processed with straight or tailored cuts to approximate target forms (Fig. 6). The straight-cut approach simplifies materialization and stores reclaimed material in a generalized form, facilitating future reuse. Connections between objects are made with finger joints, and layers are connected using wood dowels.

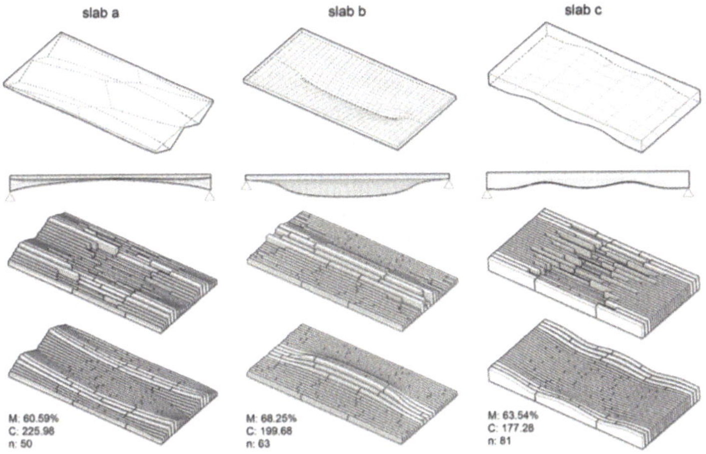

Fig. 6 The design solution using the metaheuristic nesting method for three slab units

With different inventory sizes, the study compared the material efficiency, cutting length, and number of objects using EA, first-fit, and best-fit for slabs a, b, and c (Fig. 7). Figure 7 illustrates the normalized results of these methods for each scenario. For instance, the top-left chart shows that for slab a, the EA method achieves the highest material efficiency (62%) and the minimum object count (28), while the best-fit method results in the shortest cutting length (259 m). EA consistently achieves higher material efficiency than first-fit and best-fit, demonstrating more efficient material use. Notably, slab b reaches approximately 68% efficiency with EA. Cutting lengths are generally shorter with EA, particularly in slabs b and c, reducing processing time. EA also manages fewer objects, especially with larger inventories. Overall, EA effectively balances multiple factors and optimizes material usage across varying target designs and inventory sizes.

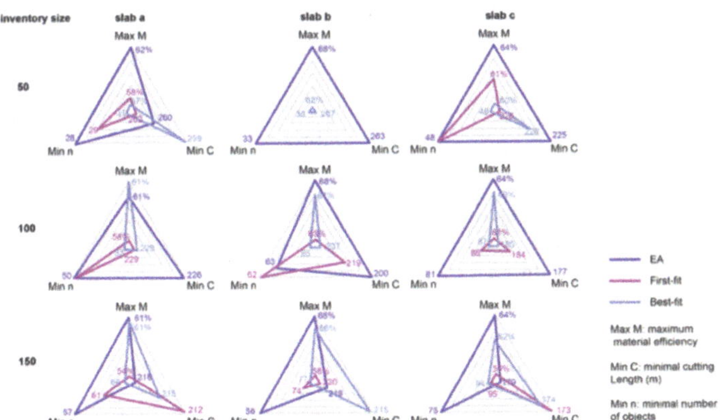

Fig. 7 The comparison of the solution quality over three methods across the case studies

5 Discussion and Conclusion

This study examines the trade-offs in nesting reclaimed CLT leftovers, focusing on material efficiency, cutting length, and object count. Most EA computations converge to optimal or near-optimal solutions within about 50 generations, typically taking around 10 min, benefiting from the well-tuned d_{min} factor. The EA achieves an average material efficiency of 64%, integrating over 60% of selected leftover materials into slab units across all case studies. It reduces cutting length by more than 30% compared to using new materials. This method offers distinct nesting patterns and unique solution spaces across different target designs despite using the same material stock. It also performs optimally with different inventory sizes over traditional heuristic methods, effectively balancing material efficiency, cutting length, and object count. Increasing the inventory size generally reduces cutting length and object count, except in a 50-piece inventory where larger pieces are used (each inventory's total area must be at least twice that of the target design). Larger inventories incur higher computational and logistical costs, especially slowing down best-fit methods. Material efficiency improvement is minimal, but could be enhanced by increasing the fitness function weight w_1. Efficiency is closely tied to the target design's form, suggesting that optimizing the form could further boost material efficiency. Furthermore, nesting preferences can be customized by adjusting objective weights: increasing w_3 favors solutions with larger pieces, whereas increasing w_2 favors smaller objects.

This research presents an inventory-informed workflow powered by computational design methodologies. The proposed optimization method addresses complex design intentions and nesting preferences, prioritizing architectural requirements and aesthetics. Although tested on 3-m by 6-m slabs with three fixed inventories, the computational framework is adaptable to dynamic inventory fluctuations and target designs with flexible dimensions. Designers and researchers can use this method to structurally and aesthetically upcycle other reclaimed planar materials into building elements, promoting circular design principles. The study opens opportunities for further research into structural simulations and prototype testing, particularly in verifying structural performance (e.g., load-bearing capacity, joint strength) and assessing fabrication feasibility (e.g., cutting costs)

References

Cousin, T., Marshall, D., Pearl, N., Alkhayat, L., Mueller, C.: Integrating irregular inventories: Accessible technologies to design and build with nonstandard materials in architecture. J. Phys. Conf. Ser. **2600**(19), 192004 (2023)

Dupas, N., Hudert, M.: Enabling the circular use of Cross-Laminated Timber by upcycling production waste. In: IASS 2024, Conference proceedings. The International Association for Shell and Spatial Structures (IASS) (2024)

Mangliar, L., Hudert, M.: Re:Shuffle. Paper presented at 5th international conference on structures and architecture, pp. 50–51. Aalborg, Denmark (2022)

Mollica, Z., Self, M.: Tree fork truss: Geometric Strategies for Exploiting Inherent Material Form. In: Adriaenssens, S., Gramazio, F., Kohler, M., Menges, A., Pauly, M. (eds.) Advances in Architectural Geometry 2016, pp. 138–153. vdf Hochschulverlag AG an der ETH Zurich, Zurich (2016). https://doi.org/10.3218/3778-4_11

Oyenuga, A.A., Bhamidimarri, R., Researcher, P.D.: Upcycling ideas for sustainable construction and demolition waste management: Challenges, opportunities and boundaries. Int. J. Innov. Res. Sci. Eng. Technol. **6**(3), 4066–4079 (2017)

Parigi, D.: Minimal-waste design of timber layouts from non-standard reclaimed elements: A combinatorial approach based on structural reciprocity. Int. J. Space Struct. **36**(4), 270–280 (2021). https://doi.org/10.1177/09560599211064091

Poteschkin, V., Graf, J., Krötsch, S., Shi, W.: Recycling of cross-laminated timber production waste. Res. Cult. Archit. **12**, 101–112 (2019)

Resnais, P., Grekis, A., Keivs, M., Gaujena, B.: Possibilities of useful use of glued wooden construction residues. Materials. **14**(15), 4106 (2021)

Robeller, C., Von Haaren, N.: Recycleshell: Wood-only shell structures made from cross-laminated timber (CLT) production waste. J. Int. Assoc. Shell and Spat. Struct. **61**(2), 125–139 (2020)

Vamza, I., Diaz, F., Resnais, P., Radziņa, A., Blumberga, D.: Life cycle assessment of reprocessed cross laminated timber in Latvia. Environ. Clim. Technol. **25**(1), 58–70 (2021)

Vessby, J., Perstorper, M., DeMonte, F., Eriksson, J.: Structural use of cut-offs from CLT-production-three examples that utilize the unique properties. In: World Conference on Timber Engineering, vol. 19, p. 22 (2023)

Younis, A., Dodoo, A.: Cross-laminated timber for building construction: A life-cycle-assessment overview. J. Build. Eng. **52**, 104482 (2022). https://doi.org/10.1016/j.jobe.2022.104482

Adaptive Bamboo Joinery: Robotic-Assisted Assembly of Traditional Indonesian Roof House Typologies

Ahmad Mansuri[1(✉)], Asterios Agkathidis[1], Davide Lombardi[2], and Hanmei Chen[1]

[1] Liverpool School of Architecture, University of Liverpool, 25 Abercromby Square, Liverpool L69 7ZN, UK
{ahmad.mansuri,a3lab,hanmei.chen}@liverpool.ac.uk
[2] Xi'an Jiaotong - Liverpool University, 111 Ren'ai Road, Suzhou, South Campus, Office DB428, Jiangsu Province 215123, PR China
Davide.Lombardi@xjtlu.edu.cn

Abstract. Traditional bamboo construction relies on a joinery system for structural integrity. Bamboo's heterogeneity and non-standardized nature challenge its integration in computational modelling and structural optimisation, particularly to address its natural flexibility, tensile strength, and lightweight properties for active bending applications. Integrated computational design and digital fabrication have the potential to modernise and enhance vernacular and heritage architecture. This study proposes adaptive joinery systems tailored to traditional Indonesian house typology to substitute conventional joints compatible with robotic assembly. The method integrates parametric design and structural optimisation for the proposed fabricable elements and joint design prototype. Key joint design criteria include mechanisms, adjustability, and angular systems. The findings highlight the feasibility of digitally fabricated joint systems in robotic assembly scenarios and digital fabrication as an alternative to traditional craftsmanship. The research provides insight into the suitability and compatibility of bamboo joint techniques for robotic assembly, bridging the gap between heritage architecture and technologies and offering solutions for contemporary bamboo architecture.

Keywords: Bamboo architecture · Bamboo joint design · Additive manufacturing · Structural optimisation · Robotic assembly · Robotic fabrication · Human-robot collaboration

1 Introduction

Developing bamboo joints compatible with robotic assembly involves understanding and integrating traditional joinery principles and advanced digital fabrication methods. Recent research advancements in digital and robotic fabrication have fundamentally linked how to overcome the main challenge in bamboo structure, including the complexity of joint design systems. Traditionally, bamboo joinery systems require human dexterity for tasks such as binding and bolting, making their adoption in robotic fabrication challenging. The significance of connection systems may limit the possible

Y. Liu et al. (Eds.): CDRF 2025, *Transindividual Intelligence*, pp. 238–254, 2026.
https://doi.org/10.1007/978-981-92-0615-5_21

geometric configuration that can be created, as their connections play a significant role in determining the structural typology and complexity of the structure [1]. Therefore, the success of the bamboo structure is highly determined by its ability to overcome the main challenge of the complexity of the jointing design [2]. The development of the tailored and adaptive joint system is essential to accommodate bamboo's unique physical properties in its hollow structure and diameter variation [3]. Recent research has addressed these limitations by integrating robotic 3D scanning technologies to obtain bamboo digitisation and investigate material mechanical properties [3, 4, 6]. developing parametric modelling and designs for bamboo joint systems [6, 7] and engineering bamboo 3D-printed joinery [1, 8–11, 13]. These efforts aim to overcome the challenges and adapt bamboo joint systems to digital fabrication. However, the literature indicates that this joint design system is intended for manually constructed structures. Only a few similar research uses of wooden sticks have been discovered in applying the joint system in a robotic assembly scenario [14, 15] which is a relevant and insightful reference for our case study? Building on our prior research on systematic literature review [16] and proposed automated fabrication techniques [17] It is suggested that no prior research has attempted to incorporate robotic technologies for bamboo assembly, especially for bending operations and joint installation in a human-robot collaborative framework.

Designing an adaptive joinery system compatible with robotic assembly requires both bamboo material constraints and automation feasibility. The human-robot collaboration (HRC) approach is a viable strategy, as bamboo joinery techniques are intricate and considered a skill that robotic fabrication is still struggling to automate fully. The joinery is critical because bamboo construction relies on well-designed connections, and highly traditional craftsmanship skill labour is required [12, 18]. Human participation remains crucial for joinery execution and depends on manual operation, particularly during material handling and bending [19]. Therefore, adaptive and compatible joinery systems are needed to make it suitable with robotic assembly and meet several criteria such as simplicity, repeatability, and a fast assembly process. This study aims to investigate and propose a design for adaptive joinery systems tailored to traditional Indonesian roof typology, compatible with robotic assembly workflows as an alternative to conventional joints. The joint systems must be a balance between structural strength, fabrication efficiency, and assembly precision while still allowing human adaptability for manual alignment, force control, and fine-tuning in a hybrid HRC scenario. Building on prior research, we aim to develop a workflow for the robotic assembly of bamboo structures based on Indonesian roof house typology within a human-robot collaboration scenario, where an adaptive joinery system will be integrated into the workflow. Adaptive bamboo joinery for robotic assembly is crucial for dynamic adjustment and sharing tasks between humans and robots, ensuring a flexible and efficient HRC assembly process.

Building on prior research and literature review, this study explores how to develop adaptive joinery for bamboo structures compatible with robotic fabrication within the HRC framework. The research explores key questions.

1. How can adaptive joinery systems support the robotic assembly of irregular bamboo structures while preserving traditional Indonesian roof typologies?
2. How do adaptive joinery systems address active bending mechanisms in bamboo joint structures for robotic assembly scenarios?

3. What are the strengths and limitations of proposed adaptive bamboo joinery systems for robotic assembly that can support their structural integrity?

1.1 Background and Context

Indonesian traditional houses feature distinctive architectural elements: the protruding roof geometry, which dominates the house figure, symbolising the head of the house and showcasing dominant proportions. The roof shape aims to show the impression of the lightness of a stilt house, which balances the heavy roof with a lighter and elevated body and is deliberately designed to be more striking and prominent compared to the body. The roof geometry is an architectural element with deep symbolic meanings and practical functionality. Key characteristics of traditional Indonesian roof shapes include: 1) high-pitched and steep roofs, 2) layered or multi-tiered roofs, 3) extended roof overhang for climate adaptation, 4) curved and symbolic roofs, 5) natural roof materials, and 6) stilt houses with raised roofs. This design element reflects local wisdom in adapting to harsh weather conditions such as heavy rainfall, intense solar radiation, and tropical climate [20]. In this study, we focus on the Bolon house of the Batak tribe in North Sumatera, Indonesia, as the object of study, and its distinctive features are the soaring and highly curved roof. It features two large cut hyperbolic paraboloid surfaces that rise steeply with the curve outward at the tips, resembling the boat or buffalo horns. The roof extends beyond the walls, providing shade and protection from the tropical climate (Fig. 1). The overall structure is symmetrical, and the curvature is carefully designed to distribute loads and withstand heavy rainfall efficiently. The steep pitch and pointed roof edges allow better airflow circulation to maintain a cool interior. Traditional roofs use lashing and binding techniques to join systems with rattan lashing or ropes. However, in modern adaptations, particularly in restoration projects or incorporating new materials, nails and screws are commonly used.

Fig. 1 Shows a Traditional Bataknese tribe house known as Bolon House, which was chosen as an object study in North Sumatera, Indonesia

Our focus on choosing the Bolon House (Fig. 1) as a case study for the prototype is due to its cultural significance, and the curvilinear shape has structural challenges in terms of its applicability for human-robotic workflows. The Bolon house, characterised by soaring, highly curved geometry, consists of a dual hyperbolic paraboloid surface, which is architecturally challenging due to its multidirectional curvatures and extended overhangs. From the technical standpoint, the roof typology requires joints for overlapping intersections of member units where curvature and angular deviation vary. The surface continuity and geometric simplicity are other reasons for the study case justification, as the roof frame consists of repeated curvilinear elements and can be translated into regular modules, simplifying the robotic task planning. For robotic applicability, the roof geometry is also more exposed and accessible for robotic arms to operate from both sides. The Bolon house roof relies on overlapping and rope joint techniques, which easily alternate with clamp-based or swivel-based adaptive joints and make it well-suited for robotic placement. The roof geometry also aligns with the scenario of human-robot collaborative assembly for robotic element handling and positioning. Human agents employ micro-adjustments during bending operations, joint installation, and post-joint fine-tuning. The layer separation between horizontal and vertical element units facilitates step-by-step robotic handling and simplified task sequencing in HRC workflows.

In terms of geometry, Bolon house curvature can be adjusted parametrically without losing the topological essence, making it ideal for simulation, optimization, and adaptive joint fitting. The behaviour compatibility with bamboo active bending and design intent can showcase the role of joint systems in a simple curvilinear roof surface where the curvature varies, demonstrating the role of joint systems in locking flexible units after deformation as a core objective of the research within the Human-Robot Collaborative framework.

The challenges in transitioning manual joinery techniques to robotic assembly in our framework are the material variability due to the uniqueness of every member and the need to balance joint precision and accuracy. Robotic systems typically work best with standardised and predictable forms. Meanwhile, developing algorithms capable of replicating this highly variable material and customising joints is challenging. Programming robots to assist humans in assembling complex joints may require real-time adjustment for fitting, which can be fastened quickly, allowing tightening and loosening. It must also account for tolerances during bending operations that may impact structural integrity. Adapting to the bamboo active bending material behaviour can also be challenging, particularly in designing a joint system that performs well under bending operation while securing the joint with good adjustability for post-assembly structure stabilisation.

1.2 Review of Digital Bamboo Joinery Systems

Based on the literature review of bamboo joinery systems utilising digital fabrication and robotic assembly techniques, we identified various types of contemporary bamboo joinery, which are summarised in the Table 1 and Fig. 2. In this study, we define the scope of our case study by focusing specifically on contemporary bamboo joinery systems that incorporate digital fabrication techniques to ensure their applicability for robotic assembly. We focus on selecting and analysing this joint system as a basis for analysis, comparison, and critical evaluation. Observing their design approach, the objective is to explore their adaptation for seamless integration with the robotic assembly workflow.

Fig. 2 Various bamboo joinery systems found in the literature review

Based on our analysis in Table 1 and Fig. 2, we identified three joint systems that are suitable for our roof prototype with overlap elements: clamp joint mechanism [15], interlocking joint [21], and rope joint [14]. The clamp joint mechanism is more favourable because it can secure and cover more friction with the tube's circular bamboo surface through clamping force and adjustable tightening systems. It allows disassembly, reassembly, and adjustability to accommodate joint fit tolerances, offering reusability and post-assembly stabilisation for final tightening and fine-tuning adjustment. In contrast, the interlocking join provides a fixed structure with a predefined direction, making it less suitable for robotic assembly due to its inability to adjust to multi-angle directions. Meanwhile, the rope join offers strong tensile strength, ideal for dynamic loads, and is sustainable and biodegradable. However, it requires skilled humans to use knot-tying techniques, is time-consuming for manual processes, and may be inconsistent in repeatability, as each joint may vary in thickness and position. Securing knots with ropes is also challenging for humans, requiring complex, dexterous, consistent movement. Considering these strengths and limitations, we conclude that clamp joints manually assembled by humans in the robotic workflow are highly compatible and practical as they provide quick, strong, and adjustable connections, enhancing the HRC construction scenario. In our case study on Indonesia roof typology, where bending elements are required for the shape, the clamping join must tolerate movement during iterative robotic assembly. Once the clamp is secured, the manual bending will continue at the subsequent overlapping joint point to achieve the hyperbolic paraboloid deformation surface.

Table 1 Case study and comparative analysis of bamboo joinery systems and digital fabrication identified in the literature

Joint Mechanism	Connection Systems	Adjustability	Angular system	Material	Technology	References
Clamp joint	Bolt	Adjustable	overlap joint	Steel/aluminium	SIP	Vasiliki et al. [15]
Clamp joint	Bolt	Fixed static	overlap joint	PLA	AM	Vasiliki et al. [15]
Interlocking joint	Mechanic	Fixed static	multiaxial joint	Nylon-Based Steel Joint	AM DMLS	Kladeftira et al. [10]
Interlocking joint	Bolt	Adjustable Fixed static	End-to-end joint	Metal	AM	Matson and Sweet [12]
Interlocking joint	Glue and a metal bolt	Fixed static	multiaxial joint	PLA Nylon and Metal	AM	Amtsberg and Raspall [13]
Interlocking joint	Glue	Fixed static	multiaxial joint	PLA	AM	Amtsberg et al. [3]
clamp joint	Bolt	Adjustable	overlap joint	Not mentioned	AM CNC machine	Qi et al. [21]
Interlocking joint	Bolt	Adjustable	multiaxial joint	PLA	AM	de Oliveira et al. (2020)
Interlocking joint	Bolt	Fixed static	multiaxial joint	PLA	AM	Di Paola and Mercurio [11]
Interlocking joint	Bolt	Fixed static	multiaxial joint	PETG filament	AM	Condezo De La Vega et al. (2024)
Interlocking joint	Bolt	Adjustable	overlap joint	PLA	AM	Qi et al. [21]
Rope Joint	Tied winding	Fixed static	overlap joint	Rope	SM	Mitterberger et al. [14]

SIP = Standardized Industrial Product, AM = Additive manufacturing (3D-printed), DMLS = Direct Metal Laser Sintering, PLA = Polylactic Acid, PETG = polyethylene Terephthalate Glycol-modified, SM = Standart Material.

2 Methods

The adaptive joinery systems methods in our object study utilised parametric design workflows for form-finding, followed by genetic algorithms with two-stage optimisation to generate the dataset intersection point of elements as the joint prototype's location (Fig. 3). Key design criteria include joint mechanisms, adjustability, and angular systems. It is tested to perform active bending structural performance to achieve the intended roof shape. In this case, the manual bending of bamboo elements enables deformation to create curvilinear structures. At the same time, joints are placed to lock the bamboo in its bent position and maintain the desired form. To showcase feasibility, we limit the scope to a reduced-scale prototype and optimise the bamboo prototype size compatible with the UR10 robotic arm's size and reachability. The physical demonstration will be tested in the Indonesian traditional roof typology with overlapping bamboo joint systems in a double surface hyperbolic paraboloid shape. The study evaluates the adaptive joint that allows elastic deformation while maintaining structural integrity. Ultimately, the study explores the integration between robotic handling and manual bending within active bending elements and observes any positional shifts in the bamboo element during joint placement within the human-robot collaboration (HRC) workflow.

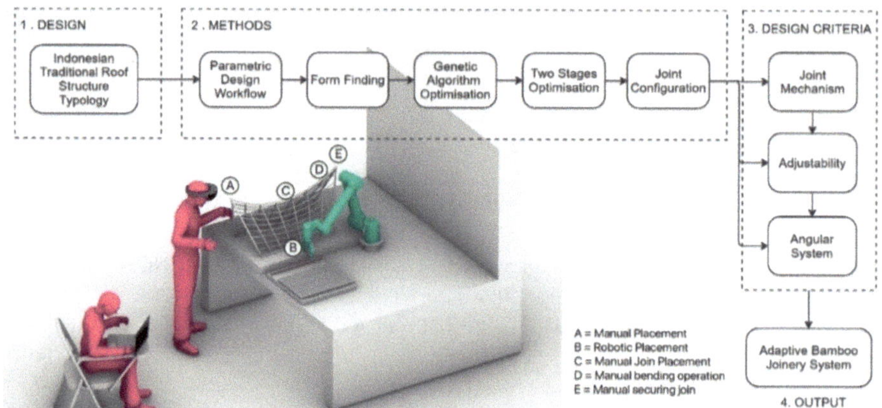

Fig. 3 Methodology, generative steps of the parametric modelling system, and joint design criteria

Our anlysis summarises the three main key design principles (Fig. 4) for adaptive bamboo joinery to overcome: 1) material variability adaptation, 2) robotic compatibility, and 3) post-assembly adaptation. It is illustrated in Fig. 4

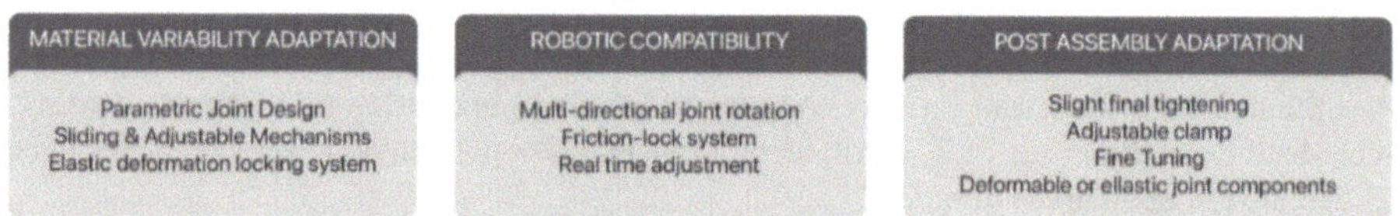

Fig. 4 Bamboo adaptive joinery key design principles for robotic assembly

3 Results and Discussion

3.1 Parametric Design Workflows, Form Finding and Optimisation

To define a joinery system compatible with our prototype and framework, the process starts by designing an Indonesian traditional roof house typology into a parametric design workflow. Parametric tools allow irregularity in the form-finding process and are addressed in the modelling process. The steps of the form-finding and optimisation process are illustrated in Fig. 5. It starts with form-finding utilising Kangaroo to generate a mesh surface. Following the design principle of roof dimensional typology, we set the proportion of the roof shape by following the original size proportional composition. To realise the roof shape, the pattern of elements is adapted from the original frame arrangement into U and V isocurves of the generated mesh. The structural bamboo element is then generated into three layers: the mainframe, vertical lines (VE), and horizontal lines (UE). The cross-sectional size of the bamboo element is gradually reduced, as the element with a smaller cross-section will be on top of the element with a bigger cross-section. In this study, we choose the overlap joint systems following the traditional roof frame rules; by generating overlapping elements between these vertical and horizontal curve lines, we define the point of intersection joint as a dataset generated to be the joint placement position.

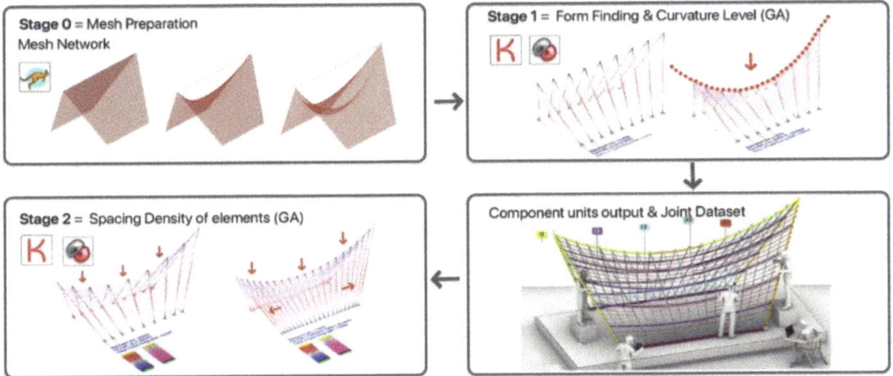

Fig. 5 The steps of the form finding and optimisation process

The post-design rationalisation process is investigated by examining the relationship between mesh shape, form-finding, and the spacing frame distance with structural performance. The first objective of optimisation is to modify the initial mesh to ensure the initial design constraints are fulfilled, to the original shape of the traditional roof. The form-finding surface optimisation seeks the level curvature of the top roof line that results in the minimum displacement value. We investigate the influence of the width curvature of the top roof line elements and see its behaviour on the structural displacement performance. By understanding the form-finding process (Fig. 6) for optimisation, we intend to investigate the original design's refined shape and better understand the balance of the top-width line curvature into the overall form surface performance. During this step, we discover that the higher or the bigger width curvature of the top line on the roof will increase the displacement value.

The mesh preparation is created by a computational logic form-finding process that starts from a simple 2D spine graph, which acts as the roof shape's conceptual frame line or layout. A base polygonal mesh is generated from this spine and defines the roof surface. To achieve a smooth and structurally viable form, the mesh undergoes subdivision, refining it into a Subdivision (SubD) mesh with more control points for detailed shaping. The process continues by applying vertical force, simulating gravity, and the expected load across the mesh surface. These forces cause the mesh to deform and settle into a minimal energy state, which optimizes the shape structurally. The resulting guided mesh represents an organic, load-responsive roof form that can be further developed for fabrication and assembly, particularly suited for flexible bamboo materials.

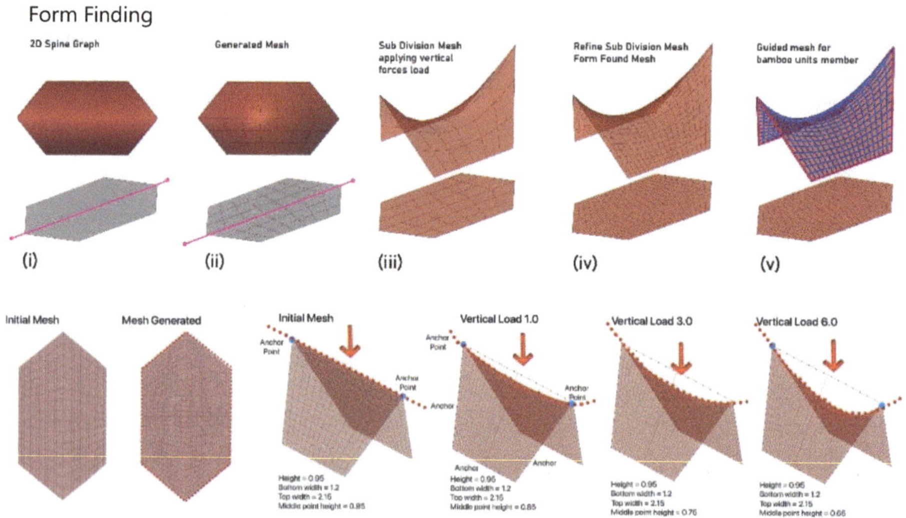

Fig. 6 The form-finding process of mesh to generate a minimal and relaxed surface

In the form-finding process, multiple vertical loads are applied to vary the magnitude and distribution of these loads on the mesh surface and explore how the roof geometry responds to the displacement simulation. Increasing or decreasing the vertical force gives the ability to control the width and curvature of the roof's upper levels. Their variation allows generations of a range of roof shapes from the same base mesh and spine graph, providing flexibility for form-finding, structural, and fabrication considerations. We define curvature level as a controllable parameter that influences the bending and shape of the roof surface. Increasing or decreasing this curvature provides the exploration of various forms, ranging from gently curved to sharply arched roofs (Fig. 7). Similarly, the cantilever distance at the roof's tip is another parameter that controls how far the roof extends beyond its supports. This affects both the spatial coverage below and the structural demands on the bamboo elements. By parametrising these two variables, we study their impact on the form finding and the structural performance.

After the first optimization stage, the number of bamboo element-type layers is determined from the mesh obtained into vertical and horizontal units from the isocurves mesh. Structural optimisation is conducted by examining the layer arrangement parameters in overlapping vertical and horizontal line elements, utilising a genetic algorithm optimisation solver with Galapagos (Fig. 8). The spacing distance range of vertical and

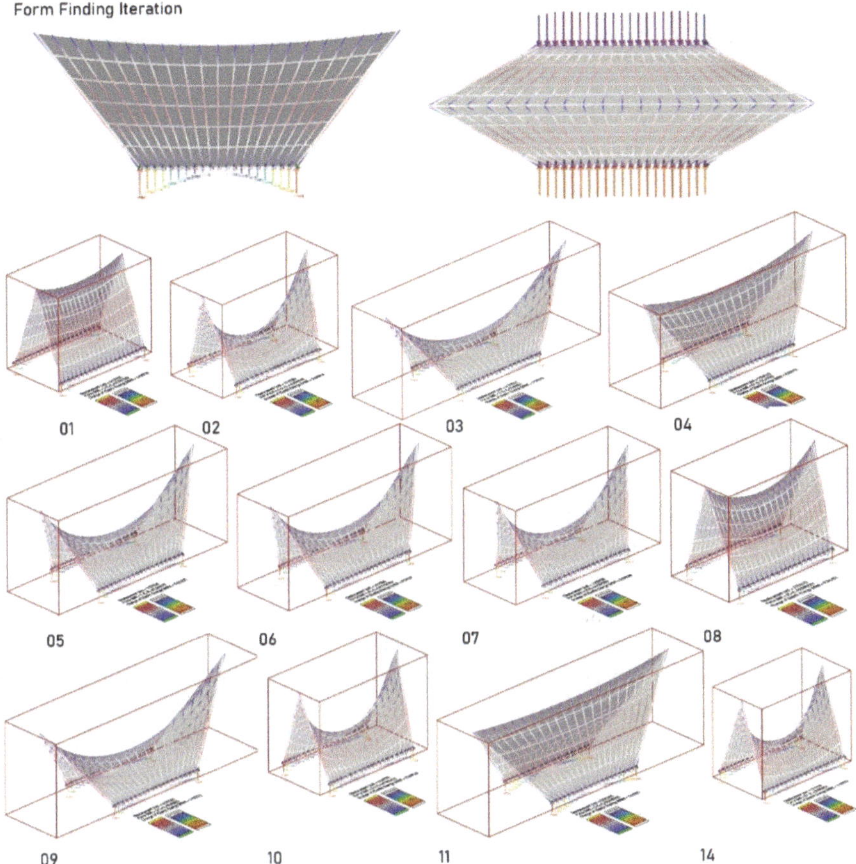

Fig. 7 The form-finding iteration process, assessing minimum displacement of roof shape based on height, top point distance, and level of curvature width with GAO and Karamba3D

horizontal element parameters is explored for optimisation to see which configuration has minimum displacement to shape as structural roof surfaces with Karamba3D for an optimal solution that satisfies the minimum target fitness of the structural displacement. Based on this optimisation, the internode of joints is then defined as a dataset for bamboo joint location. This data setpoint joint coordinate is then used to determine the joint design based on the overlapping internode joint, angle, and components. Based on this joint configuration, we categorised the type of joint based on its position.

The next stage after optimisation is translating mesh elements to curvilinear units and constructible components, consisting of horizontal units (UE) as U elements, vertical units (VE) as V elements from the mesh isocurves, and joint coordinate points. During the initial modelling, the robotic tools pick the UE units as a straight element, placing them into a targeted coordinate at the middle point. The bending curvature level varies and becomes higher or increases gradually towards the top position to achieve the desired surface. The UE level curvature is achieved by the division of equal distances of the main

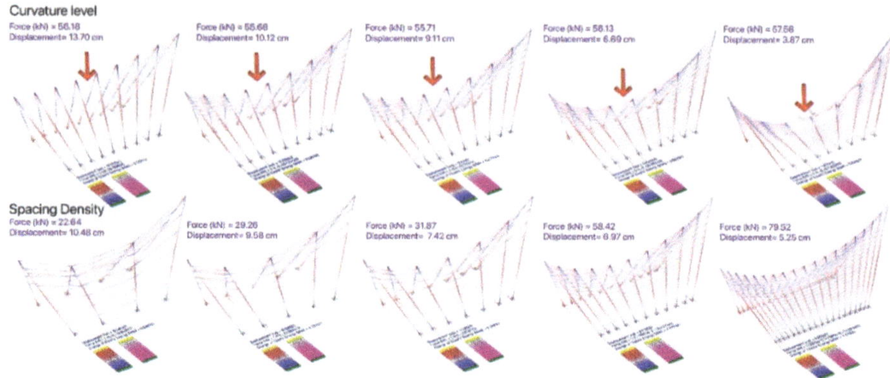

Fig. 8 Form Finding and structural optimisation process with two object optimisations: curvature level and spacing density

frame, and this will be distributed into the distance of joint placement at the end of the structure into equal distances. The joint placement operated by humans will be located in a zig-zag pattern, and the desired curvature bending is achieved by attaching and securing the joint in the overlapping point units of the UE and VE elements. The elements of the structure will maintain their position at the desired curvature level by relying on the joint connection. The joint coordinate dataset is generated into a set targeted at joint placement (Fig. 9), and the number of joints required is calculated. Based on this single unit point, the adaptive joinery design coordinates are defined. To adapt to bending operations requiring adjustability, we design a joint with a level of adjustability and fine-tuning during the robotic assembly process and post-assembly stabilisation.

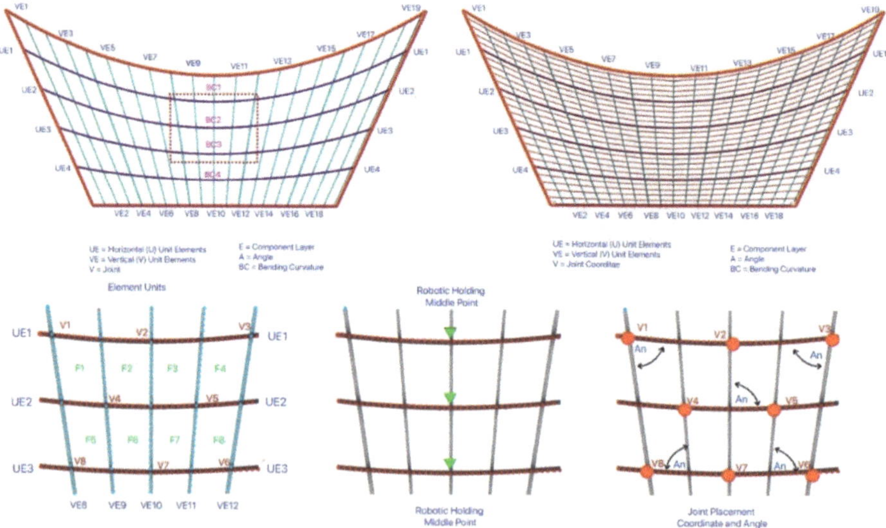

Fig. 9 Translating the component units' output and the joint dataset from mesh elements

The last stage defines and proposes the adaptive joinery systems based on the generated point coordinate. As the overlapping configuration of UE and VE elements is unique in the angular systems for the whole point, and for this case, we define the main criteria of the adaptive joints as their ability to work with the units that have a good level of real-time adjustment in direction and rotation. The second criterion is the joint with good friction to grasp and maintain the targeted joint coordinate location. During the subsequent jointing process, the connection can maintain its structural position and remain in its initial location.

Our concern for the HRC framework during the manual bending operation by human agents is that the units of the bamboo elements may shift, experiencing deformation and changing their position because of human intervention during the joining process. Finding an adaptive joint is a good balance between adjustability to achieve precision of the joint process while minimizing missed changes in the location of the joint position to achieve the desired roof surface. This study proposes two joint mechanisms adaptable for robotic assembly in bamboo structures. First are the Swivel Couples joints with clamp mechanisms with adjustable bolts, and second are the rope joint mechanisms (Fig. 10). Our future work will test the joinery system proposed to see its strengths and limitations for evaluation.

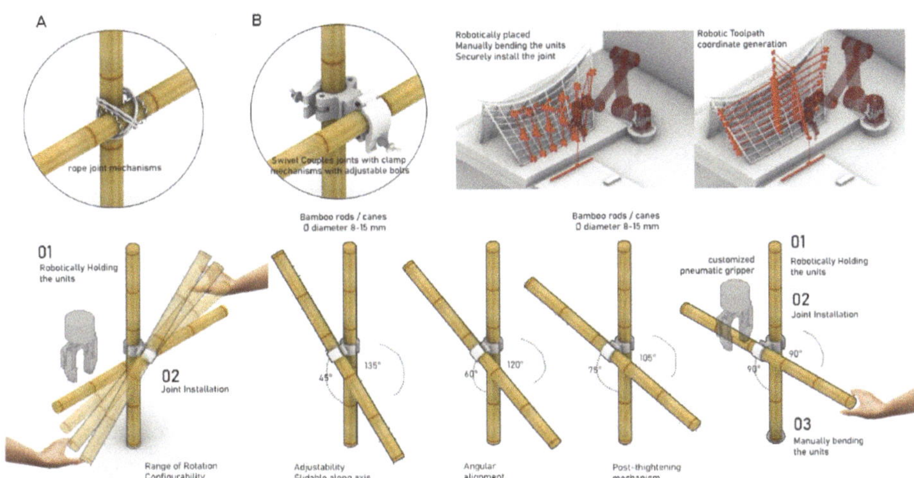

Fig. 10 Adaptive joinery system proposed with criteria

3.2 Robotic Workflows

To ensure compatibility with robotic-assisted construction, the roof prototype is scaled within the operational limits and reachability of the UR10e robotic arms. The UR10 has a payload of 10 kg and a maximum reach of 130 mm, which influenced the physical size of the roof prototype. These constraints require the overall structure to be scaled down proportionally to ensure that the joint installation point and bamboo elements are within the effective radius of the robotic setting. The robotic placement strategy is adapted by allowing sequential pickup and precise placement of the elements at the centre points of the roof for sequential robotic task planning. Their isoplanar position also generates precise alignment of different angles of vertical member units without repositioning. The horizontal units and overlapping joints could be positioned through controlled, repeatable robotic paths. A human agent will manually operate the bending task while securing the joint installation during robotic holding. By operating within this constraint, the initial test showcases the feasibility of the prototype demonstrated in the robotic system for material handling, holding the units, stabilizing the structure, joint alignment, and positioning tasks in the human-robot collaborative workflow. The robotic assembly fabrication plan and HRC workflow are illustrated in Figs. 11 and 12. The HRC framework includes the design process, model registration, material preparation for feeding the unit assembly, and the assembly process, which consists of sharing tasks between humans and robots in a sequential loop assembly..

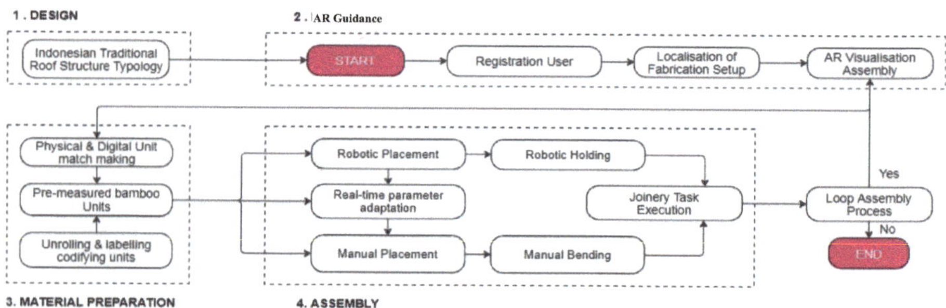

Fig. 11 Human-cobot collaboration (HRC) workflows, defining task sharing (human & robot)

3.3 Limitations

While the proposed HRC workflows demonstrate promising results in controlled and lab-scale settings, we consider several limitations for full-scale implementations. First, transitioning from a small-scale prototype to a full-size structure poses challenges regarding robot reachability, payload, workspace settings, and logistics. A robot system with a rail or mobile platform may extend or increase reachability for assembly. A mounted or movable robotic system could be used for larger assemblies and introduce new complexities in task coordination and spatial planning of robotic settings. Second, bamboo irregularities on the diameter surface and thickness potentially affect structural performance and alignment during micro-adjustment, as it is performed with active bending behaviour. Therefore, the structural integrity relies on the strength of joinery systems. Third, in the HRC workflows, coordination concerns depend on human precision during joint installation and micro-adjustment for final tightening and alignment after robotic placement. Manual bending and joint sequence deviation could lead to cumulative geometry deviations or the member units' desired curvature. These limitations highlight how

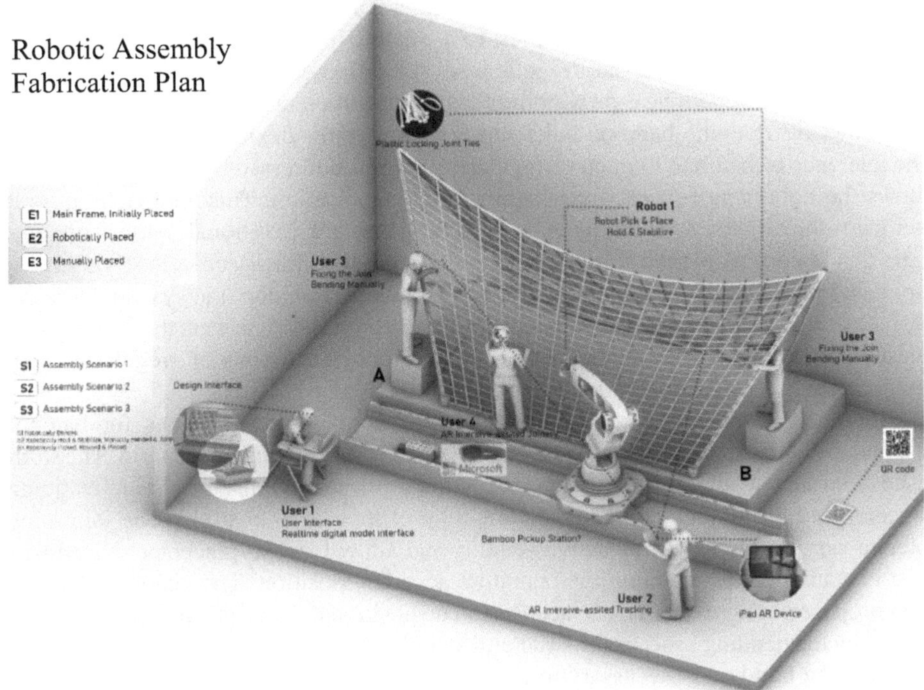

Fig. 12 Human-robot collaboration (HRC) fabrication plan and shared task scenario for the bamboo prototype

HRC could contribute to better intuitive integration between human actions and robotic operations in bamboo construction.

However, our objectives for the proposed workflow are not only to achieve perfect precision but also to explore tolerance and adaptability in Human-robot collaborative (HRC) settings. Rather than achieving exact and absolute precision outcomes, this study prioritizes the exploration of tolerance thresholds for the assembly following the irregular nature of bamboo materials. The nature of the HRC workflow with manual operations for joint tightening and bending adjustment inherently introduces variability. This study embraces that variability to allow controlled misalignment for on-site adaptability. Further work is needed to define the limits of acceptable deviation during assembly and an improved feedback mechanism that can assist human agents in fine-tuning the operation output during joint installation for improved assembly.

4 Conclusion

This research is developed to explore how adaptive bamboo joinery systems can support and be compatible with the robotic assembly while preserving traditional roof typologies. For our prototype purpose, the findings demonstrate the conclusion for adaptive joinery in a human-robotic collaboration scenario, as rather than focusing on high-precision customized joinery, the joinery systems must be adaptable for tolerance and adjustability to accommodate bamboo material irregularities. The proposed clamp-based joint mechanism to substitute repetitive rope winding can be more efficient with an adjustable system that allows overlapping bamboo units and is adaptable for flexible angular adaptation. The tolerance of bamboo diameter irregularities can be addressed by a joint system that allows the tightening process for post-structural stabilisation. This approach not only facilitates the alignment of the units but also replicates the traditional bolon roofs joint logic that is compatible with the fast pace of robotic workflows scenario. By allowing rapid assembly and the presence of human operators in the loop, the system addresses the balance of fabrication techniques and craftsmanship. These answers the first research question on how adaptive joinery systems support robotic assembly of irregular bamboo while preserving traditional indonesian roof typologies.

Additionally, the research examines how adaptive joinery systems for bamboo respond to active bending mechanisms applied in a robotic assembly scenario. The study discovered that successful robotic assembly under these circumstances requires joints capable of securing elements after deformation of their final position. The joint must tolerate varying forces during active vending and allow for post-placement adjustment by a human agent to maintain the curvature accuracy position. The clamp and swivel joint proposed in this research can accommodate positional shifts and angular deviations caused by the bending process for structural stability. Post-thickening for active bending material like bamboo is necessary to compensate for material relaxation and maintain elastic deformation. These answer the second research question on how joinery systems address the active bending mechanism in bamboo joints for robotic assembly scenarios.

In answering the third research question, we identified strengths of our approach in the human-robotic collaborative (HRC) workflows and cooperative process of assembling and bending bamboo and joint executions. These aspects facilitate intuitive interaction between the robot and manual adaptation of traditional bamboo construction. While the joint systems are a key critical point for bamboo structural integrity, it is inevitable not to integrate humans for joinery techniques, which are the skills that robots may struggle to adapt to. The adaptive bamboo joinery systems for a purpose prototype require a joint that facilitates rapid installation for the relatively fast pace of robotic assembly without disrupting the robotic sequence. While the research promotes joint system potential for scalability, our current experiment with a collaborative UR10 robot or a KUKA robot will face challenges for full-scale implementations. Our research emphasizes the human-robot collaborative process in bamboo construction and showcases its feasibility. Further development of future adaptations or full-scale implementations may require system modifications based on the same joinery principles. However, the limitation identified in our research is the absence of a real-time feedback mechanism, particularly during the manual bending execution of the human operators with robotic holding. This may introduce bending inaccuracy of the targeted curvature desired in the prototype to precisely align with the digital models. To address this limitation, further research will explore incorporating sensor-guided systems or AR-assisted tools to assess the bending deviation tolerance and ensure improved geometry assembly.

The study contributes to the growing application of hybrid human-robot collaboration that addresses natural traditional material logic like bamboo with robotic assembly workflows. Integrating computational design and digital fabrication research on bamboo promotes sustainability for contemporary applications of traditional and culturally responsive Indonesian architectural solutions.

Acknowledgements. The authors would like to express appreciation to the Indonesian Education Scholarships / *Beasiswa Pendidikan Indonesia* (BPI) Ref. Number: 0973/J5.2.3/BPI.LG/VIII/2022, Ministry of Education, Culture, Research, and Technology, and the Indonesian Endowment Fund for Education / *Lembaga Pengelola Dana Pendidikan* (LPDP), Ministry of Finance of the Republic of Indonesia, for financial support of the full scholarship.

References

1. Tanadini, D., et al.: Exploring the potential of equilibrium-based methods in additive manufacturing: The Digital Bamboo pavilion. In: Proceedings of the IASS 2022 Symposium: Innovation Sustainability Legacy (2022). https://doi.org/10.3929/ethz-b-000594566
2. Janssen, J.J.A.: Designing and Building with Bamboo. The Netherlands: International Network for Bamboo and Rattan, Eindhoven (2000)
3. Amtsberg, F., Mueller, C., Raspall, F. Di-terial – Matching digital fabrication and natural grown resources for the development of resource efficient structures. In: Proceedings of the 2021 DigitalFUTURES. Springer Singapore, pp. 330–339 (2022)
4. Lorenzo, R., Lee, C., Oliva-Salinas, J.G., Ontiveros-Hernandez, M.J.: BIM bamboo: A digital design framework for bamboo culms. Proc. Inst. Civil Eng. Struct. Build. **170**(4), 295–302 (2017)
5. Lorenzo, R., Godina, M., Mimendi, L., Li, H.: Determination of the physical and mechanical properties of moso, guadua and oldhamii bamboo assisted by robotic fabrication. J. Wood Sci. **66**(1), 1–11 (2020). https://doi.org/10.1186/s10086-020-01869-0
6. Lorenzo, R., Mimendi, L.: Digitisation of bamboo culms for structural applications. J. Build. Eng. **29**, 101193 (2020). https://doi.org/10.1016/j.jobe.2020.101193
7. Wang, T.H., Espinosa Trujillo, O., Chang, W.S., Deng, B.: Encoding bamboo's nature for freeform structure design. Int. J. Archit. Comput. **15**(2), 169–182 (2017)
8. Trujillo, O.E., Wang, T.H.: Parametric modeling of bamboo pole joints. Commun. Comput. Inform. Sci. **527**, 272–290 (2015)
9. Wu, N.H., Dimopoulou, M., Hsieh, H.H., Chatzakis, C.: A digital system for AR fabrication of bamboo structures through the discrete digitization of bamboo. In: Proceedings of the International Conference on Education and Research in Computer-Aided Architectural Design in Europe, pp. 161–170 (2019)
10. Kladeftira, M., Leschok, M., Skevaki, E., Tanadini, D., Ohlbrock, P.O., D'Acunto, P.: Digital bamboo: A study on bamboo, 3D printed joints, and digitally fabricated building components for ultralight architectures. In: 41st Annual Conference of the Association for Computer Aided Design in Architecture (ACADIA), Hybrids & Haecceities, pp. 406–417 (2022). https://doi.org/10.3929/ethz-b-000667377
11. Di Paola, F., Mercurio, A.: Design and digital fabrication of a parametric joint for bamboo sustainable structures. In: Advances in Intelligent Systems and Computing, pp. 180–189. Springer Verlag (2020)
12. Matson, C.W., Sweet, K.: Simplified for resilience: A parametric investigation into a bespoke joint system for bamboo. In: Sociedad Iberoamericana de Gráfica Digital (SIGraDI). Blucher Design Proceedings, vol. 3 (1), pp. 405–411 (2016)

13. Amtsberg, F., Raspall, F.: Bamboo^3. In: Computer-Aided Architectural Design Research in Asia (CAADRIA) 2019, vol. 1, pp. 245–254 (2018)
14. Mitterberger, D., Atanasova, L., Dörfler, K., Gramazio, F., Kohler, M.: Tie a knot: Human-robot cooperative workflow for assembling wooden structures using rope joints. Constr. Rob. **6**, 277–292 (2022). https://doi.org/10.1007/s41693-022-00083-2
15. Vasiliki, E., Joseph, A., Kenny, C., Atanasova, L., Casas, G., Dörfler, K.: Cooperative augmented assembly (CAA): Augmented reality for on-site cooperative robotic fabrication. Constr. Rob. **8**(2), 28 (2024). https://doi.org/10.1007/s41693-024-00138-6
16. Mansuri, A., Agkathidis, A., Lombardi, D., Chen, H.: Rethinking traditional indonesian roof bamboo frame structures by utilizing parametric tools and automated fabrication techniques: A systematic review. In: International Conference of the Arab Society for Computation in Architecture, Art, and Design (ASCAAD). Arab Society for Computation in Architecture, Art, and Design. 5000 Thayer Ctr Ste C, Oakland, MD 21550–1139, USA. ISSN: 2994–9777, pp. 110–133 (2023)
17. Mansuri, A., Agkathidis, A., Lombardi, D., Chen, H.: Rethinking bamboo roof-based architecture of indonesian traditional house using parametric design and automated fabrication techniques. In: Nicosia, pp. 203–212. http://papers.cumincad.org/cgi-bin/works/paper/ecaade2024_367 (2024)
18. Arce Villalobos, O.A.: Fundamentals of the Design of Bamboo Structures. Technische Universiteit Eindhoven, Eindhoven (1993)
19. Huang, Z.: Model Study on the Application of Bamboo in Building Envelope, pp. 81–118. Green Energy and Technology (2019)
20. Prasetyo, Y.H., et al.: Tropic climate form of nusantara traditional architecture's expression in regionalism. Settl. J. **12**(2), 80–93 (2017)
21. Qi, Y., et al.: Working with uncertainties: An adaptive fabrication workflow for bamboo structures. In: Proceedings of the 2020 DigitalFUTURES, pp. 265–279. Springer Singapore (2021)

BarkBeetle: Generating Conformal 3D Printing Toolpaths with a Skeleton Graph Approach

Shaoyi Wang and Simon Schleicher(✉)

University of California, Wurster Hall 94720, Berkeley, USA
{shaoyiwang,simon_s}@berkeley.edu

Abstract. Current layer-to-layer manufacturing methods face significant limitations when constructing curved forms and addressing unsupported geometries such as overhangs and bridging structures - a key bottleneck preventing the expansion of this technology to architectural scale applications and inhibiting 3D printing beyond vertical walls toward horizontal floor slabs and wide-spanning roof structures. To address these challenges, we developed BarkBeetle, a Grasshopper plugin that generates optimized toolpaths specifically for non-planar, conformal 3D printing on curved surfaces using strip-based formwork systems to create wide-spanning hybrid gridshell structures. The plugin employs a skeleton graph methodology as the primary framework for printing walls, floor slabs and roof gridshells, unifying the toolpath generation process for diverse complex geometric forms into a streamlined workflow. Additionally, BarkBeetle provides modular combinations of skeleton types, infill patterns, and layer stacking strategies, allowing designers to tailor printing parameters to specific architectural requirements and performance criteria.

Keywords: Conformal 3D printing · Continuous material deposition · Grasshopper plugin · Robotic fabrication · Toolpath generation

1 Introduction

The rise of large-scale 3D printing presents an opportunity to revolutionize conventional construction methods by improving process efficiency [1], reducing cost [2], and minimizing environmental impact [3]. However, current Layered Manufacturing (LM) techniques face significant limitations when constructing curved shapes. First, these methods often result in a stair-stepping effect that compromises surface quality and mechanical properties, particularly at larger scales. Additionally, LM workflows cannot address unsupported geometries such as overhangs and bridging structures without relying on extensive support materials. Conventional approaches to 3D printing curved or wide-spanning structures require substantial formwork, generating considerable material waste and increasing fabrication complexity. Moreover, printing on non-planar or geometrically complex surfaces – especially double-curved geometries such as synclastic (dome-shaped) or anticlastic (saddle-shaped) forms – remains challenging for existing LM techniques due to their typically coplanar slicing approach.

Y. Liu et al. (Eds.): CDRF 2025, *Transindividual Intelligence*, pp. 255–264, 2026.
https://doi.org/10.1007/978-981-92-0615-5_22

1.1 3D Printing and Conformal 3D Printing

Three-dimensional (3D) printing, also known as Additive Manufacturing (AM), is a production method that automatically creates complex geometries from 3D computer-aided design models without conventional tooling, dies, or fixtures. This technology is gaining more widespread adoption across different industries due to its ability to rapidly produce functional prototypes with minimal human labor and reduced material waste [4]. In architectural construction, 3D printing primarily employs extrusion-based processes that deposit viscous, cement-based materials along predetermined toolpaths. Current implementation strategies in the field follow two distinct approaches: some research initiatives and projects utilize prefabrication of components followed by on-site assembly [5], while others employ in-situ fabrication techniques for large-scale applications ranging from individual building walls to complete multi-story structures [6]. Conformal 3D printing is an additive manufacturing method where materials are applied directly onto a non-planar surface. In this method, materials are layered sequentially, with the Z-axis coordinates adjusting automatically to match the base surface during printing [7]. These adjustments respond to the surface's contours and complexity. Unlike traditional Layer Manufacturing (LM) methods that utilize fixed layer heights and vertical movements between layers, conformal 3D printing keeps a consistent distance between the nozzle and the base surface. This is achieved through an offset perpendicular to the base surface [8]. Recent advancements in this field have integrated vision-based sensing with conformal 3D concrete printing, enhancing its adaptability to the on-site work environment [9].

1.2 3D Printing on Bending-Active Formwork

Conformal 3D printing uses auxiliary structures or formwork as bases for depositing materials. Sub-additive Manufacturing and Non-planar Granular 3D Printing, for instance, uses granular aggregates as reusable support to efficiently create doubly curved arched concrete structures, enhancing printing speed and reducing material usage [10, 11]. These methods significantly reduces material use and minimizes waste from support structures. Additionally, there are explorations into 3D printing synclastic and anticlastic surfaces using adaptable membrane formwork, broadening the applications of conformal printing techniques [12].

Another research trajectory is the use of bending-active structures as flexible formwork and conformal 3D printing [13]. Bending-active structures are curved structures formed by the elastic deformation of initially flat elements [14, 15]. These lightweight structures offer a cost-effective alternative to traditional construction techniques and are particularly effective for creating complex curved surfaces. Bending-active structures can function as flexible formworks [16]. Research indicates that a hybrid gridshell, which combines bending-active formwork with a concrete layer on top, achieves considerable strength [13]. This demonstrates the potential of bending-active structures as lightweight formwork in the construction industry, potentially offering greater precision, accommodating more complex geometries, and reducing material usage compared to traditional casting methods.

1.3 Toolpaths from Skeleton Graphs

Several studies have applied the approach of using a skeleton graph to describe the structure of a 3D printing object. For instance, a skeleton approach can be used to partition models and minimize support structures [17]. In construction-scale 3D printing, researchers have discretized target geometry into printable layers and micro-blocks using skeleton graphs to ensure structural equilibrium throughout the fabrication process and overall shape. In this context, the skeleton graph functions as the scalar (vector) field's medial axis, facilitating essential outward radiation for precise material deposition [18]. Furthermore, recent developments in self-supporting shell structures leverage skeleton methods for geometrical and structural optimization in large-scale 3D printing applications [19].

A critical challenge in concrete 3D printing is maintaining toolpath continuity throughout the fabrication process. Continuous toolpaths are fundamental for ensuring seamless material bonding between adjacent layers and enhancing the overall structural integrity of printed components. Interruptions in the extrusion process frequently cause geometric inaccuracies due to the concrete pump's inherent delay characteristics, making consistent material flow resumption problematic after stoppage events. By employing the skeleton approach to characterize geometric properties, designers can achieve holistic control over both the printed object and the fabrication process, thereby enabling the completion of complex, large-scale structures with continuous, optimized toolpaths. Until now, however, there has been no comprehensive tool for creating 3D printing with skeleton graphs.

2 Objectives

This study proposes BarkBeetle, a Grasshopper plugin that generates conformal 3D printing toolpaths for curved surfaces using a skeleton graph approach. BarkBeetle is designed specifically for layered extrusions on non-planar surfaces as well as wide-spanning gridshell slabs and roofs on strip-composed formwork. The plugin ensures toolpath continuity to minimize interruptions during printing, maintains adequate material coverage for complex geometries, and generates high-quality curved surface printing toolpaths. Additionally, BarkBeetle adopts a user-friendly perspective, offering customized toolpath generation options to meet specific project requirements, such as various toolpath patterns and layer stacking methods. The tool seamlessly integrates existing robotic control software, facilitating efficient toolpath execution on industrial robotic arms.

3 BarkBeetle Workflow

BarkBeetle is a Grasshopper-based parametric design tool to generate conformal and continuous 3D printing toolpaths on user-specific base surfaces (Fig. 1). Implemented in Visual Studio using C# with Rhino Common and Grasshopper API integration, the plugin follows a structured workflow (Fig. 2) that progresses from input base geometries through systematic skeleton graph generation and toolpath pattern creation, culminating in the compilation of patterns into a comprehensive toolpath stack consisting of multiple printing layers. The resulting output is fully compatible with industry-standard robotic 3D printing interfaces, including KUKA PRC and the Robot Grasshopper plugin, enabling seamless implementation in digital fabrication environments.

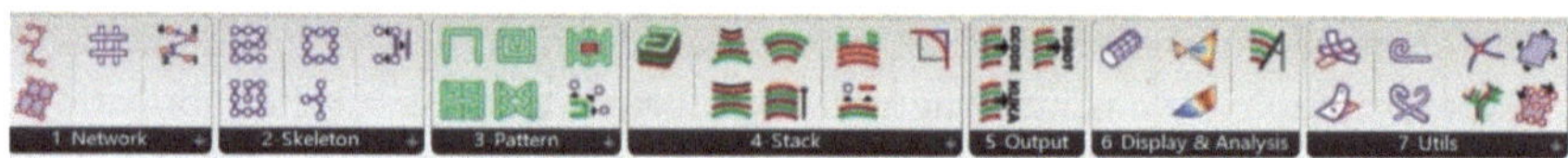

Fig. 1 BarkBeetle toolbar

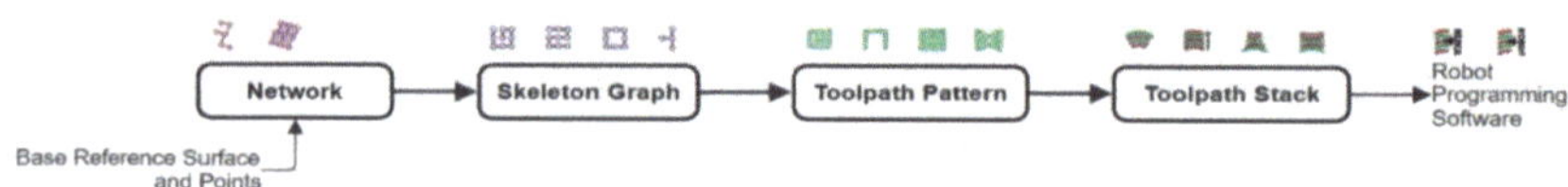

Fig. 2 Overall BarkBeetle workflow

3.1 Network: Initial Setup

The BarkBeetle toolpath generation process begins by defining a Network, which consists of a series or array of points that describes the intersections or turnings in the 3D printing process. This Network also consolidates all pertinent information required for subsequent toolpath generation, including the base reference surface, rib/wall thickness, and if needed, scanned real-world formwork in either mesh or point format.

BarkBeetle can generate a linear network from a point list, or a UV network from a point tree (Fig. 3). A linear network is ideal for fabricating layered extrusions where the input point list describes a linear geometry. This linear network can also incorporate additional branch points to create branched printings. A UV network is designed for 3D printing on gridshell formworks, utilizing an organized point array that represents the intersection points within the gridshell.

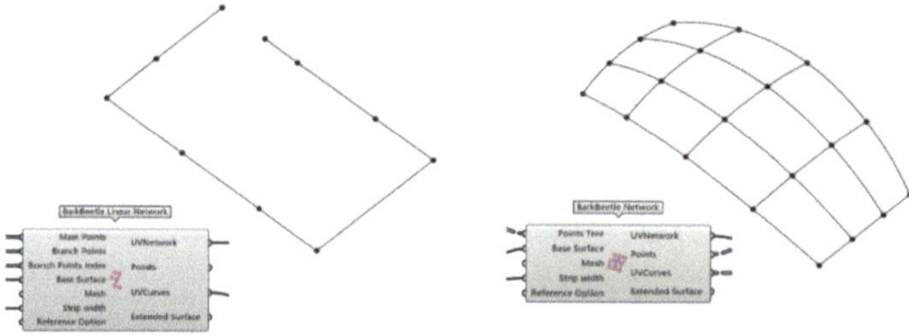

Fig. 3 A linear network (left) and a UV network (right)

3.2 Skeleton Graph: Printing Structure

As previously discussed, continuous toolpaths are crucial for ensuring seamless material bonding and enhancing the structural integrity of the printed object. To solve this challenge, we use a skeleton graph as the guiding framework for subsequent toolpath generation steps. We transform a network input into a skeleton graph, which represents the printing structure with vertices and curves. The primary curve in the skeleton graph is a single continuous, non-intersecting curve designed to ensure continuity throughout the process. Branch curves are created for branched layered extrusions or ensure sufficient material coverage on gridshell formworks.

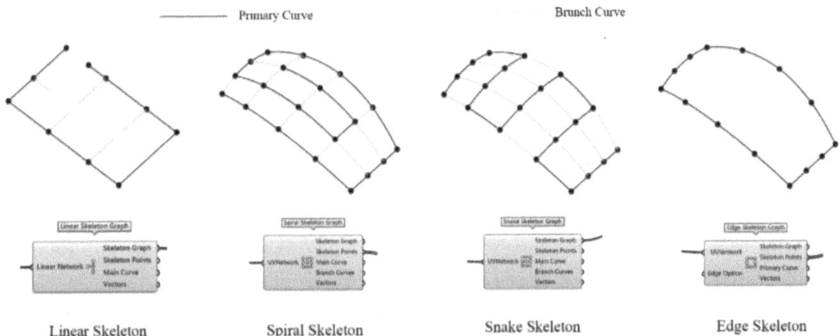

Fig. 4 Various skeleton graphs: linear-, spiral-, snake-, and edge-skeleton (from left to right)

BarkBeetle provides four distinct skeleton graphs developed from different algorithms: linear, spiral, snake, and edge skeletons (Fig. 4). The linear skeleton uses a linear network input for layered extrusions, while the other three utilize a UV network for gridshell formwork printings. The variety of skeletons offers users multiple toolpath design options and enables the generation of unique print paths across layers. This capability of BarkBeetle provides opportunities for future investigations into how varied toolpath strategies may influence material bonding—particularly in terms of interlayer contact, stress distribution, and mechanical interlock between layers.

3.3 Toolpath Pattern: Designing the First Print Layer

The toolpath pattern consists of a single-layer printing path derived from the skeleton graph. It is formed by connecting multiple closed curves around the skeleton graph into a single continuous toolpath without overlapping. Based on the four skeleton graphs described before, users can then choose between four distinct infill patterns: simple-, spiral-, snake-, and zig-zag pattern (Fig. 5).

A simple pattern consists of a single closed curve around the skeleton graph, ideal for simple walls, ribs, and gridshell structures. The spiral pattern is generated by repeatedly offsetting the skeleton graph and connecting the resultant curves to achieve a 100% infill rate. Both the snake infill and zig-zag infill patterns allow adjustments in spacing to vary

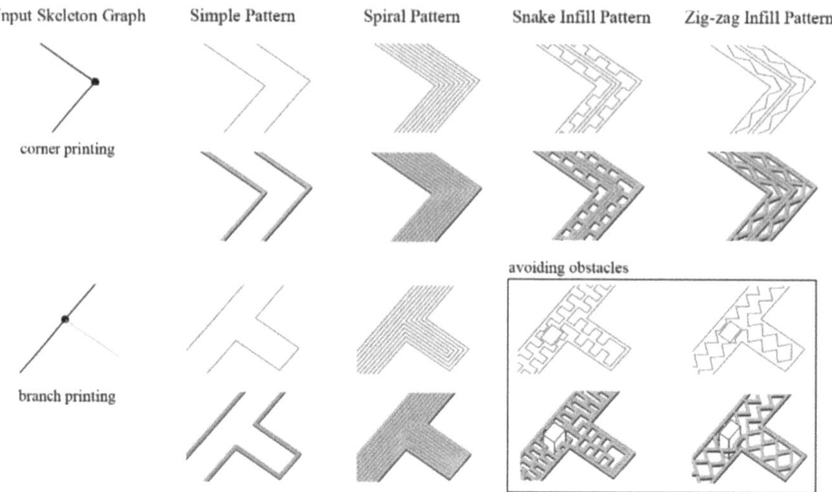

Fig. 5 Toolpath patterns generated from BarkBeetle

the infill rate. Additionally, for these two patterns, one or two print paths can be inserted in the middle to enhance material contact during printing.

Occasionally in 3D printed architecture, it is necessary to avoid specific areas, such as fasteners, building systems, pipelines within layered extrusions, or special connections between the formwork and the printing material such as shear blocks that enhance bonding at the interface. BarkBeetle features an "avoid obstacle" function, which allows toolpaths to navigate around these areas by inputting the size and location of obstacles, while still maintaining the continuity of the toolpath (Fig. 5). This functionality is applicable to both the snake infill and zig-zag infill patterns.

3.4 Toolpath Stack: Designing Multiple Layers of Printing

Traditional 3D printing typically involves slicing the geometry with flat planes to create multilayer toolpaths. However, this flat-layer approach can result in a stair-stepping effect during large-scale curved surface printing. Therefore, BarkBeetle offers various conformal 3D printing layer stacking methods (Fig. 6) to achieve uniform and smooth printing outcomes. Each toolpath stack produces a continuous toolpath curve, along with a list of planes that determine the printing direction and orientation at each location.

The vertical stack method replicates the base surface along the Z-axis to generate each layer while the offset stack is a layer stacking method where the surfaces of print layers are successively offset normal to the base surface. Both methods maintain consistent layer heights. Conversely, the vertical-tween stack and surface-tween stack methods enable the creation of variable layer height printing between a base and a top surface. Material extrusion height can be controlled by adjusting the robot's moving speed when layer heights vary.

Furthermore, BarkBeetle supports varied toolpath patterns across different printing layers. For example, the first layer might need complete material coverage, whereas

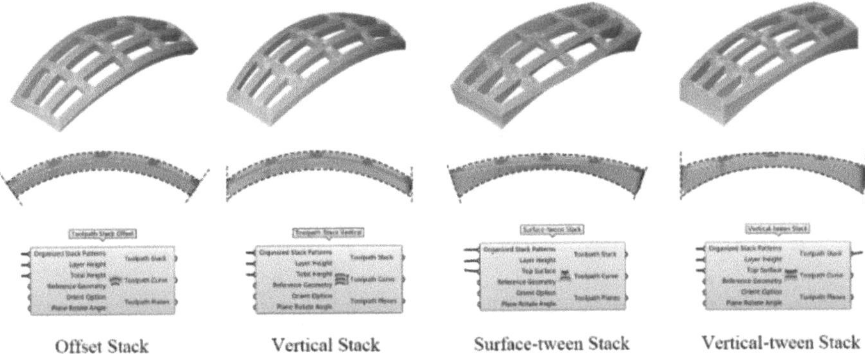

Fig. 6 Results from different toolpath stacking methods

subsequent layers could be designed with lower material density to minimize consumption. After all layer toolpaths are generated, BarkBeetle offers component interfaces for smooth integration with other robotic control plugins like KUKA PRC and Robots, transmitting data including tool tip direction and printing speed.

4 Application Examples: 3D Printing Walls, Slabs, and Roofs

BarkBeetle provides a comprehensive workflow for creating not only layered extrusions such as walls, but also lightweight curved roofs and slabs spanning long distances with minimal formwork material usage. This capability extends the potential of 3D printing in construction beyond vertical elements, enabling the efficient fabrication of horizontal and spanning structural components such as slabs and roofs.

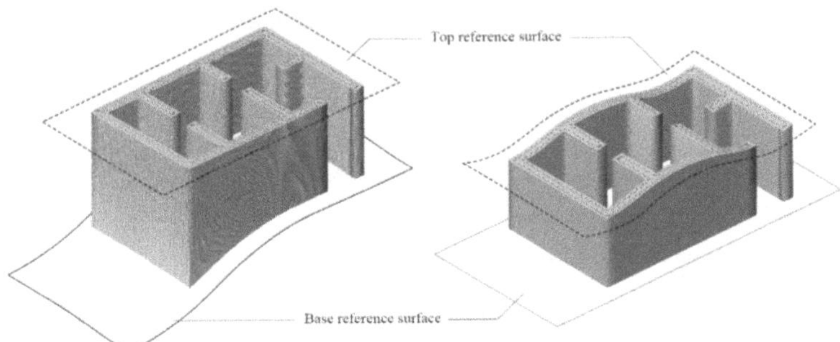

Fig. 7 Wall printings generated with BarkBeetle, each is a single continuous toolpath

Walls: BarkBeetle can generate toolpaths for non-planar layered extrusions, such as those on sloped foundations (Fig. 7, left) or when the tops of the extrusions are not flat (Fig. 7, right). Using the linear network and linear skeleton, BarkBeetle can produce interior walls and integrate the printing of all elements into a single continuous

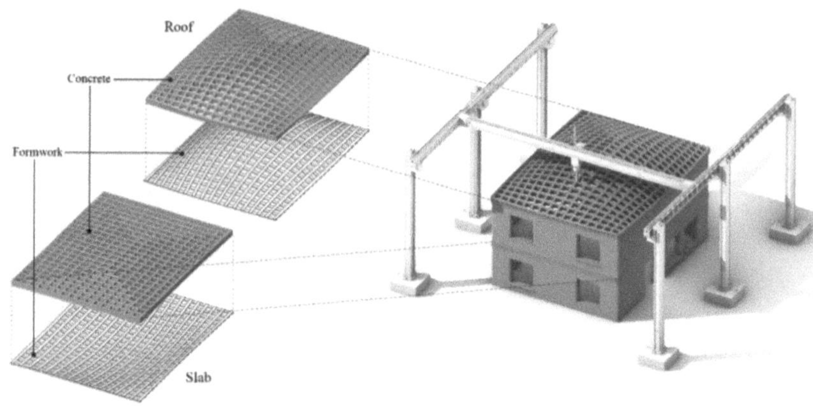

Fig. 8 BarkBeetle examples of a curved roof (left) and flat slab with curved bottom surface (right) printed on bending-active formwork, both toolpaths are continuous

toolpath. The toolpath can also accommodate designated areas for utility conduits by implementing appropriate toolpath patterns.

Slabs and Roofs: Integrating bending-active structures as a stay-in-place formwork with robotic 3D printing provides an efficient method for developing lightweight roofing and slab systems (Fig. 8). In this approach, printed materials such as concrete function as the primary compressive elements, while bending-active structures can serve as the principal tensile reinforcement for the 3D printed structure. Using the UV network and spiral/snake skeleton features of BarkBeetle, users can create a skeleton graph that connects all intersection points within a gridshell formwork. After selecting an appropriate toolpath pattern, designers can develop curved roof print paths using the offset stack function, for instance, or create material-efficient slabs by establishing multiple twining layers between a curved bottom surface and a flat top surface. The detailed mechanics of this approach will be addressed in a subsequent publication.

5 Conclusion

BarkBeetle is designed to create conformal 3D printing on curved reference surfaces. It utilizes a skeleton graph method to generate toolpaths suitable for wall-like layered extrusions as well as wide-spanning floor slabs and gridshell-like roof structures printed on a bending-active formwork. The structured workflow begins with a network created from input base geometries, transforms this network into a skeleton graph that outlines the structure of the printing object, and then proceeds to create toolpath patterns and stacks to construct the entire object. This approach ensures a continuous toolpath, minimizing interruptions during the printing process. BarkBeetle supports the approach of 3D printing on bending-active formwork, reducing the need for supporting materials when printing large-scale roofs and slabs.

Moreover, the software allows customized toolpath generation tailored to specific project requirements, such as layer stacking sequences, toolpath patterns, and obstacle avoidance, giving users precise control over print quality and material usage. The tool also seamlessly integrates with existing robotic control software, facilitating efficient execution of toolpaths on industrial robotic arms.

Future development of BarkBeetle will focus on expanding the capabilities of the skeleton generation process, making it adaptable to a wider range of geometrical topologies beyond linear extrusions and UV gridshells. Future research will integrate structural performance assessments to evaluate how different toolpath strategies influence the behavior of the resulting hybrid systems. Quantitative analysis of efficiency, material use, and structural performance will help position BarkBeetle more clearly within current construction practices. In parallel, physical 3D printing experiments will be carried out to validate and refine the generated toolpaths. These experiments will inform the iterative development of the tool, improving its reliability and performance in real-world fabrication scenarios.

References

1. Rushing, T.S., et al.: Chapter 7 - Investigation of concrete mixtures for additive construction. In: Sanjayan, J.G., Nazari, A., Nematollahi, B. (eds.) 3D Concrete Printing Technology, pp. 137–160. Butterworth-Heinemann (2019). https://www.sciencedirect.com/science/article/pii/B9780128154816000075
2. García de Soto, B., et al.: Productivity of digital fabrication in construction: Cost and time analysis of a robotically built wall. Autom. Constr. **92**, 297–311 (2018)
3. Oorschot, J.A.W.H.V., Halman, J.I.M., Hofman, E.: Getting innovations adopted in the housing sector. Constr. Innov. **20**(2), 285–318 (2020)
4. Tay, Y.W.D., Panda, B., Paul, S.C., Noor Mohamed, N.A., Tan, M.J., Leong, K.F.: 3D printing trends in building and construction industry: A review. Virtual Phys. Prototyp. **12**(3), 261–276 (2017)
5. Bhooshan, S., et al.: The Striatus bridge: Computational design and robotic fabrication of an unreinforced, 3D-concrete-printed masonry arch bridge. In: Architecture, Structures and Construction, pp. 521–543. Springer (2022). https://www.research-collection.ethz.ch/handle/20.500.11850/565076
6. PERI: PERI company history. PERI 3D printing project "Beckum" wins German Innovation Award. https://www.peri.com/en/company/history.html#&gid=1&pid=4 (2020). Accessed 17 Dec 2024
7. Alkadi, F., Lee, K.C., Bashiri, A.H., Choi, J.W.: Conformal additive manufacturing using a direct-print process. Addit. Manuf. **32**, 100975 (2020)
8. Chakraborty, D., Reddy, B.A., Choudhury, A.R.: Extruder path generation for curved layer fused deposition modeling. Comput. Aided Des. **40**(2), 235–243 (2008)
9. Çapunaman, Ö.B., Farrokhsiar, P., Bilén, S.G., Duarte, J.P., Gürsoy, B.: Vision-based sensing and digital twin technologies in conformal 3D concrete printing: Exploring operational accuracy, adaptability, and scalability, and investigating monitoring capabilities in large-scale applications. Constr. Robot. **9**(1), 4 (2025)
10. Battaglia, C.A., Miller, M.F., Zivkovic, S.: Sub-additive 3D printing of optimized double curved concrete lattice structures. In: Willmann, J., Block, P., Hutter, M., Byrne, K., Schork, T. (eds.) Robotic Fabrication in Architecture, Art and Design 2018, pp. 242–255. Springer International Publishing, Cham (2019)
11. Darweesh, B., Gutierrez, M.P., Schleicher, S.: Non-planar granular 3D printing. Constr. Robot. **7**(3–4), 291–306 (2023)

12. Lim, J.H., Weng, Y., Pham, Q.C.: 3D printing of curved concrete surfaces using Adaptable Membrane Formwork. Constr. Build. Mater. **232**, 117075 (2020)
13. Schleicher, S., Herrmann, M.: Constructing hybrid gridshells using bending-active formwork. Int. J. Space Struct. **35**(3), 80–89 (2020)
14. Lienhard, J., Alpermann, H., Gengnagel, C., Knippers, J.: Active bending, a review on structures where bending is used as a self-formation process. Int. J. Space Struct. **28**(3–4), 187–196 (2013)
15. Schleicher, S., La Magna, R., Knippers, J.: Bending-active Plates: Planning and construction. In: Fabricate: Rethinking Design and Construction, vol. 5, pp. 242–249 (2017)
16. Scheder-Bieschin, L., Van Mele, T., Block, P.: Bending-active formwork systems for concrete shells – A classification and state-of-the-art review. Structure. **67**, 106841 (2024)
17. Wei, X., et al.: Toward support-free 3D printing: A skeletal approach for partitioning models. IEEE Trans. Vis. Comput. Graph. **24**(10), 2799–2812 (2018)
18. Bhooshan, S., Van Mele, T., Block, P.: Equilibrium-aware shape design for concrete printing. In: De Rycke, K., et al. (eds.) Humanizing Digital Reality, pp. 493–508. Springer Singapore, Singapore (2018). http://link.springer.com/10.1007/978-981-10-6611-5_42
19. Zhi, Y., Chai, H., Teng, T., Akbarzadeh, M.: Local optimization of self-supporting shell structures in 3D printing: a skeleton method. In: Proceedings of the IASS Annual Symposium 2023, Melbourne, Australia (2023)

Confinement Effect of FRP Formwork by Additively Laminated Manufacturing (ALM) with UV Rapid Curing Process

Dan Luo(✉), Chenhao Wang, and Zhuoyang Xin

St Lucia, Brisbane, QLD 4072, Australia
d.luo@uq.edu.au

Abstract. Recent advancements in additive manufacturing (AM) have enabled rapid fabrication of complex construction components. Additive Laminated Manufacturing (ALM) is an emerging AM technique distinct from Fused Deposition Modeling (FDM), utilizing transverse lamination instead of traditional layer-by-layer deposition. Previous ALM systems relied on structural deoxygenation resins with fast-curing agents, which, despite high performance, required lengthy curing times. This study proposes the use of UV-curable resin in resin-based ALM, significantly accelerating curing and enhancing efficiency. The performance of FRP formwork produced with this improved system is systematically evaluated, and a strength model is established to correlate confining strength with overlapping layers. Experimental tests confirm that increased overlaps enhance the structural performance of FRP formwork in concrete structures. These findings contribute to the optimization of FRP formwork design and demonstrate that UV-cured resin-based ALM can achieve structural performance comparable to traditional chemically cured systems.

Keywords: Additive manufacture · construction automation · robotic fabrication

1 Introduction

The emergence and maturity of additive manufacturing (AM) technologies have enabled rapid fabrication of construction components with complex geometric requirements, aiming to reduce reliance on traditional concrete formwork and lower construction costs [5]. For instance, 3D Concrete Printing (3DCP) technology eliminates the dependence on concrete formwork, allowing for rapid construction of large-scale structures with complex geometries. However, concrete structures produced by this method are constrained by limited interlayer bonding strength [3], and the inability to integrate reinforcing steel further restricts structural stability [1]. Additionally, the layer-by-layer extrusion method demands materials with suitable flowability, restricting the use of coarse aggregates with larger particle sizes and complicating material selection [7].

Compared with directly printing concrete, shaping it via formwork offers certain structural advantages by optimizing the casting material constitution, increasing the

Y. Liu et al. (Eds.): CDRF 2025, *Transindividual Intelligence*, pp. 265–275, 2026.
https://doi.org/10.1007/978-981-92-0615-5_23

material integrity and providing opportunity for internal steel reinforcement. Yet traditional formwork is highly labour-intensive, drives up construction costs, and increases material waste. Additively manufactured formwork can precisely control material usage and accommodate intricate, non-standard geometries. Current formwork 3D printing mostly utilizes polymer materials, providing limited structural performance, modestly enhancing concrete's confining strength and being limited by build-volume constraints, making large-scale formwork fabrication challenging. Additive Laminated Manufacturing (ALM), a newly proposed AM formwork production technique, utilize GFRP material to addresses these shortcomings as the utilized GFRP material exhibits favourable attributes. GFRP possess notable advantages for manufacturing durable, lightweight formwork due to their high strength-to-weight ratio and corrosion resistance [2], and perform excellent ductility, strength, modulus, and deformability as well [2, 11]. In short, ALM enables the fabrication of FRP concrete formwork without size limitations, while still meeting the required strength criteria.

ALM employs industrial robots to fabricate concrete formwork. A visual programming environment is used to predefine the robot's motion path, allowing the robotic arm to exert precise control over the fabrication process. FRP formwork produced via ALM not only meets the operational demands of conventional formwork but also enhances the confining strength of concrete. Prior studies have established the feasibility of ALM for fabricating concrete formwork through a series of preliminary tests, identifying suitable manufacturing parameters such as overlap height per layer, winding speed, and resin types [6]. Additionally, the use of FRP fabric materials increases the material surface coverage area, thus enhancing manufacturing efficiency. Although structural deoxygenation resin further enhances the confining strength of the resulting formwork, its chemically based curing process extends curing times and reduces overall manufacturing efficiency. Moreover, FRP is a flexible material whose forming process depends on resin curing. During curing, the material is susceptible to deformation and collapse under gravity and the pressure exerted by the customized end-effector, thereby affecting the accuracy of the final product and potentially undermining the performance of both the formwork and the concrete–formwork assembly.

Following the initial introduction of ALM, subsequent enhancements is to replace the chemically reactive structural resin with a UV-activated resin that rapidly cures within seconds of exposure to ultraviolet (UV) light at specific wavelengths. Compared to chemical curing, this approach significantly shortens curing time and increases manufacturing efficiency. Moreover, iterative development of the customized tooling head has integrated multiple functions—including material deposition, resin application, and curing—into a single device. In this research the performance of FRP formwork produced with this improved system is sys-thematically evaluated in relationship with traveling speed, and a strength model is established to correlate confining strength with overlapping layers, as well as defect analysis. Because the FRP formwork produced by this method effectively functions as a structural form, its capacity to confine concrete is one of the most crucial metrics for evaluating the overall quality of the formwork.

2 Methodology

This study employs an ALM system with UV-curable resin to fabricate FRP formwork for concrete, investigating both the degree of confining force exerted by the formwork and the extent to which it enhances the overall performance of the concrete structure This study replicated and tested an ALM system incorporating UV-curable resin to fabricate prototype FRP formworks (Fig. 1). Tensile strength tests were conducted on ring-shaped samples extracted from these prototypes to investigate whether the formwork would fail due to hydrostatic pressure exerted by fresh concrete. Additionally, compressive strength tests were performed on concrete-filled FRP formworks to evaluate their confining strength and determine whether it matches the performance achieved by formworks fabricated with structural deoxygenation resin.

Fig. 1. ALM fabrication timelapse example

2.1 System Overview

Differing from earlier versions, the ALM system used in this study replaces the structural deoxygenation resin with a photoactivated resin. This change enables the output of FRP fabric, the application of the resin, and the curing process to occur synchronously. Consequently, the three key operations - material deposition, resin application, and UV exposure are consolidated into a single end effector (Fig. 2), thereby increasing overall manufacturing efficiency.

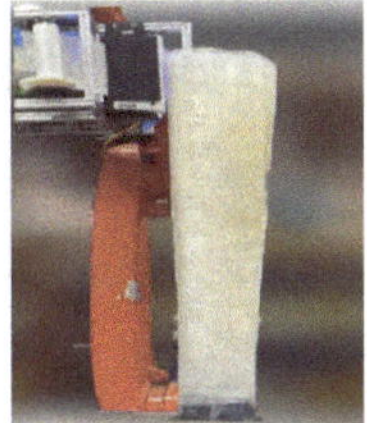

Fig. 2. Illustration of ALM Manufacturing

The general workflow of the ALM system involves importing the geometric model of the desired formwork into Rhino/Grasshopper, programming the robotic arm's motion path, uploading the path to the arm's controller, mounting the end-effector on the robotic arm, and executing the predesigned path to fabricate the formwork.

Although GFRP fabric offers favourable structural properties, it remains a flexible material. During the fabrication process, only a small base template supports the bottom portion; beyond that, the structure relies entirely on the curing of the resin to counteract gravitational forces and prevent lateral collapse of the fabric. After switching to a UV-curable resin, the ALM process no longer applies resin across the entire fabric surface. Instead, four brushes on the end-effector deposit four approximately equal strips of UV-curable resin on the fabric, thus improving interlayer adhesion and enhancing manufacturing efficiency.

Due to these characteristics of FRP materials, manufacturing parameters must be strictly controlled during fabrication to prevent uncontrollable deformation and potential failure of the FRP formwork.

2.2 Ring Tensile Test

Since FRP is a flexible material with good ductility, the hydrostatic pressure exerted by concrete during pouring can affect the structural performance of FRP formwork, potentially leading to its failure. Therefore, circumferential tensile tests must be conducted on the fabricated FRP formwork to investigate the impact of hydrostatic pressure and assess the formwork's resistance to deformation and failure.

Therefore, to investigate the ability of FRP formwork to withstand hydrostatic pressure, circumferential tensile tests were conducted on the fabricated specimens. Three sets of samples were produced using ALM, each with a different number of overlapping numbers (Fig. 3): 10, 15, and 20. Each set was then sectioned by height into five ring-shaped specimens to examine the influence of varying overlapping numbers on hydrostatic pressure (Fig. 4).

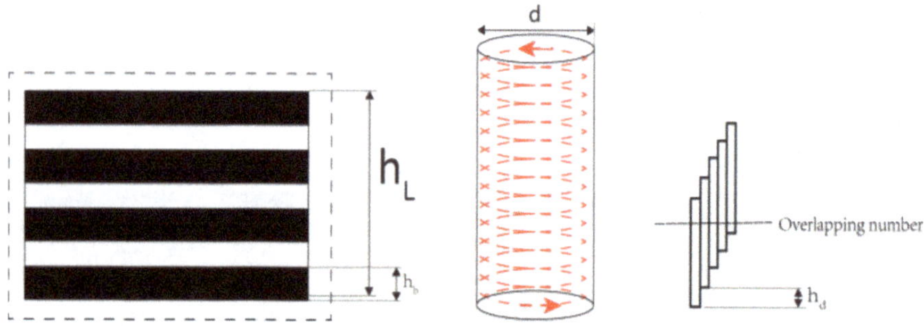

Fig. 3. Diagram of adhesive coverage on the fiber surface and layer over lapping number

Moreover, this hydrostatic pressure is influenced by the height of the concrete column. As the height of the concrete column increases, a greater volume of concrete is poured, resulting in higher hydrostatic pressure at the base of the FRP formwork. This hydrostatic pressure can be expressed mathematically as follows [10]:

$$F_{test} = d \int_0^{\pi} w_p R sin\theta d\theta = dw_p D \quad (1)$$

Fig. 4. (a) Ring-10: Ring sample which overlapping number is 10; (b) Ring-15: Ring sample which overlapping number is 15; (c) Ring-20: Ring sample which overlapping number is 20

2.3 Confining Strength of FRP Formwork

This study investigated the confinement strength provided by FRP formwork to concrete elements. Direct measurement of confinement strength poses challenges, as this confining effect is primarily reflected through variations in stress and strain under loading conditions. Therefore, compressive strength tests were conducted to indirectly evaluate the confining effect by monitoring strain variations within concrete specimens. A single FRP formwork specimen (1 m in height, 100 mm diameter) was fabricated using the ALM process and filled with concrete. After curing, the specimen was divided into three samples (each 200 mm high, 100 mm diameter) and individually tested for compressive strength. The test results were compared with those obtained from conventional concrete specimens of identical dimensions fabricated with traditional formwork, enabling evaluation of the structural enhancement provided by ALM-fabricated FRP formwork.

Furthermore, ALM-produced specimens using UV-curable resin were compared to those manufactured with structural deoxygenation resin to determine whether the UV-curable resin system could achieve comparable confinement performance. Additionally, three PLA concrete formworks of identical dimensions were fabricated using conventional 3D printing, filled with concrete, and subjected to compressive strength tests. Results were then compared with those of the FRP formwork to assess whether the ALM-produced FRP formwork offers advantages in terms of enhancing concrete strength.

Due to the intrinsic structural properties of FRP, further strengthened by resin application and curing, concrete tensile, compressive, and shear strengths are significantly enhanced. When concrete elements are subjected to axial loading, expansion activates the FRP formwork, producing a confinement effect that limits deformation and yields superior compressive strength. The relationship between confining strength and FRP thickness can be expressed as follows [7]:

$$\sigma_1 = \frac{2t_{frp}}{D}\sigma_f \tag{2}$$

3 Results Analysis

3.1 Analysis of the Tensile Strength Test Results for Ring-Shaped FRP Materials

The fabricated specimens were sectioned into ring-shaped samples, and circumferential tensile strength tests were conducted to measure their maximum tensile capacities. These data facilitated analysis of the relationship between circumferential tensile strength and the number of overlapping layers, providing insight into determining the optimal overlap number for formwork used in concrete grouting. Due to the minimal variations observed, the number of overlapping layers did not significantly affect the tensile strength of the material. Results from tensile tests on three specimen groups with varying overlapping numbers (Table 1) showed no significant correlation between circumferential tensile strength and overlap quantity. The intrinsic tensile strength of FRP fabric remained consistent regardless of layer count, suggesting that increasing the number of overlapping layers does not affect the fundamental strength properties of the material.

Table 1. The Tensile Strength and Tensile Modulus of Ring Samples

	Tensile Modulus			Tensile strength (Mpa)		
Overlapping Numbers	OL10_t	OL15_t	OL20_t	OL10_t	OL15_t	OL20_t
Sample_1	8442.5	7718.7	5894.8	108.2	128.2	105.3
Sample_2	6666.4	7263.3	5878.2	80.8	127.8	90.8
Sample_3	6296.1	5886.7	6312.0	106.6	101.7	118.2
Sample_4	6856.3	7315.0	7275.3	113.9	98.6	99.7
Sample_5	5733.4	7314.3	5264.2	94.0	108.4	137.3
Average	6798.9	7099.6	6124.9	100.7	112.9	110.3

However, it is notable that the maximum circumferential tensile load that samples could withstand demonstrated a clear positive correlation with an increase in overlapping layers (Fig. 5). Thus, when greater confinement is required for concrete structural components, increasing the number of overlapping layers in the FRP formwork provides a practical and effective solution.

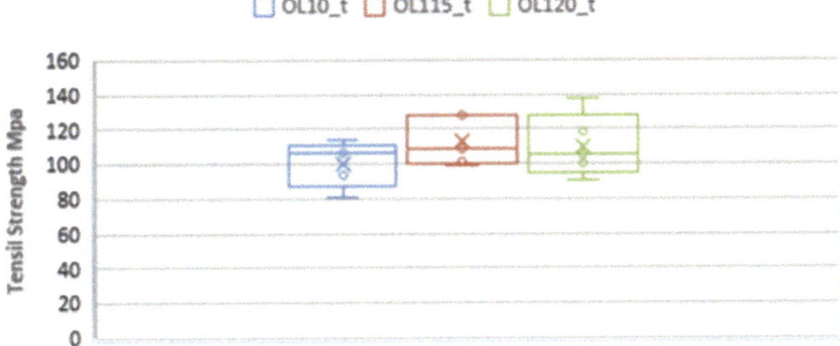

Fig. 5. The tensile strength of ring samples

3.2 Comparative Analysis of Strength Performance of FRP Formworks After Curing

An FRP formwork prototype with 15 overlapping layers was fabricated using the ALM system, filled with concrete, and allowed to cure. After curing, compressive strength tests were conducted on the resulting concrete columns. These results were then compared with compressive strength tests performed on concrete column specimens of identical dimensions produced using conventional formwork.

Figure 6 compares a standard concrete specimen (100 mm diameter) with FRP-confined and PLA formwork specimens under axial loading. The conventional specimen failed, while the FRP-confined specimen showed no visible cracks or damage, with fiber fractures limited to the outermost layer. In contrast, the PLA formwork specimen exhibited significant surface cracks and deformation. Weak adhesion between the PLA formwork and concrete, caused by curing control issues, allowed easy demolding. As a result, no effective composite system formed, limiting the PLA formwork's ability to enhance concrete compressive strength.

Fig. 6. Compressive test of samples in diameter 100 mm

Table 2 presents the specific structural performance improvements provided by the FRP formwork to concrete components. Concrete columns with FRP formwork exhibited a notable increase in compressive strength, ranging from 30% to 54% compared to standard specimens produced with traditional formwork, and between 23% and 35% compared to those using PLA formwork. However, variations in enhancement were relatively large and not confined to a narrow range, potentially due to the manufacturing process. These specimens were produced by pouring concrete into a 1-m-high FRP formwork, allowing it to cure, and subsequently dividing it into three 200 mm-high samples. During cutting, interlayer damage or voids might have formed within the FRP formwork, thereby diminishing its reinforcing effect.

Table 2. The increase of compressive strength of concrete column with the FRP formwork

Sample	Strength (MPa)	Enhancement (%)	Sample	Strength (MPa)	Enhancement (%)	Sample	Strength (MPa)
D100_1	56.6	30	St_1	43.3	35	3D_D100_1	41.7
D100_2	59.8	54	St_2	38.7	23	3D_D100_2	48.5
D100_3	38.2	-14	St_3	44.2	-10	3D_D100_3	42.2

Overall, the compressive strength of concrete specimens prepared with FRP formwork exhibited a notable improvement compared to standard specimens cast using traditional formwork, showing an increase ranging from approximately 30% to 54%. However, when compared to specimens produced with FRP formwork utilizing structural deoxygenation resin, the maximum strength increase did not reach the levels around 44% [10] obtained by the deoxygenation resin-based specimens. Nevertheless, the photopolymerized resin-based formwork demonstrated performance very close to that of the structural deoxygenation resin-based formwork. Additionally, the fabrication process using UV-curable resin significantly reduced curing time, thereby markedly enhancing production efficiency.

3.3 Mathematical Expression for the Number of Overlapping Layers

Previous research has already provided an expression for the confining strength of FRP formwork in concrete, but it did not establish a relationship with the number of overlapping layers in FRP fabric. Since the thickness of the FRP material can be represented as the product of single-layer thickness and the number of overlapping layers, the formula for the confining strength provided by FRP formwork to concrete can be expressed as follows:

$$\sigma_1 = \frac{2N_{OL}w_{frp}}{D}\sigma_f \tag{3}$$

where N_{OL} is the Overlapping number of FRP formwork, σf and wfrp is the tensile modulus and thickness of the FRP material.

This equation more accurately captures the magnitude of the confining force exerted by FRP formwork on concrete. By applying this formula, the optimal number of overlapping layers can be selected based on the required confinement level, thereby reducing material waste and enhancing manufacturing efficiency.

4 Discussion

4.1 Fabrication Failure of Formwork

The fabrication parameters are significant for high quality product making. Shown in Fig. 7, failure samples resulted from high extrusion speed and insufficient UV exposure, and insufficient resin applied. The improper parameters setting will invoke the local wrinkle for the formwork product and even damage the whole formwork without sufficient strength. Therefore, further ALM fabrication should follow a strict control to drive the robotic arm and motor in the end-effector.

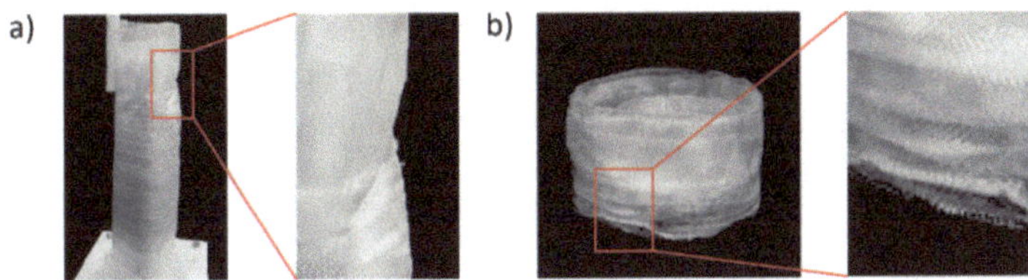

Fig. 7. (a) Surface depressions on the formwork. (b) Interlayer overlapping caused by inadequate control of manufacturing parameters

4.2 Internal Defect of Formwork

Upon demolding the grouted product, it was observed that the formwork produced by this method affected the concrete's shape, resulting in noticeable grooves on the surface with a relatively uniform distribution. In comparison to concrete components made with traditional formwork, the surface was rougher, exhibiting localized depressions that significantly reduced the overall integrity of the concrete component. These imperfections can disrupt load transfer paths within the concrete, impacting its load-bearing performance (Fig. 8).

Fig. 8. The appearance and cracks of FRP formwork on the inside surface

Analysis of the tensile test results for the annular formwork indicates that while increasing the number of layers does not enhance the intrinsic tensile strength of the material, it effectively increases the sample's maximum load-bearing capacity. Additionally, the manufacturing quality of the sample significantly influences the test results; inadequate interlayer bonding leads to reduced tensile performance. Thus, interlayer bonding strength is crucial for determining sample quality.

5 Conclusion

In summary, the analysis of compressive strength tests performed on various formwork samples after concrete grouting indicates that FRP formwork significantly enhances the strength of concrete components. However, compared to the ALM system using structural deoxygenation resin, the strength improvement achieved by FRP formwork fabricated with UV-curable resin was relatively lower, suggesting that further optimization of the manufacturing process is required. Nevertheless, the use of UV-curable resin substantially reduced production time, thereby enhancing manufacturing efficiency. Furthermore, compared to 3D-printed PLA formwork, FRP formwork exhibited superior performance, increasing the compressive strength of concrete columns by approximately 23% to 35%. This demonstrates a clear advantage of FRP formwork. Additionally, careful control of manufacturing parameters is necessary to prevent significant deformation of the FRP formwork during fabrication, which could otherwise lead to failure.

References

1. Burger, J., et al.: Eggshell: ultra-thin three-dimensional printed formwork for concrete structures. 3D Print. Addit. Manuf. **7**(2), 49–59 (2020)
2. Goyal, R., Mukherjee, A., Goyal, S.: An investigation on bond between FRP stay-in-place formwork and concrete. Constr. Build. Mater. **113**, 741–751 (2016)
3. Kristombu Baduge, S., et al.: Improving performance of additive manufactured (3D printed) concrete: a review on material mix design, processing, interlayer bonding, and reinforcing methods. Structures (Oxford) **29**, 1597–1609 (2021)
4. Kumar, B.B.M., Hsueh, Y.C., Xin, Z., Luo, D.: Process and evaluation of automated robotic fabrication system for in-situ structure confinement. In: Yuan, P.F., Chai, H., Yan, C., Leach, N. (eds) Proceedings of the 2021 DigitalFUTURES. CDRF 2021. Springer, Singapore (2022)
5. Labonnote, N., Rønnquist, A., Manum, B., Rüther, P.: Additive construction: state-of-the-art, challenges and opportunities. Autom. Constr. **72**, 347–366 (2016)
6. Ou, Y., Bao, D.-W., Zhu, G.-Q., Luo, D.: Additive fabrication of large-scale customizable formwork using robotic fiber-reinforced polymer winding. 3D Print. Addit. Manuf. **9**(2), 109–121 (2022)
7. Valasaki, M.K., Papakonstantinou, C.G.: Fiber reinforced polymer (FRP) confined circular concrete columns: an experimental overview. Buildings **13**(5), 1248 (2023)
8. Wang, W., Neaz Sheikh, M., Al-Baali, A.Q., Hadi, M.N.S.: Compressive behaviour of partially FRP confined concrete: experimental observations and assessment of the stress-strain models. Constr. Build. Mater. **192** (2018)
9. Xin, Z., Zhu, G.-Q., Hsueh, Y.C., Luo, D.: Novel additive lamination manufacturing system for rapid fabrication of large-scale reinforced structural members. Rapid Prototyp. J. **30**(10), 2161–2173 (2024)

BY NC ND

Additive Fabrication of Natural Material Through Mixed Reality

Zhao Yu and Xu Weiguo(✉)

Shenzhen International Graduate School, Tsinghua University, Beijing, China
xwg@mail.tsinghua.edu.cn

Abstract. Mixed reality construction expands the possibilities of traditional construction methods across multiple dimensions. In exploring these possibilities, various research trends have emerged, including the use of natural materials, the construction of complex geometries. Building upon these existing studies, this paper pro-poses a workflow for additive construction using natural materials within a mixed reality environment, and validates its feasibility through a practical project.

By combining discrete design rule-setting with human interaction in mixed reality, the workflow simplifies the processing, fabrication, and assembly of irregular materials. In this process, discrete design—as a form of additive manufacturing algorithm—enables natural materials to be aggregated in a structured manner, forming construction systems that are both random in appearance and operable in practice. This approach takes advantage of the small volume of offcut materials, and the high mobility of discarded natural components, while also leveraging mixed reality technology to reduce the assembly complexity typically associated with discrete design.

Keywords: Mixed reality · Natural materials · Discrete design · Additive construction

1 Background

1.1 Construction With Natural Materials in Mixed Reality

The use of natural materials in mixed reality construction is becoming increasingly common [1–7]. Unlike standardized industrial materials, natural materials tend to be irregular in shape and relatively small in size. These material characteristics have prompted changes in construction methods and led to more uncertain and variable structural forms.

A key challenge in using natural materials is managing their irregularities and deviations. Photogrammetry is widely adopted to capture precise geometric data for use in virtual-reality alignment. However, case studies have identified several limitations in current approaches:

High Equipment and Software Costs. The tools required for scanning and processing are often expensive, which limits their accessibility for low-cost or experimental projects [7].

Y. Liu et al. (Eds.): CDRF 2025, *Transindividual Intelligence*, pp. 276–287, 2026.
https://doi.org/10.1007/978-981-92-0615-5_24

Complex and Time-Consuming Scanning Processes. Accurately capturing an object's geometry typically involves 30–50 photographs and repeated scanning attempts to achieve satisfactory results [7].

Extensive Post-Processing Workload. Scanned meshes frequently contain thousands of faces, requiring manual simplification. While machine learning can assist, it still demands manual annotation of hundreds of images and hours of training to identify key points [6, 7].

Difficulty Aligning Virtual and Physical Materials. Accurately map-ping natural objects between virtual and real-world environments remains a technical challenge in mixed reality construction contexts.

These issues result in excessive data acquisition and computation, making the overall workflow time-consuming and inefficient. The complex geometry of natural materials and the need for precise alignment significantly increase operational difficulty. Together, these limitations reduce the suitability of such approaches for fast, low-tech construction scenarios where flexibility and speed are essential.

1.2 Construction of Complex Shapes in Mixed Reality

Mixed reality provides an overlaying view for construction work, and this ad-vantage has spurred a research upsurge on the construction of complex shapes [3, 8–11]. The characteristics of these complex shapes are reflected in: the complexity of components [10], the complexity of the sequence of components [11], the complexity of the final form [8], and the complexity of the construction method [9].

The roles played by mixed reality technology in the process of constructing complex shapes include: saving the time for identifying components, reducing the dependence on the layered verification path in additive manufacturing [3], and allowing for design changes during the construction process [12].

2 The Timber Tree—Mixed Reality Construction Based on Discrete Algorithms

The Timber Tree Project aims to construct a tree-shaped structure using natural materials on the lawn of the teaching building's porch. The project develops a workflow that simplifies irregular branches into geometries with a uniform appearance—elements that are connectable and compatible with standard part algorithms. This process enables the reuse of waste materials and supports additive construction of complex forms through mixed reality technologies.

Discrete design offers the potential to aggregate generic elements into complex, scalable wholes, with the combinatorial logic between individual units determining the overall form [13]. Although typically applied to standardized industrial elements, one of the core principles of discrete algorithms—that each unit has an equal number of connectable faces—can also be approximated using unprocessed natural materials.

In contrast to traditional form-finding approaches that rely on precise computation, discrete design emphasizes assembly patterns among similar elements. This focus reduces the challenge of aligning irregular natural materials with predefined forms, offering clear advantages for sustainable construction.

Structurally, discrete systems support adjustable density and follow consistent connection rules within a unified assembly logic. These characteristics simplify the construction process, even when using variable materials such as branches.

The adaptability of discrete algorithms to both the design and construction phases using natural materials aligns closely with mixed reality–based construction workflows. This synergy underpins the project's adoption of a discrete design methodology.

3 Method and Materials

3.1 Materials

The natural materials used in the project consisted of 23 wooden sticks, each with a diameter of 3.7–6 cm and a length of 18 cm.

The virtual preparation involves geometric computation and data transmission channels. In this project, Wasp in Rhino Grasshopper was used as the discrete algorithm tool.[1] The mixed reality headset employed in the project was the Meta Quest 3. The mixed reality interaction system was developed using Unity software in combination with the Meta XR SDK, and data transmission between Unity and the device was achieved via Meta Quest Link.

3.2 Simplification Treatment of Natural Materials

In order to integrate such natural materials into discrete algorithms originally de-signed for standardized elements, we developed two transformation processes to simplify the materials. These processes aim to adhere to the rules of the discrete algorithm as closely as possible while minimizing construction errors:

Visual Uniformity. Through manual selection, these irregular branches are classified by diameter and cut to uniform lengths, resulting in a set of visually similar elements.

Providing Connection Interfaces. The unprocessed surfaces of natural materials are not conducive to direct attachment or connection with other components, necessitating the use of connectors for secure assembly. The connectors were de-signed as standardized half-hexagonal elements, fabricated using 3D printing technology. They are attached to the wooden sticks with screws and can flexibly accommodate the range of diameters present in the materials used for the project. To facilitate installation and disassembly in a mixed reality environment, each connector consists of two interlocking parts based on a mortise-and-tenon structure. These standardized connectors play a tolerance-mediating role in the additive fabrication process involving irregular materials. They enable wooden sticks with varying cross-sections to conform to a uniform size, allowing the connection rules to be applied with reasonable accuracy. Furthermore, the connectors abstract away the complex geometric features of natural materials and act as intuitive linear reference

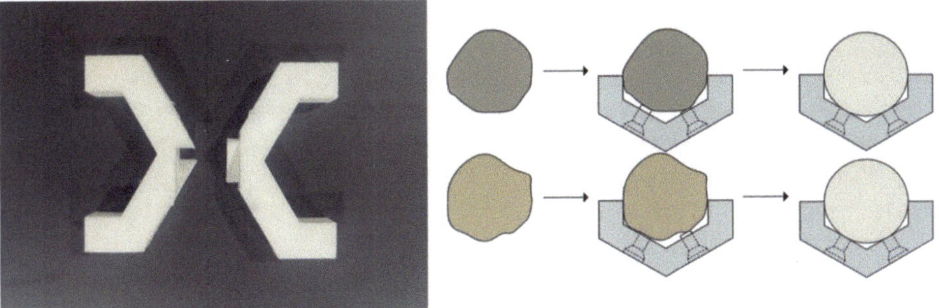

Fig. 1 Standardized connectors unifying irregular material dimensions

points within the mixed reality interface, providing clearer visual guidance for alignment and assembly.

Through these two transformation steps, the irregular wooden sticks—though similar in length and diameter—are computationally treated as uniform, reusable cylindrical elements with defined connection faces (see Fig. 1). This renders them interchangeable, eliminating the need for customized machining of either the materials or their connectors. Once standardized, the materials no longer require individual labeling or identification; they can be randomly selected during assembly, significantly reducing the time and technical effort required to identify and handle each piece. This simplification also eliminates the need for complex data input and reduces the computational load, making it possible to apply shape computation methods originally intended for standardized components to irregular ones. Additionally, in real-world construction, standardized components of the same dimensions can be freely mixed in when natural materials are insufficient.

Compared to approaches that preserve the unique characteristics of each natural element, this workflow supports a broader range of design and construction scenarios, offering greater adaptability (see Fig. 2).

3.3 Discrete Rule Setting

The connection rules followed by the components in this project significantly influenced both the visual characteristics of the generated geometry and the construction approach. Based on the target form—a tree—two contour control strategies were established: a coarse control method for the trunk and a refined control method for the branches. In terms of alignment with the target geometry, smaller connection angles were applied to the branch contours for higher fidelity, while larger angles were used for the trunk.

Regarding construction complexity, larger angles enhance the structural stability of the resulting aggregate, making the construction process easier. Therefore, an incremental reduction in angles from the base to the top was adopted to facilitate construction. Density is another crucial factor in natural material construction. Retaining larger gaps

[1] https://www.youtube.com/watch?v=IqH2xH_W4Y4&list=PLCn3-_9Z4-E5A0EFluiMldlEbDufMiN1g. Accessed 11 Jun 2025

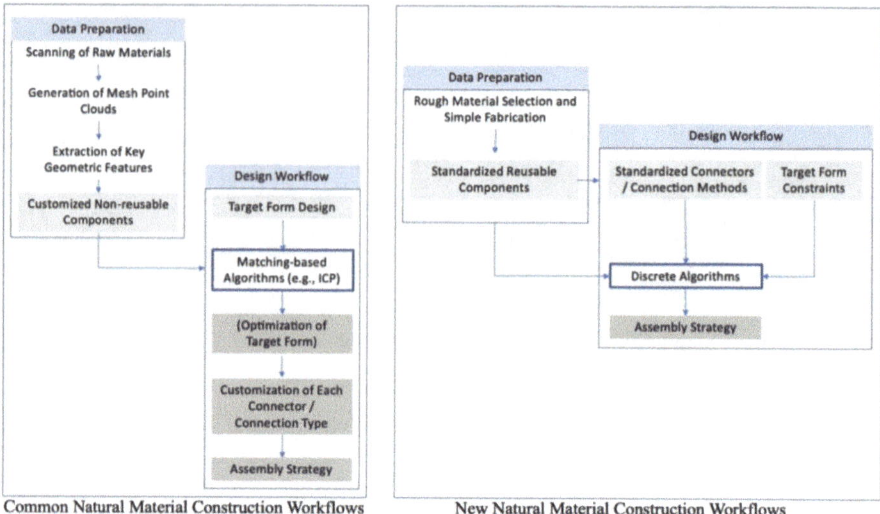

Fig. 2 Comparison of natural material construction workflows

between components during the design phase enhances the material's tolerance to natural variability (see Fig. 3).

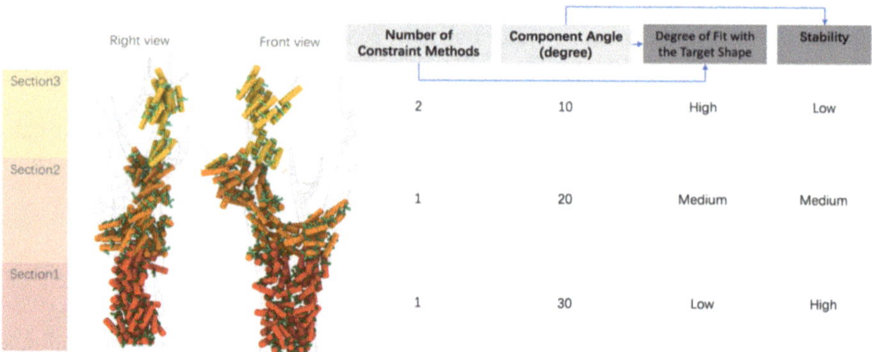

Fig. 3 Discrete design setting and aggregation results

3.4 Fabrication and Assembly

Construction in Mixed Reality. The non-standard nature of materials and the complexity introduced by algorithmic design pose significant challenges to robotic construction. However, with the assistance of mixed reality humans are provided with an additional layer of spatial perception over the physical world, making it easier to interpret the spatial relationships—such as overlaps and intersections—between components.

The interaction process in the mixed reality environment is divided into two steps in accordance with the construction sequence. First, the system displays the corresponding unit number based on user input, guiding the user to assemble the unit according to the visual instructions. Second, during the assembly process, the user can selectively display both the unit to be assembled and the already assembled partial components, depending on the current construction progress (see Fig. 4).

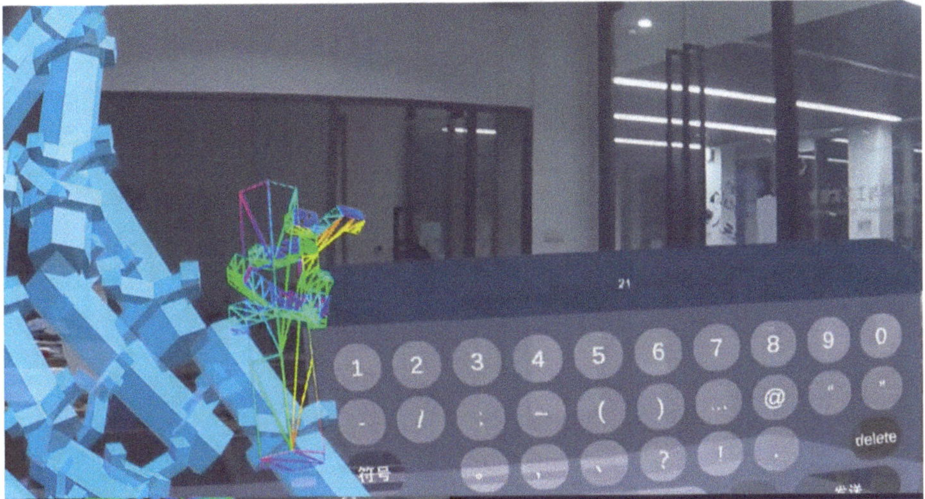

Fig. 4 Interaction in mixed reality

Assembly Sequences.

Individuals Processing. First, these pieces of wood are cut into the same 18 cm length (see Fig. 5). After preliminary screening of the diameter by manual labor, they can already be regarded as geometric components identical to standard cylinders.

For the convenience of subsequent connection with the connectors, these wooden sticks need to be pre - drilled. In virtual reality, a standard hexagonal prism with a length of 18 cm is displayed, and three hole positions are prompted on each of its lateral faces. The operator aligns the non - standard wooden stick with the hexagonal prism in the image. After completing the operation for one row, the wooden stick is rotated while simultaneously aligning the stick with the hole positions (see Fig. 6). Mixed reality and simple operations replace the cumbersome process in which a machine measures each irregular object and customizes the processing path.

Unit-Level Assembly. Each wooden stick, together with its adjacent connectors, forms a modular unit. Due to the constraints imposed by both geometry and connection rules in the algorithm, the number and positions of connectors vary from stick to stick. Displaying detailed connection information between each stick and its connectors within the mixed reality interface allows assemblers to quickly and accurately assemble these units.

Fig. 5 Processed individuals

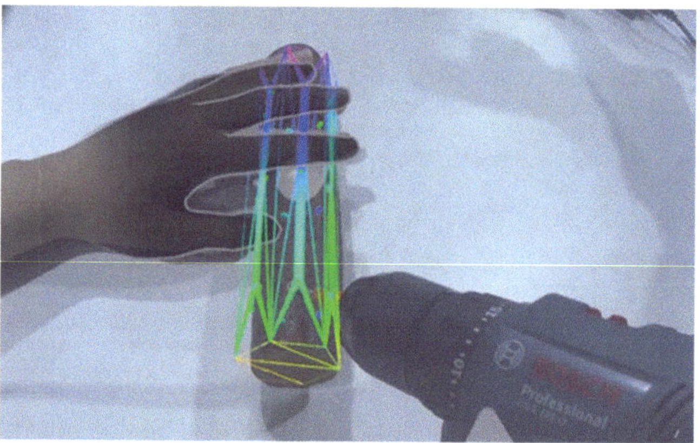

Fig. 6 Drill holes in mixed reality

During the fabrication phase, 18 holes were pre-drilled into each wooden stick. In the mixed reality environment, assemblers can match the connector positions to the corresponding holes for proper attachment (see Fig. 7). Throughout this process, fabricators can rotate or flip each stick based on its individual physical characteristics, adjusting its orientation to align the holes appropriately for the intended connections.

Partial Assembly. After the assembly of all individual units is completed, the process moves on to connecting the units into groups of two to five, forming partial assemblies. The wooden sticks do not directly contact one another; instead, they are interconnected through specially designed connectors. The angle between the upper and lower parts of each connector is predetermined during the fabrication stage. Therefore, although the angles between the sticks result in complex spatial relationships, the use of mixed reality simplifies the assembly process. Installation personnel only need to align the connectors based on the positional guidance displayed in the head-mounted device (see Fig. 8). In other words, as long as the connectors are positioned correctly, the sticks will

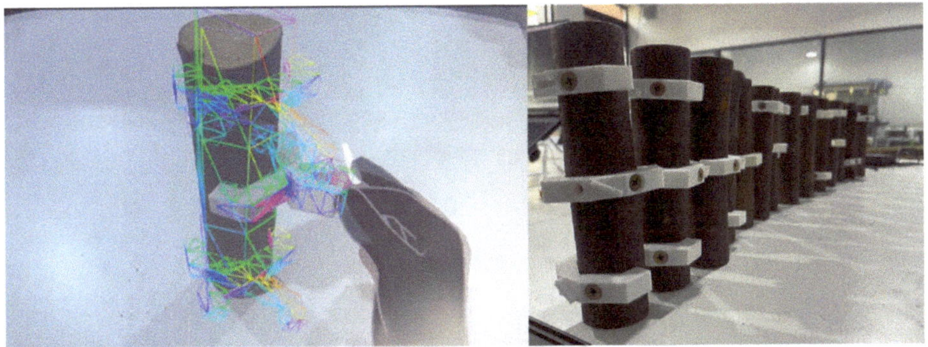

Fig. 7 Assembly individuals in mixed reality

automatically align in both position and orientation with the digital design model. This construction method eliminates the need for manual angular alignment in space, making it more suitable for operations performed while wearing head-mounted devices.

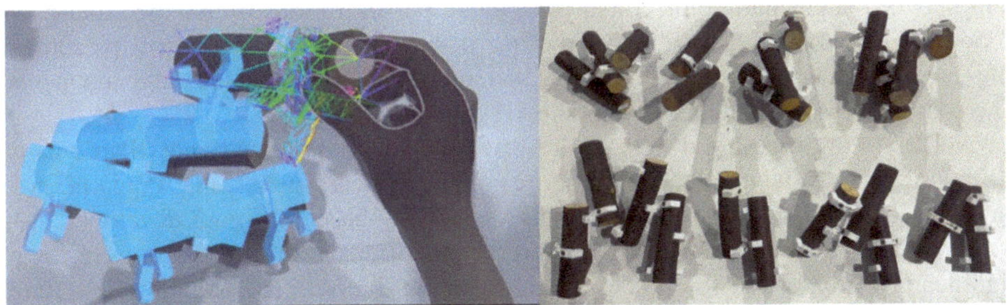

Fig. 8 Assembly partials in mixed reality

Final Assembly. After all partial assemblies are completed, the process enters the stage of overall assembly. At this stage, attention is focused on the connections between adjacent units of different partial assemblies. In the mixed reality environment, the entire constructed model up to the current progress is displayed on the screen, with the component to be installed next highlighted. This visual aid helps installation personnel to distinguish the current component from similar elements and accurately identify its orientation within the structure. Additionally, users can input a specific number via the interactive panel at any time to view the overall configuration of components at a given stage, which assists in error detection and troubleshooting.

The project tested an assembly process involving 23 wooden sticks, resulting in a structure closely aligned with the original digital model (see Figs. 9 and 10). The maximum deviation in connector placement was within 20 mm. Thanks to the adjustable aggregation density inherent to discrete design, no positional collisions occurred during the assembly process.

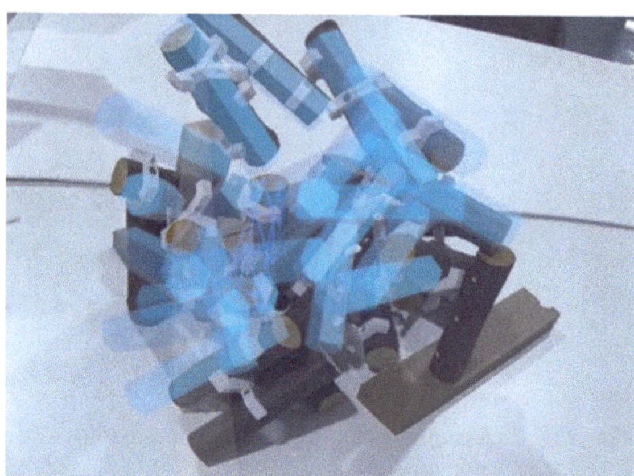

Fig. 9 Assembly whole in mixed reality

The construction was carried out by two individuals. One operator, wearing a mixed reality (MR) headset, received real-time visual guidance indicating the spatial relationship between sticks and connectors. The second person physically handled and positioned the components according to the instructions. Once the real-world placement aligned with the virtual overlay, the second person physically connected the components. The complete assembly required a total of 12 person-hours.

Notably, the process did not require prior woodworking experience. Participants did not need to evaluate material properties or acquire carpentry skills. Preparation—including material sorting, file processing, and Unity-based program development—was completed within one week. Compared to traditional workflows, the experiment significantly streamlined the process by eliminating the need for scanning, inputting, calculating, and matching individual elements to the overall geometry.

4 Conclusions

This experimental project incorporated irregular natural materials into a standardized design and construction process through simple manual selection and processing, combined with the use of standardized connectors. Compared to traditional matching-based methods that rely on scanned point clouds and target geometries, the discrete algorithm significantly reduces computational load while offering greater design flexibility. It provides a more structured approach to assembling complex forms and is better suited for manual construction in mixed reality environments. Overall, the integration of discrete algorithms with mixed reality technology presents a viable approach for constructing complex geometries using natural materials.

This experiment also allows for the identification of key principles suitable for this type of construction workflow. First, the operability of connectors is crucial: since the assembly work is carried out with head-mounted devices, the connectors must be designed for single-person, two-hand operation. Second, within the MR-assisted interactive process, the ability to flexibly visualize the assembly and disassembly of components

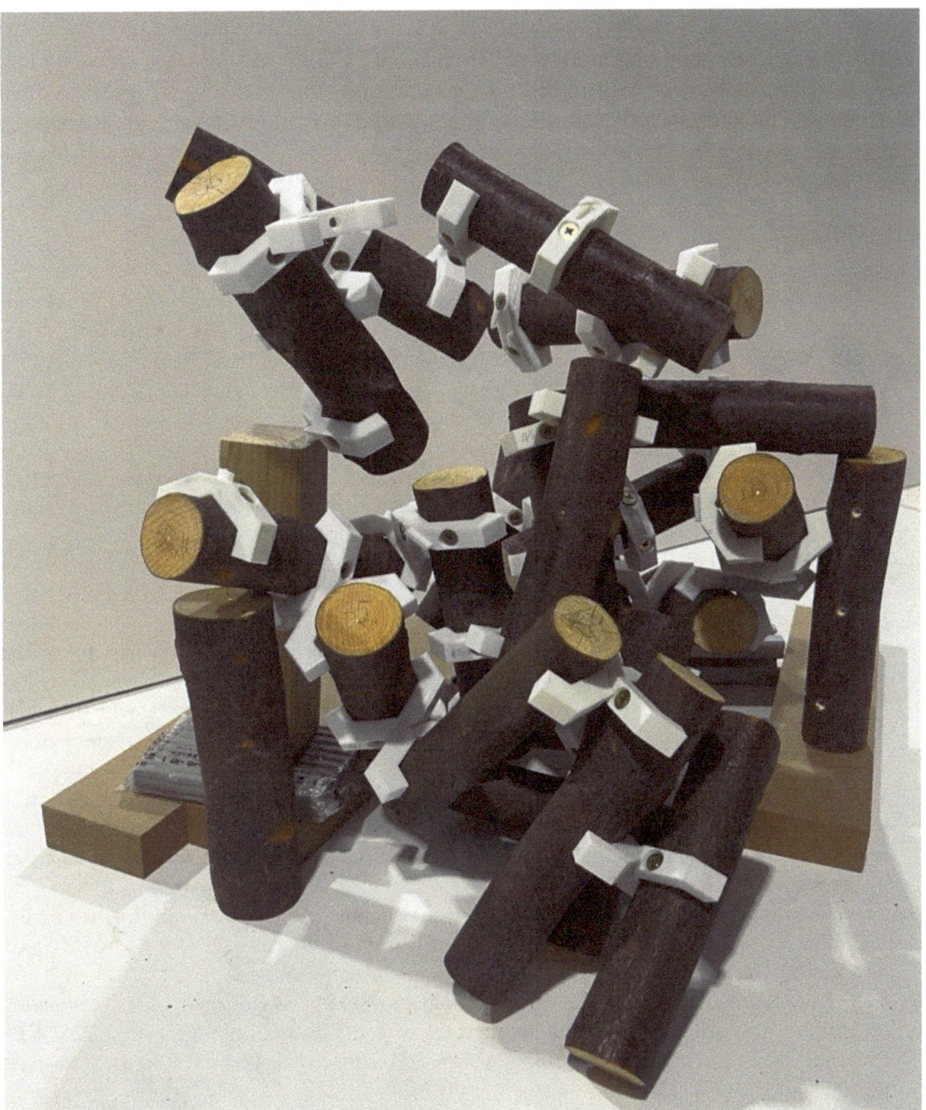

Fig. 10 Assembly result

on demand is essential. Therefore, carefully planning the assembly sequence and preparing both the corresponding models and visual cues in advance for each step are key to leveraging the benefits of mixed reality assistance.

Although this project is still in the development stage and requires further research and testing, the prototype trial demonstrates the potential of this MR-assisted construction system. Compared with other MR-based construction methods, the connection logic grounded in discrete design allows for simpler operations and facilitates easier spatial alignment between virtual and physical elements. Moreover, the data and physical components required are relatively minimal, which is crucial for enabling rapid construction. Future research will explore more complex geometric components and further simulate the irregular characteristics of natural materials, aiming to reflect their inherent adaptive capacities.

References

1. Cousin, T., Alkhayat, L., Pearl, N., Dewart, C.B., Mueller, C.: Wild wood gridshells: Mixed-reality construction of nonstandard wood. Technol. Arch. Des. **7**(2), 216–231 (2023)
2. Crolla, K., Goepel, G.: Augmented Feedback-A case study in Mixed-Reality as a tool for assembly and real-time feedback in bamboo construction. In: Realignments: Toward Critical Computation, 41th Annual Conference of the Association of Computer Aided Design in Architecture (ACADIA). Acadia Publishing Company (2021)
3. Jahn, G., Newnham, C., Berg, N.: Depth camera feedback for guided fabrication in augmented reality. In: Proceedings of the 42nd Annual Conference of the Association for Computer Aided Design in Architecture (ACADIA). ACADIA, Philadelphia (2022)
4. Lok, L., Bae, J.: Timber de-standardized 2.0: Mixed reality visualizations and user interface for processing irregular timber. In: Proceedings of the 27th CAADRIA Conference, CAADRIA, Sydney (2022)
5. Sun, C., et al.: Hybrid fabrication: A freeform building process with high flexibility and acceptable accumulative error. In: ACADIA (2018)
6. Wibranek, B., Tessmann, O.: Digital rubble compression-only structures with irregular rock and 3D printed connectors. In: Proceedings of IASS Annual Symposia. International Association for Shell and Spatial Structures (IASS) (2019)
7. Wu, N.H., Dimopoulou, M., Hsieh, H.H., Chatzakis, C.: A digital system for AR fabrication of bamboo structures through the discrete digitization of bamboo. In: Proceedings of the International Conference on Education and Research in Computer Aided Architectural Design in Europe, Rome, Italy (2019)
8. Chen, Z., Lin, D., Sun, L., Wang, S., Han, D.: A 'human-in-the-loop' workflow for realizing Taihu Rocks. In: The International Conference on Computational Design and Robotic Fabrication. Springer (2023)
9. Luo, J., Mastrokalou, E., Aldabous, R., Aldaboos, S., Lopez Rodriguez, A.: Fabrication of complex clay structures through an augmented reality assisted platform. In: CAADRIA Proceedings. Association for Computer-Aided Architectural Design Research in Asia (CAADRIA) (2023)
10. Gwyllim, J., Andrew, J.W., James, P.S.: Holographic handcraft in large-scale steam-bent timber structures. In: ACADIA, Austin, Texas (2019)
11. Zahiri, N.: Sequential augmented assembly of interlocking timber elements for circular wood construction. J. Int. Assoc. Shell Spat. Struct. **65**(4), 277–287 (2024)
12. Betti, G., Aziz, S., Ron, G.: Pop up factory: Collaborative design in mixed reality. Simul. Virtual Augment. Real. **2**(3), 115–125 (2019)
13. Köhler, D., Navasaityte, R.: Mereological tectonics: The figure and its figuration. TxA Emerg. Des. Technol., 40–52 (2016)

Controllable Deformation: A 4D-Printing Simulation Framework for Porosity-Driven Hygromorphic Bilayer Structure

Yuxing Liu[1](✉), Xuanyu Lu[2], Marjan Colletti[1], and Kostas Grigoriadis[1]

[1] The Bartlett School of Architecture, UCL, 22 Gordon Street, London WC1H 0QB, United Kingdom
yuxinglupennedu@gmail.com, {m.colletti,k.grigoriadis}@ucl.ac.uk
[2] University of Tokyo, 7 Chome-3-1 Hongo, Bunkyo City, Tokyo, Japan
luxuanyu@g.ecc.u-tokyo.ac.jp

Abstract. This research proposes a 4D-printing simulation framework that uses porosity as a design parameter to program deformation in hydromorphic bilayer structures. Grounded in Timoshenko's sandwich beam theory, the framework integrates physical experimentation, parametric modeling, and dynamic simulation (via Grasshopper), enabling inverse design of porosity distributions for target bending. Bilayers of wood-based filaments (active) and PLA (passive) demonstrate consistent humidity-driven shape changes. The study proceeds in three phases: (a) physical experiments assessing porosity–deformation relationships, (b) dynamic simulation to collect curvature data, and (c) mathematical modeling to refine design guidelines. A prototype façade panel validates the framework, with predicted bending closely matching experimental results. This confirms the method's feasibility for scalable, adaptive architectural applications.

Keywords: Porosity-driven deformation · hygromorphic bilayer · simulation framework · adaptive façade · 4D printing

1 Introduction

Hygromorphic actuation—shape changes under moisture fluctuations—is deeply rooted in biological systems, driving seed dispersal in pinecones (Song et al., 2015) and self-burial in wheat awns (Jung et al., 2014). It arises from differential swelling (governed by material properties and microscale porosity) and hierarchical structures, where active layers expand anisotropically while passive layers restrict deformation (Ruedrich et al., 2011; Wang et al., 2017). Such zero-energy actuation has inspired architecture, robotics, and adaptive systems, enabling ap-plications like self-regulating shading (Cheng et al., 2024) and humidity-driven robots (Yang et al., 2023).

Porosity, defined by void distribution, critically shapes mechanical behaviour, permeability, and environmental responsiveness. By tuning porosity across micro, meso, and macro scales, one can optimize material efficiency and deformation control. Microscale porosity is altered via material composition or particle packing (Tahouni et al., 2023),

Y. Liu et al. (Eds.): CDRF 2025, *Transindividual Intelligence*, pp. 288–296, 2026.
https://doi.org/10.1007/978-981-92-0615-5_25

and mesoscale porosity is regulated by additive manufacturing parameters (Cheng et al., 2020). In contrast, macroscale porosity ($\geq$1 dm)—a key driver of deformation magnitude, stiffness, and surface dynamics (Siéfert et al., 2019)—remains less explored despite its potential for geometry-driven design. Unlike micro/meso approaches needing specialized materials or precise printing, macroscale porosity permits structural tuning without changing composition or printer settings, offering a scalable route for adaptive façades.

Beyond porosity magnitude, pore geometry and aspect ratio strongly influence hygromorphic behaviour. Elongated or irregular pores slow deformation by creating tortuous flow paths, while rounded pores allow faster, more uniform actuation (Katagiri et al., 2015). Aspect ratios near 1 promote predictable transformations; extreme ratios induce anisotropy. These effects are amplified in bilayer systems, where differential swelling between active and passive layers—modelled by Timoshenko's bimetal theory (Krüger et al., 2021)—enables programmable bending. Porosity modulates layer stiffness and swelling, offering re-fined curvature control.

Despite qualitative insights, quantitative models linking porosity and hygro-morphic deformation remain sparse. Nonlinear material behaviour and fabrication inconsistencies often necessitate trial-and-error experimentation, which is time-intensive and unscalable. While FEA is accurate (Yogeesh, 2023), it is computationally heavy and unsuited for early-stage design loops. To address this, we introduce a Grasshopper-based solver for real-time simulation and experimental validation. Analysing solver-generated curvature data yields a mathematical model for porosity-deformation prediction, enabling inverse design without ex-tensive prototyping.

This research presents a simulation framework for controlling deformation in porosity-driven 4D-printed hygromorphic bilayers. The framework integrates: (1) physical experiments to establish porosity–deformation behaviour, (2) a para-metric solver for simulating bending outcomes under different porosity conditions, (3) mathematical models to quantify and predict deformation responses, and (4) design validation through a façade prototype. The system enables rapid, data-driven feedback for early-stage design, contributing to adaptive façade de-elopement and scalable hygromorphic applications. By linking material behaviour, digital simulation, and fabrication feasibility, this work advances programmable material strategies for responsive architectural design.

2 Methods

This study adopts a threefold methodology: physical experimentation, parametric modeling, and dynamic simulation.

2.1 Physical Experiments

Two comprehensive sets (15 samples each) and two comparative sets (2 samples each) were designed to establish qualitative porosity–deformation relationships and assess pore geometry/aspect ratio effects. Active–passive bilayer samples utilized LAYWOOD meta5 (active) and Bambu PLA (passive). LAYWOOD expands by ~6% after 24 h water immersion and contracts to ~98% on drying, swelling perpendicular to the print path. Each sample measured 200 × 1.6 × 60 mm, embedding a 160 × 0.8 × 48 mm passive layer to reduce de-lamination. Following manufacturer protocols, samples were submerged for over 24 h, then dried for 12 h at 22–25 °C and ~ 50% RH, with deformation recorded throughout.

2.2 Parametric Modelling and Digital Simulation

Parametric modeling systematically generates diverse pore geometries by adjusting porosity parameters (A, B) and subdivision (U, V), embedding nozzle constraints in a Grasshopper-based Python script to ensure 3D-printing feasibility. For instance, rectangular pore layouts (Fig. 1) are generated and exported as fabrication-ready models.

The simulation framework adapts Timoshenko's sandwich beam theory to account for hygromorphic bilayer behavior, incorporating shear deformation and rotational bending (Timoshenko, 1925). Recent refinements address unequal effective layer widths caused by 3D-printed porous mesostructures (Krüger et al., 2021). Here, porosity directly modulates these widths, allowing the model to predict curvature by calculating effective layer thickness (t) and width (w) para-metrically (1).

Two critical enhancements improve accuracy. First, material nonlinearity is addressed via a custom equation for Young's modulus (E), while the swelling ratio (α) is inversely linked to E (2). For the active layer, α is derived from the Gibson-Ashby Power-law (Gibson-Ashby Model) (3), it connects the material properties to its relative density, which could be further referred to porosity in this research. The passive layer's swelling ratio (α_p) remains constant, assuming negligible hygroscopic response. Second, empirical values for E and α from prior studies are integrated, ensuring physical relevance and accuracy (Table 1).

$$\frac{1}{\rho} = \frac{6(1+m)^2 \times \varepsilon}{H\left(3(1+m)^2 + (1+mnq)\left(m^2 + \frac{1}{mnq}\right)\right)} \tag{1}$$

$$m = \frac{t_p}{t_a}, q = \frac{W_p}{W_a}, n = \frac{E_p}{E_a}, \varepsilon = (\alpha_a - \alpha_p)(c - c_0)$$

$$E_a{}' = E_{0\_a}(1 - P_a)^{n_a}, E_p{}' = E_{0\_p}(1 - P_p)^{n_p} \tag{2}$$

$$\alpha_a{}' = \alpha_{0\_a}\left(\frac{E_{0\_a}}{E_a{}'}\right)^{\beta} \tag{3}$$

Table 1. E_{0_a} and E_{0_p} from literature. α_{0_a} is derived from E_{0_a}, while α_p is negligible, reflecting minimal response to moisture. Empirical exponents (n_a, n_p, β) require experimental calibration; here, n_a is assigned theoretically.

E_{0_a}	E_{0_p}	α_{0_a}	α_p	n_a	n_p	β
110 (N/mm^2)	2467 (N/mm^2)	0.007	0.0001	0.5	2.0	0.5

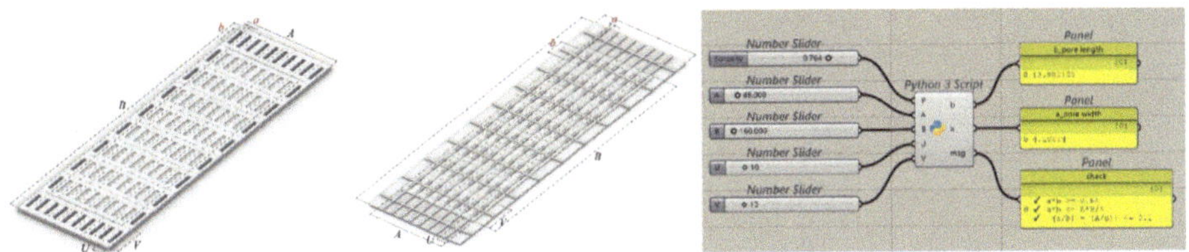

Fig. 1. Pore size (a, b) can be calculated through sample size (A, B), subdivision value (U, V) and porosity (P). On the left, A = 60 mm and B = 200 mm; on the right, A = 48 mm and B = 180 mm.

2.3 Mathematical Modelling

The effects of P_a, P_p, and R_p on deformation magnitude (bending angle, measured between start, midpoint, and end points) were independently analyzed. For each parameter, 100 samples with differential porosity were generated via parametric modeling and digital simulation. Data from these simulations informed a nonlinear polynomial regression model developed in Python.

3 Results and Discussion

3.1 Physical Experiments: Confirmation of Qualitative Relationship

3.1.1 Comprehensive Sets of Experiments

Experiments testing rectangular and hexagonal pores revealed passive layer porosity (P_p) amplifies deformation, while active layer porosity (P_a) had negligible impact. Deformation was geometry-independent. A porosity ratio ($R_p = P_a/P_p$) showed $R_p \geq 1$ produced stable, minimal bending, whereas $R_p < 1$ triggered instability. It's evident in rectangular pores with erratic bending (Fig. 2). Hexagonal pores resisted instability regardless of R_p, necessitating further comparative investigation.

3.1.2 Comparative Sets of Experiments: Pore Geometry and Aspect Ratio

Prior experiments show unexpected deformation stems from pore geometry or aspect ratio deviations. Two comparative tests were conducted: (1) rectangular pores with varied aspect ratios and (2) sinusoidal geometry at fixed ratios. Results indicate that when $R_p < 1$, instability arises if aspect ratios deviate from 1, while ratios ≈ 1 ensure stability across geometries (Fig. 3). Thus, maintaining aspect ratios near 1 or ensuring $R_p \geq 1$ prevents unpredictable deformation, providing a critical guideline for porosity-driven design to achieve controlled shape changes.

3.2 Mathematical Modelling: Establishment of Quantitative Relationship

Empirical studies confirm qualitative porosity-deformation relationships, developed into a quantitative framework via mathematical modeling. Real-time predictive feedback enables inverse design, validated against physical experiments with <4% error, facilitating reliable early-stage material programming.

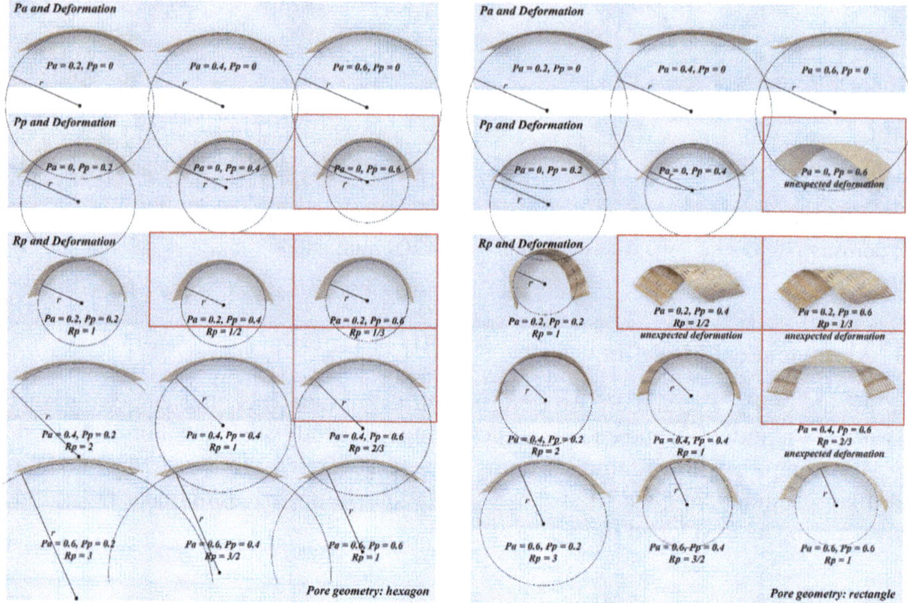

Fig. 2. Comprehensive experiments. On the left is the hexagonal set and on the right is the rectangular set. The red frame shows the range that requires comparative experiments.

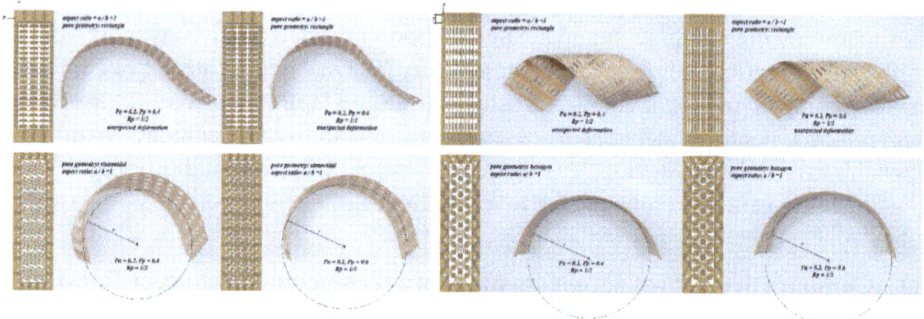

Fig. 3. Comparative experiments. On the top, the influence of pore aspect ratio on unexpected deformation was tested. On the bottom, the effect of pore geometry was tested.

3.2.1 Quantitative Relationship Between P_a and Deformation

In both rectangular and hexagonal samples, bending curvature stayed constant de-spite varying P_a. High P_a reduces active material (limiting actuation), while low P_a restricts deformation, negating porosity's influence. To capture this effect, we introduced a correction factor (4) plus non-linear E and α adjustments, greatly improving accuracy (5) (Fig. 4).

$$correction\ factor = 1 + 0.05 \times \frac{W_a}{W_a + W_p} \tag{4}$$

$$Bending\ Angle\ (Degree) = 34.255{P_a}^2 - 14.868P_a + 161.517 \quad (5)$$

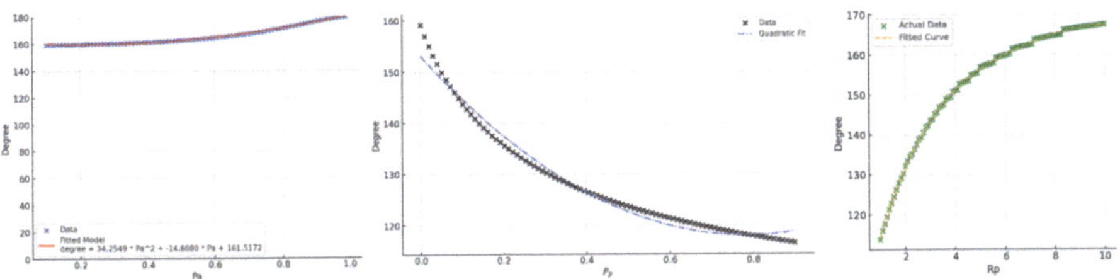

Fig. 4. From left to right are the orderly mathematical models, including P_a, P_p, R_p and deformation respectively.

3.2.2 Quantitative Relationship Between P_p and Deformation

Empirical tests demonstrate that deformation magnitude can be positively and largely influenced by P_p change. Low P_p preserves structural integrity, whereas high P_p diminishes passive layer stiffness, enabling greater deformation. Digital simulations apply amplification factors (6) derived from experimental data to enhance accuracy (7) (Fig. 4).

$$amplified\ curvature = curvature\ \times (1 + balance\ factor^2\ \times P_p) \quad (6)$$
$$balance\ factor = 4.0$$

$$Bending\ Angle\ (Degree) = 57.29{P_p}^2 - 89.755P_p + 153.205 \quad (7)$$

3.2.3 Quantitative Relationship Between R_p and Deformation

Experimental results show $R_p > 1$ prevents irregular deformation when pore ratios deviate from ≈1. The simulation workflow spans R_p values of 1–9, applying correction and amplification factors. A balancing parameter (8) handles simultaneous changes in E, α, and effective widths for P_a and P_p. Higher R_p yields smaller, stable bending, enabling data-driven models (9) (Fig. 4).

$$balance\ control = \left(\frac{{E_p}'}{{E_a}'}\right)^{control\ weight},\ control\ weight = 4.0 \quad (8)$$

$$Bending\ Angle\ (Degree) = 0.0003{R_p}^3 - 0.052{R_p}^2 + 6.973R_p - 309.64 \quad (9)$$

3.3 Design Application: Building Façade Panel

Programmable hygromorphic panels bend under humidity changes, enabling adaptive façade shading where higher porosity reduces solar blocking. Three mathematical models predict curvature from porosity, while inverse calculations allow tailoring porosity for target bending—essential for early-stage design. Such porosity-driven adaptations improve façade aesthetics, minimize material use, and support energy-efficient, weather-responsive architecture.

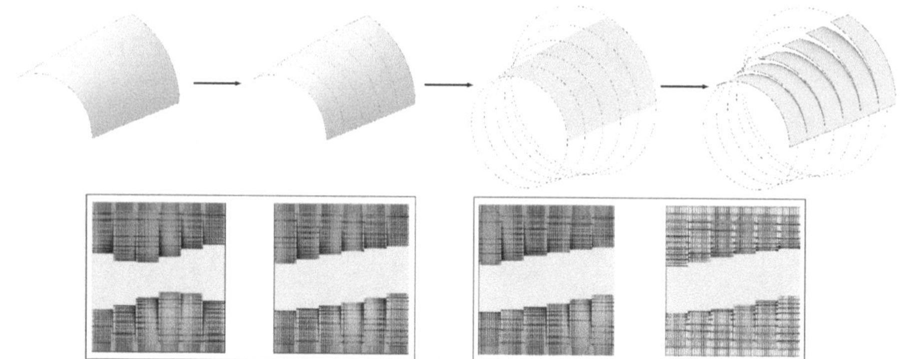

Fig. 5. (Top) The workflow includes surface generation, subdivision, angle calculation, and geometry fitting via quantitative relationships. (Bottom left) Façade aesthetics vary by adjusting P_p. (Bottom right) Identical bending can be achieved through different porosity conditions (P_p or R_p).

The application workflow (Fig. 5) involves: (1) defining a non-planar target surface, (2) segmenting it into panel-width strips, (3) calculating bending angles using the porosity-angle relationship, and (4) feeding porosity values into a parametric model to produce adaptive façade geometries.

Using the P_p-deformation relationship, a three-panel façade was fabricated (Fig. 6). Physical deformation matched simulation within ~5% (Table 2), vali-dating the numerical model. Further geometric complexity can be achieved by manipulating the original surface.

Furthermore, one deformation magnitude can correspond to multiple porosity conditions. Once both shading and aesthetic requirements are satisfied, the condition with the minimum material usage can be selected. This highlights the necessity of integrating building environment analysis and construction management with the existing workflow. Moreover, we observe that the bending curvature induced by R_p changes in rectangular and hexagonal samples differs significantly, which requires further numerical exploration.

Table 2. Cross-validation of P_p-deformation relationship.

P_p	Calculated degree (°)	Physical degree (°)
$P_{p1} = 0.492187$	122.9076	127.13
$P_{p2} = 0.167908$	133.2713	139.63
$P_{p3} = 0.103742$	144.5107	151.37

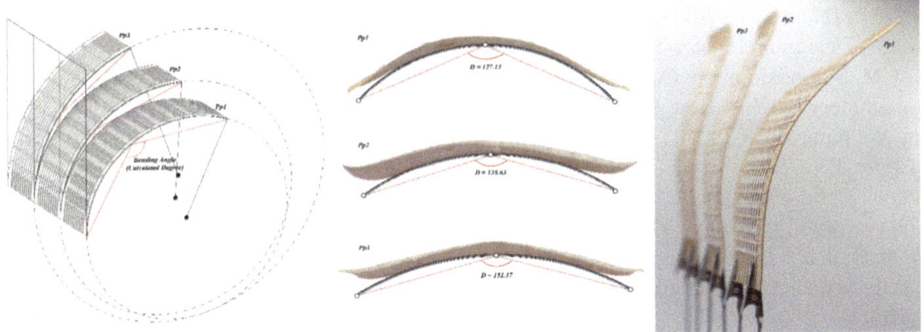

Fig. 6. Three samples are fabricated to cross-validate the quantitative relationship between porosity (P_p) and deformation. Distortion occurs due to the fluctuating environmental conditions.

4 Conclusion and Outlook

This research establishes a 4D-printing simulation framework for porosity-driven hygromorphic bilayers, enabling both predictive and inverse design of humidity-responsive deformation. The study presents three key contributions:

1. Qualitative–Quantitative Relationships: Empirical tests clarify how porosity parameters (P_a, P_p, R_p) affect deformation— P_a has minimal impact, P_p has positive impact on deformation magnitude and $R_p \geq 1$ is necessary for stable actuation when aspect ratios differ from unity.
2. Efficient Parametric Solver: A Grasshopper-based solver was developed to simulate porosity-driven deformation with low computational cost, offering rapid, design-oriented feedback for early-stage material programming.
3. Predictive Mathematical Model: A data-driven model, validated by a façade panel case, enables optimization of porosity configurations for targeted de-formation without extensive physical prototyping.

While the framework offers a scalable approach to programmable design, several areas warrant further development. These include expanding the material library, improving simulation accuracy through iterative calibration and time-dependent modeling, and integrating environmental performance criteria such as light or heat transmission. Addressing these challenges will enhance the robust-ness and applicability of porosity-driven hygromorphic systems, contributing to adaptive, sustainable architectural design through the synergy of material science and computational workflows.

References

Abdel-Rahman, A., Kosicki, M., Michalatos, P., Tsigkari, M.: Design of thermally deformable laminates using machine learning. In: Bittar, Z. (ed.) Advances in Engineering Materials, Structures and Systems: Innovations, Mechanics and Applications, pp. 1016–1021. CRC Press, Boca Raton, FL (2019)

Cheng, T., et al.: Weather-responsive adaptive shading through biobased and bioinspired hygromorphic 4D-printing. Nat. Commun. **15**(1), 10366 (2024)

Cheng, T., Tahouni, Y., Wood, D., Stolz, B., Mülhaupt, R., Menges, A.: Multifunctional mesostructures: Design and material programming for 4D-printing. In: Proceedings of the 5th Annual ACM Symposium on Computational Fabrication (pp. 1-10). Association for Computing Machinery (ACM), New York (2020)

Jung, W., Kim, W., Kim, H.Y.: Self-burial mechanics of hygroscopically responsive awns. Integr. Comp. Biol. **54**(6), 1034–1042 (2014)

Katagiri, J., Saomoto, H., Utsuno, M.: Quantitative evaluation of the effect of grain aspect ratio on permeability. Vadose Zone J. **14**(2), vzj2014–vzj2010 (2015)

Krüger, F., et al.: Development of a material design space for 4D-printed bio-inspired hygroscopically actuated bilayer structures with unequal effective layer widths. Biomimetics. **6**(4), 58 (2021)

Ruedrich, J., Bartelsen, T., Dohrmann, R., Siegesmund, S.: Moisture expansion as a deterioration factor for sandstone used in buildings. Environ. Earth Sci. **63**, 1545–1564 (2011)

Siéfert, E., Reyssat, E., Bico, J., Roman, B.: Bio-inspired pneumatic shape-morphing elastomers. Nat. Mater. **18**(1), 24–28 (2019)

Song, K., et al.: Journey of water in pine cones. Sci. Rep. **5**(1), 9963 (2015)

Tahouni, Y., et al.: Codesign of biobased cellulose-filled filaments and mesostructures for 4D printing humidity responsive smart structures. 3D Print. Addit. Manuf. **10**(1), 1–14 (2023)

Timoshenko, S.: Analysis of bi-metal thermostats. J. Opt. Soc. Am. **11**(3), 233–255 (1925)

Wang, W., et al.: Harnessing the hygroscopic and biofluorescent behaviors of genetically tractable microbial cells to design biohybrid wearables. Sci. Adv. **3**(5), e1601984 (2017)

Yang, X., et al.: Bioinspired soft robots based on organic polymer-crystal hybrid materials with response to temperature and humidity. Nat. Commun. **14**(1), 2287 (2023)

Yogeesh, N.: Fuzzy logic modelling of nonlinear metamaterials. In: Metamaterial technology and intelligent metasurfaces for wireless communication systems, pp. 230–269. IGI Global, Hershey, PA (2023)

Biobased Material Confluences: Investigating Material Formulation and 3D Printing Parameters for Biochar-Mycelium Building Composites

Raffaele Errichiello(✉) and Julio Diarte Almada

Umeå School of Architecture, Östra Strandgatan 30 C, 903 33 Umeå, Sweden
{raffaele.errichiello,julio.diarte}@umu.se

Abstract. This study explores biochar-mycelium composites for 3D printing in sustainable construction. Through four tests, we investigate material formulation, printability, and structural performance, evaluating biochar content, flour types, geometrical configuration of the 3D printing paths, and wood fiber integration. Findings suggest that biochar content, flour type and optimized geometries enhance mycelium growth by influencing antimicrobial properties and material's oxygenation while reducing shrinkage, compared to lignocellulose-based composites. These insights contribute to developing biobased composites with improved mechanical properties, supporting circular construction practices and advancing bio fabrication for architectural applications.

Keywords: Biochar-mycelium composites · 3D Printing · Bio-based construction materials · Circular architecture · Biomimicry

1 Introduction

The construction industry is undergoing a paradigm shift, seeking sustainable alternatives to conventional materials to address environmental concerns and resource depletion. Among the emerging solutions, biobased materials hold significant promise, particularly for additive manufacturing processes such as 3D printing.

This study explores the potential of biochar-mycelium composites as material for construction, using by-products from the Swedish biogas production to enhance circularity and material efficiency.

Biochar, a carbon-rich material derived from the pyrolysis of organic waste, has attracted attention for its environmental benefits, including carbon sequestration, thermal insulation, and structural reinforcement (Mohanty, et al., 2024). Its ability to regulate moisture and reduce shrinkage in composites makes it a compelling candidate for integration into new material formulations. Meanwhile, mycelium, the vegetative part of fungi, has demonstrated remarkable potential as a biobased binder, offering self-repair capabilities and the ability to form resilient, lightweight structures. Mycelium-based materials have been recognized for their thermal insulation properties and as viable

Y. Liu et al. (Eds.): CDRF 2025, *Transindividual Intelligence*, pp. 297–307, 2026.
https://doi.org/10.1007/978-981-92-0615-5_26

alternatives to fossil-based composites (Elsacker, et al., 2023), (Elsacker, et al., 2019). A striking example of mycelium's binding potential is its application in eco-asphalt, where it substitutes bitumen to create fossil-free pavements (Turrell, 2025).

Despite these promising attributes, the compatibility of biochar and mycelium in composite materials remains largely unexplored, particularly regarding their application in 3D printing. The Swedish construction industry faces an urgent need to close the circularity gap by repurposing industrial by-products into high-value materials (Circle Economy; RE:Source; RISE, 2023). Expanding biochar's application beyond soil amendment and filtration into building materials that are not only concrete presents a compelling opportunity to contribute to sustainable architecture (European Biochar Industry Consortium, 2024). However, key questions persist regarding the optimal formulation of biochar-mycelium composites and their performance under 3d printing constraints.

One major challenge in developing mycelium-based composites for 3D printing is shrinkage control. Previous studies indicate that mycelium materials incorporating lignocellulosic substrates exhibit shrinkage rates ranging from 9–20%, which compromises geometric accuracy in printed structures (Elsacker, et al., 2023). In contrast, biochar has been observed to mitigate shrinkage in cementitious composites, with Gupta et al. reporting a significant reduction in autogenous shrinkage when 5 wt% biochar was added to cement mortar (Gupta, et al., 2020). These findings suggest a potential synergy between biochar and mycelium that merits further investigation.

This study addresses these gaps through a series of empirical experiments designed to optimize biochar-mycelium formulations for 3D printing. The research examines the impact of biochar content on the material's printability and structural integrity, evaluates different flour types to enhance mycelium development, and investigates geometric patterns to optimize mycelium growth. Specifically, the geometric exploration utilizes a biomimetic approach, referencing the reaction-diffusion-based patterns found in brain corals, known for facilitating effective fluid circulation and nutrient exchange (Kaandorp, 1999). As demonstrated in previous studies, adequate oxygenation within mycelium composites enables more consistent mycelial growth (Aiduang, et al., 2024).

This biomimetic design approach ensures that the extruded material remains sufficiently interconnected to maintain structural integrity while also providing adequate porosity for optimal aeration, analogous to the natural water circulation mechanisms in coral structures. By refining these parameters, the study aims to establish a foundation for the scalable application of biochar-mycelium composites in construction.

Eventually, our research into biochar-mycelium composites for 3D printing aligns with the conference's theme "Transindividual intelligence" by integrating biobased materials and advanced fabrication technologies. This integration exemplifies the fusion of biobased waste resources with digital fabrication, promoting sustainable and intelligent construction practices.

2 Methods, Equipment, and Procedures

This study utilized a range of biobased and commercially available materials to formulate biochar-mycelium composites for 3D printing. The *biochar* sample was sourced from the Swedish energy industry; however, its precise feedstock and pyrolysis conditions remain unknown. The primary objective was to empirically evaluate its functional performance in combination with mycelium, interpreting the results based on experimental observations rather than detailed compositional analysis.

The mycelium strain used was *M9726 Ganoderma Lucidum*, provided in seed spawn form. Two types of wood fibers were incorporated: *fine sawdust* collected from the university's wood shop, with unknown wood species and possible prior treatments, and *untreated wood fiber pellets* supplied by the Research Institute of Sweden (RI.SE) in Mölndal, whose compositional properties were also unknown. To enhance mycelium development, three types of flour—*tapioca*, *full grain*, and *rice flour*—were sourced from a local food store. Additionally, *xanthan gum* and *glycerol*, serving as binders and plasticizers, were obtained from local commercial suppliers. Experimental procedures were carried out using a *WASP 40100 LDM* 3D printer equipped with a *manual-feeding extruder* and a 6 mm nozzle. Sterilization of biochar and wood fibers was conducted using a *GETINGE HS Lab Steam Sterilizer*, while kitchen utensils were used for pasteurization by boiling. The growth environment was controlled within a fermentation chamber measuring 1200 × 1200 × 800 mm, covered with transparent plastic film and equipped with a thermostat, heating mats, and humidifier. Digital modeling and toolpath generation of printed samples were performed using *Rhinoceros 3D* and *Grasshopper*.

To evaluate the potential of biochar-mycelium composites for 3D printing, four tests were conducted, each addressing different aspects of material formulation, printability, and structural performance. Table 1 summarizes the goals, experiments' setup, and data collection methods for each test.

Table 1. Research tests goals, study set-up, and data collection methods.

Goals	Experiments' set-up	Data collection
Test 1: Assess the compatibility of biochar and mycelium in different formulations for 3D printing and evaluate printability.	Four different biochar-mycelium formulations were prepared, stored in a fermentation box for 48 h, and 3D printed into small cylinders (50 mm diameter, 50 mm height). Each formulation was printed in 4 replicas to have comparable results. Samples were observed for printability and initial mycelium growth.	Printability assessment, photographic documentation of samples over two weeks, weight monitoring of formulation 1.3 to track water content.

(*continued*)

Table 1. (*continued*)

Goals	Experiments' set-up	Data collection
Test 2: Investigate the influence of different nutritional source on mycelium growth.	The most promising formulation from Test 1 (1.3) was modified by replacing tapioca flour with whole grain wheat flour (sample 2.3) and rice flour (sample 3.3). These were 3D printed into cylinders (50 mm diameter, 50 mm height), with 4 replicas for each formulation, and monitored for mycelium development.	Mycelium growth was documented photographically every 24 h over two weeks.
Test 3: Develop optimized geometries for enhanced oxygenation and structural integrity, suitable for future compressive tests according to standards (D695.15)	New geometries were designed to improve oxygenation and mycelial growth. Geometries were informed by reaction-diffusion patterns observed in brain corals, where their geometry promote fluid circulation and nutrient exchange. Twelve variations were generated using Grasshopper and Rhinoceros 3D. G3 and G7 were selected for their balance among porosity and printing path's merging points.	Structural and porosity characteristics of geometries were analyzed. Mycelium growth was observed and compared across different infill patterns.
Test 4: Enhance composite tensile resistance and lignin content by integrating wood fibers into biochar-mycelium formulations.	Biochar and wood fibers were sterilized using an autoclave (120 °C, 15 min) and mixed in five formulations (3.A to 3.E). After 14 days of initial growth at 24 °C, samples were 3D printed into cylinders (60 mm diameter, 120 mm height) using two printing paths (G3, G7). Printed samples were kept fermenting at 24 °C on 90 mm petri dishes covered with 600 ml plastic cups for a week.	Growth patterns and moist content were documented every 48 h. After 7 and 14 days, contamination, mycelium growth consistency and shrinkage were monitored across samples.

3 Results

The experimental tests provided specific observations regarding printability, mycelium growth, structural integrity, and contamination issues associated with the biochar-mycelium composites.

In Test 1 (Fig. 1), sample 1.1 had poor printability due to mycelium spawn on seeds clogging the extruder. Improved results were obtained for samples 1.2 and 1.3 due to lower mycelium and increased biochar content enhancing water management. While sample 1.4, with higher biochar content, faced extrusion difficulties, resulting in uneven and brittle extrusion. Rapid moisture loss and contamination were observed after one week in all samples.

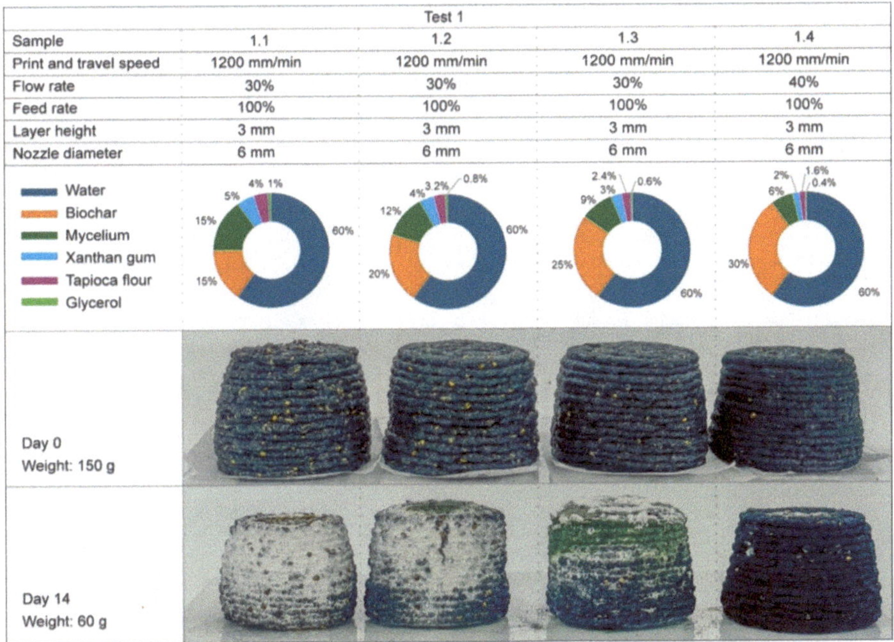

	Test 1			
Sample	1.1	1.2	1.3	1.4
Print and travel speed	1200 mm/min	1200 mm/min	1200 mm/min	1200 mm/min
Flow rate	30%	30%	30%	40%
Feed rate	100%	100%	100%	100%
Layer height	3 mm	3 mm	3 mm	3 mm
Nozzle diameter	6 mm	6 mm	6 mm	6 mm

Fig. 1. Test 1 samples composition, printing parameters, and growing documentation for 14 days.

Test 2 demonstrated the critical role of flour type in optimizing mycelium growth and maintaining antimicrobial resistance. Firstly, *tapioca flour* (Fig. 2-1.3) resulted in weaker mycelium growth and noticeable contamination. Next, *wheat flour* (Fig. 2-2.3) performed well, with slight contamination. Finally, *rice flour* (Fig. 2-3.3) provided optimal results, showing better elasticity during printing, superior mycelium growth, and no contamination. Humidity control remained challenging due to rapid water evaporation in all samples.

In Test 3, geometry optimization informed by reaction-diffusion patterns clearly influenced oxygenation and structural integrity. Figure 3 illustrates a comparative evaluation of twelve different design iterations, analyzing the relationship between material's merging points along the 3D printing paths and fill and void rate. We selected design iterations G3 and G7 for further prototyping in Test 4 because they showed a balanced relation of porosity and number of merging points.

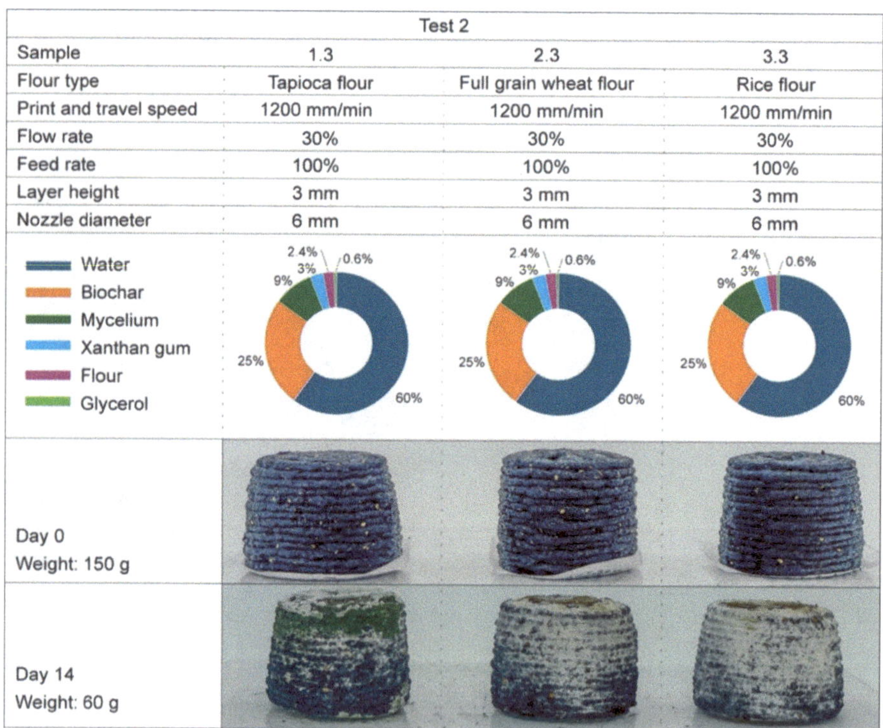

Fig. 2. Test 2 samples composition, printing parameters, and growing documentation for 14 days.

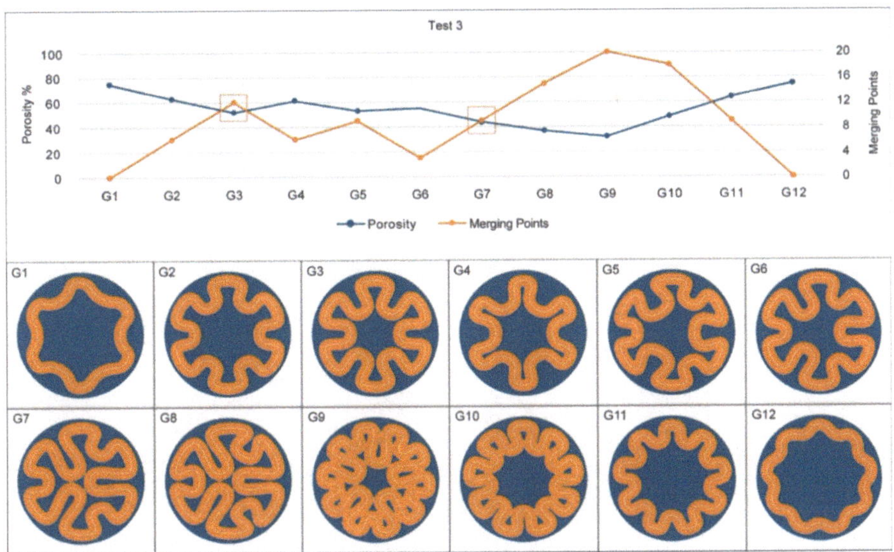

Fig. 3. Comparative geometries evaluation for Test 3

Test 4 evaluated the inclusion of wood fibers to enhance lignin content and structural behavior of the composites (Fig. 4). Geometry G3 demonstrated superior structural integrity, particularly in samples 3.A and 3.B. Sample 3.A, with slight over-extrusion, favored structural performance. Sample 3.B, 3.C, and 3.E required adjustments in water content for successful extrusion.

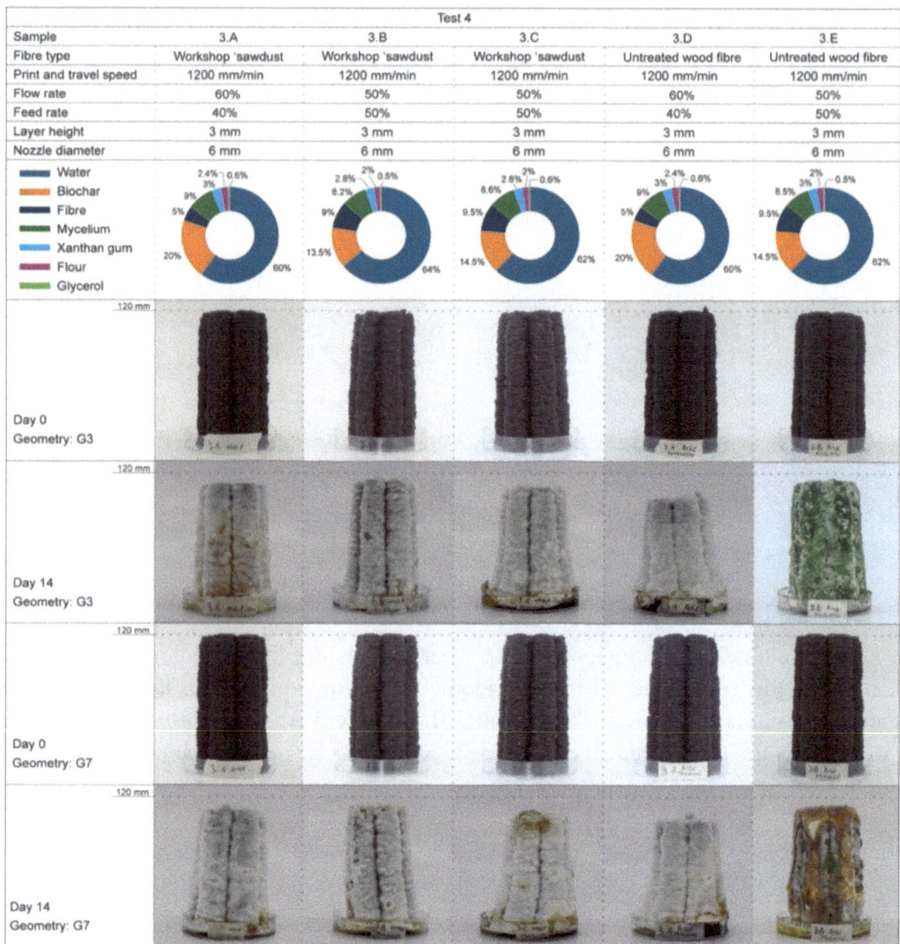

Test 4					
Sample	3.A	3.B	3.C	3.D	3.E
Fibre type	Workshop 'sawdust	Workshop 'sawdust	Workshop 'sawdust	Untreated wood fibre	Untreated wood fibre
Print and travel speed	1200 mm/min	1200 mm/min	1200 mm/min	1200 mm/min	1200 mm/min
Flow rate	60%	50%	50%	60%	50%
Feed rate	40%	50%	50%	40%	50%
Layer height	3 mm	3 mm	3 mm	3 mm	3 mm
Nozzle diameter	6 mm	6 mm	6 mm	6 mm	6 mm

Fig. 4. Test 4 samples composition, printing parameters, and growing documentation for 14 days.

Mycelial growth was uniform in all samples, with workshop sawdust producing the best growth. Sample 3.E was significantly contaminated, probably due to untreated wood fiber properties. Higher wood fiber content promoted *fruiting bodies* and *melanin* development, which is not preferable for the research's goals. Sample 3.D showed the lowest structural performance.

Moisture retention was improved in all samples compared to previous tests, losing only 2–4% of water content after one week, and 6–8% after two weeks. The best specimens showed a shrinkage of 3–5% in Z direction, measured trough photographic analysis. Overall, over extrusion and lower printing speed lead to better results.

In addition to growth and printability, compressive tests were conducted on selected samples from Test 1, Test 2, and Test 4 to assess their structural performance. These tests measured peak force, stroke, strain, compressive strength, and elastic modulus across various geometries and formulations. Results are summarized in Table 2, which presents the mechanical properties of the most representative samples, including averages for 1.1 and 3.3, and all tested 3.A–E composites.

Table 2. Compressive Strength and Elastic Modulus of the most relevant results from the different formulations. The value of samples 1.1 and 3.3 are an average among 4 replicas of each.

Samples	Diam. (mm)	Height (mm)	Cross Section (mm^2)	Peak Force (N)	Stroke (mm)	Strain (%)	Compres. Strength (MPa)	Elastic Modulus (MPa)
1.1	60	60	2827	4175	3.05	5	1.51	31.44
3.3	60	60	2827	2125	4.68	8	0.75	10.44
3.A (G3)	60	120	2748	2000	6.60	6	0.8	14.46
3.A (G7)	60	120	2827	1900	5.20	4	0.66	15.37
3.B (G3)	60	120	2748	3100	7.80	6	1.24	19.2
3.B (G7)	60	120	2827	2900	6.00	5	1.16	23.16
3.C (G3)	60	120	2748	2400	7.20	6	0.96	15.91
3.D (G3)	60	120	2748	3500	6.00	5	1.45	28.9
3.D (G7)	60	120	2827	3700	7.00	6	1.41	24.48
3.E (G7)	60	120	2827	5000	9.00	8	1.91	25.53

4 Discussion

This experimental study explored biochar-mycelium composites for additive manufacturing, highlighting their potential while identifying significant limitations. Challenges emerged primarily around printability, moisture regulation, and contamination control. Tests 1 and 2 revealed that inconsistent humidity significantly impacted mycelium growth, leading to uneven colonization and sporadic contamination during sample handling and documentation. By adjusting the fermentation chamber size in Test 4, stable humidity conditions were successfully maintained, mitigating these earlier issues. Ongoing research with the Molecular Biology Department at Umeå University aims to enhance composite's antimicrobial properties.

Each test provided insights into material formulation and geometry optimization. In Test 1, sample 1.3, with higher biochar content, showed better printability. Test 2 highlighted the critical role of flour type as nutrient; rice flour (sample 3.3) significantly improved mycelial growth and demonstrated enhanced antimicrobial resistance compared to tapioca and wheat flour. Test 3 introduced reaction-diffusion-inspired geometries, revealing that Geometry G3 facilitated better oxygenation and mycelial growth compared to Geometry G7. Integration of lignin-rich sawdust in Test 4 further strengthened mycelium development.

Mechanical performance assessments on the samples showed compressive strength values between 0.66 MPa and 1.91 MPa, and elastic modulus values ranging from

10.44 MPa to 31.44 MPa. Notably, formulations 3.A–E displayed consistent mechanical properties and progressive, foam-like compressive behavior without brittle fracture, indicating their potential as resilient structural materials. While groups 1.1 and 3.3 exhibited variability among four replicas of each, linked to issues in shrinkage, printing consistency, and uneven mycelial colonization. Compared with the performance of anisotropic mycelium-based composites found in the current literature (Rigobello & Ayres, 2022), the mycelium-biochar composites developed in this study demonstrate improved mechanical properties (Fig. 5).

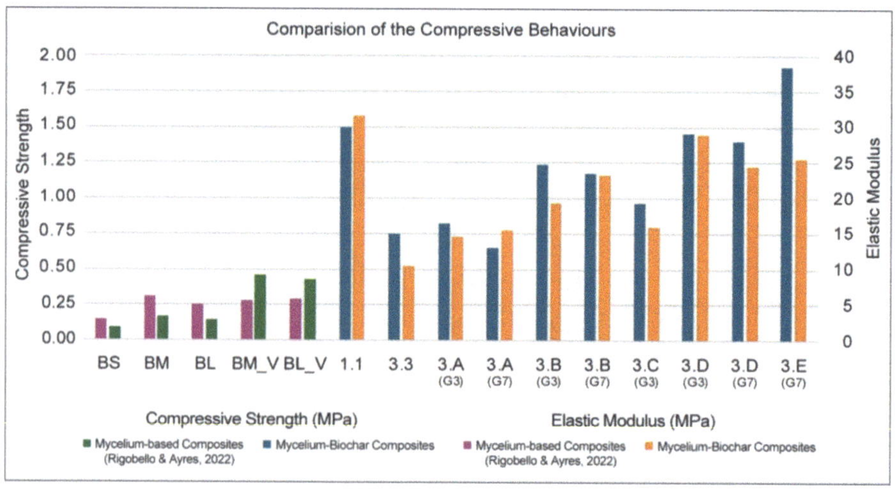

Fig. 5. Compressive behaviors of mycelium-biochar compared with mycelium-based composites.

5 Conclusion

This study provides foundational data on the mechanical and biological performance and fabrication process of biochar-mycelium composites, underscoring their viability for sustainable additive manufacturing applications.

Future investigations will address long-term environmental performance, including thermal insulation, reaction to fire test and inherent self-adherence of mycelium. These features could reduce reliance on synthetic adhesives, simplifying structural integration. Comprehensive Life Cycle Assessments (LCAs) and Life Cycle Cost (LCC) analyses will further elucidate environmental and economic benefits, particularly regarding resource utilization and waste reduction.

Acknowledgements. The authors would like to thank Lasse Kock, master student at Umeå School of Architecture, for his assistance in developing the diagrams presented in Fig. 3.

References

Aiduang, W., et al.: A review delving into the factors influencing mycelium-based green composites (MBCs) production and their properties for long-term sustainability targets. Biomimetics. **9**, 337 (2024)

Circle Economy; RE:Source; RISE: Circularity Gap Report 2023. Circle Economy, Amsterdam (2023)

Elsacker, E., et al.: Mechanical, physical and chemical characterisation of mycelium-based composites with different types of lignocellulosic substrates. PLOS ONE. **14**(7), e0213954–e0213968 (2019)

Elsacker, E., Zhang, M., Dade-Robertson, M.: Fungal engineered living materials: the viability of pure mycelium materials with self-healing functionalities. Adv. Funct. Mater. **33**, 2301875 (2023)

European Biochar Industry Consortium: European Biochar Market Report 2023–24. European Biochar Industry Consortium, Brussels (2024)

Gupta, S., Krishnan, P., Kashani, A., Kua, H.K.: Application of biochar from coconut and wood waste to reduce shrinkage and improve physical properties of silica fume-cement mortar. Constr. Build. Mater. **262**, 120688 (2020)

Kaandorp, J.A.: Morphological analysis of growth forms of branching marine sessile organisms along environmental gradients. Mar. Biol. **134**, 295–306 (1999)

Mohanty, A.K., et al.: Biocarbon materials. Nat. Rev. Meth. Primers. **4**, 20 (2024)

Rigobello, A., Ayres, P.: Compressive behaviour of anisotropic mycelium-based composites. Sci. Rep. **12**(1), 6846 (2022)

Turrell, C.: Mushroom makes tasty eco-asphalt. Nat. Biotechnol. **43**, 9 (2025)

Forming Heterogeneities: Fabricating Graded Low-Carbon CharCrete Building Elements

Nikol Kirova(✉)

Institute for Advanced Architecture of Catalonia, Carrer de Pujades, 102, Sant Martí, 08005 Barcelona, Spain
nikol.kirova@iaac.net

Abstract. As the construction industry seeks sustainable alternatives, functionally graded concrete (FGC) offers a promising approach to optimising structural performance alongside environmental performance. This paper investigates multi-material fabrication strategies for building elements with a low carbon footprint made of CharCrete, a cementitious composite material leveraging the carbon sequestration capability of biochar. The design and fabrication method "Grading Carbon" is part of a larger study that aims to balance structural integrity with embodied carbon reduction. The paper focuses on two fabrication strategies: "print and cast," using a cementitious 3D-printed formwork, and "partition and cast," where physical partitions facilitate simultaneous multi-grade casting. Experimental results highlight the critical role of bonding between grades, with sequencing and deposition timing proving essential for preventing delamination. Computational design tools and Finite Element Analysis (FEA) are used to simulate the performance of building elements with a heterogeneous materiality. Combining BCM_B10 (high-performance printable) and BCM_B30 (carbon-sequestering) grades can achieve a balance of mechanical strength and environmental impact, resulting in elements with a net neutral carbon footprint.

CharCrete 3D printing (CC3DP) has been developed and evaluated. The assessment indicates that lower biochar grades (BCM_B0–B20) are suitable for printing, while higher biochar grades necessitate casting due to increased water content and longer setting times. Mechanical testing confirms that the "print and cast" method is more efficient compared to the "partition and cast" one as it offers lower fabrication costs, greater customisation, and reduced environmental impact. Computational design workflows integrate embodied carbon considerations into the design-to-manufacturing process, providing a scalable approach for sustainable construction. The findings contribute to the advancement of graded material systems, demonstrating the viability of biochar-enhanced FGC for adaptive, high-performance building applications.

Keywords: biochar · C3DP · carbon sequestration · design optimisation · graded fabrication

Y. Liu et al. (Eds.): CDRF 2025, *Transindividual Intelligence*, pp. 308–324, 2026.
https://doi.org/10.1007/978-981-92-0615-5_27

1 Introduction

Advancements in computational design and digital fabrication have paved the way for the development of novel composite material systems that optimize both performance and sustainability. Architects and builders are increasingly interested in multi-material systems to optimize properties such as strength, thermal insulation, durability, and carbon efficiency for specific design needs. Multi-material approaches have gained significant traction in industries including aerospace and automotive engineering. However, the construction sector has predominantly utilized reinforced concrete, which effectively combines the tensile strength of steel with the compressive strength of concrete. However, traditional concrete construction remains homogeneous. This paper explores an alternative approach to concrete: graded materiality, where material properties are spatially varied to enhance both structural and environmental performance.

Functionally graded materials (FGMs) offer a promising solution to overcoming the limitations of conventional homogeneous materials by introducing controlled heterogeneities that respond to mechanical and environmental demands. In structural applications, this method facilitates increased material efficiency by reinforcing regions subjected to high stresses while introducing carbon-sequestering materials in less critical areas. By strategically incorporating biochar, cementitious materials can improve performance by minimizing mass and embodied carbon. Nonetheless, the application of graded materiality in construction has faced limitations, primarily due to difficulties in fabrication.

The literature on Functionally Graded Concrete (FGC) strategies explores the optimisation of cement and lightweight aggregates for enhanced structural and thermal performance (Herrmann and Sobek, 2017, 2015; Oxman, 2010; Oxman et al. 2011). Advances in casting and spraying techniques have enabled layered and continuously graded FGC, improving performance while reducing cement content (Aylie et al. 2015; Bajaj et al. 2013). Recent developments at MIT, Stuttgart University, and Cambridge's CIRG have explored material porosity, density grading, and lightweight aggregates to enhance efficiency and sustainability (Maier and Lees, 2022; Torelli et al. 2020; Torelli and Lees, 2020, 2019). However, challenges in material compatibility, layer bonding, and manufacturing complexity persist, and the integration of biochar for carbon sequestration in FGC remains underexplored.

Existing research on Concrete 3D Printing (C3DP) has primarily focused on homogeneous cementitious mortars for structural optimisation, as seen in the stress-line-informed beams developed at the University of Southern Denmark (Breseghello et al. 2023). Some advances in multi-material C3DP, typically by printing a high-performance material and subsequently casting a lower-density filler, an approach explored by Xtree (Gaudillière et al. 2019, 2018). Despite advancements, manufacturing techniques for functionally graded concrete remain underdeveloped, with stepwise grading being the most practical method currently available.

2 Methodology

This study explores the fabrication of graded CharCrete, a carbon-sequestering biochar-based cementitious composite. A key challenge in incorporating biochar into cementitious materials lies in the trade-off between mechanical performance and environmental benefit: as biochar content increases, compressive and flexural strength decrease, while carbon sequestration improves. These opposing optimisation criteria highlight the need for a novel approach to material distribution that balances structural integrity with embodied carbon reduction.

To address this, the research builds on the Grading Carbon design method, developed as part of a broader study, where a method for strategically varying the biochar content across building components is used to optimise carbon footprint (Kirova et al. 2023a, 2023c). However, realising these digitally designed material gradients requires precise and feasible fabrication strategies. This paper focuses on the experimental evaluation of two stepwise grading techniques:

- *Partition and Cast*: simultaneous multi-grade casting using physical partitions to separate and control material zones;
- *Print and Cast*: casting low-strength material into a performative C3DP formwork.

The methodology comprises two main phases. First, small-scale prototypes are fabricated to investigate interfacial bonding and mechanical stability between grades under both fabrication strategies. Second, full-scale structural probes, designed using the *Grading Carbon* computational method, are produced and tested under controlled laboratory conditions to assess their mechanical performance and environmental impact. By combining computational material design with digital and conventional fabrication methods, this research aims to demonstrate the potential of functionally graded CharCrete for low-carbon, high-performance construction. The outcomes contribute to the evolving discourse on sustainable architecture, emphasizing the critical role of fabrication in enabling advanced material systems for carbon-conscious building design.

2.1 Interface and Bonding

The transition between constituents significantly affects bonding and overall performance, with poor adhesion leading to delamination and potential failure. Concrete casting, the dominant construction method, relies on precise timing between pours to ensure optimal bonding. Studies, including those by CIRG, highlight that wet-on-wet casting within two hours achieves superior adhesion, whereas wet-on-dry requires bonding agents (Torelli and Lees, 2020, 2019).

The material interface and bonding resulting from the "partition and cast" and the "print and cast" strategy are assessed in terms of stability, casting time, grade sequencing, and deformation. By addressing these factors, the research aims to refine fabrication techniques for stepwise graded cementitious materials, mitigating risks of delamination and structural weakness.

The "*partition and cast*" strategy involved casting cuboid specimens (160 × 40 × 40 mm) using two CharCrete grades (BCM_A20 and BCM_A50 (Kirova et al. 2023b), separated by 3D-printed PLA partitions. The experiments assessed both partition removal timing and interface geometry to determine their influence on bonding and mechanical performance. The study revealed that removing partitions too early (within 5–10 min) led to deformation due to hydrostatic pressure, while waiting 15–20 min ensured strong bonding and retained interface geometry. Delays beyond 30 min resulted in poor adhesion and delamination. The second study examined interface geometries, showing that increased surface area and smooth curves improved bonding, whereas sharp corners caused material accumulation, weakening adhesion. These findings highlight the critical role of timing and interface design in achieving stable multi-grade composites. While effective, this method requires precise control over casting sequences and material behaviour.

The "*print and cast*" approach leveraged 3D printing to create formwork before casting a secondary grade inside. CharCrete 3D Printing (CC3DP) builds upon C3DP principles but introduces additional complexities due to the hydrophilic nature of biochar, which affects the mixture's rheology. Standard cementitious mixes for 3D printing require reduced water content and plasticisers to achieve print stability, maintaining a water-to-binder ratio below 0.3. However, biochar-rich CharCrete grades exceed this threshold, impacting mechanical performance and printability. The research evaluated various CharCrete mixtures, revealing that those with higher biochar content (BCM_B20 to BCM_B70) exhibited excessive shrinkage and workability challenges.

The CharCrete grades were evaluated using a CC3DP setup, with BCM_B10 and BCM_B20 emerging as the most suitable for additive manufacturing, owing to their print stability and mechanical performance. Cylindrical specimens (160 mm Ø, 100 mm height) were printed with BCM_A5, followed by casting BCM_A50 at different time intervals. Immediate casting led to buckling, deformation, and leakage due to hydrostatic pressure on the fresh print. A 10-min delay reduced deformation but still caused expansion at the base. The optimal casting window, observed at 25–45 min post-printing, allowed the print to stabilize while remaining wet enough to ensure strong bonding. Casting after over an hour resulted in poor adhesion and separation due to shrinkage. A benefit of this method in comparison to the "partition and cast" one is that the textured surface of the CC3DP formwork enhances bonding through better interlocking.

2.2 Design for Manufacturing

This study investigates the integration of computational design and digital fabrication to produce functionally graded concrete elements using CharCrete, a biochar-based cementitious material system. The objective is to optimise structural elements for reduced embodied carbon without compromising mechanical performance. To achieve this, a custom computational workflow was developed, combining Finite Element Analysis (FEA) with Karamba3D and parametric modelling in Grasshopper.

The design process begins with structural simulation using a mono-grade material (BCM_B10), which informs the spatial allocation of different material grades. Structural parameters such as utilisation, principal stress, and proximity to boundaries or connections are used to drive material grading. Each parameter is weighted using the Wallacei plug-in to generate a performance-based distribution. A continuous gradient is then discretised into a stepwise grading scheme for fabrication. This approach allows the material composition to vary spatially, reducing carbon emissions while retaining structural integrity (Fig. 1).

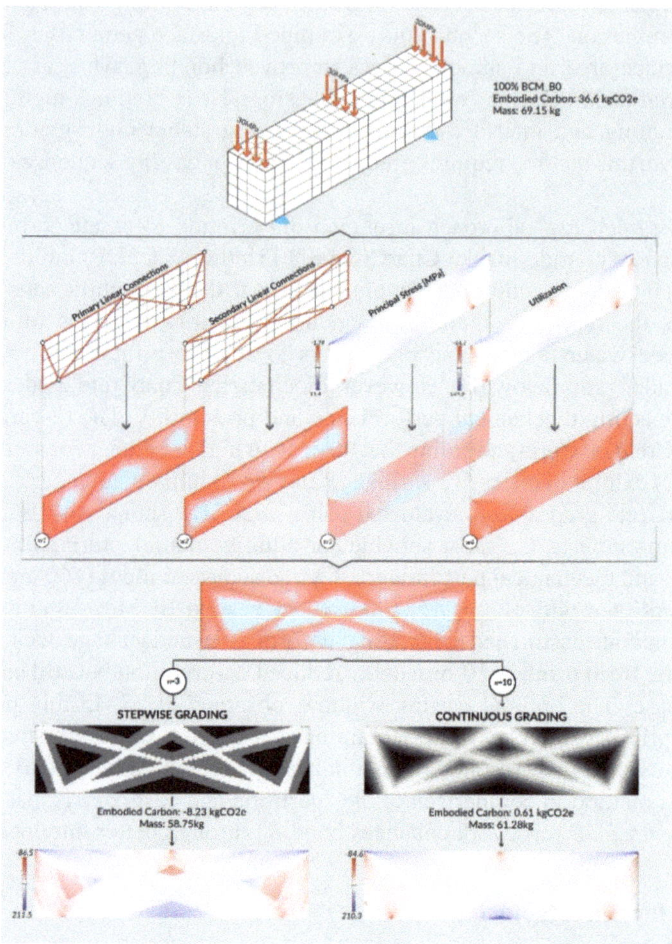

Fig. 1. The computational generative process for the design of a probe with a graded materiality

The developed method is used to design a probe of 1000 × 250 × 200 mm, optimised to fit the available gantry-based printing setup at Swinburne University's Digital Construction Laboratory. Comparative simulations between continuous (ten grades) and stepwise (three grades) grading revealed only marginal differences in structural utilisation. However, the stepwise approach demonstrated significantly better environmental performance, achieving a net negative embodied carbon of −8.23 kgCO_2e compared to 0.61 kgCO_2e for the continuous version and 36.6 kgCO_2e for a mono-grade control (Fig. 1). It also resulted in a 15% reduction in material mass compared to 11.4% with continuous grading, highlighting its efficiency in both material use and carbon reduction.

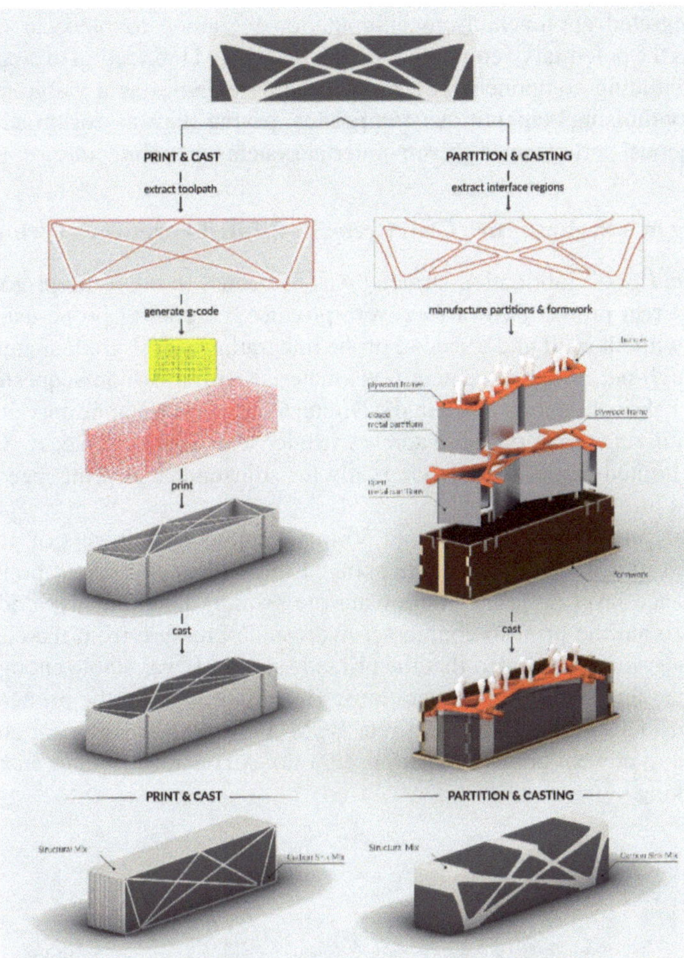

Fig. 2. Manufacturing process of the two fabrication strategies: print and cast; partition and cast

Stepwise grading was selected for physical prototyping due to its practical advantages and compatibility with available fabrication technologies. While continuous grading has aesthetic appeal, it poses significant manufacturing challenges, especially with cementitious materials of varying density. This research focuses on the fabrication of the digitally simulated and optimized graded CharCrete topology by exploring the "*print and cast*" and the "*partition and cast*" methods. Both approaches aim to control spatial material variation in a wet state, enabling the formation of a monolithic, graded element (Fig. 2).

Three probes were fabricated based on the same digital design: a control probe made entirely from a 3D-printed high-performance mix; a "*print and cast*" probe combining printed and cast materials; and a "*partition and cast*" probe with dual casting separated by physical partitions. These prototypes allow for a direct comparison of how fabrication strategies affect structural performance, material use, and environmental impact.

This integrated approach, from computational grading to physical prototyping, demonstrates the potential of combining digital design and fabrication to create efficient, low-carbon building components. Stepwise grading emerges as a viable and scalable method for optimising cementitious composites, paving the way for broader adoption of heterogeneous, performance-driven material systems in architecture.

2.3 Casting in a Performative CharCrete 3D Printed Formwork (Print and Cast)

The "*print and cast*" fabrication strategy was evaluated through the prototyping of a graded CharCrete probe. Two probes were produced: a control probe using standard 3D printing without infill and a second probe integrating a dual-grade material system, combining high-performance cementitious material with a carbon-sequestering CharCrete grade. Fabrication took place in the Digital Construction Laboratory at Swinburne University, utilizing a gantry-based screw extruder with a manual feeder. The printing process had limited flexibility, allowing only for adjustments to print speed, extrusion feed rate, and first-layer height.

The printing process lasted 1 h and 30 min, producing 25 layers of 10 mm layer height and 20 mm weight. Water was sprayed throughout to prevent premature drying and enhance layer adhesion. Following the printing phase, a plywood formwork was added around the print to counteract hydrostatic pressure from the cast material, although observations indicated that the printed formwork was stable enough to potentially eliminate this step in future iterations. The cavities within the printed formwork were then manually filled with CharCrete BCM_B30 mix, with casting completed in under 10 min to prevent premature setting (Fig. 3). A concrete vibrator ensured proper material packing.

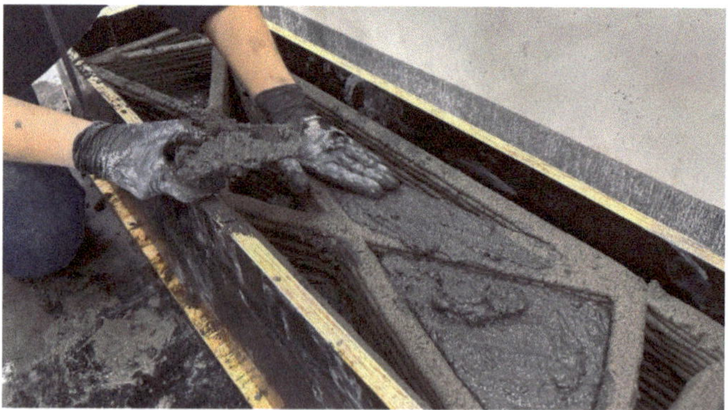

Fig. 3. Process of manually filling in the designed cavities with the CharCrete BCM_B30 mix during the production of the probe manufactured with the print and cast method

After 24 h of curing, the formwork is removed, revealing a clear distinction between the printed and cast materials, evidenced by the colour contrast (Fig. 4). No deformation was observed due to hydrostatic pressure, and the printed texture remained intact except for the top surface, which was slightly flattened due to manual packing. The probe was then left to fully cure before undergoing mechanical testing, where its performance was compared against both the control probe and the partition and cast probe.

Fig. 4. Front view of the probe manufactured following the print and cast method 24 h after manufacturing

2.4 Simultaneous Multi-grade Casting with Physical Partitions (Partition and Cast)

The "*partition and cast*" strategy was evaluated through the prototyping of a graded CharCrete probe. This method uses physical partitions to control the spatial distribution of multiple material grades during casting. The formwork and partitions were fabricated over three days at Swinburne University's ProtoLab, requiring woodworking, metal-working, and digital fabrication expertise. Though labour-intensive, the partitions are reusable, enabling multiple casting variations with consistent typology but different material compositions.

Fig. 5. A snapshot of the manufacturing process of the probe after both grades were cast in their respective regions and prior to the removal of the partitions

The formwork and frames were CNC-milled from laminated plywood and included 2 mm grooves to hold the metal partitions in place during casting. Ten partitions were made from thin metal sheets, five closed and five open, pre-cut, bent, and fixed to plywood frames for support and easy removal (Fig. 2).

Casting followed a central-to-peripheral sequence: CharCrete was poured into the closed partitions first, followed by the high-performance cementitious mix, and finally the open partitions near the formwork. This order reduced material displacement. While access limitations in the closed partitions presented challenges, all mixes were successfully placed within 15 min (Fig. 5). A concrete vibrator was used in accessible zones to remove air pockets; however, a vibrating table may improve results in future applications.

Fig. 6. The final probe produced with the "partition and cast" method after demolding

Partitions were removed immediately after packing, with precision and vertical motion to avoid material deformation. Due to the pressure exerted by the cast material, light vibrations were applied to facilitate removal, a process that took over 30 min. After 24 h of curing, the formwork was removed, revealing the graded material composition with a clear distinction between CharCrete (dark) and high-performance concrete (light) (Fig. 6).

3 Results

The probes were subjected to mechanical testing to quantify the impact of fabrication strategies on structural performance. Conducted at Swinburne's Smart Structures Laboratory, the tests assessed mechanical capacity, fracture behavior, and load-bearing differences between the two fabrication methods. Two key aspects were observed: whether the elements exhibited monolithic behavior and whether CharCrete integration within cavities enhanced load-bearing capacity.

The three fabricated probes were tested using the Instron 5MN Four-Column Static Testing Machine under a three-point bending setup, replicating the loading conditions from digital simulations. Each probe underwent an identical testing procedure, with load applied until failure, while deformation and load data were automatically recorded.

Fig. 7. The control probe produced with C3DP while undergoing the mechanical testing. Two distinct moments are depicted: the first appearance of cracks (top), and the moment of failure (bottom)

The control probe, composed entirely of 3D-printed high-performance material, failed at 13.9 kN, with cracks propagating from the bottom to the side, rather than in the expected central region (Figs. 7 and 8). This deviation in failure mode may be attributed to misalignment of the supports, minor deformations during printing, or geometric factors such as the proximity and length of structural elements. The first crack emerged in a side lateral element connecting the bottom to the top corner, where the density of elements was lowest. Further tests need to be conducted to verify the results.

Additionally, the axial displacement of the control probe measured 0.93 mm, the highest among the three samples (Fig. 8). This indicates that the probe exhibited higher flexibility and lower stiffness, likely due to its open structure with air cavities, allowing for increased flexural behaviour. These findings suggest that while fully 3D-printed elements provide structural integrity, their open configurations may lead to localized weaknesses under applied loads.

Following the "print and cast" probe underwent mechanical evaluation, reaching a maximum load capacity of 19.6 kN before failure. The failure mode was characterized by a crack propagating from the bottom center to the top (Figs. 9 and 10). Notably, this probe exhibited the highest mechanical strength. The high-performance mix forming the perimeter likely contributed to its reinforcement, enhancing structural performance. The symmetrical material layout, combining C3DP high-performance mix with Char-Crete cast within the cavities, resulted in a central failure point, aligning with expected behavior. Additionally, compared to the control probe, the axial force distribution was

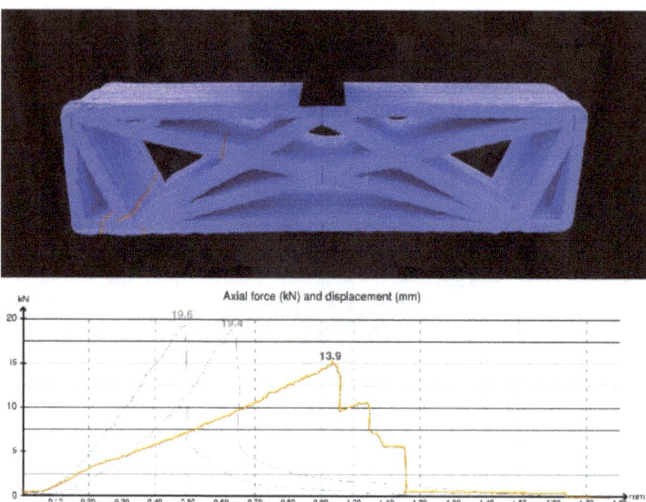

Fig. 8. The testing results from the control probe: i) the cracking process prior to failure is depicted on the top with the cracks highlighted in orange, ii) the results from the control probe's the load-test and specifically the axial force in kN is depicted in orange in comparison to the other two probes (bottom)

more uniform, demonstrating monolithic behavior, a critical indicator of a successful graded material composition achieved through the "print and cast" fabrication strategy.

The graded probe demonstrated higher mechanical performance, with a maximum axial force of 19.6 kN compared to 13.9 kN for the control probe, representing a 41% increase. This enhancement in strength can be attributed to the addition of the CharCrete mix in the cavities. Although this results in increased mass, it does not lead to a higher embodied carbon footprint, as the CharCrete mix offsets the embodied carbon of the high-performance mix through the carbon sequestration process associated with the biochar it contains. Examining the axial displacement at the moment of failure, the control probe outperformed the "print and cast" probe, exhibiting a displacement of 0.93 mm compared to 0.49 mm for the latter. This indicates a significant reduction in flexural capacity and higher stiffness in the graded probe.

Finally, the "partition and cast" probe was subjected to testing. As shown in Figs. 9 and 11, the probe cracked very similarly to the "print and cast" probe, with a crack initiating in the bottom central region and propagating upwards. The "partition and cast" probe failed at a load of 19.4 kN, marginally lower than the other graded probe produced using the "print and cast" method, which sustained a maximum axial force of 19.6 kN (Figs. 10 and 12).

The axial displacement was recorded at −0.49 mm for the "print and cast" probe and −0.64 mm for the "partition and cast" probe. Interestingly, the "partition and cast" probe demonstrated higher flexural capacity compared to the "print and cast" probe. This difference can be attributed to the variation in geometry, where the bottom layer of the 'print and cast' probe comprised a high-performance mixture with lower flexural strength, whereas the 'partition and cast' probe featured CharCrete mix in the bottom

Fig. 9. The "print and cast" probe produced with C3DP and casting while undergoing the mechanical testing. Two moments are depicted: first appearance of cracks (top), and failure (bottom)

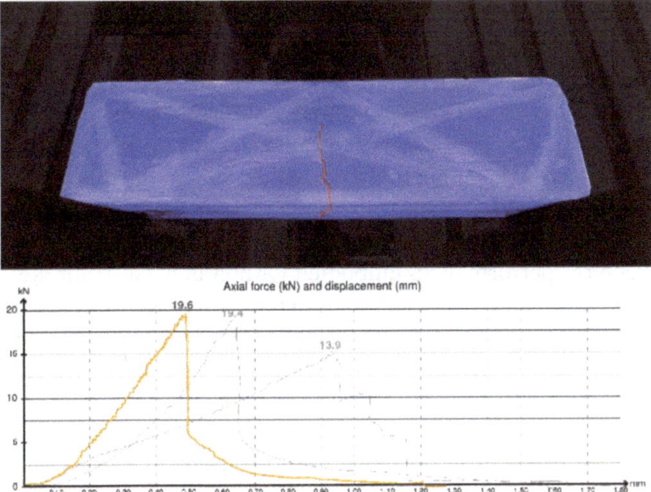

Fig. 10. The testing results from the 'print and cast' probe: i) the cracking process prior to failure is depicted on the top with the cracks highlighted in orange (top), ii) the results from control probe's the load-test and specifically the axial force in kN is depicted in orange in comparison to the other two probes (bottom)

Fig. 11. The 'partition and cast' probe produced with dual-material casting while undergoing the mechanical testing at the Smart Structures Laboratory. Two distinct moments are depicted (from top to bottom): first appearance of cracks, and failure

central area. This discrepancy in geometry arises from the distinct fabrication strategies employed.

The mechanical testing revealed that the "partition and cast" probe exhibited slightly higher strength than the "print and cast" probe, though the difference was marginal. Both graded probes showed consistent performance, with similar maximum axial force capacities and no signs of delamination or weak bonding at the material interfaces. The "partition and cast" probe demonstrated slightly greater flexural capacity, suggesting a limited structural advantage.

Beyond structural performance, embodied carbon and fabrication efficiency were also assessed. The "print and cast" method used more high-performance mix, increasing embodied carbon due to the solid printed outline. However, it eliminated the need for separate formwork, offering flexibility and adaptability for customized elements. In contrast, the "partition and cast" strategy required labor-intensive formwork and partitions, contributing to higher initial embodied carbon. Yet, the reusability of these components may offset their impact in repetitive manufacturing scenarios, especially where specialized equipment is unavailable.

Overall, the "print and cast" method proved more efficient, offering lower embodied carbon and greater design flexibility. Aesthetic differences also emerged—C3DP produced a rough, layered texture, while casting delivered a smooth, formwork-defined surface. These considerations become critical at the building scale, where assembly

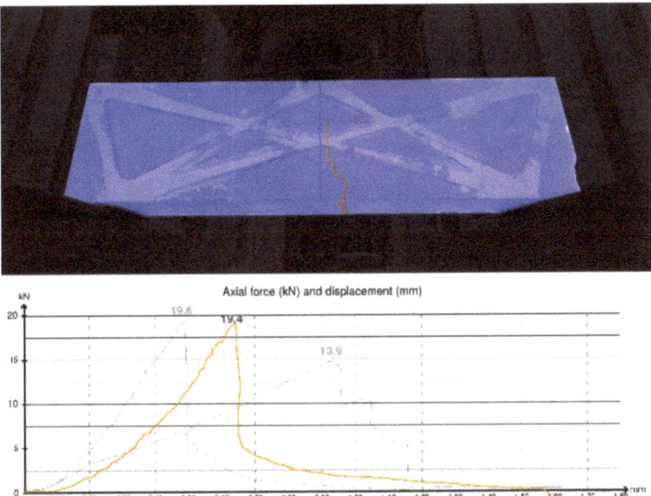

Fig. 12. The testing results from the 'partition and cast' probe: i) the cracking process prior to failure is depicted on the top with the cracks highlighted in orange (top), ii) the results from the control probe's load-test and specifically the axial force in kN is depicted in orange in comparison to the other two probes (bottom)

and finish quality matter. Given the comparable structural performance of both methods, decisions at scale should weigh embodied carbon, fabrication costs, and visual integration.

4 Conclusions

This study investigated two fabrication strategies: "print and cast" and "partition and cast"; for producing graded CharCrete elements, designed to balance structural performance and carbon reduction (Fig. 13). By comparing their mechanical, environmental, and fabrication outcomes, the research offers a comprehensive understanding of how material grading can enhance both performance and sustainability in architectural components.

Mechanical testing revealed that both graded probes exhibited comparable structural performance, with no evidence of delamination or poor bonding between material interfaces. The "partition and cast" probe showed slightly greater flexural strength and higher geometric accuracy, closely replicating the digital design. This marginal structural advantage was counterbalanced by the method's higher labour demands and initial embodied carbon due to complex formwork and partitions.

In contrast, the "print and cast" method emerged as the more efficient and scalable approach, eliminating the need for additional formwork while enabling customised design through 3D-printed performative zones. Although it uses more high-performance mix, which slightly increased embodied carbon, the overall process was simpler, quicker,

Fig. 13. The two probes manufactured following the "print and cast" (left), and "partition and cast" method (right) alongside the fabrication equipment for each strategy: metal partitions and C3DP extruder.

and more adaptable for custom or one-off elements. Moreover, the use of digital fabrication techniques allowed for precise material placement and reduction of material waste, aligning with sustainability goals.

The study also identifies an optimal dual-grade material system: BCM_A10 (high-performance), and BCM_A30 (carbon-sequestering); which offers a balanced combination of strength and carbon reduction. While simulations suggested that two grades could suffice structurally, physical testing confirmed that including an intermediate grade improves bonding, reduces shrinkage differentials, and enhances overall material stability.

Fabrication timing proved to be a critical parameter. Managing the workability window of high-biochar CharCrete grades is essential to avoid premature curing and to ensure strong inter-grade bonding. For larger components, simultaneous printing and casting may be required, underscoring the need for integrated planning in scaled-up applications.

Finally, aesthetic outcomes also differ between methods: 3D-printed elements featured a textured, layered finish, while cast elements reflected the smoothness of their formwork. These visual characteristics become particularly relevant at the architectural scale, where assembly, surface finish, and visual continuity impact the overall quality of construction. Additionally, reinforcement integration has to be considered when scaling up.

By integrating computational design, material grading, and digital fabrication, this research demonstrates a clear path toward more sustainable, low-carbon, and high-performance concrete construction. The findings support the use of functionally graded materials as a viable strategy for future architectural applications, contributing to the shift towards environmentally responsible design and construction practices.

Acknowledgments. The author acknowledges the support of the Smart Structures Laboratory and the Digital Construction Laboratory at Swinburne University of Technology for providing access to fabrication and testing infrastructure essential to the experimental components of this research. The author also gratefully acknowledges the guidance and critical feedback provided by the supervisory team: Dr. Jane Burry, Dr. Mehrnoush Latifi, and Dr. Areti Markopoulou. Their expertise and ongoing support significantly contributed to the development and refinement of this work.

References

Aylie, H., Gan, B.S., As'ad, S., Pratama, M.M.A.: Parametric study of the load carrying capacity of functionally graded concrete of flexural members. Int. J. Eng. Technol. Innov. **5**, 233–241 (2015)

Bajaj, K., Shrivastava, Y., Dhoke, P.: Experimental study of functionally graded beam with fly ash. J. Inst. Eng. (India) Ser. A. **94**, 219–227 (2013). https://doi.org/10.1007/s40030-014-0057-z

Breseghello, L., Hajikarimian, H., Jørgensen, H.B., Naboni, R.: 3DLightBeam+. Design, simulation, and testing of carbon-efficient reinforced 3D concrete printed beams. Eng. Struct. **292**, 116511 (2023). https://doi.org/10.1016/j.engstruct.2023.116511

Gaudillière, N., et al.: Large-scale additive manufacturing of ultra-high-performance concrete of integrated formwork for truss-shaped pillars. Robot. Fabr. Archit. Art Des. **2018**, 187–198 (2018)

Gaudillière, N., et al.: Building applications using lost formworks obtained through large-scale additive manufacturing of ultra-high-performance concrete. In: 3D Concrete Printing Technology, pp. 37–58. Elsevier, Amsterdam (2019). https://doi.org/10.1016/b978-0-12-815481-6.00003-8

Herrmann, M., Sobek, W.: Functionally graded concrete: numerical design methods and experimental tests of mass-optimized structural components. Struct. Concr. **18**, 54–66 (2017). https://doi.org/10.1002/suco.201600011

Herrmann, M., Sobek, W., Development of weight-optimised, functionally graded precast slabs. In: Proceedings of the International Association for Shell and Spatial Structures (IASS) Annual Symposium 2015: Future Visions – Amsterdam. International Association for Shell and Spatial Structures (IASS), Amsterdam, pp. 1–11. (2015)

Kirova, N., Burry, J., Latifi, M., Markopoulou, A., Functionally graded CharCrete slabs for embodied carbon optimisation in architecture. In: Proceedings of IASS Annual Symposia, IASS 2023 Melbourne Symposium: Optimisation methods and applications. (2023a)

Kirova, N., Markopoulou, A., Burry, J., Latifi, M., A study on carbon-neutral biochar-cementitious composites. In: Design for Rethinking Resources, UIA 2023. Sustainable Development Goals Series. Springer, Cham, pp. 37–58. (2023b). https://doi.org/10.1007/978-3-031-36554-6_33

Kirova, N., Markopoulou, A., Bury, J., Latifi, M., Grading CharCrete: embodied carbon optimization of load-bearing walls. In: Proceedings of the Association for Computer Aided Design in Architecture (ACADIA) 2023: Habits of the Anthropocene – Scarcity and Abundance in a Post-Material Economy, ACADIA, pp. 130–139. (2023c)

Maier, M., Lees, J.: Interlayer fracture behaviour of functionally layered concrete. Eng. Fract. Mech. **271**, 108672 (2022). https://doi.org/10.1016/j.engfracmech.2022.108672

Oxman, N.: Structuring materiality: design fabrication of heterogeneous materials. Archit. Des. **80**, 78–85 (2010)

Oxman, N., Keating, S., Tsai, E., Functionally graded rapid prototyping. In: Proceedings of the 2011 International Conference on Computer-Aided Architectural Design Research in Asia (CAADRIA 2011). Association for Computer-Aided Architectural Design Research in Asia (CAADRIA), Hong Kong, pp. 1–10. (2011)

Torelli, G., Fernández, M.G., Lees, J.M.: Functionally graded concrete: design objectives, production techniques and analysis methods for layered and continuously graded elements. Constr. Build. Mater. **242**, 118040 (2020). https://doi.org/10.1016/j.conbuildmat.2020.118040

Torelli, G., Lees, J.M.: Interface bond strength of lightweight low-cement functionally layered concrete elements. Constr. Build. Mater. **249**, 118614 (2020). https://doi.org/10.1016/j.conbuildmat.2020.118614

Torelli, G., Lees, J.M.: Fresh state stability of vertical layers of concrete. Cem. Concr. Res. **120**, 227–243 (2019). https://doi.org/10.1016/j.cemconres.2019.03.006

BY NC ND

Graphene-Enhanced Timber Systems: Real-Time Monitoring and Lifecycle Data Integration for Maintenance and Adaptive Reuse

Clara Hernández, Maria Alejandra Morales, Nitsan Mor(✉), Sunida Sailasuta, Areti Markopoulou, Nikol Kirova(✉), and Daniil Koshelyuk

Institute for Advanced Architecture of Catalonia, C. de Pujades, 08005 Barcelona, Spain
nitsan.mor@students.iaac.net, nikol.kirova@iaac.net

Abstract. This study explores the integration of graphene nanoplatelets (GNP) within glue-laminated timber (glulam) to develop an intelligent composite system for real-time structural monitoring in mass timber architecture. Amid growing interest in sustainable construction, timber is increasingly used to replace carbon-intensive materials. However, the industry lacks affordable tools to assess timber performance during its service life and non-destructive methods to evaluate its potential for reuse afterward. This study addresses that gap by embedding GNP ink between glulam lamellas to detect deformation through changes in electrical conductivity. A series of scaled prototypes was fabricated and tested under load conditions. Electrical resistance data was collected and processed using machine learning algorithms to predict deformation patterns and stress distribution.

A custom digital interface was developed to visualize structural behavior, enable predictive maintenance, and generate a Service Life Passport (SLP) linked to a digital twin for each timber element, allowing professionals to track performance, detect material displacement, and integrate reuse strategies. The results confirm the feasibility of this graphene-enhanced system as a low-cost, scalable solution for monitoring and extending the life cycle of timber components. This innovation supports circular construction practices by transforming mass timber into a sensor-enabled material that retains value beyond its initial application.

Keywords: Graphene-Embedded Timber · Real-Time Structural Monitoring · Life-Service-Passport · Smart composite · Predictive Maintenance · Performance optimisation · Digital Material Twin

1 Introduction

The mass timber industry presents a sustainable alternative to conventional anthropogenic construction materials, aiming to lower pollutant emissions from the building sector and meet emerging standards for environmentally responsible construction. Projects such as Mjøstårnet (Arquitekturmæsteren AS, 2019) have demonstrated the feasibility of high-rise buildings using industrial timber components, while more free-form approaches, such as the Cambridge Central Mosque (Marks Barfield Architects,

Y. Liu et al. (Eds.): CDRF 2025, *Transindividual Intelligence*, pp. 325–337, 2026.
https://doi.org/10.1007/978-981-92-0615-5_28

2019), have demonstrated the structural versatility, aesthetics, and potential of free-form glulam in architecture. The industry continues to evolve, with future projects such as The World's Largest Wooden City in Brumunddal, Norway (Lund Hagem Architects, 2022), scheduled for construction in 2025, highlighting the growing role of timber and technological advances in urban development.

Despite advances in engineered timber construction, significant research gaps remain in areas such as increasing building height, managing the material cycle, extending component life, monitoring structures, and enabling reuse. Timber is a natural material that, to some, even if only to a limited extent, exhibits unpredictable behaviors influenced by natural conditions. Currently, timber monitoring relies on simple sensors, limited to humidity and temperature, or highly complex structural stability sensors, used in high-budget projects. Additionally, assessing the condition of timber components after their service life requires invasive and destructive testing, the results of which are extrapolated to estimate the condition of other components. There is an urgent need for affordable monitoring systems for mass timber projects and non-destructive methods to assess the condition of timber during and after use.

At the same time, graphene, a novel carbon-based material discovered in 2004, has been widely studied for its exceptional mechanical, electrochemical, and thermal properties, advancing fields such as material science, electronics, biology, and energy (Zhong, Zhen, & Zhu, 2017; Yu et al., 2017). The responsiveness of graphene and its derivatives to light, electricity, chemicals, and mechanical stimuli enables the creation of gas, humidity, and strain sensors (Yu et al., 2017; Koshelyuk et al., 2019, p. 665).

Although graphene is a remarkable material, its most impactful architectural applications may not be in isolation but rather in combination with other materials and systems, particularly as an affordable, sustainable, and highly effective sensor. The Pro Skin project (Ramirez et al., 2016) explored the integration of graphene nanoplatelets (GNPs) into flexible silicon membranes, creating a responsive surface capable of real-time data feedback and adaptive interactions. Similarly, the Alive project (Koshelyuk et al., 2019) used graphene to monitor deformation in real time, using changes in conductivity to create a digital twin that deforms in sync with the physical membrane, acting as both a sensor and an actuator.

Building on these studies, this research explores the use of a graphene ink composite within glue-laminated timber (glulam) components as a sensor. As loads are applied and deformations occur, changes in electrical resistance within the graphene are monitored. These measurements are processed by an AI algorithm to predict deformations and structural stresses based on the resistance changes.

This study aims to establish the feasibility of embedding graphene-based sensing technology in timber components to track their structural history. Additionally, it seeks to expand the range of measured variables and increase model complexity, ultimately developing a predictive system for evaluating glulam components without destructive testing. This innovation could enable design for reuse by creating a service-life passport for timber components, extending their life and turning them into valuable assets beyond their initial use in buildings.

2 Methodology

2.1 Material System

Timber and graphene composite The developed composite system integrates graphene with glulam timber to create a sensor-enabled structural element. Although single graphene sheets offer high electrical conductivity, their fabrication is challenging due to their nanometric thickness (Singh et al., 2011). Instead, graphene nanoplatelets (GNPs) were chosen to overcome these processing challenges. GNP powder was combined with other materials to form a liquid ink, which simplifies the coating process while maintaining the electrical properties required for sensing (Kirova et al., 2018; Luque Mayor et al., 2017; Shi et al., 2016). Glulam was selected as the base material because it is widely used in the timber construction industry, and its engineered flexibility makes it ideal for demonstrating deformation under heavy loads. In this system, both the top and bottom layers of the glulam beams were coated with graphene ink to monitor changes in conductivity under tension and compression. In the following stages of the experiment, graphene was embedded in the wood layers to protect the conductive layer without compromising measurement accuracy. Additionally, the study systematically investigated how variations in the width of the graphene layers, the number of layers applied, and the interaction with the adhesive used in the glulam manufacturing process affected the electrical resistance of the graphene and thus the overall system performance (see Fig. 1).

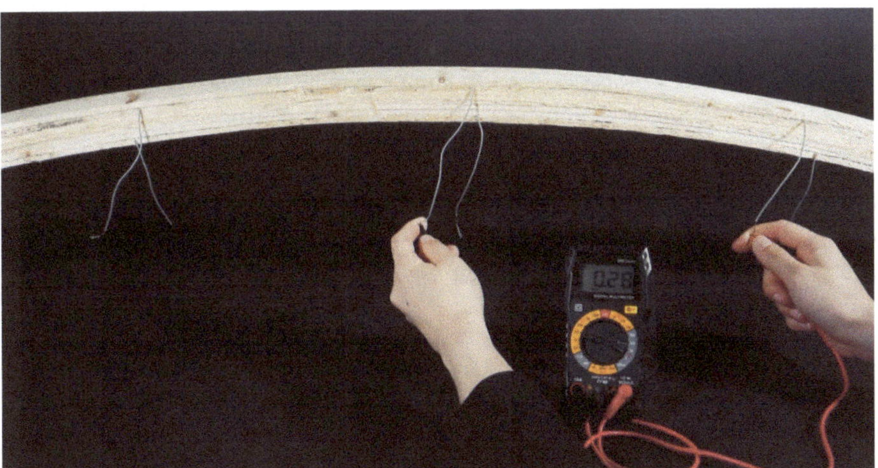

Fig. 1. Measuring the impact of different parameters on the electrical resistance of graphene, including glue type, number of layers, length, and width, using an electrical multimeter.

2.2 Material Performance

The first experiment was designed to understand how deformations within the timber element affect the performance of graphene as a conductive medium. A single-layer timber beam 2500 mm long, 14 mm thick, and 40 mm wide was used with a 1 cm wide graphene ink coating applied along the center of its surface. The sample beam was placed on a base, and a 5 V electrical pulse was applied from one end to enable resistance measurement at the midpoint. The difference between the pulse sent and received corresponded directly to the electrical resistance of the graphene between these points. Initial measurements showed a resistance shift between 200 and 900 Ω, with a baseline of 3,900 Ω in the straight state of the beam, increasing to a range of 4,100–4,900 Ω as the beam flexed (see Fig. 2). This behavior aligns with the mechanical stress-strain response of timber, which initially follows a linear elastic phase where stress and strain are directly proportional. Within this elastic zone, structural elements are required to remain within allowable deflection limits defined by timber construction standards. If these limits are exceeded, the element must be reassessed for another purpose. Once the strain exceeds the elastic zone, permanent deformation may occur, requiring maintenance before the element can be reused. A similar pattern was observed in the graphene layer, where electrical resistance increased as mechanical strain grew. This is because stretching the material reduces the continuity of its conductive pathways, leading to higher resistance. Consequently, both the timber and graphene exhibit strain-sensitive behavior, enabling the electrical signals to serve as a reliable proxy for tracking structural deformation in real time.

Since the structural deformation of the timber beam directly affected the electrical properties of the integrated graphene, this relationship allowed the electrical signals to be translated into geometrical data, which led to the second experiment. Consequently, the data acquisition process was designed to simultaneously translate the electrical resistance into a simulation of the physical deformation in real-time. However, the initial measurements showed significant noise in the resistance readings, making it difficult to obtain a stable correlation. To address this issue, a moving average of 10 samples was used. This approach reduced fluctuations in the data and enabled a smoother transition of information into a digital twin representation within Grasshopper, allowing it to continuously respond to the changing geometric state of the beam, whether straight or bent (see Fig. 3).

Beyond instantaneous deformation tracking, this setup also opens the opportunity to monitor time-dependent behaviors such as creep. Creep in wood refers to the gradual, permanent deformation that occurs under sustained load, even if the load remains constant and within the elastic range. Traditionally, deformation is measured using sensors such as strain gauges, which capture geometric changes over time. However, these methods make it difficult to distinguish whether the measured deformation results from creep or other load-induced behaviors. In contrast, the graphene-based sensing system continuously records changes in electrical resistance, offering a more direct correlation with internal strain. These long-term resistance variations can serve as indicators of cumulative stress exposure. Since creep is influenced by environmental factors such as humidity and temperature, recording them alongside resistance data enables a more reliable long-term deformation analysis within the digital twin.

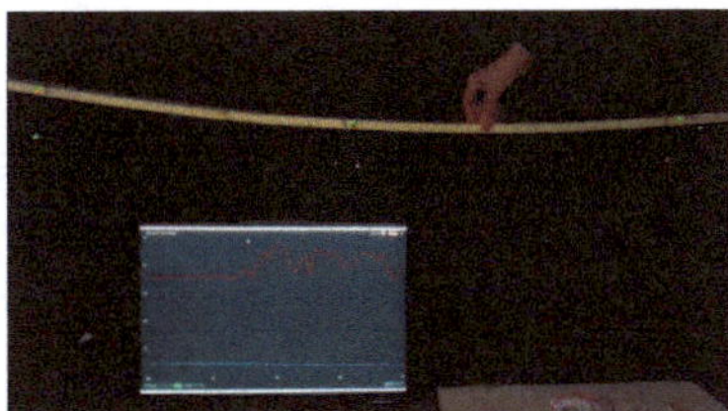

Fig. 2. Experiment 1: Electrical resistance variations captured during the composite beam's deformation.

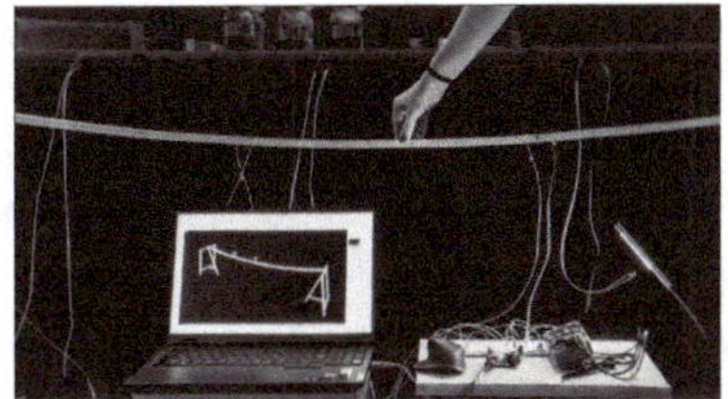

Fig. 3. Experiment 2: A manually bent composite beam's electrical resistance changes are translated into a digital geometrical representation using Grasshopper and Firefly plugin.

2.3 Material Processing and Computation

In order to develop a dataset for training the machine learning model, samples of laminated timber with integrated graphene coatings were fabricated. This chapter outlines the fabrication process of these samples and the digital workflow established to process and analyze the collected data.

The glue lamination fabrication process consists of the following steps:

1. Lamella preparation: Timber lamellae with thicknesses of 4 mm, 5 mm, and 7 mm were fabricated to ensure a smooth surface for optimal bonding and coating adhesion.
2. Graphene coating: A 1 cm wide stripe of graphene nanoplatelet (GNP) ink is applied to the center of each lamella. The coating process is repeated 5 to 7 times at 30-min intervals to achieve uniform conductivity.
3. Wiring integration: Electrical wiring is integrated every 400 mm along the graphene-coated lamellae to facilitate data extraction and enable real-time monitoring of conductivity changes under load.
4. Glulam lamination: The graphene-coated layers are positioned at the top and bottom of the glulam assembly, with structural adhesive applied between the layers. The laminated structure is then subjected to controlled pressure and curing conditions to ensure strong interlayer bonding and structural integrity (see Fig. 4).

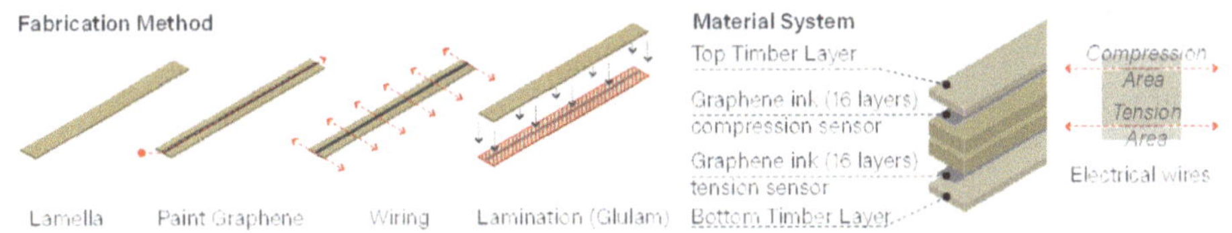

Fig. 4. Diagram illustrating the fabrication process of the glulam prototype and the isometric layers of the material system.

Several prototypes were fabricated for testing:

1. Straight beam 1 – height 28 mm, width 42 mm, length 2500 mm

Fig. 5. Composite glulam-graphene scaled prototypes.

2. Straight beam 2 – height 44 mm, width 42 mm, length 2500 mm
3. Straight beam 3 – height 30 mm, width 50 mm, length 2100 mm
4. Straight beam 4 – height 26 mm, width 44 mm, length 1700 mm
5. Curved beam – height 30 mm, width 44 mm, length 2100 mm, radius 1145 mm (see Fig. 5)

Experiment 3 Setup Given the results of the previous experiments, which showed that physical deformations of the timber beam affected the electrical behavior of the embedded graphene, the goal was to use an AI algorithm to interpret and record the changes in electrical behavior and record them as physical geometry-related data. The physical setup of the beams remained the same as in previous experiments, but the received electrical pulses were measured in five segments along the beam in separate circuits for the tensed and compressed layers. Simultaneously, frontal camera images were collected and processed with computer vision tools for physical deformation data collection. The collected data was processed into a database for the machine learning training process (see Fig. 6).

Machine Learning After post-processing to ensure data quality, approximately 45,000 clean samples were prepared for training. The input data consisted of 19 pre-scaled features, including voltage/stress changes, material properties, and environmental conditions, while the output data represented deformation vectors, load positions, and weights. The dataset was divided into training, validation, and test sets in an 80:10:10 ratio. A dense neural network (DNN) was constructed using TensorFlow's Sequential API for the regression task. The model features an input layer of 19 neurons, two hidden layers of 64 and 32 neurons using sigmoid activations, and a linear output layer of 17 neurons. The model was trained via back-propagation using the Adam optimizer with an exponentially decaying learning rate starting at 0.001 for up to 300 epochs with a batch

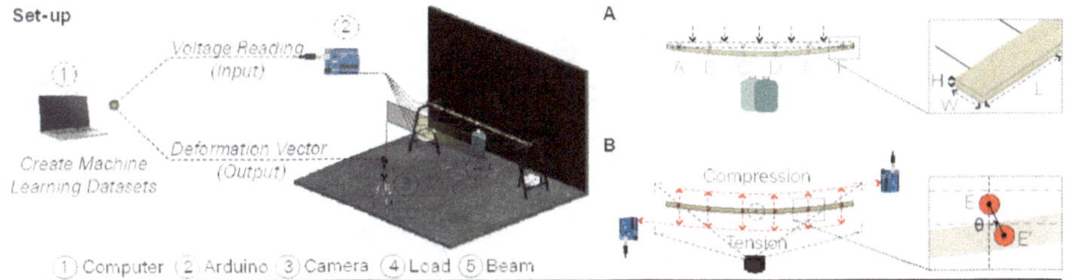

Fig. 6. Experiment 3 setup. **A** Load application positions, at the midpoints between the resistance measurement points. **B** Resistance measurement points in the tension and compression layers, and computed deformation mapping from camera images.

size of 16, while early stopping based on validation loss was applied to prevent overfitting. Training logs showed a progressive decrease in mean loss. Although high noise levels in the voltage readings posed challenges, the final model demonstrated reliable performance for real-time structural monitoring and digital twin integration (see Fig. 7).

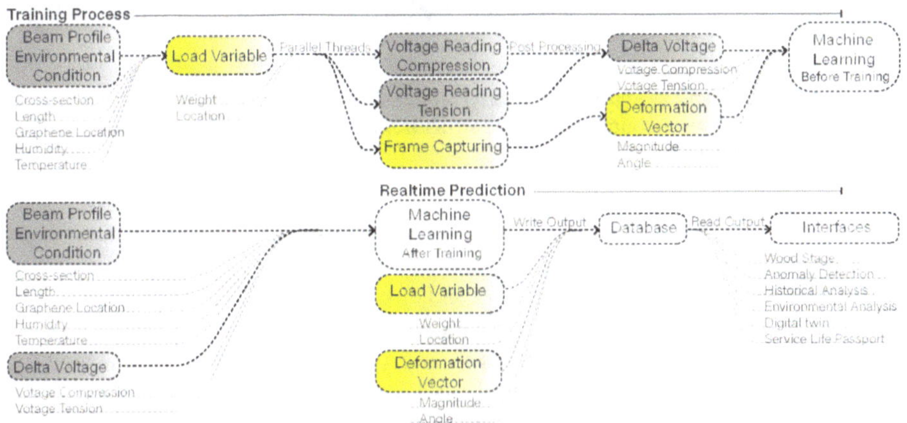

Fig. 7. Digital flow

2.4 Digital Interface & Lifecycle Assessment

To present and visualise the multi-layered data, an interface with multiple dashboards was developed using Unity software to facilitate use by different stakeholders. Key features include real-time deformation tracking to detect internal stresses in the wood (acting as a strain gauge) and a digital twin model of the timber for predictive maintenance and performance optimization. Furthermore, a Service Life Passport (SLP) provides a digital record of the timber's history, promoting reuse, minimizing waste, and enhancing classification after disassembly.

The real-time data dashboard features an anomaly detection section and a graph that monitors the strain gauges on the timber, ensuring quick and efficient checks for

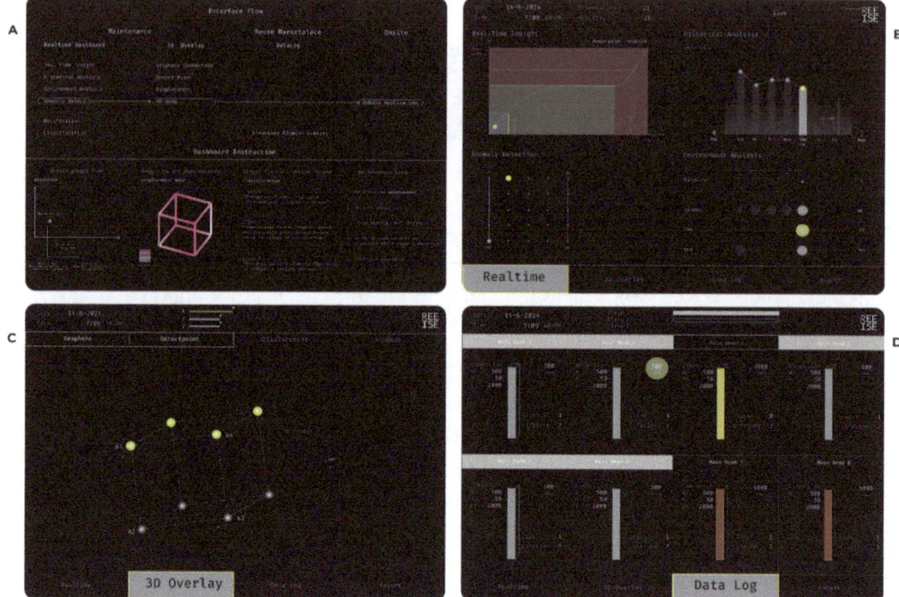

Fig. 8. Interface Data Dashboards. **A** "InterfaceFlow"- Dashboard Instruction. **B** "Real-Time Data Dashboard" - Displays anomaly detection alongside historical and environmental analysis. **C** "3D Overlay" - Shows a digital twin of the structure. **D** "Data Log" - Serves as a Service Life Passport for the element.

immediate action. The same dashboard and interface also provide weekly data records along with environmental factors such as weather and humidity to optimize maintenance performance, prevent timber deformation, and ensure structural safety.

The second dashboard allows users to monitor each graphene-enhanced element to observe displacements and deformations caused by loads. This data is managed by a digital twin for accurate and efficient analysis.

The final dashboard presents a Service Life Passport for the material. This passport adds value by using digital twins to extend the lifespan of timber. The digital twin of each element and the Service Life Passport are categorized based on the material's condition as indicated by the strain gauge. (see Fig. 8)

Class 1: The material is in good condition and ready for reuse (white zone) Class 2: The material shows some damage and requires refurbishment before it can be reused or repurposed (yellow zone) Class 3: The material is critically damaged and needs to be downcycled (red zone) (see Fig. 9).

3 Results

3.1 Validation of Graphene-Enabled Sensing System

To demonstrate the effectiveness of the graphene-enabled sensing system, a bridge prototype was designed and fabricated at a scale of 1:5. The structural design was optimized using Grasshopper Karamba analysis to ensure stability and adequate flexural strength.

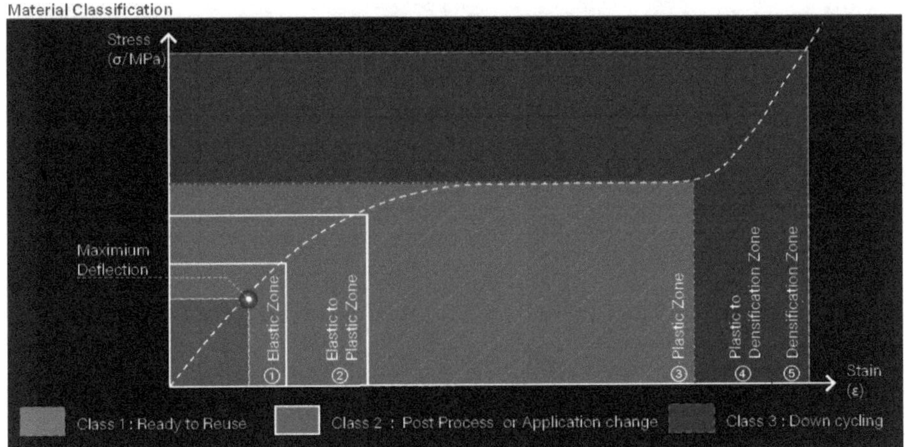

Fig. 9. Strain-stress graph displaying the range for each composite component class

The bridge followed a glulam fabrication technique, consisting of four layers of 7 mm laminated timber (lamellas). Graphene nanoplatelet (GNP) ink was applied as strips along the bridge, positioned on the top and bottom layers of the lamellas to monitor tension and compression. One pair of strips was placed on the left half of the bridge and another on the right half to ensure full-width coverage.

Voltage readings were taken at six predefined locations along the bridge and processed by the ML-trained model at 100 millisecond intervals. The system reliably predicted the applied load and its location in real-time. When the detected load exceeds a predefined threshold, an integrated NeoPixel LED strip lights up red at the corresponding location, providing an immediate visual representation of the structural stress. Simultaneously, the ML-derived results were transmitted to a digital interface that displayed real-time deformation data on a digital twin of the bridge, along with recorded environmental conditions such as temperature and humidity. Additionally, the interface displayed the Service Life Passport (SLP) of the bridge, consolidating all recorded data over its lifespan (Fig. 10).

The experimental results confirmed the reliability of the embedded sensing system in detecting micro-deformations in the timber structure. The correlation between applied loads and voltage fluctuations demonstrated the feasibility of graphene-based conductivity monitoring as a non-invasive method for real-time structural assessment.

4 Discussion and Future Steps

Throughout the experimental process, the ML model demonstrated a consistent correlation between applied loads and variations in voltage across the glulam timber (GLT) components. However, several factors were identified as critical for optimizing the training accuracy:

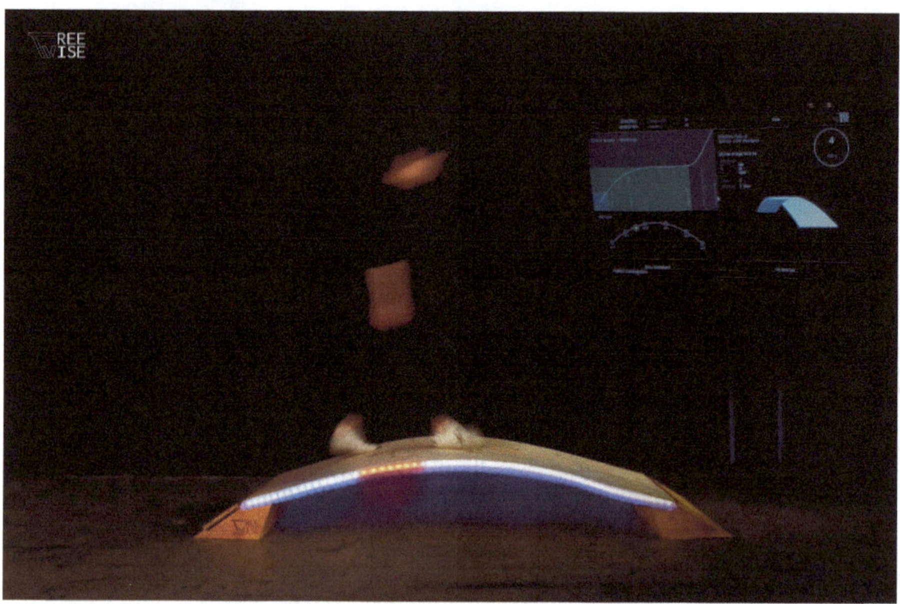

Fig. 10. Scaled composite bridge and the "Real-Time Data Dashboard"

- Increasing the number of samples collected, as the effectiveness of the ML model relies on a sufficiently large dataset. Expanding both the quantity and variety of samples will also enable direct comparison between the neural network model and simpler baseline approaches, clearly highlighting the benefits of the ML method through standard evaluation metrics such as Root Mean Square Error (RMSE) and Mean Absolute Error (MAE).
- In the present study, all prototypes were physically manufactured, and the time-intensive nature of the manufacturing process limited the variety of samples.
- The integration of graphene nanoplatelets (GNPs) into timber was monitored over six months to evaluate their impact on mechanical performance and adhesive bonding stability. Throughout this period, the prototypes showed no signs of delamination or degradation in structural behavior. While these results indicate short-term consistency, further investigation is necessary to determine the long-term stability of GNPs under varied environmental conditions and extended use.
- Temperature and moisture were included in the dataset due to their direct impact on the mechanical behavior of wood. Although these variables showed minimal fluctuation during the short-term tests in the current experiment, their inclusion is essential for future scalability. Expanding the technology will require long-term tests under both constant and variable environmental conditions, such as sustained humidity/temperature and simulated daily or seasonal cycles, to better understand how wood responds to these factors over time. To support this, the system may need to be coupled with temperature and humidity sensors not only during the data collection and model training phase, but also as part of the finalized sensing setup, so that the digital twin continuously reflects environmental influences in real time..

- Prolonged testing of each beam could further clarify the effect of material fatigue on electrical resistance variations, strengthening the predictive capabilities for long-term structural health monitoring and SLP assessment.

To move beyond the proof of concept established in this research, the next phase of experimentation will focus on controlled testing to further validate the material system's performance. These experiments will be conducted in laboratory environments such as Matter Labs, where key parameters like temperature, humidity, and moisture can be systematically controlled. Additionally, a robotic arm will be employed to apply controlled multi-axial loads such as shear, torsion, and lateral forces. This approach will increase the reliability of the data sets and support the validation of the material system behavior for real-world structural applications.

The fabrication process demonstrated in this research presents a promising avenue for automation and large-scale adoption within the GLT industry. The integration of graphene sensing technology into mass timber construction is feasible without significant modifications to existing manufacturing workflows. Potential automation strategies include:

- Using laser technology to precisely deposit graphene ink between GLT lamellas could improve the efficiency and repeatability of the manufacturing process.
- Embedding automated voltage data loggers would enable seamless transmission of real-time measurements to monitoring platforms, ensuring continuous structural assessment without manual intervention.
- Developing a wiring system for resistance-based sensing in timber elements requires careful cable routing, strategies to minimize electrical noise, and consistent contact between the conductive surface and the measurement devices. To resolve these issues, collaboration with electrical engineers will be essential for circuit design and system calibration, as well as with data engineers to define robust acquisition and signal processing protocols.

These advances suggest that graphene-enhanced GLT structures could be adopted by the mass timber industry as an intelligent, self-monitoring material system.

References

Zhong, Y., Zhen, Z., Zhu, H.: Graphene: Fundamental research and potential applications. FlatChem. **4**, 20–32 (2017). https://doi.org/10.1016/j.flatc.2017.06.008

Novoselov, K.S., et al.: Electric field effect in atomically thin carbon films. Science. **306**(5696), 666–669 (2004)

Yu, X., Cheng, H., Zhang, M., Zhao, Y., Qu, L., Shi, G.: Graphene-based smart materials. Nat. Rev. Mater. **2**(9), 17046 (2017). https://doi.org/10.1038/natrevmats.2017.46

Singh, V., Joung, D., Zhai, L., Das, S., Khondaker, S.I., Seal, S.: Graphene-based materials: Past, present and future. Prog. Mater. Sci. **56**(8), 1178–1271 (2011). https://doi.org/10.1016/j.pmatsci.2011.03.003

Koshelyuk, D., et al.: Alive: a multi-layered, flexible and elastic shape-aware graphene-based interface. In: *ACADIA 2019: Ubiquity and Autonomy*. Conference presentation, The University of Texas at Austin, October, pp. 664–673 (2019). https://www.researchgate.net/publication/339987914

Ramirez, I., Staples, R., Paksoy, B., Naagendran, C. (n.d.). Pro_Skin [Programmable Skin]. Institute for Advanced Architecture of Catalonia (IAAC). http://www.iaacblog.com/projects/pro_skin-programmable-skin/. Accessed 3 Sept 2019

Kirova, N., Jooste, E., Mahdi, H., Jalodia, S.: SYNAPSE: an interactive medium for spatiotemporal behavioral data collection and analysis. Institute for Advanced Architecture of Catalonia (IAAC) (2018). http://www.iaacblog.com/projects/synapse-2/. Accessed 3 Sept 2019

Luque Mayor, R., López-Alascio Hervás, J., Van Limburg Stirum, G., Zervos, T.: Graphene composites. Institute for Advanced Architecture of Catalonia (IAAC) (2017). http://www.iaacblog.com/programs/graphene-composite/. Accessed 3 Sept 2019

Shi, G., et al.: Highly sensitive, wearable, durable strain sensors and stretchable conductors using graphene/silicon rubber composites. Adv. Funct. Mater. **26**(42), 7614–7625 (2016). https://doi.org/10.1002/adfm.201602619

Dreiling, J., Smith, D.: The role of mass timber in sustainable construction. J. Green Build. **15**(1), 143–156 (2020)

Choi, W., Lahiri, I., Seelaboyina, R., Kang, Y.S.: Synthesis of graphene and its applications: A review. Crit. Rev. Solid State Mater. Sci. **35**(1), 52–71 (2010)

Kuzmenko, A.I., van Heumen, E., Carbone, F., van der Marel, D.: Universal optical conductance of graphite. Phys. Rev. Lett. **100**(11), 117401 (2008)

Wimmers, G.: Wood: Nature's Construction Material. Routledge, Abingdon (2017)

Wilkinson, S., Remøy, H., Langston, C.: Sustainable Building Adaptation: Innovations in Decision-Making. John Wiley & Sons, Chichester (2014)

Espinoza, O., Buehlmann, U., Bumgardner, M.: Mass timber: Industry perceptions and willingness to adopt. BioProd. Bus. **1**(6), 48–57 (2016)

Anthony, R.W., Gould, P.L.: Performance of mass timber structures in earthquakes. J. Struct. Eng. **142**(9), C4015004 (2016)

Jones, C., Williams, J.: The application of graphene in modern construction. Mater. Today Commun. **24**, 101097 (2020)

Van de Kuilen, J.W.G., Ceccotti, A., Xia, Z., He, M.: Very tall wooden buildings with cross laminated timber. Procedia Eng. **14**, 1621–1628 (2011)

Mallo, M.F.L., Espinoza, O.: Awareness, perceptions, and willingness to adopt cross-laminated timber by the architecture community in the United States. J. Clean. Prod. **94**, 198–210 (2015)

Liu, Y., Chen, X., Xie, Q., Yang, J.: Graphene-based materials for strain sensors. J. Mater. Sci. Technol. **35**(4), 674–684 (2019)

Harmsworth, S., Whyte, J.: Smart buildings and digital twin technologies. Int. J. Build. Pathol. Adapt. **37**(3), 297–316 (2019)

Hossain, M.A., Kundu, A.: Advancements in graphene-based sensors for structural health monitoring. Sensors. **20**(10), 2996 (2020)

Montero, J., Teibert, H.K.: Structural health monitoring of timber structures: A review. Struct. Control. Health Monit. **25**(8), e2184 (2018)

Anaf, W., Cabal, A., Robbe, M., Schalm, O.: Real-time wood behaviour: The use of strain gauges for preventive conservation applications. Sensors. **20**(1), 305 (2020)

Upcycling Cork: Design and Fabrication of a Binder-Free Cork Component Architectural System

Michael Sinzinger(✉), Shrey Kapur, Anish Hatekar, Nikol Kirova(✉), and Areti Markopoulou

Institute for Advanced Architecture of Catalonia, C. de Pujades, 08005 Barcelona, Spain
michael.sinzinger@students.iaac.net, nikol.kirova@iaac.net

Abstract. This paper presents a binder-free, upcycled cork material system for architectural use, proposing a sustainable alternative to conventional insulation and structural components. Using post-industrial cork waste and suberin's natural adhesive properties, activated through controlled hot compression, the study develops modular blocks without synthetic binders. Material experiments, including compressive strength and thermal insulation testing, demonstrate that the fabricated cork components meet or exceed the performance of conventional cork products, including polyurethane-bound agglomerates. The system features friction-based interlocking geometry and post-tensioning for dry assembly and disassembly, enabling reuse. A parametric design protocol further tailors material density to structural and thermal needs. The findings highlight suberin-based binding as both an environmentally and functionally viable approach and demonstrate the potential of processed cork waste as the basis for a scalable, digitally integrated construction system.

Keywords: cork · binder-free materials · waste-derived materials · bio-renewable materials · modular construction · circular construction · digital design

1 Introduction

1.1 Cork: Overview

In 2023, the architecture and construction sector accounted for 32% of global energy demand, 34% of CO_2 emissions, and approximately one-third of global waste [1]. To meet sustainability targets, the adoption of circular building practices - such as design for disassembly, modularity, and reuse - is becoming increasingly critical. Equally important is a shift in how resources are perceived and utilized within architectural design [1]. Within this context, cork has emerged as a promising bio-renewable material due to its favourable physical properties and environmental profile [2, 3].

On a cellular level, cork is composed of hexagonally packed prismatic cells with air-filled interiors, growing radially outward from the tree's core. Its tissue contains

Y. Liu et al. (Eds.): CDRF 2025, *Transindividual Intelligence*, pp. 338–352, 2026.
https://doi.org/10.1007/978-981-92-0615-5_29

approximately 43 ± 6% suberin and 22 ± 3% lignin by weight, alongside various polysaccharides and extractives [4, 5]. Suberin, a glyceride biopolymer, is released when cork is exposed to heat - an effect naturally observed during wildfires, where the bark provides a fire-resistant barrier [6]. In this thermally activated state, suberin exhibits adhesive properties, which enables its use as a natural binding agent. It is also the primary contributor to cork's elasticity and its low permeability to gases and liquids, which makes cork suitable for use as a sealant [7].

Currently, around 70% of global cork production is dedicated to the bottle stopper industry [6–9]. This sector generates a hierarchical waste stream, with offcuts from stopper production largely processed into agglomerated cork products. Nevertheless, between 25–31% of cork production is waste material, a significant portion of this by-product is cork powder with particles smaller than 1 mm in diameter, which is either incinerated for biomass energy or used in wastewater treatment [9–12]. Literature suggests that extending the lifespan of cork materials prior to decomposition or energy recovery is key to increasing material circularity within the sector [13, 14].

The architecture and construction industry, while secondary to the bottle stopper market, accounts for roughly 22% of cork usage [3, 10, 15]. Cork's performance characteristics, renewability, and distinct tactile and aesthetic qualities have contributed to a growing interest in cork within contemporary architectural practice, where it has emerged as a sustainable alternative to fossil-based insulation materials and façade systems [2, 3, 9, 10]. The most common architectural applications involve agglomerated cork boards and expanded cork boards. Expanded cork, with a lower density (~120 kg/m^3 compared to ~180 kg/m^3 for agglomerated cork), offers superior thermal insulation (0.038 W/m·K vs. 0.045 W/m·K), though it is generally more brittle [3, 10].

While agglomerated cork boards typically rely on binding agents like polyurethane, expanded cork boards use suberin, released during the autoclaving process, as a natural adhesive [14].

1.2 Research Background and Relevant Work

Suberin-based binding for high-density cork components has been the subject of research since the early 1990s, notably by Flores et al. in 1992 [17] and Gil (1994, 1996) [18, 19], who focused on the densification of cork dust. Flores et al. produced consolidated cork materials with densities between 950 and 1250 kg/m^3, concluding that the resulting material retained favourable thermal insulation, as well as fire and chemical resistance. Despite exhibiting lower mechanical strength, it was considered a viable alternative to wood-based materials and polymers. Gil similarly identified the potential for this process to enhance material utilization within the cork industry.

While these early studies did not result in widespread industrial adoption, recent research has renewed interest in densified cork as a means to expand its architectural applications. Knapic et al. demonstrated that increasing the density of expanded cork boards up to 290 kg/m^3 led to improved structural performance [20]. In 2024, Sergi et al. further explored the consolidation of agglomerated cork boards and advocated for their use in decking systems, positioning them as sustainable alternatives to conventional wood products [21].

In parallel, the architectural application of cork as a structural material has been advanced by Oliver Wilton and Matthes Barnett Howland, who proposed a monolithic wall system composed of expanded cork blocks. Their work culminated in the construction of a fully inhabitable housing unit, effectively reintroducing the concept of load-bearing cork architecture into contemporary discourse [22].

1.3 Research Objective

Building on this precedent of load-bearing cork components and the demonstrated performance improvements through densification, this research investigates the potential of suberin-based binding to upcycle industrial cork waste into high-performance architectural elements. The objective is to develop a modular, binder-free building system that adheres to circular design principles and eliminates the need for synthetic adhesives. The proposed system uses interlocking geometries and post-tensioning strategies to allow for efficient, dry assembly and complete, non-destructive disassembly. This is supported by a performance-driven digital design process, which optimizes material distribution through computational simulations to meet localized thermal and structural performance criteria.

2 Material Development

2.1 Material Experimentation

Prototyping was carried out at the Institute for Advanced Architecture of Catalonia (IAAC), using a hot press setup consisting of a rectangular aluminium mold with heated top and bottom plates measuring 250×250 mm, and height-adjustable side frames mounted to a conventional hydraulic press. Cork granulates were sourced locally, primarily through ICSURO, the Catalan Cork Institute [23].

The experimentation focused on binding cork granulates through hot compression, without any additional binders or adhesives. A wide range of granule sizes (0.5-40 mm) and post-industrial cork waste types were tested to evaluate their compressibility and bonding behaviour under different thermal conditions. The supplied bulk material required no further preparation before being loaded into the pre-heated press, which was prepared with a silicone-based mold release beforehand. Parameters such as temperature, pressure, and compression time were iteratively refined to optimize adhesion.

This process led to the development of a catalogue of material samples (Fig. 1), with the post-industrial waste material, ranging between 250–700 kg/m^3 of density and between 0.5–10 mm in granule size.

The densification limit for cork, the state of complete collapse of the cell structure and elimination of airgaps, is reported to be around 1200 kg/m^3. Flores et al. [17] compressed samples of non-specified initial size to 950–1250 kg/m^3, with a reported optimum result at 180 °C, 32 MPa, for 25 mins to achieve a density of 1200 kg/m^3. Gil et al. [18] compressed samples of agglomerated cork boards initially measuring 1000 × 1000 × 50 mm and reported an optimum by preheating for 70 min at 180 °C, compressing for 10 min at 230 °C and 3 MPa for a density of 760 kg/m^3. Sergi et al. [21] compressed

Fig. 1. Material catalogue: waste derived samples marked in lower left corner

polyurethane-bound agglomerated cork boards initially measuring 30 × 30 × 15 mm to 740–940 kg/m^3, with a reported optimum result at 180 °C, 10 MPa for 31.5 min for a density of 940 kg/m^3.

For optimal adhesion, a pressing time of 60 min at 200 °C was established for 40 mm-thick samples, independent of the target density. For samples with a thickness of 80 mm, the parameters were adjusted to 90 min at 210 °C.

Regardless of the initial applied pressure, a reduction of approximately 50% was observed within the first 15 min of compression, likely due to an initial adaptive deformation process of cork cells. A notable increase in pressure occurred only at temperatures exceeding 240 °C, due to cell expansion similar to the behaviour observed in expanded cork board production, though this resulted in less favourable material characteristics for the intended application.

Table 1 summarizes the hot compression parameters for samples used in material testing, and Table 2 presents samples of the same granule size with different pressing parameters that lead to inferior binding, as well as additional samples of different composition. Samples B1, B3, B5, B7 with 180 °C displayed weaker adhesion at the core that improved only slightly with longer pressing time (B3, B7), whilst a temperature increase to 210 °C (B2, B4, B6, B8) resulted in no improvement compared to 200 °C. No significant discrepancy in adhesion was observed between different granule sizes below 20 mm, but larger granule composition proved to produce airgaps weakening the integrity of the samples (B9, B10). A mix of larger with finer granularity examined adhesion equal to finer granule samples (B11).

The densification limits of this process were not explored entirely, as the lower end (<250 kg/m^3) is covered by commercially available products and the higher end (>700 kg/m^3) by the referenced research, although neither investigate waste-derived material sources. Future work should also address variation based on chemical composition, origin and age of material.

Table 1. Hot compression matrix of samples used in material testing

Sample	Granule size	Height	Initial pressure	Temperature	Pressing time	Density
M1	3–7 mm	55 mm	0.628 MPa	200 °C	70 min	345 kg/m^3
M2	3–7 mm	50 mm	0.941 MPa	200 °C	65 min	565 kg/m^3
M3	<1 mm	45 mm	0.612MPA	200 °C	60 min	350 kg/m^3
M4	<1 mm	50 mm	0.951 MPa	200 °C	65 min	600 kg/m^3

Table 2. Hot compression matrix of additional samples

Sample	Granule size	Height	Initial pressure	Temperature	Pressing time	Density
B1	3–7 mm	50 mm	0.621 MPa	180 °C	60 min	325 kg/m^3
B2	3–7 mm	50 mm	0.654 MPa	210 °C	60 min	390 kg/m^3
B3	3–7 mm	52 mm	1.01 MPa	180 °C	90 min	590 kg/m^3
B4	3–7 mm	55 mm	0.955 MPa	210 °C	85 min	615 kg/m^3
B5	<1 mm	55 mm	0.612 MPa	180 °C	60 min	345 kg/m^3
B6	<1 mm	50 mm	0.613 MPa	210 °C	65 min	370 kg/m^3
B7	<1 mm	45 mm	0.934 MPa	180 °C	90 min	575 kg/m^3
B8	<1 mm	45 mm	0.944 MPa	210 °C	85 min	610 kg/m^3
B9	10–30 mm	40 mm	0.644 MPa	200 °C	60 min	310 kg/m^3
B10	10–30 mm	40 mm	0.945 MPa	200 °C	60 min	525 kg/m^3
B11	mix 1–30 mm	45 mm	0.701 MPa	200 °C	65 min	415 kg/m^3

2.2 Compressive Strength Testing

Based on the hypothesis of increasing performance through densification, the study's experimentation investigated the relationship between density and compressive strength. A series of comparative tests were conducted using a hydraulic press equipped with a load cell to ensure accurate measurement of the applied force. The experimental setup provided a controlled and repeatable environment and was validated using a control material with known compressive strength.

Tests were carried out on cubes a = 40 mm, with four sample cubes for each material, by continuously applying vertical pressure through the hydraulic press. The resulting pressure was recorded through the load cell and vertical compression of the sample cubes through the press extension and additionally photographically. The results of the four iteratively tested samples per material were averaged and nominal stress (σ_n) and nominal strain (ε_n) were calculated by the following Eqs. (1) and (2), with A_0 the section area in mm^2 of the sample the force F in N was applied to, and L_0 and ΔL the initial vertical length respectively relative compression in mm.

$$\sigma_n = \frac{F}{A_0} \tag{1}$$

$$\varepsilon_n = \frac{\Delta L}{L_0} \tag{2}$$

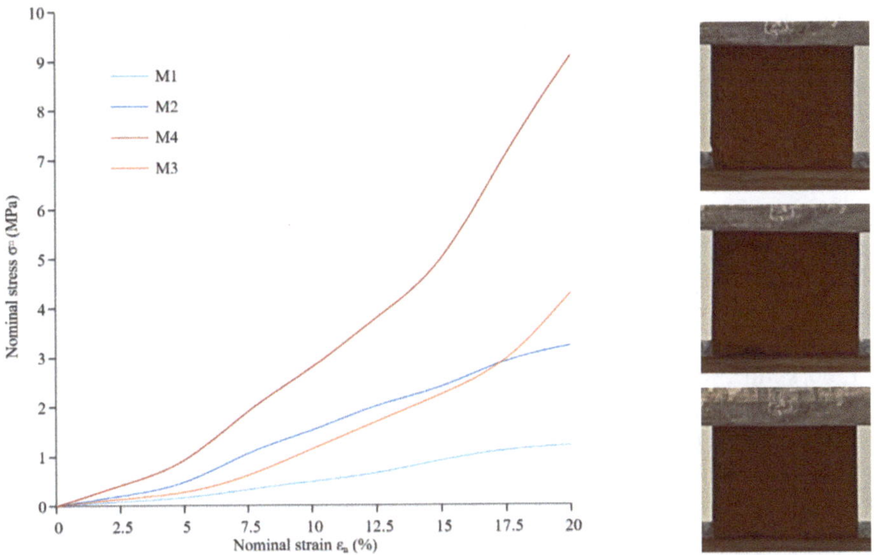

Fig. 2. Left: Stress-strain curve for samples: M1, M2, M3, M4: initial elastic domain followed by viscoelastic behaviour. Right: photograph series of M3 under compression

To assess the influence of granule size and material density on compressive performance, four cork-based materials were tested across two density and two granule size categories: M1 (345 kg/m^3, 3–7 mm granule size), M2 (565 kg/m^3, 3–7 mm granule size), M3 (350 kg/m^3, <1 mm granule size), M4 (600 kg/m^3, <1 mm granule size). These samples were chosen due to their most consistent fabrication reproducibility during the pressing process.

As shown in Fig. 2, all samples exhibited an initial low-resistance elastic phase, followed by a prolonged period of linear, delayed elasticity. The duration of this second phase decreased with increasing material density, while the time required for dimensional recovery increased. Samples with finer granule sizes (M3 and M4) displayed clearly defined elastic and post-elastic phases, whereas this behaviour was less discernible in the coarser granule samples (M1 and M2), which also demonstrated overall lower compressive strength. The results indicate a positive correlation between density and compressive strength across both granule size domains.

Cork's typical compressive behaviour is characterized by an initial elastic region (5–7% strain), followed by a plateau (50–70% strain) associated with cell wall buckling, ending with densification and a rise in stress [24–26]. The observed stress-strain curves in this study deviate from this standard pattern but align more closely with the behaviour documented by Flores et al. [17] for consolidated cork dust, where large strains lead to collapsed cellular structures. However, due to the rapid dimensional recovery observed, even in the denser samples, it is likely that complete cell collapse did not occur.

Young's Modulus in compression was calculated from the elastic domain (0–5%), yielding the following values: M1 3.56 MPa (sd 0.56 MPa), M2 9.21 MPa (sd 1.37 MPa), M3 5.66 MPa (sd 0.49 MPa), M4 16.92 MPa (sd 1.37 MPa). For comparison, expanded cork boards typically exhibit values between 2–3.5 MPa, while polyurethane-bound agglomerated cork boards range between 3.5–6 MPa [27]. These results indicate that the newly manufactured, binder-free cork materials demonstrate superior compressive performance compared to suberin-bound expanded cork boards and meet or exceed the mechanical strength of agglomerated cork boards. Additionally, the lower standard deviation of the finer granule samples further highlights their performance.

2.3 Thermal Insulation Testing

As one of cork's key material properties, thermal insulation is strongly influenced by its density. To evaluate this relationship, the same 4 test materials used in the compression tests, M1 (345 kg/m^3, 3–7 mm granule size), M2 (565 kg/m^3, 3–7 mm granule size), M3 (350 kg/m^3, <1 mm granule size), M4 (600 kg/m^3, <1 mm granule size), were subjected to thermal resistance testing.

The tests were performed using a custom-built, two-compartment thermal chamber equipped with an infrared heat source on one side and a thermal camera on the opposite side. Each sample, cut to a uniform thickness, was positioned between the insulated compartments to function as a thermal barrier. As heat was applied, the temperature rise over time was recorded by averaging measurements from nine designated points across the surface of each sample. Although not sufficient to confidently report insulation values, this controlled setup allowed for a consistent, comparative assessment of the thermal resistance properties of the various cork materials.

As illustrated by the temperature curves in Fig. 3, both lower material density and finer granule size correlate with improved thermal resistance. The thermal imaging results (Fig. 4) further support this observation: samples M1 and especially M2—both composed of larger granules—exhibited more irregular and patchy heat distribution. These visual inconsistencies suggest higher internal variability in material structure, which, as indicated by the testing, contributes to reduced performance in both compressive strength and thermal insulation. In contrast, the finer granularity of M3 and M4 produces a more homogeneous material composition that leads to improved performance. The relative difference between densities correlates with thermal resistance for both granule sizes.

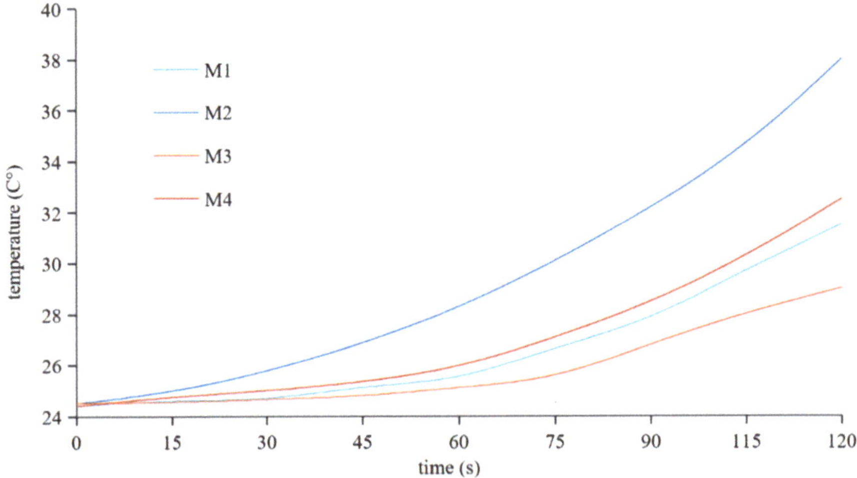

Fig. 3. Increase of temperature over time when acting as thermal barrier against an infrared heat source in an enclosed environment. Samples: M1 (345 kg/m^3, 3–7 mm granule size), M2 (565 kg/m^3, 3-7 mm granule size), M3 (350 kg/m^3, <1 mm granule size), M4 (600 kg/m^3, <1 mm granule size)

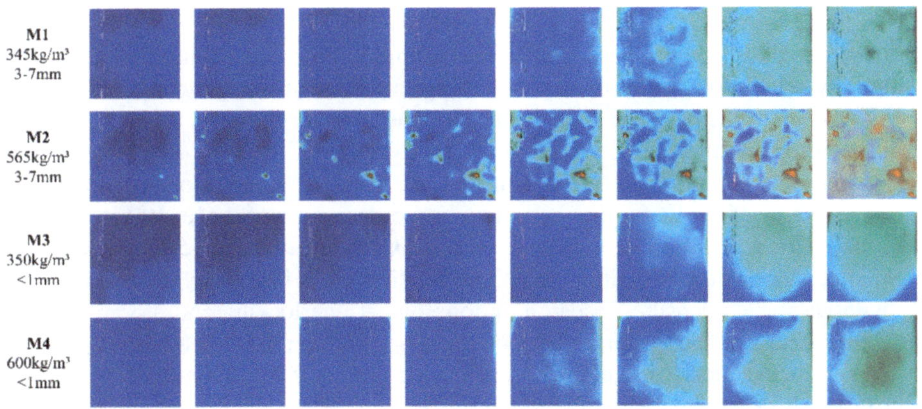

Fig. 4. Thermal imagery of the cold side of the tested materials when acting as thermal barrier against an infrared heat source in an enclosed environment

The results confirm known correlations of density and insulation for cork, but further testing is needed to validate these observations.

3 Design Development

3.1 Block Component System

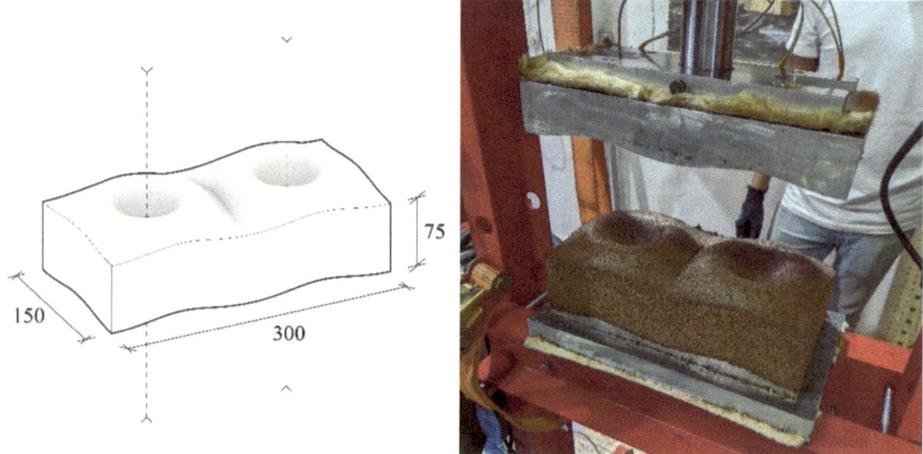

Fig. 5. Block component with measurements and post-tensioning axes (left), Manufacturing process of the block with the custom-built hot press (right)

Aligned with the research objective, the design development focused on creating a modular system that prioritizes ease of assembly and disassembly, while accommodating the integration of the developed material catalogue with varying structural and thermal performance properties.

Geometrical limitations observed during the material exploration were the total height and angle of repose for the bulk material. Component height over 100 mm disproportionately increases production time and results in inadequate adhesion. The repose angle affects possible inclination, eliminating vertical walls inside the geometry boundary without uneven density. Research on topological interlocking and mortarless wall structures identifies friction as primary material property for these geometries [28–30]. Cork exhibits excellent friction coefficients, even though the hot compression slightly reduces this by smoothening surfaces.

The developed interlocking block, measuring 300 × 150 × 75 mm (Fig. 5), maximizes friction area through its undulating rotational surfaces based on the repose angle, implementing a convex-concave curvature for optimal transmission of forces from block to block. With this geometry, various wall thicknesses based on conventional brickwork bonds are achievable and corners can be formed freely between 90–270° (Fig. 6).

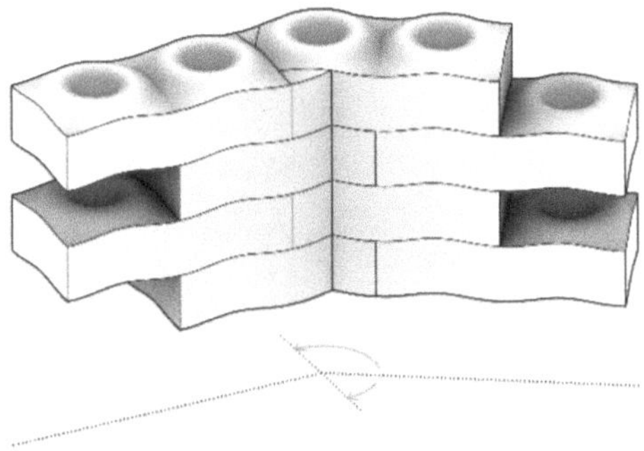
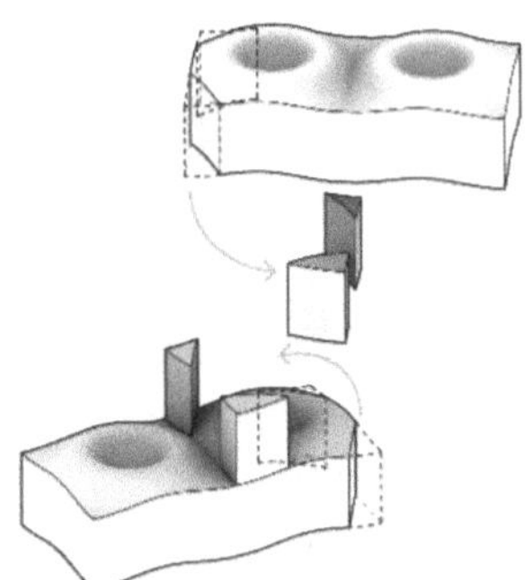

Fig. 6. Geometric sequence of the developed block: brickwork bond and corner formation

As tensile strength in cork is limited, the geometry was also designed to be fitted with holes for post-tensioning cables (Fig. 5). This post-tensioning system enhances the overall structural integrity by anchoring the interlocking blocks at key points, specifically at the roof and floor connections, effectively locking the assembly in place and enabling stability under vertical and lateral loads.

3.2 Digital Design Protocol

As the developed manufacturing process allows an individual adaptation of density for each block, a digital design protocol was developed that conceptually utilizes this adaptability to further increase material utilisation. Based on the results of the insulation and compression tests, a linear correlation of these properties with the density of the material can be assumed. To increase the performance range of the components, seven different material densities were extrapolated from the testing data, ranging between 250–700 kg/m^3 with corresponding insulative and compressive properties based on the material testing. This set of component properties informed the digital design protocol to enable a precise adaptation to locally varying requirements.

The algorithm was designed inside the visual scripting environment Grasshopper and evaluates structural and insulation requirements through the Karamba3d [31] respectively LadyBug Honeybee Plug-Ins [32, 33]. It informs each individual block by assigning the optimal out of the seven density values to minimize material consumption. Figure 7 illustrates this protocol: 1) A minimum density based on compressive loads is obtained through finite element analysis. 2) Insulation benefits are calculated as the coefficient of indoor thermal comfort in an annual thermal load analysis. This desired density value is maximized within the previously determined structural requirement boundary, resulting in an individual mapping of component properties throughout the whole structure.

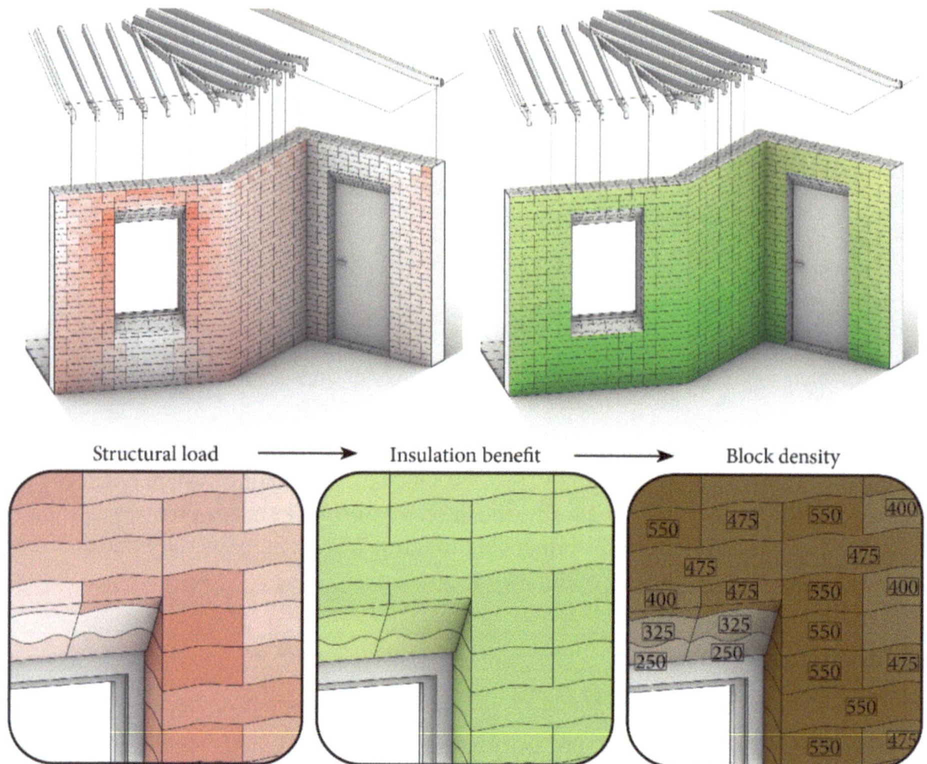

Fig. 7. Digital design protocol: Insulation benefit maximized within structural load boundary to individually map component densities

3.3 System Scaling-Up: 1:1 Prototype

The research process, levered by extensive material experimentation and testing, culminated in the creation of a 1:1 scale prototype (Fig. 8), showcasing part of a building section, with the aim of investigating interlocking, post-tensioning as well as tactile and aesthetic qualities. Blocks were pressed at a temperature of 210 °C for 90 min to varying densities between 300–600 kg/m^3.

The prototype implemented a CNC-milled acoustic pattern to investigate the post-processing capabilities of the material and reveals its intriguing haptic properties, adding to the versatility of the developed system by reinforcing cork's sound absorbing properties. The integration of a simple sill and presentation of the post-tensioning system showcases possible solutions of common connection systems and its implementation into an architectural building. The execution of these connections points to different material systems reinforced the capability of the chosen interlocking and post-tensioning system. The friction-based, high-surface-area connections proved to enable sufficient transmission of forces and ease of assembly. Possible limitations of the material were observed at local, high-impact points such as exposed block corners during the assembly process.

Fig. 8. Prototype: showcase of the modular block, connections and post-tensioning system

In line with the circular design principles of the process, the re-pressing of the developed material was investigated in a short experiment. Milling waste of the post-processed blocks was collected and successfully repressed into a small material sample. Although mechanical properties of this piece were not investigated, the sample exhibited similar behaviour, indicating that adhesive properties of Suberin can be activated repeatedly.

4 Conclusion

The study highlights the potential of binder-free cork materials from a performance perspective, the upcycling potential of cork industry waste streams and advocates for digital design methods to increase efficiency and utilization in material-related design tasks.

The tests of compressive and thermally insulative properties underline the potential of the newly developed material system and demonstrate that higher performance in cork products doesn't necessitate the use of binding agents or composites. It advocates for suberin-based binding not only from an environmental but also a performative standpoint, by combining structural and insulating properties in one component. The exhibited stronger compressive performance than polyurethane-bound counterparts challenges the necessity of binders in the first place, although standardised testing of the presented material properties is necessary, as is additional assessment of behaviour under water and moisture, long-term stress and UV exposure.

The additional energy demand of the hot compression process needs to be evaluated for scalability and industrial feasibility. Even though further market assessment is necessary, based on the interaction with supplier ICSURO, the availability of better performing, finer waste material predominates, which indicates good compatibility with existing industrial processes. The collaboration with industry partners and cork research institutions emphasizes this real-world potential outside of an academic interest. With cork waste making up 25–31% of total production, the proposed upcycling process presents a compelling case for the cork industry. Despite its limitations in component dimension, the hot compression process facilitates the utilization of these heterogenous resource streams and enables a precise adaptation of density and composition of these components.

Whilst the 1:1 prototype represents a significant step towards actual implementation, large-scale feasibility of monolithic cork architecture remains unresolved. The proposed dry-joint block system, as the developed material itself, requires further testing to be fully adopted. Nevertheless, the research extends the application of cork by combining structural and insulating performance with modularity and introduces additional design value.

This design value of tactile and aesthetic characteristics is leveraged through the digital design process. The developed protocol also emphasizes the capability of advanced digital methods to address the multifactorial challenges of working with waste-derived and heterogeneous material streams. Through the close relationship between manufacturing, testing and implementation, the proposed system can incorporate and exploit these anomalies, increasing material utilisation and efficiency and presents itself as a case study for the wider adoption of cork as a building material to increase circularity in architecture and the cork industry.

References

1. United Nations Environment Programme GA for B and C. Not just another brick in the wall: The solutions exist - Scaling them will build on progress and cut emissions fast. Global Status Report for Buildings and Construction 2024/2025. https://wedocs.unep.org/20.500.11822/47214 (2025). Accessed 4 May 2026
2. Wilton, O., Barnett, H.M.: Cork: An historical overview of its use in building construction. Constr. Hist. **35**(1), 1–22 (2020)
3. Miranda, I., Pereira, H.: Cork façades as an innovative and sustainable approach in architecture: a review of cork materials, properties and case studies. Materials. **17**(17), 4414 (2024)
4. Pereira, H.: Variability of the chemical composition of cork. Bioresources. **8**(2), 2246–2256 (2013)
5. Jové, P., Olivella, M.À., Cano, L.: Study of the variability in chemical composition of bark layers of Quercus suber L. from different production areas. BioRes. **6**(2), 1806–1815 (2011)
6. Pereira, H.: Cork: Biology, Production and Uses, 1st edn. Elsevier, Amsterdam; London (2007) 336
7. Correia, V.G., et al.: The molecular structure and multifunctionality of the cryptic plant polymer suberin. Mater. Today Bio. **1**(5), 100039 (2020)
8. Rives, J., Fernandez-Rodriguez, I., Gabarrell, X., Rieradevall, J.: Environmental analysis of cork granulate production in Catalonia – Northern Spain. Resour. Conserv. Recycl. **58**, 132–142 (2012)
9. Zhai, W., Zhong, Y., Xu, M., Wei, X., Cai, L., Xia, C.: Transforming wastes into functional materials: natural cork-based physical structural components and polymers. Green Chem. **26**(15), 8615–8641 (2024)

10. APCOR. O SETOR DA CORTIÇA [Internet]. https://apcor.pt/uploads/Media/Estudos/Caracterizacao-diagnostico-tecnologico-estrategia/1_O%20Setor%20da%20Cortic%CC%A7a_Caracterizac%CC%A7a%CC%83o_Diagnostico_Estrategia.pdf (2023). Accessed 29 Dec 2024
11. Knapic, S., Oliveira, V., Machado, J.S., Pereira, H.: Cork as a building material: A review. Eur. J. Wood. Prod. **74**(6), 775–791 (2016)
12. Gil, L.: Cork powder waste: An overview. Biomass Bioenergy. **13**(1–2), 59–61 (1997)
13. Gürgen, S.: Sustainability of cork and protective systems. In: Gürgen, S. (ed.) Guarding with Cork: A Sustainable Solution for Protection [Internet], pp. 1–17. Springer Nature Switzerland, Cham (2025). https://doi.org/10.1007/978-3-031-72882-2_1
14. Demertzi, M., Paulo, J.A., Arroja, L., Dias, A.C.: A carbon footprint simulation model for the cork oak sector. Sci. Total Environ. **1**(566–567), 499–511 (2016)
15. Rives, J., Fernandez-Rodriguez, I., Rieradevall, J., Gabarrell, X.: Environmental analysis of the production of natural cork stoppers in southern Europe (Catalonia – Spain). J. Clean. Prod. **19**(2–3), 259–271 (2011)
16. Tártaro, A.S., Mata, T.M., Martins, A.A., Esteves da Silva, J.C.G.: Carbon footprint of the insulation cork board. J. Clean. Prod. **1**(143), 925–932 (2017)
17. Flores, M., Rosa, M.E., Barlow, C.Y., Fortes, M.A., Ashby, M.F.: Properties and uses of consolidated cork dust. J. Mater. Sci. **27**(20), 5629–5634 (1992)
18. Gil, L.: Effect of hot pressing densification on the cellular structure of black agglomerated cork board. Holz Roh Werkst. **52**(2), 131–134 (1994)
19. Gil, L.M.C.C.: Densification of black agglomerate cork boards and study of densified agglomerates. Wood Sci. Technol. **30**(3), 159–167 (1996) http://link.springer.com/10.1007/BF00231635
20. Knapic, S., Santos, C.P.d., Pereira, H., Machado, J.S.: Performance of expanded high-density cork agglomerates. J. Mater. Civ. Eng. **29**(2), 04016198 (2017)
21. Sergi, C., Sarasini, F., Bracciale, M.P., Russo, P., Tirillo, J.: Cork consolidated by hot compression as a viable bio-based alternative to polyolefines in decking boards: A preliminary study. Constr. Build. Mater. **22**(420), 135541 (2024)
22. Wilton, O., Howland, M.B.: Cork construction kit. J. Archit. **25**(2), 138–165 (2020)
23. Icsuro [Internet]. Catalan Institute of Cork. [online]. https://icsuro.com/en/catalan-institute-of-cork/
24. Gerometta, M., Gabrion, X., Lagorce, A., Thibaud, S., Karbowiak, T.: Towards better understanding of the strain-stress curve of cork: A structure-mechanical properties approach. Mater. Des. **235**, 112376 (2023)
25. Saadallah, Y.: Modeling of mechanical behavior of cork in compression. Fract. Struct. Integr. **14**(53), 417–425 (2020)
26. Oliveira, V., Rosa, M.E., Pereira, H.: Variability of the compression properties of cork. Wood Sci. Technol. **48**(5), 937–948 (2014)
27. Jardin, R.T., Fernandes, F.A.O., Pereira, A.B., Alves de Sousa, R.J.: Static and dynamic mechanical response of different cork agglomerates. Mater. Des. **5**(68), 121–126 (2015)
28. Estrin, Y., Krishnamurthy, V.R., Akleman, E.: Design of architectured materials based on topological geometrical interlocking. J. Mater. Res. Technol. **15**, 1165–1178 (2021)
29. Dyskin, A.V., Pasternak, E., Estrin, Y.: Mortarless structures based on topological interlocking. Front. Struct. Civ. Eng. **6**(2), 188–197 (2012)
30. Weizmann, M., Amir, O., Grobman, Y.J.: Topological interlocking in architecture: A new design method and computational tool for designing building floors. Int. J. Archit. Comput. **15**(2), 107–118 (2017)
31. Preisinger, C.: Linking structure and parametric geometry. Archit. Des. **83**(2), 110–113 (2013)

32. Sadeghipour Roudsari, M., Pak, M.. Ladybug: A parametric environmental plugin for grasshopper to help designers create an environmentally-conscious design. In: Proceedings of BS 2013: 13th Conference of the International Building Performance Simulation Association, 1 January, pp. 3128–3135. (2013)
33. Sadeghipour Roudsari, M., Subramaniam, S. Automating radiance workflows using Python [Internet]. https://www.radiance-online.org/community/workshops/2016-padua/presentations/213-SadeghipourSubramaniam-AutomatingWorkflows.pdf (2016)

Responsive 3D Printing: A Stimulus-Driven Approach Inspired by Phototropism

Lucas Helle, Joseph Naguib, and Roberto Naboni(✉)

SDU CREATE - University of Southern Denmark, Campusvej 55, Odense, Denmark
ron@iti.sdu.dk

Abstract. This paper proposes a responsive 3D printing framework inspired by phototropism, enabling additive manufacturing processes to adjust material deposition iteratively based on environmental stimuli. The system employs an agent-based model (ABM) that interprets light conditions captured through computer vision, dynamically influencing deposition trajectories. Discrete Deposition Modelling (DDM) is used to facilitate bead-wise material placement, allowing reassessment of environmental inputs at each cycle. The methodology is implemented through an integrated feedback loop, coordinating sensing, behavioural computation, and on-the-fly G-code generation. Validation experiments, supported by digital simulations, demonstrate that individual and collective agent behaviours effectively respond to changing light conditions, producing directional growth patterns. These results confirm the feasibility of stimulus-responsive fabrication and suggest new directions for developing context-aware additive manufacturing workflows informed by environmental inputs.

Keywords: Responsive 3D Printing · Discrete Deposition Modelling · Bio-inspired design · Behavioral Fabrication · Phototropism

1 Introduction

Typical 3D printing processes rely on predetermined toolpaths that divide digital models into discrete layers, operating under the assumption of stable and predictable fabrication environments (Zhou et al., 2024). Closed-loop error correction systems have been introduced to address deviations such as layer misalignments or extrusion inconsistencies, yet they typically adjust the process after deviations occur (Narupai et al., 2022). Research in cyber-physical construction has expanded this approach by integrating sensing and robotic control to introduce context-aware fabrication workflows (Naboni, 2022). Recent advances in environment-aware manufacturing, such as terrain-adaptive concrete printing (Breseghello et al., 2023) and robot-vision systems for deposition control (Naboni, et al., 2022), have further demonstrated the potential of integrating environmental feedback into additive processes. Model-free approaches integrating environmental feedback have been proposed, such as the reinforcement learning framework by Felbrich et al. (2022). These developments provide a foundation for exploring adaptive fabrication systems where deposition decisions are driven by real-time environmental inputs. In this

Y. Liu et al. (Eds.): CDRF 2025, *Transindividual Intelligence*, pp. 353–362, 2026.
https://doi.org/10.1007/978-981-92-0615-5_30

context, nature offers additional inspiration; phototropism, for example, enables plants to reorient growth towards light through local stimulus responses (Liscum et al., 2014). Building on this principle, we present a stimulus-responsive 3D printing framework that dynamically adjusts deposition based on live light sensing, influencing global geometry and local material placement.

2 Methodology

The project explores the feasibility of a responsive 3D printing process inspired by phototropism, where deposition decisions dynamically respond to external light conditions. The approach employs Discrete Deposition Modelling (DDM), a technique distinct from conventional Fused Deposition Modelling (FDM) in depositing discrete polymer beads rather than continuous filaments. This allows for flexibility in the deposition sequence (Andreen *et al.*, 2019) and extended geometry control, enabling condition reassessment and dynamic adjustments at each printing cycle. Material placement decisions are guided by an Agent-Based Model (ABM) that processes sensed light stimuli. Each printing cycle involves three sequential stages: *sensing*, where external light conditions are captured; *response*, where the ABM evaluates conditions and determines the structure's deposition behaviour; and *action*, where these vectors inform precise bead deposition. A digital simulation representing print cycles is implemented to test the agent interactions and anticipate potential issues before physical printing. The system's performance is then assessed by comparing simulation predictions with physical experimental outcomes.

2.1 Experimental Setup and System Architecture

The setup for running the experiments consists of a Prusa i3 MK3 3D printer, a 220-lm LED lamp for the light source, and an Intel RealSense D455C depth camera, all integrated within a ROS workspace managed by a Raspberry Pi 4. The light source is positioned at different locations during printing to evaluate the printing behaviour emerging from varying light conditions. The camera is positioned above the build area, capturing RGB images, which are subsequently processed in OpenCV to prepare them for analysis within Grasshopper 3D. The ROS2 Workspace acts as the central coordinator of the system, facilitating communication between hardware components and computational models (Fig. 1). It employs a finite state machine to manage the iterative printing process across the *sensing, response,* and *action* stages. The ABM is hosted within Grasshopper 3D, where it computes the vectors that guide the printing behaviour and generates G-code commands, which are sent on the fly to the 3D printer.

2.2 Stimulus-Driven 3D Printing Workflow

The responsive 3D printing workflow operates through an iterative cycle structured around three primary stages: *Sensing, Response*, and *Action.* Each cycle begins by capturing environmental data in the *Sensing stage (1)*, using camera vision to preprocess images, analyse shadows, and accurately determine the position of the light source. This processed information, along with predefined startup conditions, including system

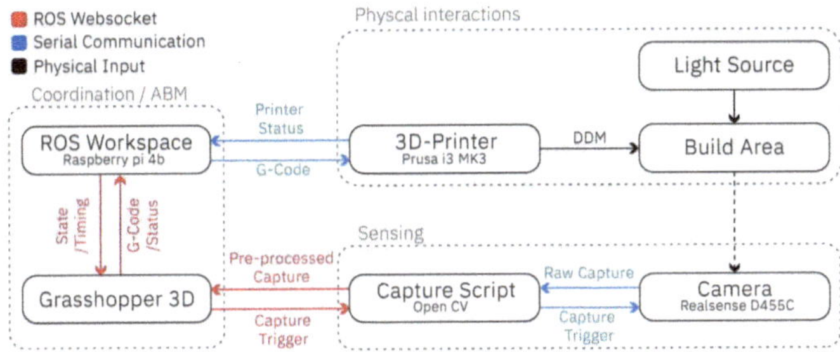

Fig. 1. System architecture of the responsive 3D printing setup, showing communication and physical interactions

initialisation, capture boundary definition, and selection of startup points, informs the *Response stage (2)*, where an Agent-Based Model (ABM) calculates local interactions, generates response vectors, and updates relevant parameters. Subsequently, these parameters guide the *Action stage (3)*, during which print settings are established, G-code is generated, and discrete deposition modeling (DDM) is executed. Complementing this physical workflow, a simulation loop is implemented as a design and testing environment to explore and refine agent behaviours before their physical implementation. The interactions between these two loops, along with supporting conditions such as startup settings and light stimuli, are detailed in Fig. 2.

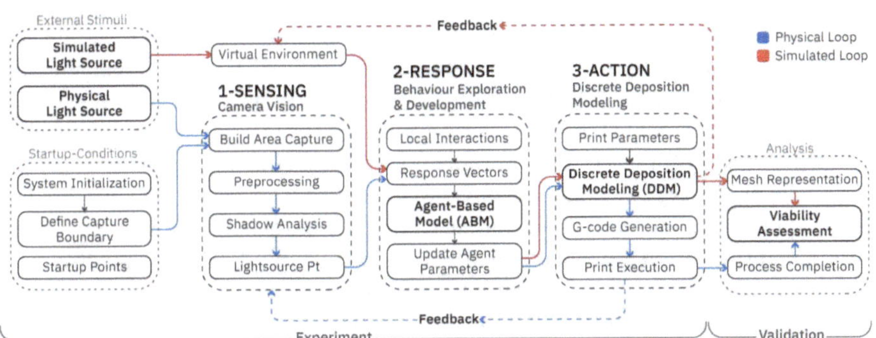

Fig. 2. Experimental and validation framework integrating sensing, agent-based response, and discrete deposition modelling, with interconnected physical and simulated feedback loops

Sensing: Camera Vision - The purpose of the sensing stage is to detect the external light on the build area via brightness analysis and determine the source position on the XY plane. On each cycle, the print bed is moved towards a predefined position to maintain a consistent viewpoint for each capture. The camera raw images are then cropped and corrected for perspective distortion by defining the capture boundaries of the print bed in OpenCV during a startup operation. The preprocessed image is then sent to Grasshopper 3D and converted to grayscale to enhance shadow contrast (Fig. 3b). A light direction vector is obtained by analysing the shadow cast by the printed structure.

The system defines a search radius around the initial bead deposition (*P0*) to identify a predefined number (*n*) of the darkest pixels based on the grayscale value. The centroid of the darkest pixels (*Ps*) represents the average position of *n*. A vector from *P0* to *Ps* is then projected onto a 20 mm offset perimeter surrounding the print bed; this projected point is defined as the light source position (*Pl*) (Fig. 3b,c). This vector calculation abstracts the sensed light conditions into a precise point in space used for the subsequent stage.

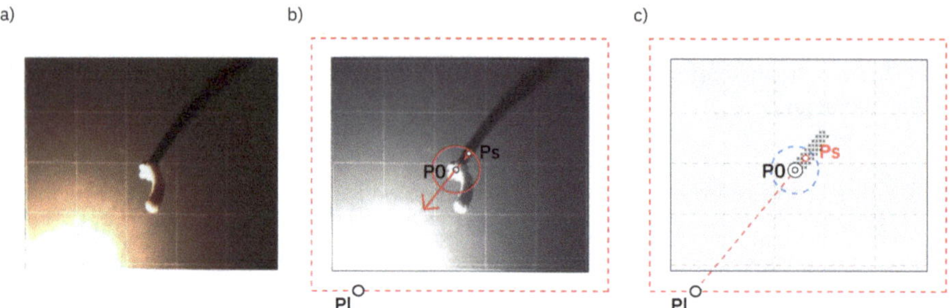

Fig. 3. a) The resulting capture after the preprocessing script, b) View from Grasshopper 3D after shadow analysis, c) A simplified diagram for the shadow algorithm.

Response: Behaviour Exploration and Development - The ABM translates sensed light conditions into response vectors, determining deposition coordinates. Inspired by the Boids flocking model (Olfati-Saber, 2006), the system smoothly combines external attraction to the light source *Vl(n)* with local agent interactions *Vi(n)*. The resulting vector *V(n)*, defining each agent's XY positional shift per cycle, is set to a predefined maximum magnitude, and a fixed rate of change limits how quickly an agent can adjust its displacement. An increment in the Z-axis is then added to *V(n)* to obtain the next deposition coordinate *P(n + 1)* (Fig. 4a). Four weighted local behaviours integrate internal agent conditions: merging (agents combine into one), branching (an agent splits into two based on XY displacement amount), cohesion (agents being attracted to neighbours' average position), and separation (agents repelled until a threshold radius (*r*)) (Fig. 4b).

Action: Discrete Deposition Modeling - The action stage utilises DDM to convert the print targets generated by the ABM into precise material deposition commands. During each cycle, the system defines a linear path connecting the current deposition point *P(n)* to the next target position *P(n + 1)*. Material extrusion begins at a normalised position along this path, defined by a parameter *(t)*, to ensure proper adhesion. The extrusion volume rate is calibrated proportionally to the distance between *P(n)* and *P(n + 1)* based on preliminary tests. This method guarantees consistent material flow, avoiding over- or under-extrusion and ensuring structural integrity (Fig. 4a). These print parameters are then translated into G-code commands, which are executed by the printer at each cycle.

2.3 Digital Simulation

The digital simulation is a key component of the ABM framework, allowing the exploration of print behaviours without requiring an entire physical print cycle. This simulation

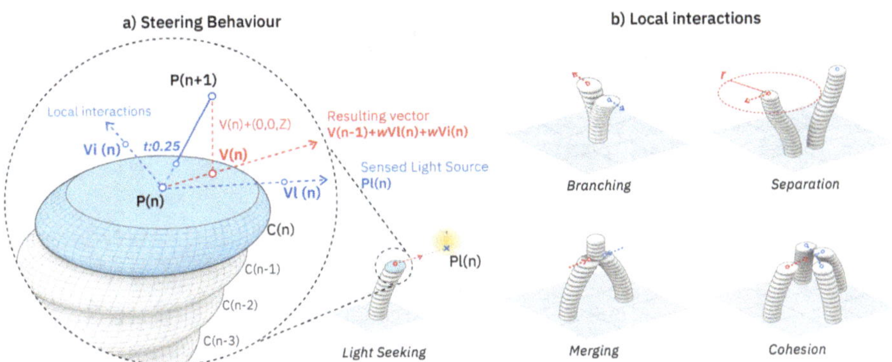

Fig. 4. a) The steering behaviour combines multiple conditions to get the next deposition point. b) Local interactions conditioning agent behaviour

replicates the ABM algorithm used in the actual printing process, generating deposition targets by simulating how agents respond to a virtual light source represented as a predefined 3D point in the CAD environment. The simulation maintains control over the light source position and agent parameters, allowing the dynamic environment to be accurately modelled and adaptive behaviours to be tested effectively. After calculating the deposition coordinates, the simulation employs a fast, mesh-based representation of the plastic beads, with volumes and sizes derived from previously collected physical samples. This approach provides a rapid visualisation of anticipated material behaviour while highlighting significant volume deviations and potential overlaps between beads.

3 Validation Experiments

The responsive 3D printing system was evaluated through printing experiments. The experimental campaign was based on two series. The first experiment focused on validating the essential light-seeking behaviour of an individual agent. The second experiment explored the interaction among multiple light-seeking agents to evaluate emergent responsive behaviours. Throughout the printing process, agent positions, distances to the sensed light source, and captured images were logged for further validation. A time-lapse of each experiment was recorded from a fixed position to assess the results over time. To ensure reliable and consistent results in the physical experiments, specific printing parameters were first established and fine-tuned through iterative trials within the simulation environment. These calibrated parameters were then directly applied to the printing process. Each experiment began by capturing an initial baseline image and generating the first agents based on user input. Subsequent cycles applied a fixed vertical increment of 3 mm, while horizontal (XY) adjustments were computed dynamically by the Agent-Based Model (ABM). This iterative process continued until the printed structure reached the predefined height limit.

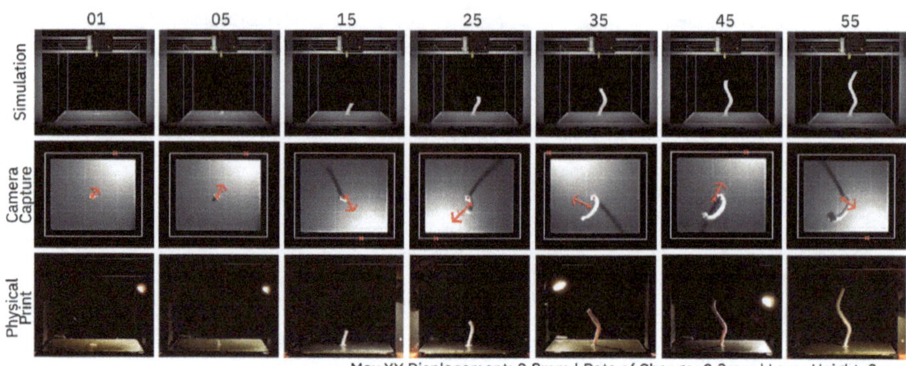

Fig. 5. Iterative cycles of the light-seeking experiment showing simulation predictions, camera captures, and corresponding printed outputs.

3.1 Experiment 1: Light Seeking Behaviour

In this experiment, the system was configured to test a phototropic response by adjusting material deposition based on light stimuli. The starting point is set at the centre of the print bed. The light source was repositioned across four different locations, one at each corner of the print bed, rotating clockwise at predefined cycle intervals. The print parameters and light positions were tested in simulation before being replicated in the physical setup. Figure 5 presents printing parameters and snapshots from various cycles representing simulation outputs, processed camera captures, and actual printed results.

3.2 Experiment 2: Incorporating Local Interactions

Two additional experiments were conducted to examine how multiple agents, while responding to light stimuli, modulate their behaviour through local interactions, collectively contributing to deposition patterns. These experiments were run in the simulation and printing environment and compared (Fig. 6). Experiment 2A investigated separation forces and branching, starting with a single agent at the center of the print bed. When the cumulative XY displacement exceeded 2.6 mm, the agent branched into two separated by 50% of the bead XY displacement from the center, while the XY displacement was reduced by 40%. Agents then move apart due to a separation force. Experiment 2B examined cohesive forces and merging, beginning with a number of randomly distributed agents influenced by a cohesion force. When two or more agents come within 5 mm of each other, they merge into a single agent, averaging their movement until only one agent remains.

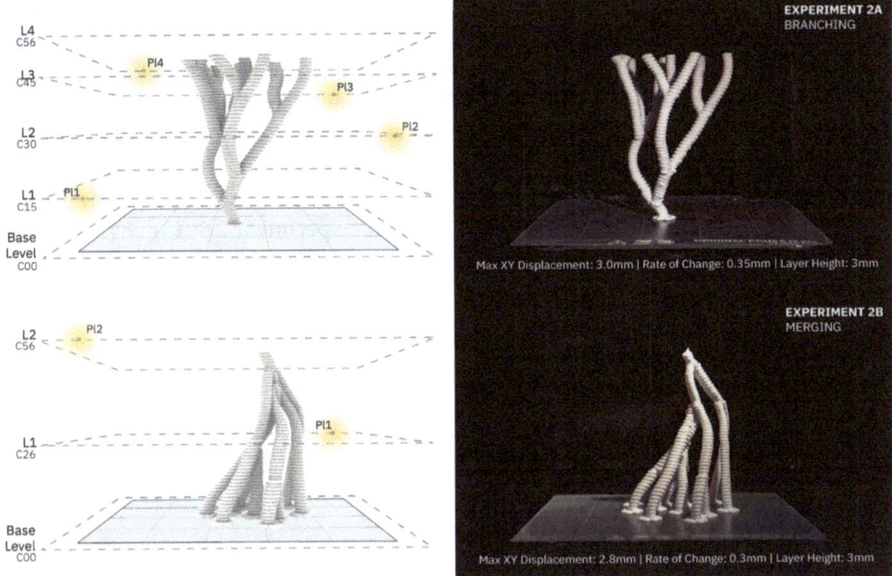

Fig. 6. Experiment 2 exploring agent interactions: branching (top) and merging (bottom), demonstrating representative simulation outputs and corresponding printed results.

4 Results

The results of the validation experiments demonstrate distinct behavioural patterns emerging from the implementation of stimulus-driven deposition, both at the individual and collective levels. In the first experiment, the three printed specimens consistently exhibited light-seeking behaviour, with a measurable reduction in the distance between deposition points and the light source over successive cycles. Each displacement of the light source leads to an initial increase in distance, followed by a clear reorientation of deposition trajectories towards the new stimulus location. While the precise distance profiles vary among specimens, the overall trend is robust and aligns closely with simulation predictions, indicating that the light-seeking mechanism alone is sufficient to produce directional adaptation (Fig. 7).

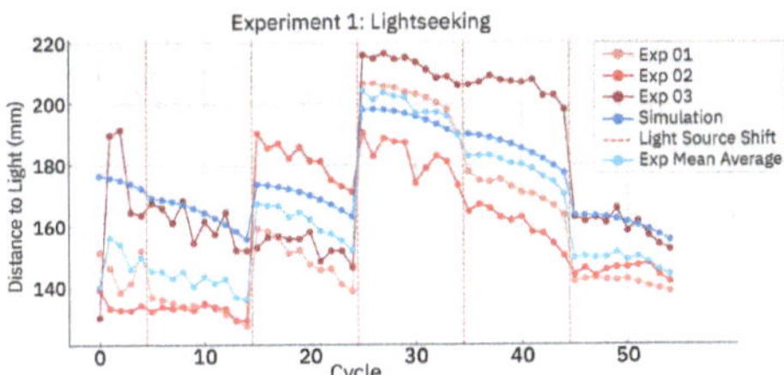

Fig. 7. Experiment 1 distance to light comparison on each cycle

The second experiment introduces local interaction rules that modulate agent behaviour beyond stimulus response alone. In the first series (branching and separation), agents divide when displacement thresholds are exceeded, resulting in bifurcating trajectories that maintain phototropic orientation while maximising spatial dispersion. In contrast, the second series (cohesion and merging) exhibits an opposing dynamic: agents progressively reduce their spatial separation, converging into a single trajectory over time. This difference is evident in the resulting morphologies, branching patterns in Experiment 2A versus consolidated structures in Experiment 2B (Fig. 8).

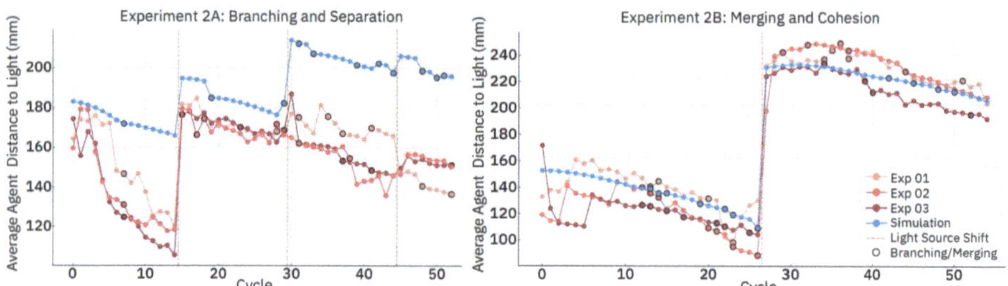

Fig. 8. Experiment 2 distance to light comparison on each cycle

While both behaviours remain driven by the light stimulus, introducing local interactions increases variability in trajectory and final form. Deviations between simulation and physical specimens are more pronounced in Experiment 2B, particularly during merging phases, suggesting that proximity-based interaction rules introduce greater geometric unpredictability (Fig. 9). Nevertheless, the overall agreement between predicted and printed outcomes confirms that the agent-based model effectively governs individual and collective deposition behaviours in response to dynamic stimuli.

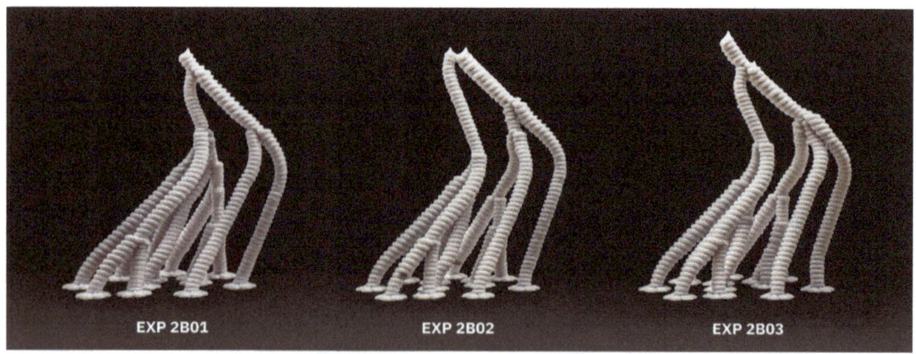

Fig. 9. Printed samples of experiment 2B

5 Discussion and Conclusion

This research has demonstrated the feasibility of embedding stimulus-responsive behaviours within additive manufacturing workflows, inspired by plant phototropism. By integrating environmental sensing and decentralised agent decision-making, the system redefines material deposition as an adaptive process that evolves in response to external stimuli. The experimental results confirm that deposition trajectories systematically reorient towards dynamic light sources and that local interaction rules effectively shape emergent deposition patterns. These findings establish a novel paradigm in additive manufacturing, shifting from predefined, model-driven toolpaths to behavioural fabrication informed by external conditions. Rather than treating deviation and variation as errors to be corrected, the proposed framework embraces local responsiveness and interaction as drivers of material organisation. The introduction of local agent interactions further enhances the system's capacity for situated problem-solving, allowing collective behaviours to negotiate between environmental inputs and spatial relationships without centralised control. While the present implementation operates within a controlled laboratory setting, with a single stimulus type and predefined interaction rules, these boundaries outline a proof of concept rather than intrinsic limitations. Future work may expand the framework to integrate multiple stimuli, continuous feedback mechanisms, and adaptive learning strategies, enabling fabrication systems to engage more fully with unstructured environments. Scaled to architectural components, the approach could fabricate structurally informed shapes that embed material efficiency while passively responding to their environment. This research contributes to a rethinking of additive manufacturing as an open, behavioural practice. The demonstrated capacity for emergent, context-aware adaptation departs from deterministic fabrication models and opens new pathways towards self-organising, bio-inspired production systems.

Acknowledgements. This work was supported by the Villum Foundation through the Villum Young Investigator Programme, as part of the research project Bio-Inspired Behavioural Additive Construction [grant number 53094].

References

Andreen, D., Goidea, A., Johansson, A., Hildorsson, E., Swarm Materialization Through Discrete, Nonsequential Additive Fabrication. In: IEEE 4th International Workshops on Foundations and Applications of Self Systems, pp. 225–230. (2019). https://doi.org/10.1109/FAS-W.2019.00059

Breseghello, L. et al. (2023) Shape-Env - Camera-enhanced robotic terrain-shaping for complex 3D concrete printing. In: eCAADe 2023: Digital Design Reconsidered, Graz, Austria, pp. 539–548. https://doi.org/10.52842/conf.ecaade.2023.1.539

Brion, D. A. J., & Pattinson, S. W. (2022). Generalisable 3D printing error detection and correction via multi-head neural networks. Nature Communications, **13**(1), 4654. https://doi.org/10.1038/s41467-022-31985-y

Felbrich, B., Schork, T., Menges, A.: Autonomous robotic additive manufacturing through distributed model-free deep reinforcement learning in computational design environments. Constr. Robot. **6**(1), 15–37 (2022). https://doi.org/10.1007/s41693-022-00069-0

Liscum, E., Askinosie, S.K., Leuchtman, D.L., Morrow, J., Willenburg, K.T., Coats, D.R.: Phototropism: Growing towards an understanding of plant movement. Plant Cell. **26**(1), 38–55 (2014). https://doi.org/10.1105/tpc.113.119727

Naboni, R.: Cyber-physical construction and computational manufacturing. In: Bolpagni, M., Gavina, R., Ribeiro, D. (eds.) Industry 4.0 for the Built Environment, pp. 515–540. Springer, Cham (2022)

Naboni, R., Breseghello, L. and Sanin, S. (2022) Environment-aware 3D concrete printing through robot-vision. In: eCAADe 2022: Co-creating the Future – Inclusion in and through Design. Ghent, Belgium, pp. 409–418. https://doi.org/10.52842/conf.ecaade.2022.2.409

Olfati-Saber, R.: Flocking for multi-agent dynamic systems: Algorithms and theory. IEEE Trans. Autom. Control. **48**(7), 988–1001 (2006)

Zhou, L., et al.: Additive manufacturing: A comprehensive review. Sensors. **24**(9), 2668 (2024). https://doi.org/10.3390/s24092668

Structural Nodes from Composite Tree Bifurcations: Investigation and Development of an Adaptive Construction System from Naturally Grown Raw Wood Members

Raman Suliman(✉), Kevin Moreno Gata, and Martin Trautz

Department of Structures and Structural Design, RWTH Aachen, Aachen, Germany
raman.suliman@rwth-aachen.de

Abstract. The growing demand for sustainable construction has intensified interest in timber as a renewable material. However, current industry practices favor standardized straight lumber, overlooking naturally occurring irregular timber as a resource. These irregular wood elements, specifically tree branch bifurcations, have the potential to develop sustainable structural systems. This research addresses resource efficiency by leveraging the unique properties of angular timber elements (ATEs) combined with computational design methods. The geometry and grain alignment of raw timber components are examined using 3D scanning to provide digital libraries documenting key properties. These libraries inform parametric design procedures, optimizing the geometric arrangement of 3D trusses based on data obtained from branch bifurcations. This optimization directly affects the configuration and geometry of structural nodes and joints. Through case studies on typologies such as walls, towers, ceilings, and bridges, the integration of these natural materials into node-based systems was examined. The findings highlight how the inherent characteristics of the material align with structural requirements, aiming for a balance between robust load-bearing performance and design flexibility. This study demonstrates the feasibility of combining computational design with raw timber variability to reduce waste and promote sustainability, opening new avenues for architectural expression.

Keywords: Node-based structures · Spatial structures · Raw timber elements · Computational design methods · Inventory-based design

1 Introduction

Wood is becoming increasingly important in construction, but rising demand may soon outpace production capacity. In 2022, Germany's forests covered 29.9% of the land and produced 78.7 million cubic meters of wood annually.

Harvested wood is classified into categories, such as stem wood, industrial wood, energy wood, and non-utilizable, based on various factors, with straight timber prioritized for construction [1]. Irregularly shaped wood, such as branch forks and curved branches,

Y. Liu et al. (Eds.): CDRF 2025, *Transindividual Intelligence*, pp. 363–372, 2026.
https://doi.org/10.1007/978-981-92-0615-5_31

is often underutilized, typically left to decompose or used in thermal energy production [2]. The growth of a tree, influenced by species, sunlight, and climatic factors, affects the density and orientation of the wood grain, hence determining its shape and strength [3].

These natural junctions redirect loads while maintaining grain continuity, making them suitable for structural applications that involve force redirection and bending stiffness [4]. To make use of these characteristics, samples of raw timber were analysed to identify their unique occurrence tendencies. Combining this data with current production techniques could allow for the integration of natural wood components into spatial knot structures and evaluate the scalability of this concept.

The two main goals of the resulting design process are to create spatial structures that adapt to the properties and availability of the raw material while also maintaining the alignment between the function and natural grain flow of each timber component. By implementing an adaptive building system that responds to the base material, this approach promotes resource efficiency by harmonizing designs with the natural diversity of raw wood.

2 State of Research

The development of structural nodes in timber construction involves understanding material properties and manufacturing constraints. Current research investigates the development of irregular spatial trusses, the value of grain and geometric continuity, and advanced fabrication techniques to improve structural efficiency and adaptability.

The subsequent research examples analyse significant contributions in these domains, establishing a basis for the incorporation of harvested timber components into computationally generated systems.

2.1 Irregular Spatial Trusses

Irregular node structures and their formation require examination, along with suitable manufacturing methods and their limitations. The work "3D Printed Space Frames" by F. Raspall, F. Amtsberg, and C. Banon [5] highlights the use of 3D printing in structural design. By integrating parametric modeling, structural optimization, and additive manufacturing, their approach creates bespoke, material-efficient, and adaptable space frame nodes. Four full-scale projects demonstrate flexibility and enhanced performance achievable with this method, paving the way for highly customized structural nodes.

2.2 Fibers, Grain and Geometric Continuity

The role of grain orientations within a timber node is pivotal for achieving desired performance. M. Tamke's "The Rise: Building with Fiberous Systems" [6] explores thread-like materials in construction, emulating biological systems.

This project emphasizes continuity within components, where nodes serve as material boundaries. this generative process incorporates assembly logic, creating three-dimensional structures from actively bent strand bundles. This highlights the potential of integrating similar systems to achieve continuity between structure and node design

2.3 Fabrication of Timber Junctions/Nodes

The structural and changeable nature of timber nodes must be supported by manufacturing procedures. A new column-beam connection modelled after tree stem-branch junctions is presented in F. Hawasly's" Designing and Testing of a Wood-Only Timber Frame Connection" [7].

This design does not require steel reinforcement because it uses a tenon-mortise interface and layers of laminated veneer. Elastic qualities and continuous reinforcement from curved timber elements were demonstrated in testing, which showed bending performance comparable to standard joints. This technique shows promise for efficient and natural load rerouting in timber structures.

3 Methodology

Wood specimens from a forest near the city of Aachen and from the Aachen municipal gardening centre made up a sample set of raw material. as these samples were acquired, key features such as wood species, cross-section and crack patterns were recorded. A digital library containing accurate 3D scans of the sample set was then generated.

Characteristics, such as element curvature, were extracted from the library and expressed numerically. The formation of node configurations and spatial patterns was ultimately guided by these entries. To refine the design, a multi-objective optimization (MOO) algorithm was applied. followed by a matchmaking process using the Hungarian algorithm [8], where each node segment was paired with the most suitable raw wood element. Finally, fabrication data for each segment were generated and stored, along with identifiers for efficient tracking and production.

3.1 Raw Timber Geometry Analysis and Documentation

Depending on the tree species, branch forks can exhibit significant variation in the number of branches. The most common type consists of a single fork, where a parent branch splits into two.

However, these can also take on unique 3D shapes, which can pose a challenge for modular applications. Therefore, only the relationship between each combination of two branches were examined, rather than the entire bifurcation. This allowed for a reduction in the complexity of the raw material while still utilizing the properties of the continuous wood grains.

The following analysis focused on identifying strictly curved i.e. angled geometries, excluding straight sections, which are more suitable as struts. Each wood element's angular value and corresponding cross-section were documented.

3.1.1 Data Visualization

In addition to bifurcations, bent branches have been included as a separate category for analysis. Custom Grasshopper scripts were created for each of these two categories to gather the necessary information. The primary goal was to acquire an initial overview of the geometries within the library, which would serve as the foundation for the subsequent design phases.

For the bifurcation, the process begins by importing the meshes as OBJ files and smoothing them through multiple iterations. This Step simplified the extraction the centerlines of the geometries.

Next, the plugin "Cockroach" [9] is employed to derive the centerlines for each branch within the bifurcation. The centerlines of paired branches are then connected, and centre point of the bifurcation is defined, allowing the angle between the branches to be calculated. This way for each tree bifurcation at least three potential use cases can be identified.

For bent wood elements the number of use cases is dependent on the number of useable partitions. Taking this into consideration, a single continuous centerline is generated using the previous method. The centreline is divided into short subsegments, and the radii of curvature are extracted at each division point. Because a single radius value does not fully capture the curvature of a timber partition, multiple lists are created—each containing an origin point on the centerline along with several adjacent points in both directions. Averaging the radii in each list creates overlapping regions that provide better description of the curvature along the timber while reducing the impact of outlier values.

Finally, segments with less curvature are filtered out to ensure that only the sections with the strongest curvature are retained for integration into the design. The angles between the vectors from the point of greatest curvature to each end are then documented. These entries help identify and numerically describe the angular characteristic of the timber segments. Groupings were created based on the angular sizes of the individual raw timber segments, with intervals of 5° across the range from 0 to 180°. The data is then plotted on a two-axis diagram, where the x-axis represents the groupings and the y-axis shows the number of entries in each group (see Fig. 1).

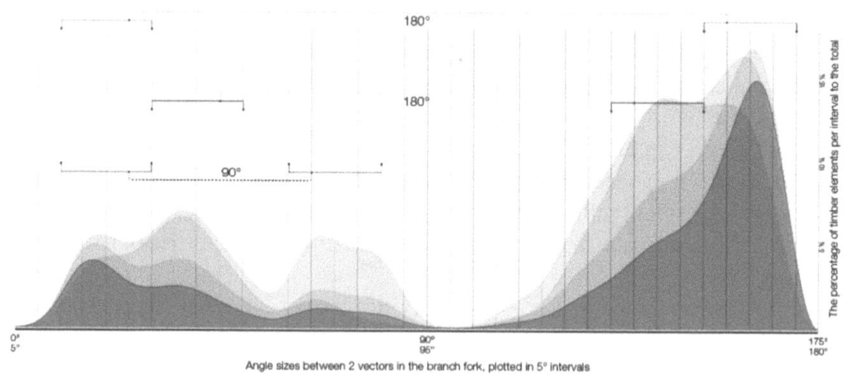

Fig. 1. Data plotting on an A/S diagram

3.2 Node Definition

In a structure composed of nodes and struts, the nodes are points in 3D space, while the struts or bracing elements are lines connecting pairs of nodes. To understand the generation of the final geometry of a node, A GH algorithm was initially developed to generate node geometries based on these spatial relationships Fig. 2.

This process started with a random set of points distributed in space, minimizing the influence of any predetermined arrangement. From this initial set, one point is chosen as the center of a primary node (Pz), with several other points designated as neighboring nodes. Struts then connect these nodes to form a partial structure. New reference points (Sr) are then positioned in the direction of each strut but are set at a specific distance from Pz. From (Sr), a convex hull was generated, creating a triangulated mesh. The edges of the mesh faces serve as reference curves (Cr), which will determine the position and orientation of the node segments (Ns). From each Cr a node segment is derived as a polyline and defined by the two endpoints of Cr and Pz.

Additionally, two vectors were isolated, originating from PZ and aligning towards the endpoints. These vectors were categorized into primary ($\overrightarrow{vp}$) and secondary vectors ($\overrightarrow{vs}$) based on the composition of the node. The angle between the two vectors defines the desired angular characteristic from the raw wood piece for that specific Ns. For each Ns, a base plane (π) was assigned, which is crucial for manufacturing. This plane is defined by Pz and a normal vector, that is the cross product of $\overrightarrow{vp}$ and $\overrightarrow{vs}$. Subsequently, the plane is oriented so that $\overrightarrow{X}$ of π is parallel to $\overrightarrow{vp}$. Finally, a plane (β) is defined as the basis for generating the interface —i.e., the bonding surface— between each pair of segments. This, in turn, informs the assembly of the complete node.

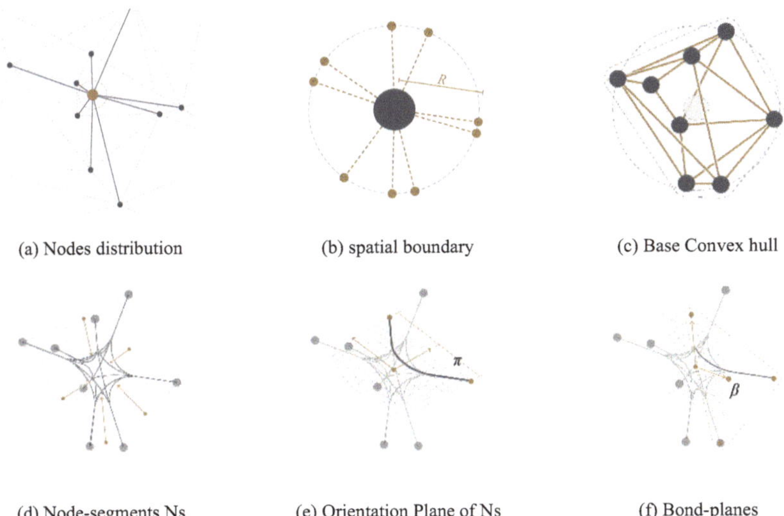

Fig. 2. Node-segment geometry based on its relative positioning. (a) Nodes distribution. (b) Spatial boundary. (c) Base Convex hull. (d) Node-segments Ns. (e) Orientation Plane of Ns. (f) Bond-planes

3.3 Parametric Spatial Frame

During the development of the node composition, multiple iterations were conducted to approximate the angular values observed in the material library. Key formative angles, identified as peaks in the A/S diagram, were integrated into the initial spatial arrangement to achieve structural coherence and flexibility.

The Truss structure followed a 2.5D geometry (curved plane), starting with a uniform curvilinear distribution of nodes that have angular relationships close to 90°. By shifting the Struts, a parallelogram configuration emerges, in the case of nodes uniformly distributed along the U and V directions, the change forms 135° and 45° angular relationships. These could be further adjusted by varying the number of nodes along each direction, enabling greater design adaptability. To fine-tune the system, specific angles derived from the library's maximum values served as guidelines, including:

$$20^\circ - 70^\circ(\text{sum } 90^\circ)/20^\circ - 160^\circ(\text{sum } 180^\circ)/40^\circ - 140^\circ(\text{sum } 180^\circ)$$

A parametric skeleton model of discrete lines with a corresponding pattern was created. The positions of individual nodes along the U and V axes of the surface can be varied within a defined sub-domain. Each node can be moved up to a maximum distance, determined by the midpoint between adjacent nodes minus a buffer to prevent collisions.

For structures with a large number of nodes, individually manipulating the position of each node is impractical. Instead, regular groupings in both the U and V directions are defined, with only the edge nodes of each group being adjusted. The displacement of nodes within each group was then interpolated, significantly reducing the number of input parameters (see Fig. 3).

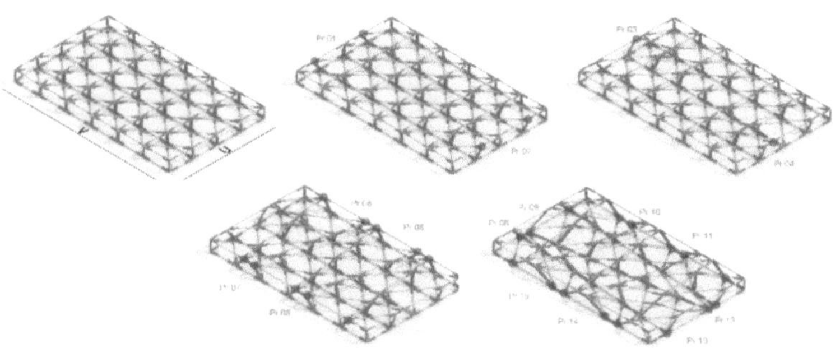

Fig. 3. Input parameters for the spatial structure

3.3.1 Multi Objective Optimization

For This step the Plugin Octopus [10] was used. While the optimization criteria encompassed load-bearing behavior, angular value comparison and genetic variety, the constraints relied on the repositioning of nodes within their designated subdomain of a UV surface.

For angular value comparison, a descriptive curve was plotted on the A/S diagram, similar to the material library. Node segments were measured and grouped in 5° increments, with a direct comparison made between the library and structure values. The vertical deviation between these curves formed the first optimization objective.

3.3.2 Matchmaking

A fitness factor is calculated to assess the compatibility between each raw wood element to each node segment and based on multiple criteria, generating a value matrix where one axis represents the node segments and the other the raw wood elements. The Hungarian algorithm is then applied to assign raw wood elements while minimizing the total evaluation sum, ensuring the best possible match between the structure and raw material.

4 Results

This Design process was observed across three key steps to assess its effectiveness. The first being the uniform distribution of nodes and standard curvilinear interconnections, serving as the baseline. This was followed by the introduction of the structural pattern, derived from the initial evaluation of the library, and concluded with multi-objective optimization to refine the design and align it with target library values.

This process was applied to a range of typologies, including both vertical structures (e.g., walls, facades, towers) and horizontal elements (e.g., ceilings, bridges). In all cases, the initial geometry played a significant role in shaping the optimization outcomes. Starting with a non-uniform configuration often facilitated better results in the later stages, enhancing the overall optimization (see Fig. 4).

5 Discussion

This study presents an optimization-based method for the integration of naturally grown timber into structural systems, focusing on fabrication challenges and establishing a systematic approach for node selection and refinement. Key factors include assessing the fitness of various timber dimensions for machining and structural connections.

The use of an A/S diagram or similar approaches could assist designers in selecting the most suitable 3D-truss topology early in the design process while ensuring compatibility with the existing material library. This could be supported by establishing a comprehensive catalog of node configurations and truss arrangement.

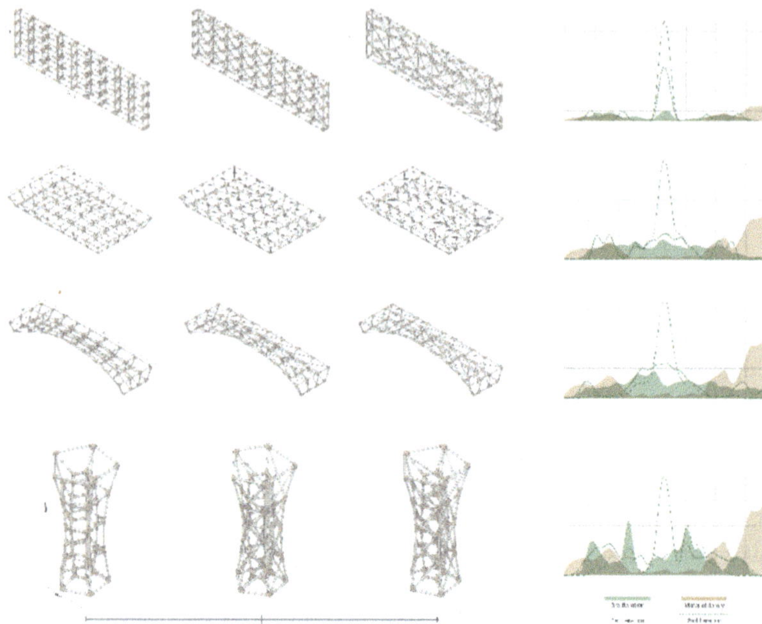

Fig. 4. Typologies and iterations

Although the potential for realizing nodes from wood segments is promising, several aspects require further investigation. Preliminary prototypes revealed that fabrication speed was suboptimal, and mechanical connections need further development to ensure durable bonds Fig. 5.

The bonding surfaces between node segments are critical for proper assembly, and fine-tuning the placement of a node segment within a Node composition is necessary to minimize material losses during subtractive manufacturing.

While the proposed method demonstrates promise, it currently serves only as a preliminary roadmap. Future work should involve extensive testing with a larger sample size, studying the influence of further characteristics. Moreover, refining the compatibility evaluation process—by expanding the initial A/S diagram to include additional dimensions for each evaluation characteristic—will be essential for practical implementation.

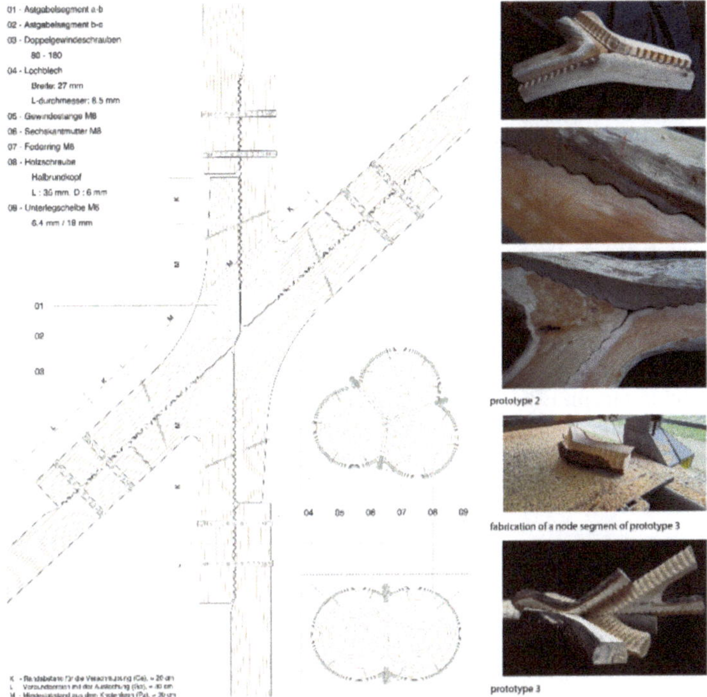

Fig. 5. Node-strut and in-node connections: detailed drawing.

Acknowledgement. This publication is based on the master's thesis of Raman Suliman at RWTH Aachen, supervised by Kevin Moreno Gata M.Sc. The thesis was evaluated by Univ.-Prof. Dr.-Ing Martin Trautz and Univ.-Prof. Dr. Jakob Beetz, whose guidance and feedback were greatly appreciated.

Special thanks go to the forestry department of the Aachen city region and the city nursery for their valuable support.

References

1. Ramage, M.H., Burridge, H., Busse-Wicher, M., Fereday, G., Reynolds, T., Shah, D.U.: The wood from the trees: the use of timber in construction. Renew. Sust. Energ. Rev. **68**, 333–359 (2017)
2. Rohholzmarkt I: Holzmarktbericht 2022. https://www.bmel.de/DE/themen/wald/holz/holz_node.html (2022)
3. Zhao, L., Zhou, H., Jin, S., Cui, W., Wang, M.: Mechanical properties of poplar (NL-3412) branches. Wood Mater. Sci. Eng. **20**(1), 1–11 (2024)
4. Moreno Gata, K., Seiter, A., Grizmann, D., Trautz, M.: Development of construction methods with naturally grown timber and bending-resistant joints. In: World Conference on Timber Engineering (WCTE 2023), World Conference on Timber Engineering (WCTE 2023), Oslo, Norway, pp. 705–710 (2023)

5. Bañón, C., Raspall, F.: 3D Printing Architecture: Workflows, Applications, and Trends (SpringerBriefs in Architectural Design and Technology). Springer Singapore, Singapore (2021)
6. Tamke, M., Stasiuk, D., Thomsen, M.R., Gramazio, F., Kohler, M., Langenberg, S.: The rise: building with fibrous systems. In: Fabricate. DGO-Digital Original, pp. 136–145. UCL Press, London (2017)
7. Hawasly, F., Matsumoto, N., Koshihara, M., Adachi, K.: Designing and testing of a wood-only timber frame connection inspired by the growth formation of trees' stem-branch junction. In: World Conference on Timber Engineering, Oslo, Norway, pp. 1365–1374 (2023)
8. Huang, Y., Alkhayat, L., De Wolf, C., Mueller, C.: Algorithmic circular design with reused structural elements: Method and tool. In: Fib Symposium Proceedings, vol. 55, pp. 457–468 (2021)
9. Vestartas, P., Settimi, A.: Cockroach: a plug-in for point cloud post-processing and meshing in Rhino environment. EPFL ENAC ICC IBOIS (2020)
10. Vierlinger, R.: Multi objective design interface. PhD Thesis, University of Applied Arts Vienna (2013)

Computational Analytics and Design for Architectural Heritage

Research on the Restorative Perception of Jiangnan Garden Scenes Based on Computer Vision and Generative Artificial Intelligence

Jianan Qin(✉) and Xu Li(✉)

School of Architecture and Planning, Hunan University, Changsha, China
{qjn-2000,leexu_2004}@hnu.edu.cn

Abstract. Jiangnan garden scenes represent the ideal human habitat in traditional Chinese culture, highlighting the harmonious coexistence of people and their environment while providing transindividual restorative experiences. Previous studies have not focused on the quantitative visual characteristics of Jiangnan garden scenes that enhance restorative perceptions. Additionally, research on the restorative perception of the Jiangnan garden scenes often faces challenges regarding its applicability in design practice, as well as issues concerning credibility and interpretability. To this end, this study uses social media images and self-captured images to create a dataset of Jiangnan garden scenes, obtains quantitative features of the visual elements within them by Semantic Segmentation through the Pyramid Scene Parsing Network. Subsequently, Deepseek-R1 is used to design descriptive prompts for Jiangnan garden scenes based on the identified features. And the Low-Rank Adaptation of Stable Diffusion XL is trained using the image dataset, which is then paired with the designed prompts to generate Jiangnan garden scenes that have restorative effects. Finally, the restorative perception of these scenes is evaluated through a perceptual scale. The results show that by using Computer Vision and Generative Artificial Intelligence, it is possible to effectively extract visual features and intelligently generate healing scenes. This method provides a reference for the paradigm shift in architectural design during the era of Artificial Intelligence. It also provides insights for improving the quality of the built environment and promoting emotional recovery in the public sphere.

Keywords: Jiangnan garden scenes · Restorative perception · Pyramid Scene Parsing Network · Deepseek-R1 · Stable Diffusion XL · Low-Rank Adaptation

1 Introduction

With the acceleration of urbanization and changes in the economic environment, people are increasingly facing significant life pressures, leading to a growing demand for physical and mental recovery and healing (Helliwell et al. 2024; Wu et al. 2024). Jiangnan garden scenes symbolize the harmonious coexistence of humans and nature in traditional Chinese culture, representing an ideal living environment that fosters a sense of restoration (Zhou 2004; Zhang et al. 2024). Applying the experiential qualities found

Y. Liu et al. (Eds.): CDRF 2025, *Transindividual Intelligence*, pp. 375–387, 2026.
https://doi.org/10.1007/978-981-92-0615-5_32

in Jiangnan garden scenes to modern urban and architectural design—thereby enhancing environmental quality and addressing people's emotional needs—has long been a focus of academic research. However, the historical influence of Chinese literati culture on Jiangnan gardens has resulted in design concepts and dissemination modes that are primarily expressed through non-systematic literary and artistic works. This overly subjective interpretation can undermine the credibility and interpretability of research concerning the restorative perception of garden scenes (Chen et al. 2024).

Computer Vision (CV) is a field of science and technology that uses machine learning techniques for image processing, pattern recognition, and image understanding. It has been extensively applied in architecture and the built environment, including tasks such as measuring environmental greenness, extracting building colors, and detecting damage to building facades. The effectiveness of CV in analyzing and representing visual elements within physical environments has been validated (Ito et al. 2024; Zhang and Pan 2024). Furthermore, in recent years, Generative Artificial Intelligence (GAI) has garnered significant attention for its ability to utilize algorithms and multimodal models to generate diverse content that meets user needs. Among the most advanced tools in this domain are Large Language Models, such as ChatGPT, and Diffusion Models, such as Stable Diffusion. These technologies provide a wide range of possibilities for schemes designing and performance evaluation in architecture and building engineering (Ghimire et al. 2024; Lyu et al. 2024; Ma and Zheng 2024; Yuan, P.F. et al. 2023).

Therefore, in response to current social needs, this study proposes a research method that combines CV and GAI, which can intuitively understand the restorative perception provided by Jiangnan garden scenes and facilitates the rapid generation of renderings with their characteristics. By doing so, architects will develop a more direct design strategy and create numerous scenes that evoke restorative perceptions through intelligent workflows, thereby enhancing the ability to fulfill their social responsibilities.

2 Materials and Methods

The conceptual framework of this study is illustrated in Fig. 1. To analyze the restorative perception provided by Jiangnan garden scenes, a comprehensive image dataset was constructed, incorporating widely popular content from social media as well as content recognized by professionals. Semantic segmentation based on CV technology was performed to extract the visual elements and their quantitative features of the scene. In addition, to reproduce the restorative features of Jiangnan garden scenes, Large Language Model (LLM) and Stable Diffusion (SD) based on GAI, were used for deep learning of the scene images and their visual elements, and Low-Rank Adaptation (LoRA) was trained to fine-tune the generated image effects. Finally, by referencing the structure and content settings of the Perceived Restorative Scale (PRS), Artificial Intelligence Generated Content (AIGC) was evaluated to verify the feasibility of the proposed method in this study.

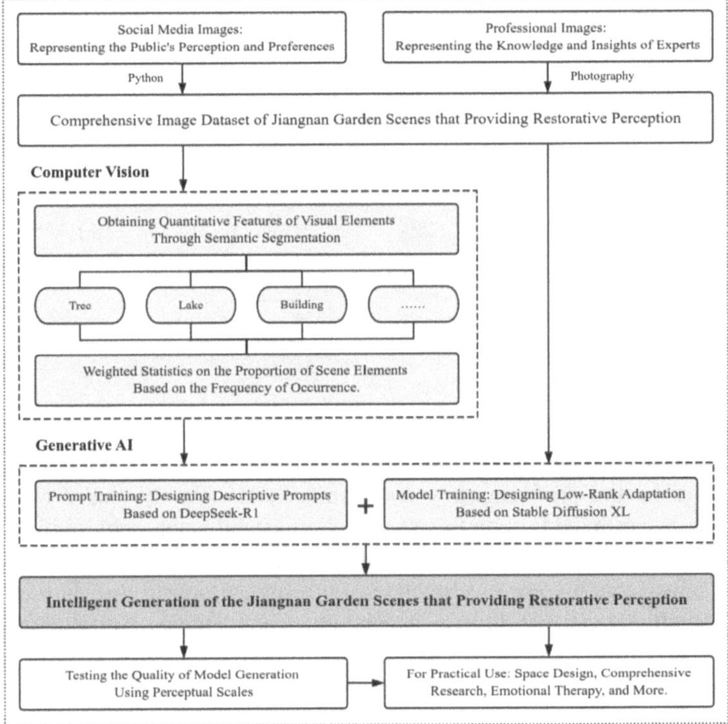

Fig. 1. Conceptual framework and research methods

2.1 Acquisition of Multimodal Data

2.1.1 Analysis of Social Media Text Data

Social media information is characterized by its large volume, dynamic changes, and authentic content. This study accessed mainstream social media platforms in China, such as Weibo and Rednote, using Python and set search keywords including "Jiangnan Garden", "Jiangnan Classical Garden", "Suzhou Garden". The content and web links posted by users from January 1, 2022, to December 31, 2024, were collected in bulk based on popularity and subsequently conducted manual data cleansing. The text content was analyzed for word frequency and sentiment using ROST CM6 software, with the results presented in Table 1. These textual data support the premise of this study, which posits that people can experience transindividual restorative perceptions in Jiangnan garden scenes. Additionally, the findings will serve as a reference for developing perception scales in follow research.

Then, based on this, search keywords of perception class such as "healing", "recovery", "cultivation", and "tranquility" were added to obtain user posted content in bulk. The public's expression of restorative perception in Jiangnan garden scenes was analyzed to summarize the descriptive features that evoke a high level of emotional satisfaction. Subsequently, a vocabulary set was formed to serve as a reference for the subsequent optimization of LLM.

Table 1. High-frequency words and their emotional tendencies

Emotional types	Number of texts	Text proportion	Emotional level
Positive	2633	72.98%	General: 1422 (54.01%), Moderate: 425 (16.14%), High: 786 (29.85%)
Neutral	28	0.77%	——
Negative	947	26.25%	General: 361 (38.12%), Moderate: 332 (35.06%), High: 254 (26.82%)
Total	3608	100%	——

2.1.2 Creation of a Comprehensive Image Dataset

The Comprehensive Image Dataset established in this study comprises two parts: social media image data and professionally taken image data. Firstly, based on predefined search criteria, we utilized Python to access web links and batch-acquire highly popular image content from social media. These images were manually screened to eliminate low-quality ones, which were characterized by a high proportion of human portraits, limited informational content, blurriness, and other deficiencies. Subsequently, Python was used to uniformly center-crop the images to a size of 1024 × 1024 pixels and convert their formats, resulting in a social media image dataset containing 300 images. All images in this dataset represent the intuitive perceptual preferences of the general public.

Additionally, we compiled a professionally selected and photographed collection of Jiangnan garden scenes. Specifically, relying on the previous research findings of our research group and the learning outcomes from the undergraduate course "Suzhou Traditional Gardens: An Oriental Aesthetic Lifestyle" cognitive internship at the School of Architecture and Planning, Hunan University (http://arch.hnu.edu.cn/info/1126/5055.htm), we constructed a dataset of 200 professionally taken images that have also been standardized in size and format. All images in this dataset reflect the evaluation criteria of professionals in architecture and related fields, including urban and rural planning, landscape architecture, design, and more.

2.2 Scenes Element Recognition Based on Computer Vision

2.2.1 Image Semantic Segmentation

This study utilizes the Pyramid Scene Parsing Network (PSPNet) for image recognition and understanding. PSPNet extracts features from the original image using a Convolutional Neural Network (CNN), subsequently employs the PSPModule to perform multi-scale pooling, convolution, and upsampling on the feature maps. The processed feature maps are then concatenated with the original feature maps. Finally, a prediction mask is generated through convolution and softmax operations. This model is recognized for its satisfactory accuracy in scene semantic segmentation while maintaining reasonable hardware requirements. This step will enable the extraction and recognition of visual elements from the comprehensive image dataset of Jiangnan garden scenes that we have created.

2.2.2 Classification and Statistics of Visual Elements

The dataset labels adopted are from the ADE20K dataset, a large-scale dataset trained and released by MIT comprising 20,210 images. This dataset categorizes image content into 150 classes, covering three visual concepts: discrete objects (items with well-defined shapes, such as cars and people), stuff (amorphous background areas like grass and sky), and object parts (components of existing object instances with functional significance, such as heads or legs). These classifications align well with individuals' visual experiences in Jiangnan gardens, thereby establishing a connection with restorative perception. Semantic segmentation is conducted on the comprehensive image dataset using PyCharm to obtain quantitative data for each element within the images, as illustrated in Fig. 2.

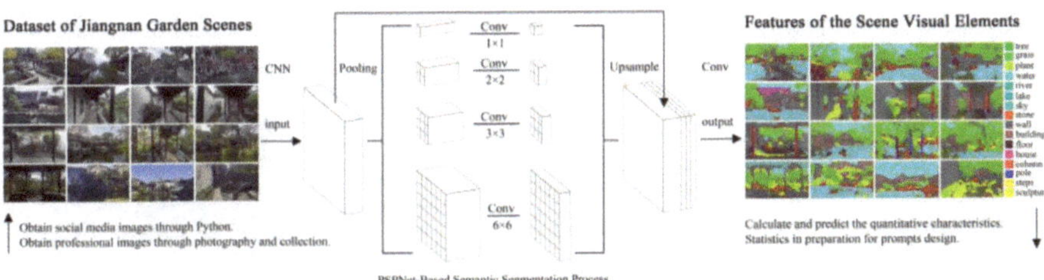

Fig. 2. Analyzing the visual elements of Jiangnan garden scenes by Computer Vision

2.3 Model Training Using Generative Artificial Intelligence

2.3.1 Developing Prompts Using LLMs

LLMs are deep learning models pre-trained on extensive datasets of text, capable of mimicking human language cognition and generation processes to a certain extent. The combined application of the advantages of LLM-generated prompts with text-to-image large models has garnered attention from researchers (Ma et al. 2024). Released at the beginning of 2025, Deepseek-R1 currently leads in performance among LLMs for knowledge-based tasks and long-text processing.

In this study, based on a quantitative analysis of visual elements in Jiangnan garden scenes and a collection of descriptive terms compiled from social media text data, we utilized Deepseek-R1 to generate standardized prompts and data labeling standards for these scenes. The aim was to accurately and effectively convey information about the visual elements in the images and align with public perception preferences. These standardized prompts will be used for LoRA training and to test the effectiveness of AIGC.

2.3.2 Training LoRA Based on SDXL

Stable Diffusion XL (SDXL), released in 2023, comprises a Base model equipped with fundamental capabilities such as txt2img, img2img, and inpainting, as well as a Refiner model that enhances the latent features of images generated by the Base model, essentially optimizing image-to-image generation (Podell et al. 2023). SDXL has demonstrated impressive generation results in the field of architecture and spatial design and has established a comprehensive application ecosystem (it is one of the versions with the most compatible models and plugins available on platforms like Civital).

LoRA represents a efficient and flexible model fine-tuning technique suitable for various situations. In the application of SD, LoRA serves as a plugin that allows users to control the generated image results by adjusting its weights while preserving the performance of the original SD model. The advantages of LoRA include its small file size, low computational requirements, and significant impact. It enables the training of models with specific artistic styles or characteristics using a limited amount of data (Liu et al. 2024). The process of training LoRA for designing restorative garden scenes based on Deepseek-R1 and SDXL is illustrated in Fig. 3.

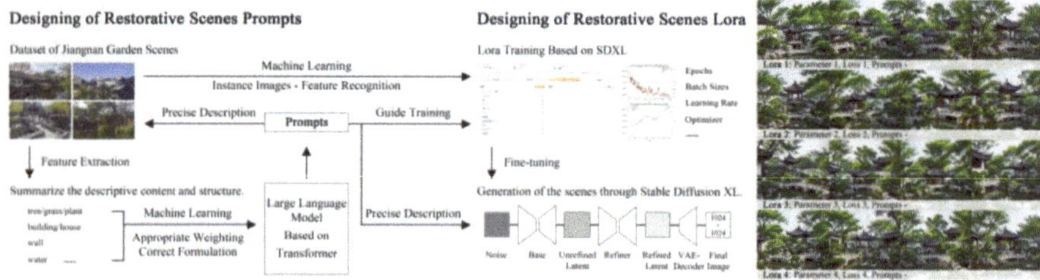

Fig. 3. Designing restorative scenes Lora by Generative Artificial Intelligence

2.4 Testing and Comparing the Effectiveness of Generated Content

2.4.1 Quality Analysis of Generated Images

In this study, the validation of the training model's generation effect encompasses two pivotal dimensions: objective and subjective. Firstly, the objective aspect focuses on the loss rate during the training phase, which acts as a proxy for the adaptation of the machine learning process to the initial dataset. Specifically, this metric quantifies the discrepancy between the model's predicted outputs and the actual target values, thereby providing a direct indication of the model's accuracy and efficiency in learning image features. Secondly, the subjective dimension revolves around image quality, reflecting the intuitive outcomes of the generated images. This evaluation involves a comprehensive assessment of visual quality, detail richness, color accuracy, and overall naturalness, ultimately determining the applicability of the generated images for real-world scenes. These dual considerations collectively contribute to a comprehensive understanding of the model's performance and its suitability for practical applications.

2.4.2 Questionnaire Survey on Restorative Perceptions

Due to the specific age range and preferences of the online user group, there are certain limitations in the social media data, so this study supplements the questionnaire with more targeted and content-specific questions to obtain a more accurate evaluation of perceptions. Specifically, the evaluation indicators of the PRS (Kaplan 1995, a survey questionnaire designed to collect participants' subjective experiences through a series of questions to quantify the environmental recovery potential) were referenced, and the high-frequency content from the previously mentioned social text data was utilized to construct the scale. A shorter and more focused perception scale was developed, as shown in Table 2. After generating images of Jiangnan garden scenes using the self-developed LoRA, 30 students majoring in architecture and related fields, along with 30 tourists visited Jiangnan gardens from various regions of China, were invited to complete the questionnaire on-site to evaluate the effectiveness of the AIGC in this study.

Table 2. Items in the perception scale

1. How would you evaluate the visual quality of the image?			
V1	Unreal	(very much) 1__2__3__4__5__ (very much)	Naturally
V2	Broken	(very much) 1__2__3__4__5__ (very much)	Aesthetic
2. How would you describe the scene of the image?			
V1	Messy	(very much) 1__2__3__4__5__ (very much)	Attractive
V2	Polluted	(very much) 1__2__3__4__5__ (very much)	Healthy
3. What are your physical reactions to the scene of the image?			
V1	Breathing Acceleration	(very much) 1__2__3__4__5__ (not at all)	
V2	Heart Rate Acceleration	(very much) 1__2__3__4__5__ (not at all)	
4. What do you prefer to do in the scenes of the pictures?			
V1		(not at all) 1__2__3__4__5__ (very much)	Stay here longer.
V2		(not at all) 1__2__3__4__5__ (very much)	Visit here more often.
5. What emotions do you feel in the scene of the picture?			
V1	Anxious	(very much) 1__2__3__4__5__ (very much)	Relaxed
V2	Bored	(very much) 1__2__3__4__5__ (very much)	Lively

3 Results

3.1 Obtaining Key Information About the Restorative Scenes

3.1.1 Visual Element Statistics

Due to the diverse spatial characteristics of the scene image dataset from Jiangnan gardens, there are notable variations in the layout and design of both buildings and landscapes. Consequently, all images are classified into three categories: forest landscapes, waterfront landscapes, and architectural landscapes, which are used to distinguish the extent to which different image contents influence restorative perception. The proportion of visual elements in each category is obtained through semantic segmentation and then weighted and normalized based on the frequency of each scene. The quantification data of visual elements in Jiangnan garden scenes, which have the potential to provide restorative perceptions, is calculated as shown in Eqs. 1 and 2, as well as in Table 3.

$$P_i = \frac{S_i}{S_{all}} \tag{1}$$

$$P_{w(i)} = \frac{P_i - P_{min}}{P_{max} - P_{min}} \tag{2}$$

Table 3. Quantitative statistics of visual elements in Jiangnan garden scenes

	Main objects	Weighted proportions		Main objects	Weighted proportions
1	tree/grass/plant	0.999	8	mountain	0.027
2	building/house	0.421	9	fence/railing	0.015
3	wall	0.349	10	road/sidewalk	0.011
4	water	0.3451	11	door	0.001
5	rock	0.152	12	signboard	0.001
6	floor/ground	0.069	13	window	0.001
7	sky	0.035	... ...		

3.1.2 Design of Descriptive Prompts

By integrating quantitative data from highly popular textual content on social media and visual elements in specific garden scenes, we use Deepseek-R1 for feature learning. The focus is on distinguishing the significance of visual elements, allocating appropriate weights to each, and aligning descriptive methods with public preferences. This approach is applied in the following steps: (1) After annotating a comprehensive image dataset using the Deepbooru algorithm, we utilize Deepseek-R1 to optimize the tag labels, thereby enhancing the model's accuracy in learning from images. (2) We design standardized prompts to optimize training directions and facilitate subsequent standardized model testing. The resulting standardized prompts form a set of formulas with practical value, as illustrated in Table 4.

Table 4. Standardized prompt words with unified structure and content

	Overall description	Visual elements	Image quality
Prompt	Restorative scene, Jiangnan garden, landscape design, ...	vibrant greenery, traditional buildings, reflective water, ...	realistic, best quality, masterpiece, ...
Example	Restorative Jiangnan garden scene, landscape design, (lush plants:1.2), (vibrant greenery:1.2), (traditional buildings), (historical architecture), moss-covered walls, reflective water, streams, natural textured rocks, subtle paths, soft sky, diffused daylight, distant mountains, bamboo fences, a circle door, signboards, windows, people, natural style, peaceful, harmonious, restorative atmosphere, (realistic:1.8), best quality, masterpiece.		

3.2 Evaluating Generated Effects of the LoRA

3.2.1 Analysis of the Training Results

To train the LoRA model, we utilized the lora-scripts developed by Akegarasu (https://github.com/Akegarasu/lora-scripts). This training process employed a selection of 100 high-quality images from the comprehensive image dataset, ensuring no repetition of content and a close alignment with the visual features relevant to the restorative scenes mentioned in this study. Multiple rounds of training were conducted with varying parameters, and Fig. 4 presents the loss rate visualizations obtained after the completion of each training round. Theoretically, a lower loss rate indicates a better learning outcome. Based on this criterion, an initial screening of the LoRA was conducted, leading to the determination of setting the *save precision* to bf16, *max_train_epochs* to 10, *train_batch_sizes* to 4, *learning rate* to 2e-4, *unet_lr* to 2e-4, *text_encoder_lr* to 5e-5, *lr_scheduler* to cosine, *optimizer_type* to prodigy, *network_dim* to 32, *network_alpha* to 16.

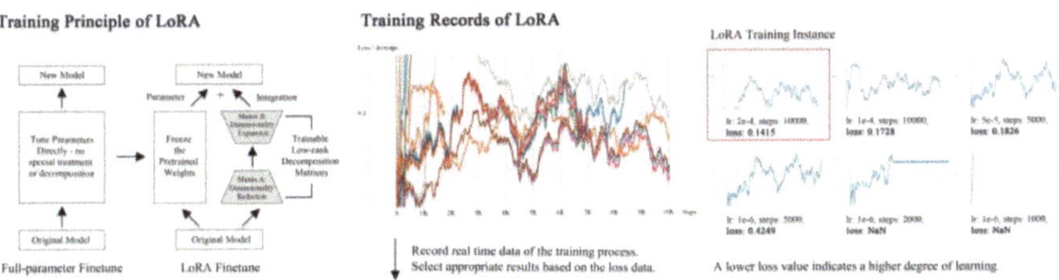

Fig. 4. Parameters during model training process

The study controlled for various variables. After completing the self-developed LoRA, we exclusively used the base model of SDXL, standardized prompts, and consistent parameters to conduct multiple rounds of image generation tests. Based on evaluation criteria such as the completeness of generated image details, logical coherence of content, and the absence of overfitting, we selected the LoRA that produced the highest quality of image generation.

Furthermore, we compared the generation performance of the self-developed LoRA with that of mainstream, highly popular LoRA of the same type. Under identical parameter settings, we generated images in various situations. The results, as shown in Fig. 5, demonstrate that the self-developed LoRA exhibited comparable performance to the benchmarked LoRA models. Notably, when using prompts related to "restorative" themes, the self-developed LoRA outperformed its counterparts, suggesting a superior capability in generating scenes that evoke restorative perceptions.

Fig. 5. Comparison test of AIGC effects

3.2.2 Analysis of Restoration Perception of Generated Images

Finally, an individual evaluation through perceptual scale is conducted on the generative performance of the self-developed LoRA. As shown in Table 5, the results obtained from 60 questionnaires in this study indicate that the proposed method for transindividual intelligent extraction of scene visual features and the self-developed AI scene generation model demonstrate effectiveness in reproducing Jiangnan garden scenes. The AIGC provide restorative perceptions and fulfill people's emotional needs. This finding contributes to a deeper understanding of the restorative perceptions evoked by the visual elements of Jiangnan gardens, offering promising insights for advancing design in this field.

Table 5. Evaluation of the restorative perceptions provided by AIGC

	1-V1	1-V2	2-V1	2-V2	3-V1	3-V2	4-V1	4-V2	5-V1	5-V2
Average	3.9	3.8	4.1	3.6	4.1	3.9	4.4	4.4	4.5	4.0
Max	5	5	5	4	5	5	5	5	5	5
Q3	4	4	5	4	5	4	5	5	5	5
Q2	4	4	4	4	4	4	5	4.5	5	4
Q1	4	3	4	3	4	3	4	4	4	4
Min	2	1	3	3	3	2	3	3	3	1

4 Conclusion and Discussion

This study proposes a method for understanding and reproducing the restorative perception of Jiangnan gardens based on CV and GAI, capable of providing transindividual healing experiences at a relatively low computational cost and short design cycles. The research findings indicate: 1. The application of PSPNet and Deepseek-R1 can effectively extract the visual element features that provide restorative perception in Jiangnan garden scenes in a natural language manner. 2. The application of SDXL and Deepseek-R1 facilitates the precise training of LoRA for the intelligent generation of restorative scenes. This process has been demonstrated to be rational (based on training process data) and feasible (based on participant evaluation data). 3. This study also allows architects to utilize AI tools more conveniently. Although AI is often considered a "black box" (Ghimire et al. 2024), this method provides an interpretable application approach for AI-assisted design.

However, this study does have certain limitations. Firstly, in future research, further subdividing and examining different types of Jiangnan garden scenes (such as courtyard scenes, indoor scenes, forest scenes, rockery scenes, among others) individually will enhance our understanding and reproduction accuracy of restorative perceptions of these scenes. Additionally, in practical operations, integrating the LoRA with SD plugins such as ControlNet, along with parametric design software like Grasshopper, will generate more controllable effects, aligning the design of restorative scenes more closely with architects' intentions and facilitating implementation. This will promote the application and dissemination of this technology in practice. Lastly, current laws provide clear and increasingly comprehensive regulations regarding data privacy and copyright issues. It is imperative to consider how to conduct the information acquisition process necessary for the use of intelligent technologies responsibly, while also ensuring the open-source sharing of AI-generated content and preventing unequal monopolies. This aspect is often overlooked in the current research.

This study contributes to the vertical application of advanced technology in the field of architecture by addressing the persistent challenges of prolonged project cycles and low levels of intelligence, which are significant constraints impeding the development of the current construction industry. Furthermore, this public data-driven design methodology bridges a crucial gap between design and practice—specifically, the perception that architectural proposals are often disconnected from genuine public needs and instead cater solely to architects' subjective whims. As a result, it significantly enhances architects' capacity to address societal problems. Additionally, this study also foreshadows the future direction of AI agent intervention in architectural design. Existing design frameworks based on CV and GAI have already achieved the perception of environmental information and the formulation of design decisions. In the future, the further integration of robotic fabrication will make the complete intelligence of the design process a reality. It is also important to note that due to the unimaginable speed of updates and iterations in AI, the method proposed in this study can only ensure a limited degree of timeliness and innovation. Continuous monitoring of technological advancements is critical to effectively implement transindividual intelligence in practice.

Acknowledgement. This research is supported by the Humanities and Social Science Foundation of Ministry of Education, China (Grant No.24YJAZH071).

References

Helliwell, J.F., Layard, R., Sachs, J.D., De Neve, J.-E., Aknin, L.B., Wang, S. (eds.): World Happiness Report 2024. University of Oxford: Wellbeing Research Centre, Oxford (2024)

Wu, Y., Liu, Q., Hang, T., Yang, Y., Wang, Y., Cao, L.: Integrating restorative perception into urban street planning: A framework using street view images, deep learning, and space syntax. Cities. **147**, 104791 (2024). https://doi.org/10.1016/j.cities.2024.104791

Zhou, W.: Cultural background for the development of Chinese classical landscape architecture. Chin. Landsc. Archit. **2004**(09), 62–65 (2004)

Zhang, Z., Jiang, M., Zhao, J.: The restorative effects of unique green space design: Comparing the restorative quality of classical Chinese gardens and modern urban parks. Forests. **15**(9), 1611 (2024). https://doi.org/10.3390/f15091611

Chen, X., Yu, H., Xiong, R., Ye, Y.: Construction of an analytical framework for spatial indicator of Chinese classical gardens based on space syntax and machine learning. Landsc. Archit. **2024**(03), 123–131 (2024)

Ito, K., Kang, Y., Zhang, Y., Zhang, F., Biljecki, F.: Understanding urban perception with visual data: A systematic review. Cities. **152**, 105169 (2024). https://doi.org/10.1016/j.cities.2024.105169

Zhang, X., Pan, W.: A review of architectural design methods in the age of artificial intelligence. New Archit. **2024**(5), 66–72 (2024). https://doi.org/10.12069/j.na.20231022

Ghimire, P., Kim, K., Acharya, M.: Opportunities and challenges of generative AI in construction industry: Focusing on adoption of text-based models. Buildings. **14**(1), 220 (2024). https://doi.org/10.3390/buildings14010220

Lyu, Z., Li, Z., Wu, Z.: Research on image-to-image generation and optimization methods based on diffusion model compared with traditional methods: Taking façade as the optimization object. In: Yan, C., Chai, H., Sun, T., Yuan, P.F. (eds.) Phygital Intelligence. CDRF 2023. Computational Design and Robotic Fabrication, pp. 35–50. Springer, Singapore (2024). https://doi.org/10.1007/978-981-99-8405-3_4

Ma, H., Zheng, H.: Text semantics to image generation: A method of building facades design base on stable diffusion model. In: Yan, C., Chai, H., Sun, T., Yuan, P.F. (eds.) Phygital Intelligence. CDRF 2023. Computational Design and Robotic Fabrication. Springer, Singapore (2024). https://doi.org/10.1007/978-981-99-8405-3_3

Yuan, P.F., Xu, X., Wang, Y.: Toward an AI-augmented generative design era. Archit. J. **2023**(10), 14–20 (2023). https://doi.org/10.19819/j.cnki.ISSN0529-1399.202310003

Ma, B., Zong, Z., Song, G., Li, H., Liu, Y.: Exploring the role of large language models in prompt encoding for diffusion models. arXiv. (2024). https://doi.org/10.48550/arXiv.2406.11831

Podell, D., et al.: SDXL: Improving latent diffusion models for high-resolution image synthesis. arXiv. (2023). https://doi.org/10.48550/arXiv.2307.01952

Liu, C., Takikawa, T., Jacobson, A.: A LoRA is worth a thousand pictures. arXiv. (2024).https://doi.org/10.48550/arXiv.2412.12048

Kaplan, S.: The restorative benefits of nature: Towards an integrative framework. J. Environ. Psychol. **15**, 169–182 (1995). https://doi.org/10.1016/0272-4944(95)90001-2

cc
BY NC ND

Architectural Style Recognition Using Multisource Data and Deep Learning: A Case of Historical Districts in Shanghai

Ma Shengxin[1], Huang Heng[1], Zhang Xiyuan[1], and Luo Linxi[2(✉)]

[1] School of Architecture and Urban Planning, Tongji University, Shanghai, China
[2] School of Architecture, Southeast University, Nanjing, China
220220116@seu.edu.cn

Abstract. Architectural style plays a crucial role in urban heritage preservation and urban landscape management. However, the diverse architectural styles shaped by the fusion of Eastern and Western cultures in Shanghai pose challenges for style classification. This study proposes an identification framework that integrates multi-source data and CNNs. First, 11 primary architectural styles in Shanghai were identified based on a review of classical literature. A cross-cultural architectural style dataset comprising 4,584 expert-verified images was constructed using web and social media data, covering 11 categories, such as Lilong-style Houses and Art Deco. A comparative analysis of six CNN architectures, including VGG16, ResNet50, and ConvNeXt-Tiny, revealed that ConvNeXt-Tiny achieved the highest accuracy (84%), outperforming the baseline model by 5.13%. An architectural style map was generated using ConvNeXt-Tiny and 10,313 street view images(SVIs) collected from 12 Historical Districts. This study effectively addresses cross-cultural architectural style recognition and provides insights applicable to other cities.

Keywords: Architectural style · Deep learning · Image recognition · Historical districts · Transfer learning

1 Introduction

As a tangible representation of cityscape, architectural style is essential for heritage preservation, district delineation, and cultural continuity. However, as a city where Eastern and Western cultures blend, the diverse buildings in Shanghai pose challenges for architectural style recognition.

Advances in big data and artificial intelligence have expanded the possibilities for architectural style analysis. First, new data sources, including urban vector data, Points of Interest (POI), and SIVs, provide a robust foundation for large-scale urban research. Second, emerging deep learning techniques, particularly CNN architectures such as DeepLabV3, ResNet, and VGG, enable the precise extraction and classification of urban elements. The development of these field has laid a solid foundation for this study.

Y. Liu et al. (Eds.): CDRF 2025, *Transindividual Intelligence*, pp. 388–399, 2026.
https://doi.org/10.1007/978-981-92-0615-5_33

1.1 Related Work and Research Objectives

The research on the architectural styles of Shanghai has been thoroughly documented. Zheng Shiling, Luo Xiaowei and Wu Jiang had sorted out the architectural styles of Shanghai from a diachronic perspective (Luo Xiaowei 1990; Wu Jiang 1997; Zheng Shiling 1999). Additionally, some studies have focused on the architectural style research of specific stylistic types (Xu Yihong 2023) and specific regions (Chai Xingkang et al. 2022; Zhang Ying 2009). Following the chronological evolution of styles, Shanghai's architecture can be classified into four major categories: (1) Chinese Traditional Architecture since Shanghai's establishment as a county, (2) Classicism in the foreign concessions since the 1840s, (3) Eclectic architecture blending Western and Chinese influences since the 1910s, and (4) Modernism from the 1940s onward. These classifications highlight the cross-cultural nature of Shanghai's architectural heritage and provide a foundation for dataset construction and annotation.

Existing architectural style recognition datasets are insufficient for cross-cultural style classification, and the performance of different recognition algorithms remains unclear. Current datasets include mainstream Western architectural style datasets (Z. Xu et al. 2014), British architectural dataset (Lindenthal and Johnson 2021), datasets on Chinese official and vernacular architecture (Zhang et al. 2024), and modern architecture datasets for individual cities (Shan and Zhang 2022). Although few datasets cover architectural styles from both the East and the West (H. Xu et al. 2023), these datasets are not annotated by professional architects. In terms of recognition algorithms, CNN-based models such as VGG (Folken 2023), DenseNet (Zhang et al. 2024), ResNet, and Fast-R-CNN (H. Xu et al. 2023) have been extensively used. Some studies have employed the method of model comparison, significantly improving recognition performance (Folken 2023).

Historical districts play a crucial role in preserving cityscape. While deep learning has been applied to analyzing building color (Zhou et al. 2023), materials, street styles (Jiang Haobo et al. 2022), and street perception (Wu et al. 2023), research on architectural style recognition remains limited. Nonetheless, related studies offer valuable insights, such as the method of image preprocessing (Zhou et al. 2023) and mapping techniques (H. Xu et al. 2023). This study extends style recognition to historical districts and incorporates data preprocessing methods from prior studies to enhance model robustness and generalization.

Therefore, the objectives of this study are:

(1) Construct a cross-cultural architectural style dataset that reflects Shanghai's situation from the professional perspective of architects.
(2) Evaluate CNN-based models to determine the most effective architectural style recognition algorithm.
(3) Generate a architecture style map of historical districts.

2 Materials and Methods

This study is conducted in three steps: (Fig. 1) (1) Data Collection and Processing: Define architectural style types, collect architectural images, and construct the dataset.(2) Model Performance Comparison and Evaluation: Select six typical CNN models, compare their recognition performance, and identify the best model for this task.(3) SVIs Analysis and Map Generation: Collect SVIs, recognize the styles, and ultimately generate an architectural style map.

This study focuses on 12 historical districts in central Shanghai, as designated in 2003. These districts encompass nearly all architectural typologies in the city while exhibiting stylistic variations (Chen Fei and Ruan Yisan 2008). Such diverse styles enables a comprehensive evaluation of the model's generalization ability and provides an intuitive assessment of its recognition performance.

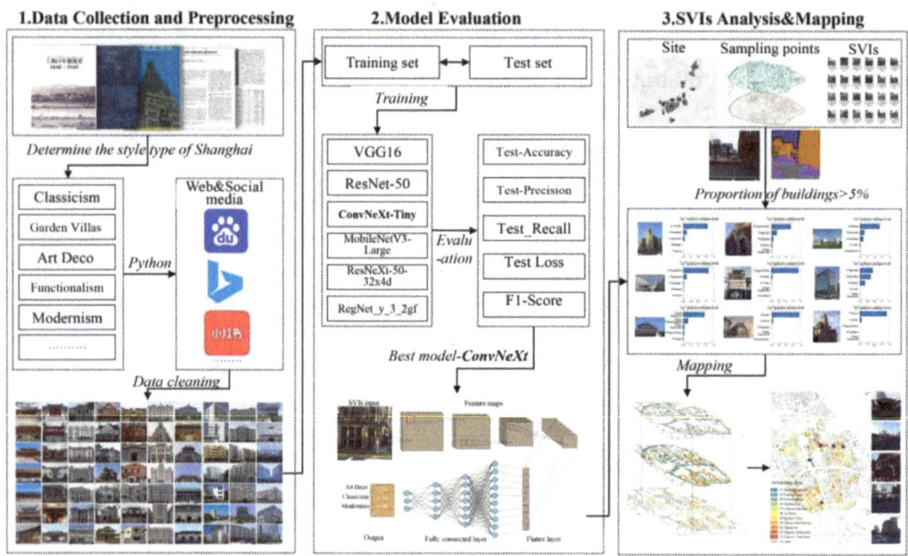

Fig. 1. Technical route

2.1 Data Collection and Preprocessing

Based on a series of studies on Shanghai's architecture (Chai Xingkang et al. 2022; Luo Xiaowei 1990; Wu Jiang 1997; Xu Yihong 2023; Zhang Ying 2009; Zheng Shiling 1999), 11 major architectural styles have been identified. These styles encompass nearly all buildings while also reflecting the city's architectural evolution since its opening in 1842 (Table 1).

Python scripts were developed to crawl image data from websites and social media using the 11 labels. Additional tags, such as "exterior," "façade,"and "perspective" were incorporated into the crawling labels (e.g., "Classical Architecture Exterior" and "Classical Architecture Façade") to capture images from various angles.

Data cleaning was performed based on the following criteria to filter out non-conforming images: (1) Images where the building occupies a small portion of the frame. (2) Aerial views, interior images, or others unsuitable for street view recognition. (3)

Table 1. Classification of architectural styles in Shanghai

Style tpye	Characteristic	Example
01-Chinese Traditional Architecture	Traditional Chinese public buildings with timber structures and sloping roofs, including pagodas, and pavilions	
02-Chinese Vernacular house	Traditional Chinese courtyard-style residential architecture, such as Suzhou-style,Huizhou-style, Siheyuan , and others	
03-Classicism	Emphasizing proportions and the application of column orders, arches, domes, and pediments, with the use of stone materials	
04-Lilong-style Houses	The integration of traditional Chinese courtyard-style residential architecture with Western-style housing	
05-Garden Villas	Western-style villa residences, verandas, pointed roofs	
06-Art Deco	Emphasizing geometric shapes and simple lines, focus on proportion, symmetry and balance, and decorative components	
07-Chinese Classical Revival	The mixed use of Chinese architectural elements and Western elements, stone materials, the three-section style, an emphasis on proportion	
08-Modernism	Simple building volume, emphasizing the abstractness of the building's geometric form, removing decorations, and focusing on the unity of form and function.	
09- Functionalism	The most common architectural styles include modern ordinary office buildings, commercial buildings and residential buildings, which generally adopt the window-wall system.	
10-Expressionism	Emphasize the free creation of architectural forms, volume combinations, and be adept at using combinations of materials such as steel, bricks, and glass	
11-Deconstructivism	Create distorted and displaced spaces through the combination of fragmented architectural forms, non-linear architectural forms, and building skins	

Images not depicting buildings. (4) Images mismatched with the specified architectural style label. To ensure accurate classification of images, 9 trained architects and PhD students participated in the data cleaning and annotation. For labels with limited data, additional image crawling was performed, and supplementary datasets, such as Arch28 and arcDataset (Z. Xu et al. 2014), were used.

2.2 Model Training and Evaluation

The study first selected VGG16, a classic image recognition model, as the baseline for evaluation. VGG16 is widely used in computer vision tasks due to its stability and compatibility. Additionally, ResNet, ConvNeXt, MobileNetV3, ResNeXt, and RegNet were chosen for comparison. VGG and ResNet represent traditional CNN architectures, while ResNeXt is a variant of ResNet. MobileNetV3 is a lightweight model, ConvNeXt and RegNet are currently the most advanced models. To ensure a fair comparison, models with similar parameter counts were selected (Table 2).

Table 2. Select models with similar parameter counts for comparative experiments

Model name	Number of parameters
VGG16	138M
ResNet-50	25.6M
ConvNeXt-Tiny	28M
MobileNetV3-Large	5.4M
ResNeXt-50-32x4d	25M
RegNet_y_3_2gf	21M

Due to the small scale of the dataset, this study employed transfer learning to train the model. The batch size was set to 32, with Adam as the optimizer, a learning rate of 0.001, and StepLR as the learning rate scheduler. The learning rate was halved every 5 epochs (step_size = 5) with a gamma of 0.5. The model was trained for 60 epochs. Test loss, precision, accuracy, recall, and F1-score were used as evaluation metrics. The model with the best overall performance on the test set was selected as the style recognition model.

2.3 Architectural Style Recognition in Historical Districts

SVIs was crawled using the Baidu Map API, primarily focusing on street views perpendicular to the streets for building recognition. A sampling point was set every 40 m, resulting in a total of 13,502 SVIs. DeepLab_V3 was then used to clean the data, filtering out street views with less than 5% of the image occupied by buildings. In the end, 10,313 valid street views were obtained.

The label with the highest confidence was selected as the architectural style data for each sampling point. The architectural style map was then generated in GIS. First, a circle with a radius of 20 m was drawn around each street view point to represent the architectural style of that area. The data was then mapped to individual buildings, resulting in the presentation of the architectural style map.

3 Results

3.1 Dataset Construction

A dataset containing 4,584 images of 11 architectural styles was constructed. The dataset was split into a training set and a test set at a ratio of 2:8(Fig. 2a). To enhance the model's generalization ability and robustness, the dataset underwent preprocessing including random cropping and horizontal flipping. The VGG16 model was trained using this dataset, and the test-recall, test-accuracy, and F1-score were all around 0.78 (Table 3), indicating that the dataset quality is high and can support the model in making accurate predictions. The model's train-accuracy was 0.7892, while the test-accuracy was 0.7398 (Table 3), showing that the dataset has comprehensive coverage, reasonable proportions, and quantity, which supports training a model with high generalization capability. For individual style types, Six styles achieved accuracy rates exceeding 80%, while the recognition accuracy for closely similar styles, such as Chinese Classical Revival and Chinese Traditional Architecture, as well as Classical Architecture and Art Deco, also exceeded 70% (Fig. 2b). This indicates that the annotation quality of the dataset is high, allowing the model to effectively recognize the features of different styles.

Table 3. The performance of VGG16 baseline

Test-accuracy	Test-loss	Test-recall	F1-score	Train-accuracy
0.7892	0.6156	0.7891	0.7869	0.7398

3.2 Comparison and Evaluation of Deep Learning Models

ConvNeXt-Tiny is the best performing model. ConvNeXt draws inspiration from the design principles of Transformers and introduces large convolution kernels to better capture global contextual information (Liu et al. 2022). It's architecture is stable, and its classification accuracy even surpasses that of the ViT model. Experiments show that ConvNeXt is highly suitable for architectural style classification tasks, with all of its parameters significantly outperforming other models (Table 4).

Compared to VGG16, accuracy improved by 5.13% (0.84), F1-score increased by 5.12% (0.842), loss decreased by 25.64% (0.438), precision rose by 4.07% (0.844), and recall improved by 5.51% (0.842). This indicates that ConvNeXt not only provides more accurate predictions but also has a lower misclassification rate, lower omission rate, and better generalization ability (Table 4). In addition to its superior classification capabilities, ConvNeXt also has a smaller parameter size. ConvNeXt-Tiny (28M parameters)

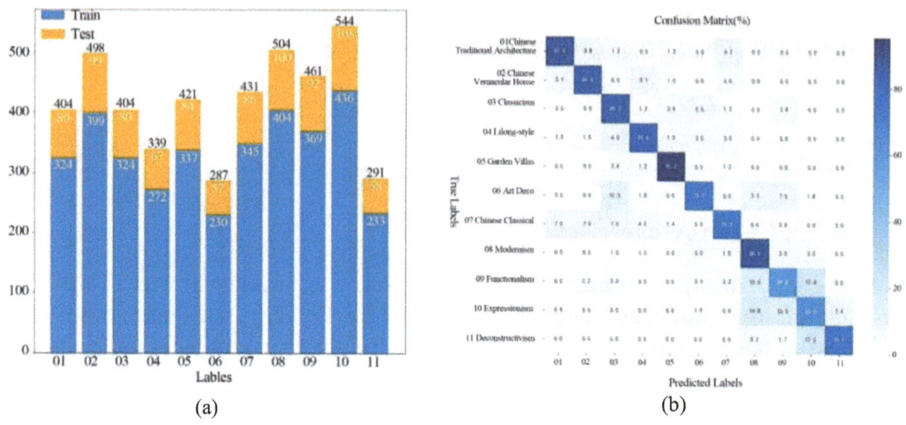

(a) (b)

Fig. 2. The performance of dataset

has only 20.2% of the parameters of VGG16 (138M parameters), yet its accuracy is 5.13% higher than that of VGG16.ConvNeXt's performance on recognizing each architectural style. ConvNeXt performs exceptionally well in recognizing each architectural style. Among the 11 architectural styles, 7 of them achieved an accuracy above 85%. For styles with lower recognition accuracy in VGG16, such as Chinese Classical Revival and Functionalist architecture, ConvNeXt improved accuracy by 15.4% and 13.0%, respectively (Fig. 3a).

The model trained with ConvNeXt also performs well in distinguishing between similar architectural styles, such as Chinese Traditional Architecture vs. Chinese Classical Revival and Classical Architecture vs. Art Deco (Fig. 3b). Therefore, ConvNeXt-Tiny was ultimately chosen as the model for the architectural style classification task in historical districts.

Table 4. Data comparison of several CNN models

Step	RegNet	ResNeXt	MobileNet	ConvNeXt	ResNet	VGG
test_accuracy	0.813	0.789	0.806	**0.840**	0.799	0.779
f1-score	0.817	0.791	0.809	**0.842**	0.801	0.781
test_loss	0.537	0.595	0.585	**0.438**	0.589	0.627
test_precision	0.821	0.797	0.812	**0.844**	0.811	0.784
test_recall	0.819	0.792	0.809	**0.842**	0.798	0.782

3.3 Architectural Style Recognition and Mapping

Figure 4 shows the style types and distribution of architectural styles across the 12 historical districts. The Old City, Shanyin Road, Hengfu, and Tilanqiao areas mainly feature

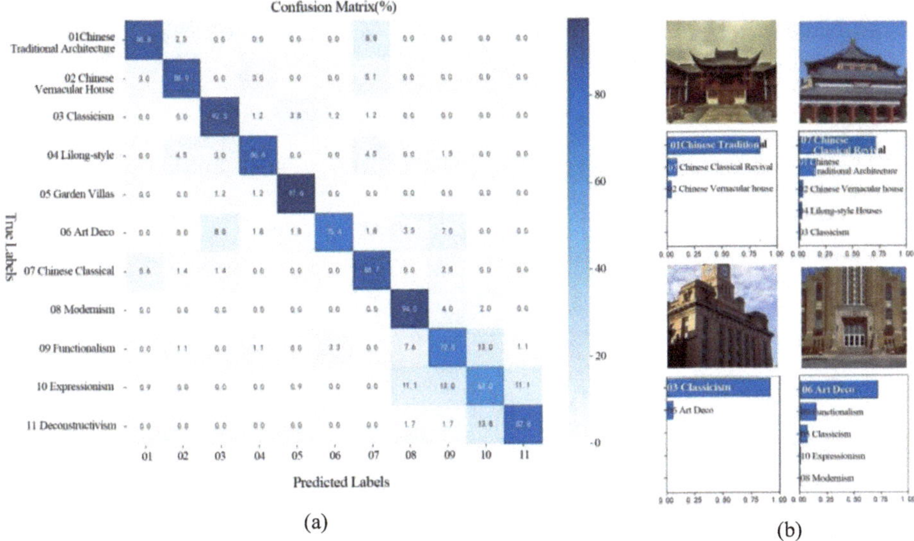

Fig. 3. The performance of ConvNeXt

linong-style house, while the Bund is dominated by classical architecture, Art Deco style, and linong-style house. The buildings in Jiangwan and Hongqiao are primarily modernism, functionalism, and Expressionism, which aligns with the general public's perception of architectural styles.

The Bund and Old City areas have been selected as typical regions for the study (Fig. 5). The Bund is renowned as the "Exhibition of World Architecture" and, since Shanghai opened its port in 1842, it has become a center for classical architecture and Art Deco. Classical architecture such as the No. 1 China Ever bright Bank and No. 5 Asia Building, as well as Art Deco structures like the No. 3 Bank of China Building and No. 4 Shanghai Federation of Trade Unions, were all correctly identified. On the other hand, the Old City Area, located in contrast to the Bund, preserves much of Shanghai's traditional architecture (Fig. 5a). From the style map of the Old City, it can be seen that the architecture is primarily characterized by Chinese traditional architecture and linong-style houses. Additionally, with the development of the Old City, a certain number of modern buildings have been constructed along Henan Road and Fuxing Road. Buildings such as No. 1 Xuanhutai, No. 3 Yu Garden Department Store representing Chinese traditional architecture, No. 2 typical linong-style houses, No. 4 Hong Kong Mindu functionalist architecture, and the Art Deco-style Huangpu District Market Supervision Administration were all correctly recognized (Fig. 5b).

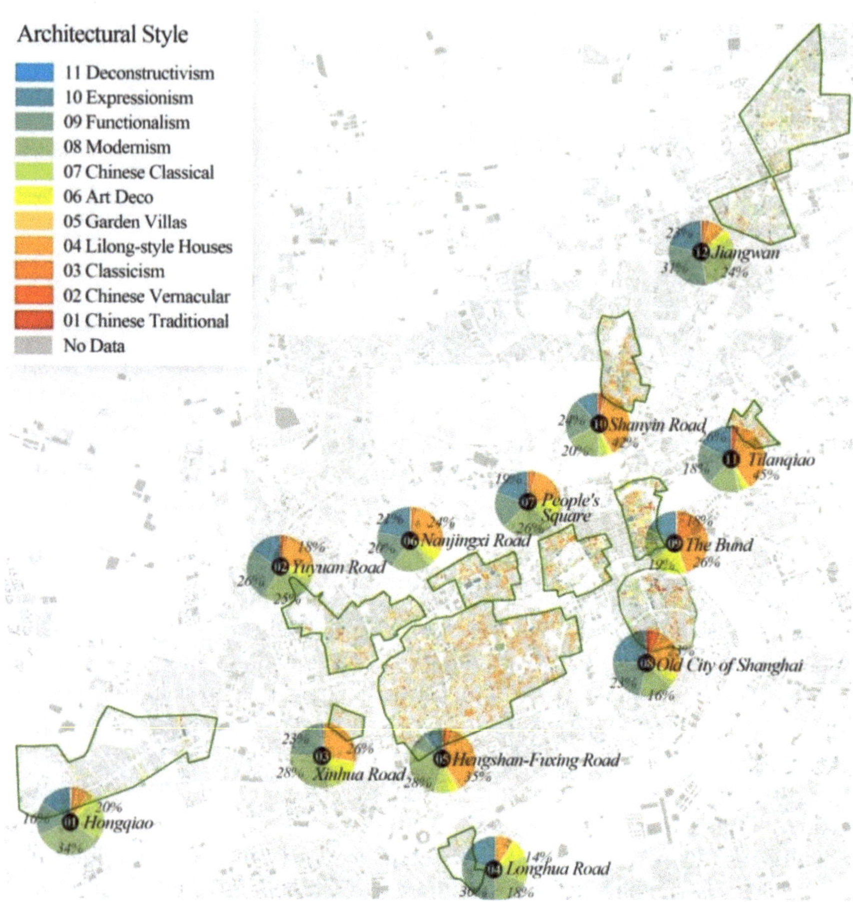

Fig. 4. The style map of the 12 historical districts

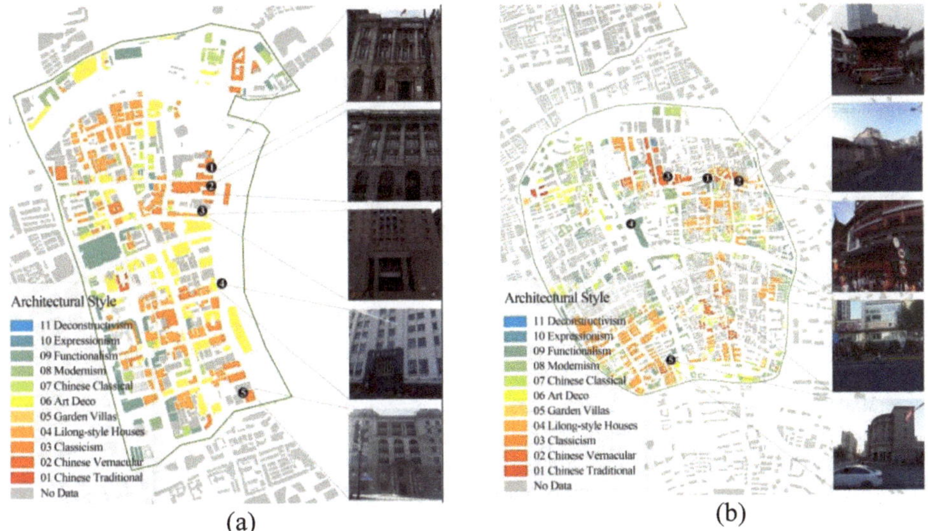

Fig. 5. Architectural style maps of the Bund (a) and Old City (b).

4 Discussion and Conclusion

A cross-cultural architectural style dataset containing 11 architectural styles and 4,584 images was constructed. The study comprehensively compared the recognition performance of mainstream CNN architectures on the architectural style task. An application experiment was also conducted in the Historical Districts of Shanghai. The research indicate that: (1) The constructed dataset exhibits high annotation quality, achieving a 78% recognition accuracy with the VGG16 model.(2) ConNeXt-Tiny is the best-performing CNN model for style recognition, reaching an 84% recognition accuracy with a loss of 0.438 after training, which represents a 5.13% improvement in accuracy and a 25.64% reduction in loss compared to VGG16.(3) A total of 10,131 SVIs were collected to implement architectural style recognition in the Historical Districts of Shanghai, achieving high precision. Architectural style recognition can be further supported by research on building materials and colors to assist in delineating urban landscapes and detecting historical street blocks and current building conditions. This study can be adapted to cities such as Wuhan and Tianjin. However, for cities with stronger traditional architectural identities (e.g., Xi'an, Suzhou) or highly modern urban contexts (e.g., Shenzhen, Hong Kong), more granular dataset partitioning may be necessary.

However, the study also has some limitations. Firstly, the current street view data do not cover all streets, which restricts the model to recognizing only buildings with available street view data along the streets. Secondly, occlusions from trees in some SVIs affect the recognition accuracy. In future work, remote sensing data (Zhang et al. 2024) may be incorporated to assist in architectural style recognition.

References

Xingkang, C., Jiahao, Q., Chengpan, Y., Shan, J., Xiaoxiao, R.: Research on the facade style of the bund buildings in modem Shanghai. Urban. Archit. **19**(21), 119–122, 152 (2022). https://doi.org/10.19892/j.cnki.csjz.2022.21.25

Fei, C., Yisan, R.: A comparative study on conservation planning of historic and cultural areas in Shanghai and the response of conservation planning. Urban Plan. Forum. **2**, 104–110 (2008)

Folken, M.: Architectural style classification using convolutional neural networks and transfer learning. M.S. Thesis, Truman State University (2023). https://www.proquest.com/docview/2909473755/abstract/C61B6975967243F8PQ/1

Jiang Haobo, L., Shan, & Xiao Yang.: Typology study on streetscape through street view technology-evidence from Shanghai's historical district. Hous. Sci. **42**(3), 24–31 (2022). https://doi.org/10.13626/j.cnki.hs.2022.03.006

Lindenthal, T., Johnson, E.B.: Machine learning, architectural styles and property values. J. Real Estate Financ. Econ. (2021). https://doi.org/10.1007/s11146-021-09845-1

Liu, Z., Mao, H., Wu, C.-Y., Feichtenhofer, C., Darrell, T., Xie, S. A ConvNet for the 2020s. 2022 IEEE/CVF Conference on Computer Vision and Pattern Recognition (CVPR), 11966–11976, (2022). https://doi.org/10.1109/CVPR52688.2022.01167

Xiaowei, L.: Shanghai's architectural styles and Shanghai culture. Time Archit. **1**, 4–6 (1990)

Shan, L., Zhang, L.: Application of intelligent technology in facade style recognition of Harbin modern architecture. Sustainability. **14**(12), 7073 (2022). https://doi.org/10.3390/su14127073

Wu, C., Ye, Y., Gao, F., Ye, X.: Using street view images to examine the association between human perceptions of locale and urban vitality in Shenzhen, China. Sustain. Cities Soc. **88**, 104291 (2023). https://doi.org/10.1016/j.scs.2022.104291

Jiang, W.: A History of Shanghai Architecture: 1840–1949. Tongji University Press, Shanghai (1997)

Xu, H., Sun, H., Wang, L., Yu, X., Li, T.: Urban architectural style recognition and dataset construction method under deep learning of street view images: A case study of Wuhan. ISPRS Int. J. Geo Inf. **12**(7), 264 (2023). https://doi.org/10.3390/ijgi12070264

Yihong, X.: Shanghai urban culture and art deco architecture. Shanghai Art Rev. **6**, 17–20 (2023)

Xu, Z., Tao, D., Zhang, Y., Wu, J., Tsoi, A.C.: Architectural style classification using multinomial latent logistic regression. In: Fleet, D., Pajdla, T., Schiele, B., Tuytelaars, T. (eds.) Computer Vision-Eccv 2014, Pt I, vol. 8689, pp. 600–615. Springer International Publishing Ag, Cham (2014) https://webofscience.clarivate.cn/wos/alldb/full-record/WOS:000345524200039

Zhang, X., Li, S., Chen, C.: Classification of architectural styles in Chinese traditional settlements using remote sensing images and building facade pictures. J. Geogr. Sci. **34**(12), 2457–2476 (2024). https://doi.org/10.1007/s11442-024-2300-5

Zhang, Y.: On modern architecture style of Shanghai bund. Master's Thesis, Soochow University (2009). https://kns.cnki.net/KCMS/detail/detail.aspx?dbcode=CMFD&dbname=CMFD2011&filename=2010056558.nh

Shiling, Z.: The Evolution of Shanghai Architecture in Modern Times. Shanghai Educational Publishing House, Shanghai (1999)

Zhou, Z., Zhong, T., Liu, M., Ye, Y.: Evaluating building color harmoniousness in a historic district intelligently: An algorithm-driven approach using street-view images. Environ. Plan. B-Urban Anal. City Sci. **50**(7), 1838–1857 (2023). https://doi.org/10.1177/23998083221146539

cc
BY NC ND

Research on the Renovation of Industrial Architectural Remains in Rural Areas of Southern Anhui Province Based on Generative Artificial Intelligence and Human Perception Experiments

Hanwen Yu[1(✉)], Zao Li[2(✉)], Rui Zeng[1], and Shaotong Yan[1]

[1] Hefei University of Technology, Hefei, China
473033797@qq.com
[2] Anhui Jianzhu University, Hefei, China
277240440@qq.com

Abstract. Southern Anhui Province retains a significant number of rural industrial architectural remains, which face widespread abandonment and deterioration due to functional decline and insufficient preservation. Traditional design methods suffer from inefficiency and a lack of public perception validation, struggling to balance historical authenticity with modern functionality. This study innovatively integrates generative artificial intelligence and human perception experiments to propose an efficient renovation approach. By constructing a database of industrial architectural remains in Southern Anhui and combining the Stable Diffusion model with lightweight finetuning (LoRA), we generated renovation schemes that harmonize regional style and functional requirements. Experiments demonstrated that adjusting model parameters and textual prompts enables spatial renovation while preserving architectural heritage styles. Further validation through eye-tracking and Semantic Differential (SD) methods revealed a significant increase in heatmap focus concentration for updated images. SD scores indicated high evaluations for renovation effectiveness, though optimization of cultural characteristics remains necessary.

Keywords: Industrial architectural remains · Generative artificial intelligence · Eye-tracking · SD method

1 Introduction

Under the context of rural revitalization and heritage conservation in China, the renovation of modern industrial architectural remains (e.g., tea factories, granaries, and workshops) in rural Southern Anhui Province faces dual challenges: preserving historical authenticity while adapting to modern functional demands. This study proposes a "generative artificial intelligence (GAI) + human perception experiment" approach. A database of Southern Anhui industrial architectural forms is constructed, and diffusion

Y. Liu et al. (Eds.): CDRF 2025, *Transindividual Intelligence*, pp. 400–409, 2026.
https://doi.org/10.1007/978-981-92-0615-5_34

models are employed to generate renovation schemes. Eye-tracking experiments and Semantic Differential (SD) questionnaires are integrated to establish a "form-cognition-experience" evaluation framework. Focusing on rural industrial architectural remains, the research develops an "AI generation-perception verification-model optimization" closed-loop system, addressing the design limitation of prioritizing form over perception. This methodology provides a quantifiable paradigm for rural heritage renewal.

2 Methodology

This study employs generative artificial intelligence (GAI) tools for digital renovation of industrial architectural remains. The core methodology trains deep neural networks to learn data distributions, autonomously generating human-aligned creative content. By analyzing statistical patterns in large-scale multimodal data (e.g., images, text), generalized generative models are built for intelligent image synthesis and text generation[1].In human perception experiments, eye-tracking[1–3]—a well-established method for evaluating visual experiences in built environments—is applied to assess AI-driven renovation effects and guide model optimization [5–14].

3 Experiments

The study's experiments were divided into two phases. Phase 1 involved sample selection, database construction of rural industrial heritage in Southern Anhui, and comparative analysis of LoRA model training performance across optimizers via parameter adjustments. Image quality was controlled by tuning Stable Diffusion's weight parameters. Phase 2 recruited participants for eye-tracking experiments to compare pre and post-renovation physiological metrics for indoor/outdoor spaces. Next, SD method is adopted to conduct a psychological evaluation of the generated images, guiding the optimization design of the artificial intelligence model. This two-part experiment enables cross-integration of generative AI and human perception.

3.1 Field Surveys and Sample Collection

Industrial heritage buildings serve as historical markers of rural industrial development, with sites like old granaries, tea factories, oil mills, and cocoon stations increasingly obsolete. To systematically document these structures, researchers conducted field surveys in Shexian, Yixian, Jixi, and Qimen counties from March to June 2024. Over 70 sites were visited, yielding interior and exterior images of rural industrial architecture in Southern Anhui.

Analysis of the samples identified shared characteristics among Southern Anhui's rural industrial heritage structures. Predominantly brick-wood with tiled, double-pitched roofs, these buildings typically feature standalone layouts but can also form clustered complexes. Their interiors are defined by spacious, rectangular floor plans with minimal partitions, reflecting historical industrial process requirements. Larger in scale than traditional local dwellings, these structures are highly recognizable within villages and demonstrate significant potential for adaptive reuse (Fig. 1).

Fig. 1 Selected surveyed case images of industrial heritage in southern Anhui

3.2 Database Construction and LoRA Model Training

Prior to LoRA model training, a database of Southern Anhui's industrial heritage was developed. Training data was split into two subsets: field-surveyed sample images and renovation case studies. Taking interior spaces as an example, 30 well-preserved structural photos (Training Set 1) were selected for strong representativeness and typical features of Southern Anhui's industrial heritage. Dataset 2 sourced 30 renovation examples from the Gooood Design website (www.gooood.cn) and Baidu Images using keywords like "rural architecture renovation," "timber structures," and "interior design," ensuring high-quality reference cases.

Text annotation aligned with image data was critical for precise model training. Training Set 1 and 2 images were annotated using BooruDatasetTagManager software to enhance feature learning and detail control during renovation image generation [15]. Shared trigger words like "Chachang, indoors" were retained to activate the LoRA model. Inherent genetic features (e.g., truss structures, sloped roofs) required no annotation, while modifiable traits (e.g., column types, window materials, flooring) were tagged to enable detail adjustments via textual prompts (Fig. 2).

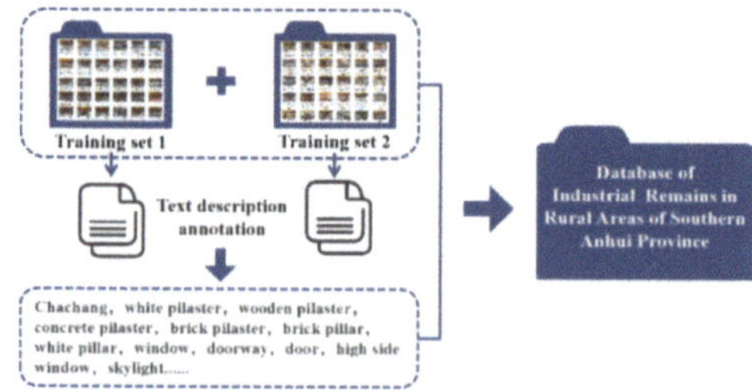

Fig. 2 Process of industrial heritage database construction in southern Anhui

Training was conducted using the RealisticVision V2.0 base model over 10 epochs, comparing AdamW8bit and Lion optimizers. Synthesis results were saved per epoch for comparison. The Lion optimizer achieved loss convergence to 0.08, meeting deep learning criteria and demonstrating balanced style abstraction and generalization.

3.3 Image Generation Experiment Workflow

The generative AI tool used was the open-source Stable Diffusion WebUI (https://github.com/AUTOMATIC1111/stable-diffusion-webui), deployed locally with Python setup. Models were accessed via browser for real-time parameter tuning and image generation. Experiments utilized an NVIDIA GeForce RTX 4060 Ti 8GB GPU.

For the blank control group, images were generated using Stable Diffusion alone without LoRA Four widely used SD models were tested with prompts like "Southern Anhui architectural style, indoor, wood truss structure" in the text-to-image module (other parameters fixed) [16].. Outputs relied on SD's default image database, yielding results inconsistent with expected heritage styles due to insufficient dataset alignment (Fig. 3).

Fig. 3 Image generation results of four SD models. Prompts: Southern Anhui architectural style, indoors, wood truss structure, no humans, window, door, white pilaster, white wall, concrete floor.

Despite prompting key features of Southern Anhui's industrial heritage, the four SD models produced inconsistent results deviating from expected styles, primarily due to insufficient dataset alignment with rural industrial architecture.

Further experiments incorporating the LoRA model while keeping prompts/parameters constant revealed weight-dependent outcomes: low weights retained SD's default style with minimal heritage influence; increasing weights progressively aligned outputs with training data but risked content deterioration beyond 0.8. Optimal results (0.5–0.8 weight) balanced similarity and quality, demonstrating successful style transfer of Southern Anhui's industrial heritage characteristics (Fig. 4).

Subsequent experiments used the img2img module to renovate heritage images by adding functions like libraries, restaurants, and exhibition halls via prompts. Testing CFG Scale and denoising strength impacts revealed optimal results at CFG Scale 6–8 and denoising strength 6–7, balancing style preservation and structural integrity (Fig. 5).

Finally, applying optimal parameters derived from prior experiments, renovations were generated for 8 interior spaces of Southern Anhui's industrial heritage using modified text prompts. Exterior image restoration and generation followed the same LoRA

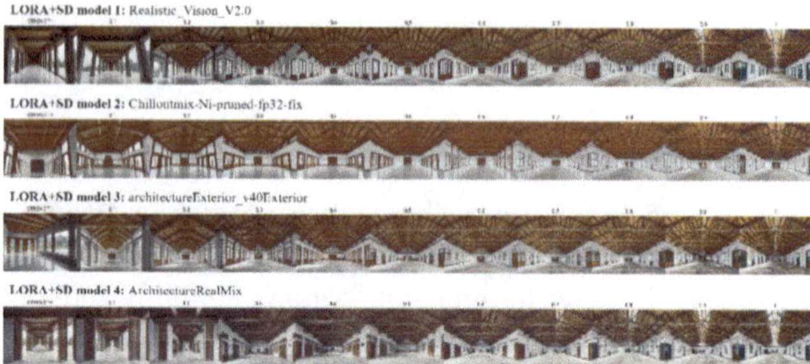

Fig. 4 Image generation results controlled by four base models with LoRA: LoRA model weights progressively increasing from 0, 0.1, 0.2…1 left to right

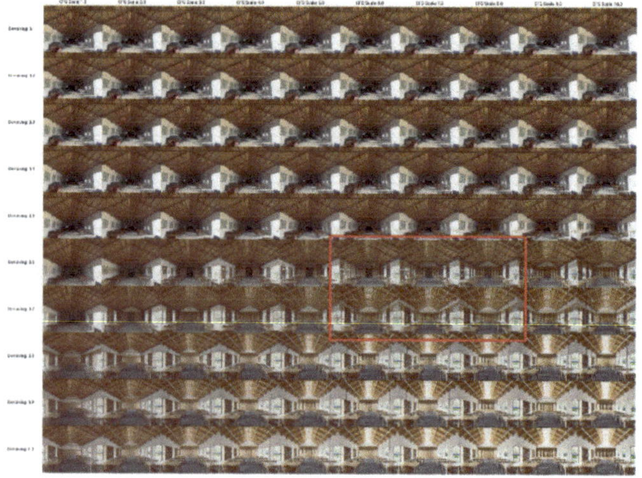

Fig. 5 Horizontal axis: CFG Scale from 1 to 10; vertical axis: denoising strength from 0.1 to 1.0. Other parameters fixed. Prompts: Chachang, library, indoors, white wall, gray tile floor, no humans, glass window, glass door, wooden bookshelf, wooden table, wooden chair, white pilaster.

database construction and training process, producing updates for 4 original photos. Synthesized images will serve as stimuli for subsequent human perception experiments (Fig. 6).

3.4 Eye-Tracking Experiment and Result Analysis

The eye-tracking experiment was conducted at the Architecture and Environment Behavior Research Lab of the School of Architecture and Art from March 22–24, 2025, 15:00–18:00 daily. The windowless room featured soft lighting to ensure image clarity on displays. A Tobii Glasses 3 head-mounted eye tracker with adjustable prescription lenses enabled precise gaze recording. Eighteen architecture master's students (gender ratio ≈1:1) participated (Fig. 7).

Twenty-four paired renovation images (12 pre- vs. 12 post-upgrade) from Fig. 9 were selected for consistent clarity and aspect ratio. Participants viewed images sequentially

Fig. 6 AI-generated indoor and outdoor renovation results of industrial heritage

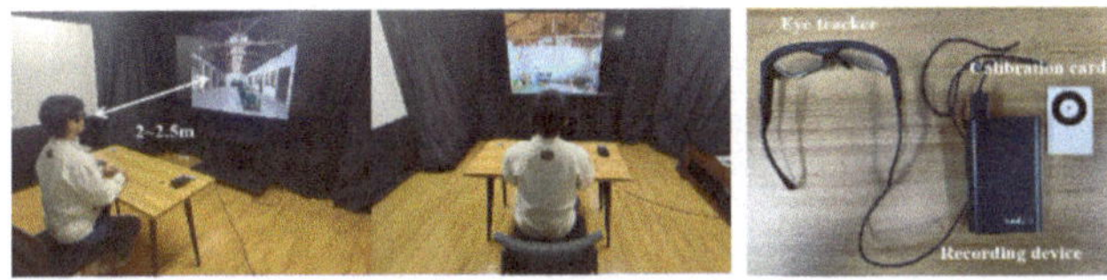

Fig. 7 Experiment setup and equipment

(10 s/image + 10 s rest) in pre-to-post order. Data were analyzed in Tobii Pro Lab using 60-pixel heatmap radii.

Post-experiment, data were imported into Tobii Pro Lab software for visual analysis. Heatmaps showing average fixation durations of 18 participants (with a fixed 60-pixel fitting radius) revealed significant differences between pre- and post-upgrade images. Pre-upgrade heritage photos elicited scattered attention with short fixation times across indoor/outdoor elements, lacking sustained engagement [17]. Post-upgrade images demonstrated concentrated gaze in red heatmap zones, indicating stronger visual interest in renovated features.

Heatmaps from pre- and post-upgrade images also revealed shared patterns: participants' attention concentrated significantly in central areas, particularly pronounced in indoor scenes—a finding corroborated by saccade data analysis. This phenomenon stems from both natural viewing habits and the strong visual guidance of spatial vanishing points. Across all 12 image pairs, gaze fixation clustered on building facades, walls/roofs, and furniture/fixtures, with minimal attention to flooring. These shared patterns provide critical guidance for AI-driven image renovations (Fig. 8).

To explore in greater detail the impact of renovations on participants, the study selected two indoor and two outdoor image pairs (pre- and post-upgrade) for Area of Interest (AOI) analysis. Images were first divided into zones (e.g., roofs, doors/windows, furniture, clutter), followed by software-based quantitative analysis of average Total Fixation Duration (TFD) data from 18 participants [18] (Table 1).

Comparing pre- and post-upgrade AOI data revealed distinct trends:

Indoor spaces: Roof truss structures maintained high fixation durations post-renovation, though reduced from pre-upgrade levels, underscoring their significance

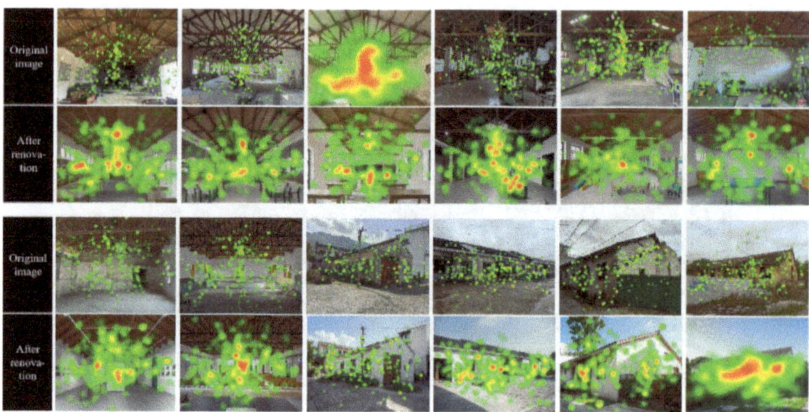

Fig. 8 Eye-tracking experiment heatmaps of 12 experimental samples

Table 1 Pre- and post-upgrade AOI date

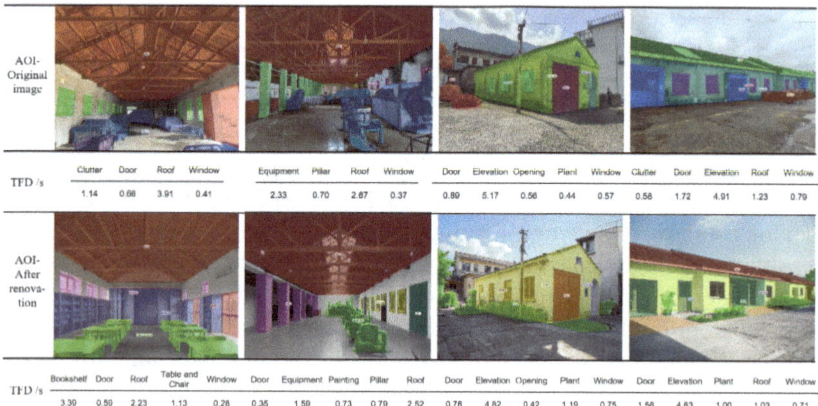

AOI-Original image

TFD /s	Clutter	Door	Roof	Window
	1.14	0.68	3.91	0.41

TFD /s	Equipment	Pillar	Roof	Window
	2.33	0.70	2.67	0.37

TFD /s	Door	Elevation	Opening	Plant	Window
	0.89	5.17	0.56	0.44	0.57

TFD /s	Clutter	Door	Elevation	Roof	Window
	0.58	1.72	4.91	1.23	0.79

AOI-After renovation

TFD /s	Bookshelf	Door	Roof	Table and Chair	Window
	3.30	0.50	2.23	1.13	0.28

TFD /s	Door	Equipment	Painting	Pillar	Roof
	0.35	1.59	0.73	0.79	2.52

TFD /s	Door	Elevation	Opening	Plant	Window
	0.76	4.82	0.42	1.19	0.75

TFD /s	Door	Elevation	Plant	Roof	Window
	1.56	4.63	1.00	1.03	0.71

in industrial heritage interiors. Furniture/equipment additions significantly increased attention, while doors/windows remained low-interest zones.

Outdoor spaces: Building facades with vegetation showed stronger engagement, suggesting greenery integration could enhance visual appeal during renovations.

3.5 SD Questionnaire Result Analysis

Following the eye-tracking experiment, participants completed a Semantic Differential (SD) questionnaire to psychologically evaluate AI-generated renovation images. Twelve pre/post-upgrade image pairs from the eye-tracking study served as SD samples, paired with 16 adjective scales using a 7-point Likert scale (-3 to $+3$, higher scores indicating stronger agreement).

Analysis of SD questionnaire results for 12 image pairs revealed average scores of $\geq$1.3 for overall renovation effectiveness, indicating high participant approval of AI-generated updates. The lowest scoring metric—lack of greenery elements in indoor images—was the only negative-scoring category due to perceived barrenness.

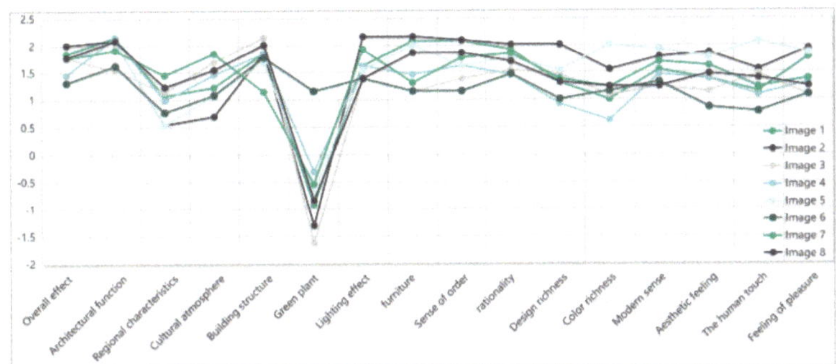

Fig. 9 SD method scores for indoor renovation images of industrial heritage buildings

Line chart analysis revealed that images with low overall renovation scores correlated strongly with unclear regional identity, weak cultural atmosphere, and monotonous design/color. These factors exerted stronger influence on exterior renovation scores. In the future iterative optimization of AI models, significant emphasis should be placed on improving the above-mentioned low-scoring aspects.

3.6 Form-Cognition-Experience Model

Human perception experiments identified key renovation priorities and critical metrics for Southern Anhui's industrial heritage buildings. Image generation should focus on preserving indoor truss integrity, optimizing central hotspot design, and refining text prompts (e.g., adding greenery/cultural artifacts like landscape paintings and traditional furniture) based on SD questionnaire feedback. Regional ambiguity in exterior renovations can be addressed by augmenting LoRA databases with local architectural examples (e.g., horse-head walls) and modern design case studies.

This study proposes a "Form-Cognition-Experience" framework: LoRA databases capture heritage style, latent diffusion models generate AI learning systems, human experiments validate outcomes, and feedback refines models—establishing a closed-loop process from generation to optimization (Fig. 10).

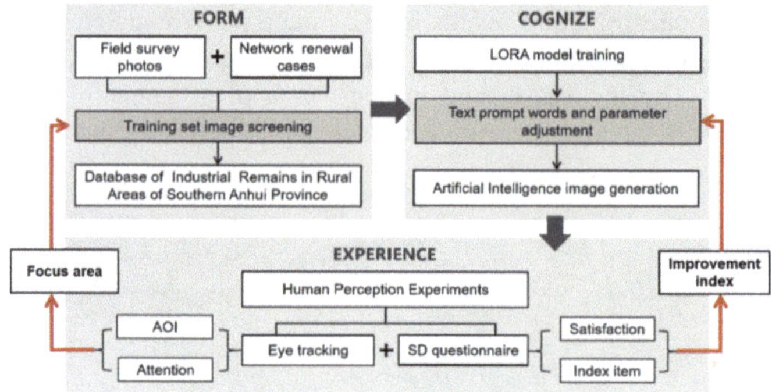

Fig. 10 Form-cognition-experience model

4 Conclusion and Discussion

This study introduces a novel renovation method for architectural heritage, establishing a reliable experimental workflow that significantly boosts design efficiency to address resource shortages in rural areas. The approach leverages AI's potential to learn and expand architectural styles, despite structural instability in generated outputs—a limitation mitigated through reinforcement learning optimization of training datasets. AI's rapid generation of diverse design options empowers designers to prototype concepts efficiently for iterative refinement.

However, generative AI remains a supplementary tool rather than a replacement for human designers, lacking the capacity to deeply understand user needs and emotional contexts. While current technical complexity and equipment requirements pose challenges, advancements in AI will likely enhance model intelligence and accessibility, democratizing design capabilities. Perhaps one day in the future, the idea of everyone being able to become their own designer will no longer be just a dream.

Funding Information. This research was funded by the National Natural Science Foundation of China (No. 52378001); the Collaborative Innovation Project for Colleges and Universities in Anhui Province (No. GXXT-2023-047); and the Open Project of Anhui Provincial Key Laboratory (No. 2024HPJZ-KF05).

References

1. Chaillou, S.: Artificial Intelligence and Architecture: From Research to Practice. Birkhauser Verlag GmbH, Basel, Switzerland; Boston, MA (2022)
2. Elsadek, M., Liu, B., Xie, J.: Window view and relaxation: Viewing green space from a high-rise estate improves urban dwellers' wellbeing. Urban For. Urban Green. **55**(1), 126846 (2020)
3. Liu, R., Neisch, P.: Measuring the effectiveness of street renewal design: Insights from visual preference surveys, deep-learning technology, and eye-tracking simulation software. Landsc. Urban Plan. **256**, 105291 (2025)
4. Xing, Y., Leng, J., Zhou, H.: Cognitive preferences for architectural renovation strategies in traditional villages combining subjective evaluation and eye tracking. Front. Archit. Res. **14**(4), 1017–1034 (2025)

5. Goodfellow I J, et al.: Generative adversarial nets. In: Proceedings of the 28th International Conference on Neural Information Processing Systems, MIT Press, Cambridge, pp. 2672–2680, 2014
6. Kingma D P, Welling M: Auto-encoding variational bayes. In: Proceedings of the 2nd International Conference on Learning Representations, ICLR Press, Banff, pp. 1–14, 2014
7. Larochelle H, Murray I: The neural autoregressive distribution estimator. In: Proceedings of the 14th International Conference on Artificial Intelligence and Statistics, JMLR Press, Fort Lauderdale, pp. 29–37, 2011
8. Ho J, Jain A, Abbeel P.: Denoising diffusion probabilistic models. In: Proceedings of the 34th International Conference on Neural Information Processing Systems. Red Hook, Curran Associates Inc: Article No. 574, pp. 6840–6851, 2020
9. Rombach R, Blattmann A, Lorenz D, Esser, P., Ommer, B.: High-resolution image synthesis with latent diffusion models. In: Proceedings of the IEEE/CVF Conference on Computer Vision and Pattern Recognition, IEEE Computer Society Press, Los Alamitos, pp. 10674–10685, 2022
10. Zhang, Lvmin, Maneesh Agrawala: Adding conditional control to text-to-image diffusion models. arXiv. https://doi.org/10.48550/arXiv.2302.05543 (2023). Accessed 10 Feb. 2023
11. Hu E J, et al.: LoRA: Low-rank adaptation of large language models. arXiv 2021
12. Jarodzka, H., Brand-Gruwel, S.: Tracking the reading eye: Towards a model of real-world reading. J. Comput. Assist. Learn. **33**(3), 193–201 (2017)
13. Hwang, J.W., Hong, C.U., Chong, W.S.: SD and EEG evaluation of the visual cognition to the natural and urban landscape. J. Environ. Sci. Int. **15**(4), 305–310 (2006)
14. Cottet, M., Vaudor, L., Tronchère, H., Roux-Michollet, D., Augendre, M., Brault, V.: Using gaze behavior to gain insights into the impacts of naturalness on city dwellers' perceptions and valuation of a landscape. J. Environ. Psychol. **60**, 9–20 (2018)
15. Radford A, et al.: Learning transferable visual models from natural language supervision. arXiv 2021
16. Hertz A, Mokady R, Tenenbaum J, Aberman, K., Pritch, Y., Cohen-Or, D.: Prompt-to-prompt image editing with cross-attention control. In: Proceedings of the 11th International Conference on Learning Representations, ICLR Press, Kigali Rwanda, pp. 1–19, 2023
17. Korpela, K.M.: Perceived restorativeness of urban and natural scenes—photographic illustrations. J. Archit. Plan. Res., 23–38 (2013)
18. Li, J., et al.: An evaluation of urban green space in Shanghai, China, using eye tracking. Urban For. Urban Green. **56**, 126903 (2020)

Knowledge-Augmented Stable Diffusion Model for Semantic Representation of Architectural Heritage: A Case Study of Huizhou Traditional Dwellings

Zhe Guo[1(✉)], Qingyi Xiong[1], Zihuan Zhang[1], and Yuhang Kong[2]

[1] Department of Architecture and Art, Hefei University of Technology, Hefei 230000, Anhui, China
guogal@hotmail.com

[2] Department of Architecture, Tianjin University, 300350 Tianjin, China

Abstract. Efficiently generating traditional architectural models with regional characteristics is a growing challenge in digital heritage preservation. Existing approaches often lack semantic depth and struggle to integrate construction logic, stylistic rules, and multi-source data. This study addresses these issues through a Knowledge-Augmented Stable Diffusion model, focusing on Huizhou traditional dwellings as a representative case. We propose a hybrid framework that combines Graph-Based Reasoning (GBR) and Stable Diffusion to enable semantic representation and controlled façade generation. First, the compositional features of Huizhou traditional dwellings (HzTD), hierarchical layout, materials, and ornaments, are analyzed and encoded into a shape grammar. Then, using the Property Graph (PG) data model standard and Neo4j, we construct a formal knowledge base that supports reasoning, retrieval, and prompt generation. A Multimodal Diffusion Transformer (MMDiT) is introduced to incorporate KAR-guided ControlNet instances, improving generation accuracy and semantic alignment. Experiments on the façades of Xidi Ancient Village show that the proposed model outperforms baseline generative methods in style consistency, efficiency, and structural fidelity. This research demonstrates the potential of integrating domain knowledge into generative AI for architectural heritage, offering new possibilities for cultural preservation, design assistance, and virtual reconstruction. Future work will explore multimodal 2D-to-3D generation and dynamic scene synthesis.

Keywords: Knowledge-Augmented Reasoning (KAR) · Semantic enrichment · Stable Diffusion Model · Architectural heritage · Huizhou traditional dwellings

1 Introduction

1.1 Background

With the rapid advancement of digital preservation of cultural heritage and intelligent 3D model reconstruction technologies, efficiently and accurately generating traditional architectural models with regional construction characteristics has become a critical

Y. Liu et al. (Eds.): CDRF 2025, *Transindividual Intelligence*, pp. 410–419, 2026.
https://doi.org/10.1007/978-981-92-0615-5_35

issue in the fields of architectural heritage research and architectural design. Particularly in the digital representation of architectural heritage with unique regional historical and cultural value, such as Huizhou traditional dwellings (HzTD), challenges remain in achieving multi-perspective, precise, and rapid façade generation while preserving traditional styles. Existing digital research on the stylistic features of traditional residential building mainly focuses on aesthetic expression and structural analysis, relying on static data analysis and phenomenological interpretation, which presents two issues: (1) a lack of in-depth research on the dynamic generation of architectural forms and styles, especially in relation to construction knowledge semantics; (2) insufficient attention to the correlation between the construction logic of traditional buildings with regional characteristics, such as structural methods and construction techniques, resulting in generated models lacking realism and rationality.

To address this background, this study firstly adopts graphical analysis and ontological reasoning to parse the features of Huizhou building façades and construct shape grammar rules. Subsequently, based on Property-Graph Structure (PGS), knowledge modelling using Neo4j is performed to provide support for G-AI generation. In terms of generative modelling, Stable Diffusion Model, LoRA and ControlNet are optimised with Knowledge-Augmented Reasoning (KAR) to improve the sample compression and multi-view generation accuracy. Taking the traditional dwellings in Huizhou, China as a case study, we propose a method for constructing a semantically enriched model of traditional architectural knowledge based on KAR and combine it with the stable diffusion generative model to realise the automatic generation of regional façade styles.

1.2 Related Works

For the research on automatic generation of building façades, scholars have been exploring different approaches in recent years. Yin et al. (2020) used Mask R-CNN to identify elements in architectural drawings and automatically generate BIM models, laying the foundation for building element identification and modelling. Sun et al. (2022) optimised the automatic façade generation process in historic urban renewal using generative adversarial networks (GANs). Fan et al. (2021) proposed a method based on Layout Graph combined with Gibbs sampling to improve the accuracy of semantic façade modelling from point cloud data. Wang et al. (2020a, 2020b), on the other hand, classified different elements of building façades and repaired damaged façades based on topological graph semantic modelling. In the development of generative AI, Stable Diffusion and its enhancement methods have gradually become an important technological tool for façade generation. Related works (Shi et al. 2024) developed a generative AI-based façade design method by combining the regional identity training model and achieved the design of façade from the topology map semantic modelling by combining Stable Diffusion and ControlNet. Further research (Xu et al. 2024) combined Grasshopper with Stable Diffusion to ensure that the generated façade meets the style, size and form requirements.

Meanwhile, knowledge graph and semantic modelling have gradually gained attention in architectural space research, promoting the structured representation and automated analysis of architectural information. Acierno et al. (2017) systematically defined the information structure of architectural heritage's lifecycle based on the CIDOC-CRM ontology modelling. Study (Wang et al. 2020a, 2020b) used knowledge graph embedding technology to optimise the structured query of architectural data, which improved the accuracy and efficiency of retrieval. HBIM also played an important role in cultural heritage preservation, which focusing on geometric modelling, data storage and standardisation issues (Rocha and Tomé 2021). In the field of architectural semantic segmentation, studies (Hou et al. 2024) combined machine learning algorithm to improve the accuracy of façade segmentation in architectural heritage reconstruction.

2 Method

2.1 Overall Framework

Our proposed Knowledge-Augmented Stable Diffusion Model (KA-SDM) consists of three major components:

Base Stable Diffusion Backbone (pretrained on LAION-5B)

Domain Knowledge Embedding Module

Prompt-Conditioned Semantic Controller (PCSC) (Figs. 1 and 2)

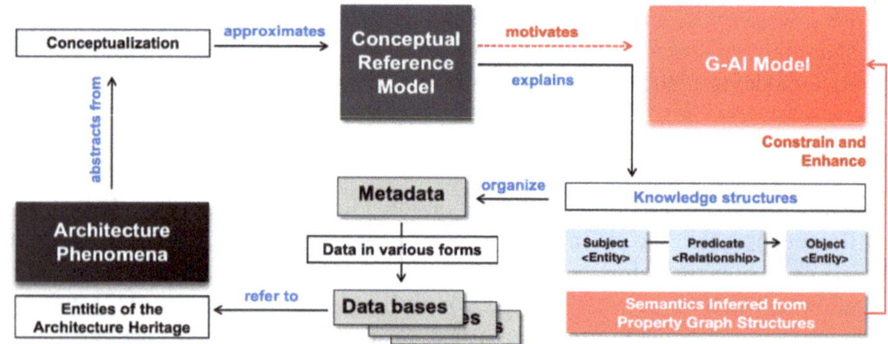

Fig. 1 Framework of conceptual reference model

2.2 Ontology Modeling

2.2.1 Architectural Heritage Construction Knowledge Structuring

As a prominent representative of Chinese vernacular architectural heritage, HzTD are primarily distributed in the southern Anhui Huangshan, Jixi, and Shexian. These dwellings originated during the *Song* dynasty and flourished in the *Ming dynasty* embodying the profound integration of regional geography, cultural traditions, and clan-based social structures, which are organized into organically formed settlements.

At the individual building level, spatial organization centres around an open-air courtyard (*'TianJing'* in Huizhou context), with front and rear halls arranged along a central axis. The overall layout emphasizes the principles of *'Ting(厅)' in front and*

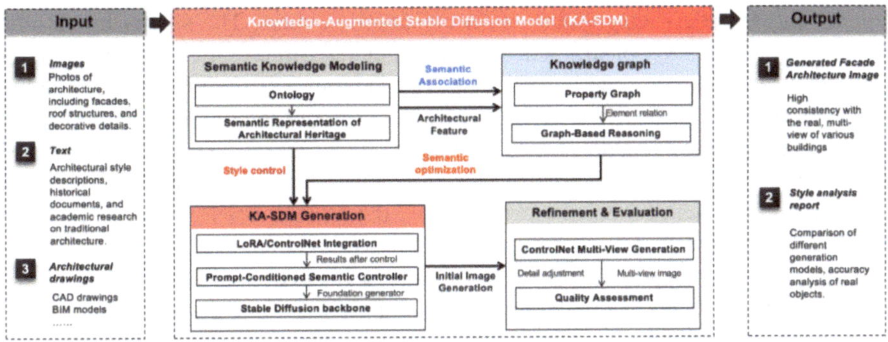

Fig. 2 The application workflow of knowledge-augmented stable diffusion model

'Tang(堂)' in back, which means *'front halls and rear chambers'* and axial symmetry. Structurally, these buildings adopt timber frame construction systems, with primary materials including locally sourced timber, stone, grey bricks, and tiles. The dwellings feature enclosed courtyards surrounded by high walls on all four sides, forming a spatial typology of *'solid exterior, void interior'*. The interior facades are articulated with intricately carved wooden windows, balustrades, and ornamental components, while the exterior facades are characterized by the iconic *'white walls and grey tiles'* aesthetic. A distinctive feature is the *'horse-head gable walls'* (*'MaTouQiang(马头墙)'* in Huizhou context), which visually articulate the roofline and reflect regional identity (Fig. 3).

We constructed the knowledge ontology based on PGS and set specific to architectural heritage of HzTD, encompassing four construction attributes:

Function & Type *(Space Type)*
Layout & Composition *(Plan Element, Façade Element)*
Material & Tectonic *(Structure, Component, Tectonic Detail)*
Culture & Symbolism *(Ritual, Hierarchy, Motif, Narrative, Regional Style)*

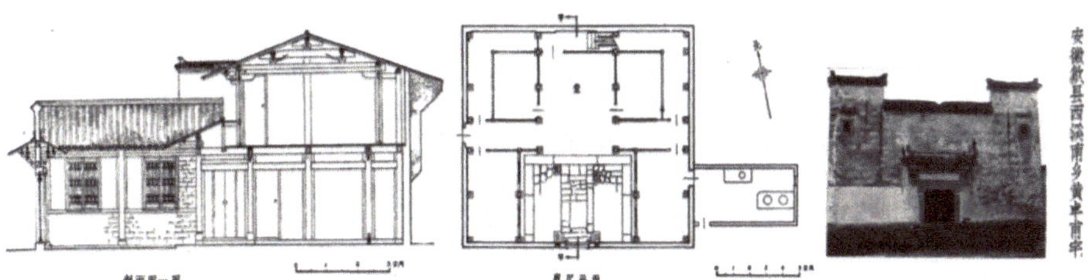

Fig. 3 Huizhou traditional dwelling resource *(张仲一 et al. 徽州明代住宅 [M]. 建筑工程出版社. 1957)*

2.2.2 Semantic Extraction and Graph Schema Design

Based on these structured four categories of construction attributes in architectural heritage, we focus specifically on entity extraction related to Space Type, as well as Plan

Elements and Façade Elements based on construction rules. The spatial layout characteristics of HzTD are constrained by a fundamental structural framework, namely, the horizontal coordinates and heights of posts. This reflects the timber-frame organizational model, which is *'DaMuZuo' (大木作)'* in Chinese construction system and exhibits strong regional characteristics, making it a key focus of extraction in this study. In the direction facing the entrance, the *'KaiJian(开间)'* module determines the compositional characteristics of the front façade, while the *'JinShen(进深)'* module, which is perpendicular to the entrance, constrains the compositional features of the side façade's *'MaTouQiang'*. The relationship extraction pattern between entities is based on construction logic, with a focus on the organization of architectural elements, the sequence of component connections, and the spatial layout patterns. The method shows in Fig. 4.

It is important to note that in this case study, we focus solely on the fundamental prototypical characteristics of the HzTD type. Variations in spatial layout derived from this architectural prototype are not discussed at this stage.

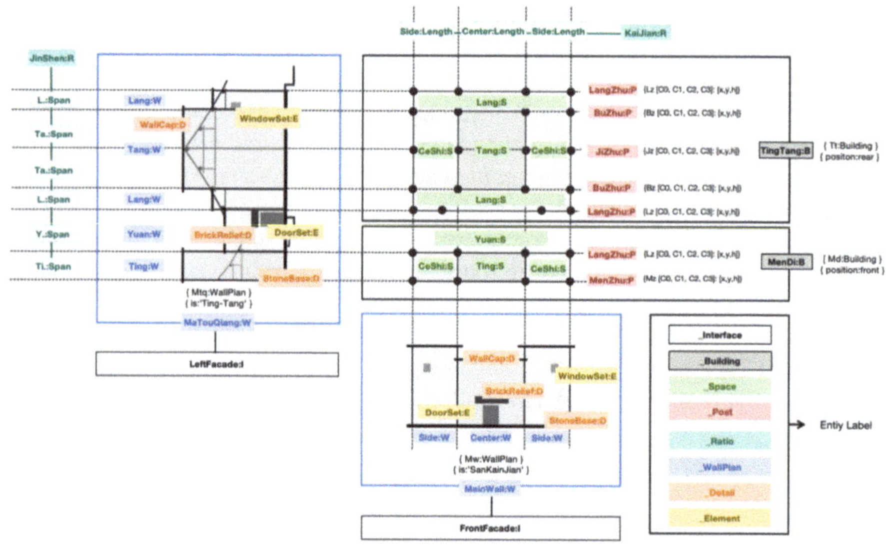

Fig. 4 Entity extraction of Huizhou traditional dwelling

After establishment of HzTD conceptual modeling, the graph schema was designed by abstracting entities and relationships. We abstracted eight types of entities, represented as key-value pairs in the format *<entity, property>*, and further model the primary-level entity properties as attribute entities. Simultaneously, we extract four types of relationships, which are stored in a triple format *<entity, relationship, entity>*. All *Entities* and *Relationships* are listed in Fig. 5.

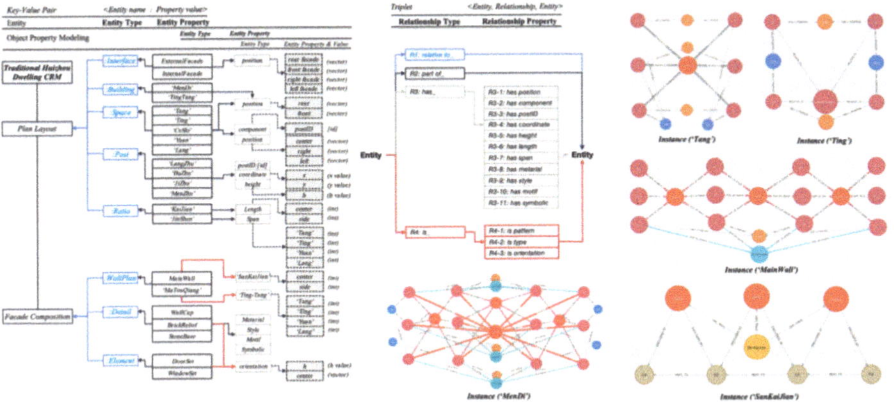

Fig. 5 Entity and relationship ontology structure with Neo4j

2.3 Graph-Based Reasoning of Huizhou Traditional Dwellings

This study utilizes Neo4j for structured knowledge storage, visualization, and reasoning. As a database based on the property graph data model, Neo4j does not natively support formal logic reasoning as enabled by ontology languages such as OWL. However, through graph-based data modelling and rule-based query programming, it can achieve a range of quasi-reasoning functions suitable for architectural semantic interpretation and component relationship analysis. By performing reasoning based on Cypher queries, rule-based descriptions and structural path analysis over the Neo4j database, relevant data can be exported in CSV format which could be accessed by subsequent processes (shows in Table 1).

Table 1 Graph-based reasoning methodology and instance in case study

Inference type	Application instance in this case study: *Plan layout* -to- *façade composition*	
Class-Hierarchies-Based Structural Inference	input:	<component: Space {postID:value}> <space: SpaceType {name: 'Tang'}>
	logic:	inheritance semantics
	output:	< part: WallPlan {name: 'Tang'} > = < height: h {h value}>
Attribute and Label-Based Rule Inference	input:	If <position: 'name' > = 'front' 'rear' & < space: 'name' > = 'Ting' 'Tang'
	logic:	if A then B
	output:	<façade composition: WallPlan {type: 'Ting-Tang'}>
Component-Level Spatial Logic Inference	input:	<post: 'Tang' {x value} > = < post: 'Lang' {x value}>

(continued)

Table 1 *(continued)*

Inference type	Application instance in this case study: *Plan layout* -to- *façade composition*	
	logic:	topological & design logic
	output:	create relationship: (Tang) - [: NEXT_To] - (Lang)

2.4 Integration with Stable Diffusion Model (KA-SDM)

LoRA enables Stable Diffusion to learn specific styles or fine-grained visual patterns through lightweight tuning, while ControlNet enhances controllability by guiding image generation with structural inputs such as masks. By reasoning over the graph, we can dynamically generate ControlNet constraint masks and context-aware data to guide LoRA training, ensuring that both processes are grounded in semantically rich and structurally accurate architectural knowledge.

The goal of this step is to automatically generate semantic segmentation masks from specific viewpoints and to export descriptive prompts as supplementary constraints for the AI generator. A Python script can then be used to map this data, via coordinate transformation, onto a 512 × 512-pixel canvas, thereby computing the segment mask layout required to represent the target architectural view. Taking the generation of the front façade in this study as an example, the aim is to generate *'SanKaiJian(三开间) Huizhou traditional dwelling main entrance façade'*. Reasoning from PGS, the geometric blocks corresponding to its 3 semantic regions directly map to related entity attributes and their values under the façade composition type (Fig. 6). This process generates semantic segments by directly drawing 2D pixel fragments, eliminating the need for 3D modelling and significantly improving generation efficiency by focusing solely on pixel-level outputs.

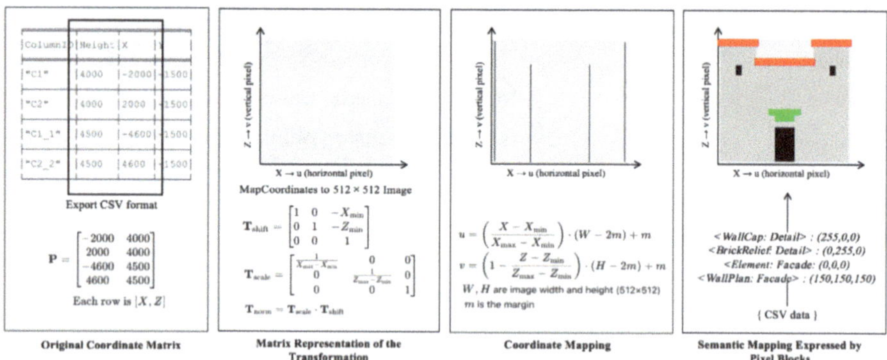

Fig. 6 Data mapping base on PGS inference: From number to pixel

3 Experiment and Results

To fine-tune the generative model for the specific visual and semantic characteristics of HzTD, we prepared the small-scale dataset containing approximately 30 high-quality images, which emphasizes the external spatial characteristics of Huizhou dwellings, such as wall-enclosed village edges, roof silhouettes, and street-facing interfaces. These datasets were curated from architectural photography and were used to fine-tune LoRA (Low-Rank Adaptation) modules for controlled generation. The base model employed for LoRA fine-tuning is the pretrained Stable Diffusion v1.5 released by CompVis, which offers a robust foundation for style transfer and spatial composition learning in architectural contexts.

We compare our KA-SDM with Original Stable Diffusion v1.5, SD + LoRA (no knowledge guidance) and SD + ControlNet (no semantic control). Evaluation metrics imply Visual Quality via FID and Human Ranking. Then we designed two tasks including Front-perspective and Contextual-perspective façade generation.

The result shows that our KA-SDM outperformed all baselines in terms of FID (Fréchet Inception Distance) and user-based aesthetic evaluation. Notably, the generated images preserved the hierarchical layout and material textures characteristic of HzTD (shows in Table 2). We invited five experts in the fields of architectural history and Huizhou vernacular studies to evaluate a random sample of 40 generated images, using a 5-point Likert scale on the following criteria (Table 3). Each image received scores from all five experts, and inter-rater agreement was analyzed using Cohen's kappa ($\kappa = 0.76$), indicating substantial agreement (Fig. 7).

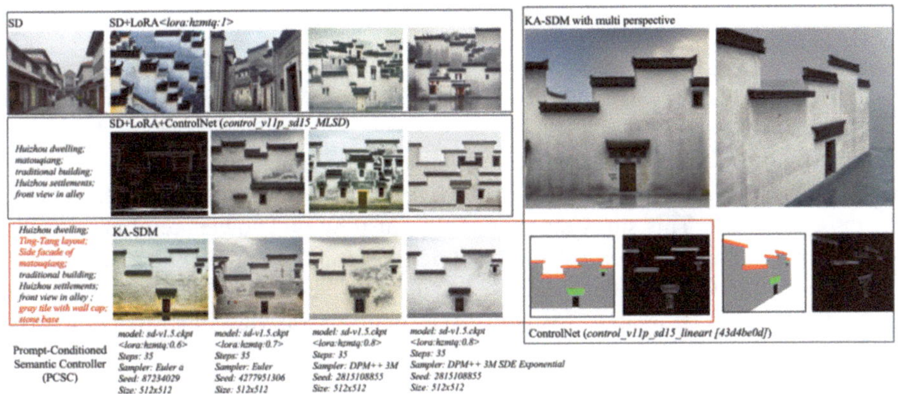

Fig. 7 Multi-model experiment with multi-perspective generation results

Table 2 Visual and structural performance

Model	FID	Expert semantic score	Layout accuracy
SD	39.2	2.8/5	61.5%
SD + LoRA/ControlNet	24.7	3.6/5	73.3%
KA-SDM (ours)	16.9	4.5/5	89.4%

Table 3 Human expert evaluation criterion and average human scores (out of 5)

Evaluation criterion	Description
Typological Fidelity	Correctness and clarity of typological form
Spatial Composition Logic	Coherence in layout and scale
Cultural Symbolism Expression	Appropriateness of decorative and compositional symbolism
Material Construction Plausibility	Whether materials, textures and joinery correspond to real Huizhou style

Model	Typology	Symbolism	Spatial	Materials
SD	2.8	4.2	2.3	3.7
SD + LoRA/ControlNet	3.5	3.9	3.6	3.8
KA-SDM (ours)	4.6	4.5	4.8	4.6

4 Conclusion

In this study, the PGS provides a semantic foundation for organizing architectural entities and their properties, enabling precise knowledge representation, visualization and inference through Neo4j. This study structures the construction knowledge of architectural heritage, enabling multidimensional reasoning, constraint enforcement, and automatic completion within G-AI models. Experimental results show that this method performs well in rapid image generation for both fixed architectural views and dynamic views (e.g., perspective views with defined vanishing points), demonstrating advantages such as high image consistency, fast processing speed and lightweight computation, thus laying a solid foundation for further research.

References

Acierno, M., Cursi, S., Simeone, D., Fiorani, D.: Architectural heritage knowledge modelling: An ontology-based framework for conservation process. J. Cult. Herit. **24**, 124–133 (2017)

Bloch, T., Sacks, R.: Comparing machine learning and rule-based inferencing for semantic enrichment of BIM models. Autom. Constr. **91**, 256–272 (2018)

Fan, H., Wang, Y., Gong, J.: Layout graph model for semantic façade reconstruction using laser point clouds. Geo-spat. Inf. Sci. **24**(3), 403–421 (2021)
Hou, J., Zhou, J., He, Y., Hou, B., Li, J.: Automatic reconstruction of semantic façade model of architectural heritage. Herit. Sci. **12**(1), 400 (2024)
Htet, A., Liana, S.R., Aung, T., Bhaumik, A.: The intersection of façade engineering and building information modeling: Opportunities and challenges. J. Technol. Innov. Energy. **2**(3), 94–110 (2023)
Rocha, J., Tomé, A.: Multidisciplinarity and accessibility in heritage representation in HBIM Casa de Santa Maria (Cascais)—A case study. Digit. Appl. Archaeol. Cult. Herit. **23**, e00203 (2021)
Shi, M., Shi, M., Seo, J., Cha, S.H., Xiao, B., Chi, H.-L.: Generative AI-powered architectural exterior conceptual design based on the design intent. J. Comput. Des. Eng. **11**(5), 125–142 (2024)
Sun, C., Zhou, Y., Han, Y.: Automatic generation of architecture façade for historical urban renovation using generative adversarial network. Build. Environ. **212**, 108781 (2022)
Wang, R., Wang, M., Liu, J., Cochez, M., Decker, S.: Structured query construction via knowledge graph embedding. Knowl. Inf. Syst. **62**(5), 1819–1846 (2020a)
Wang, Y., Fan, H., Zhou, G.: Reconstructing façade semantic models using hierarchical topological graphs. Trans. GIS. **24**(4), 1073–1097 (2020b)
Xu, S., Zhang, J., Li, Y.: Knowledge-driven and diffusion model-based methods for generating historical building façades: A case study of traditional Minnan residences in China. Information. **15**(6), 344 (2024)
Yin, M., Meng, X., Zhang, H., Song, S., Du, S.: Automatic layer classification method-based elevation recognition in architectural drawings for reconstruction of 3D BIM models. Autom. Constr. **113**, 103082 (2020)
Zhang, Z., et al.: Ming Dynasty Residences in Huizhou. China Architecture & Building Press, Beijing, China (1957)

A Quantitative Study on Urban Fabric of Historical Areas with Graph Analysis: Case Studies of Datong and Nanjing, China

Yijia Zhu and Lian Tang(✉)

School of Architecture and Urban Planning, Nanjing University, Nanjing, China
502023360032@smail.nju.edu.cn, tanglian@nju.edu.cn

Abstract. The exploration of quantitative methods for urban fabric, grounded in urban morphology theory, offers digital tools to accurately characterize the physical attributes of urban fabric. While exiting quantitative approaches primarily focus on general morphological features such as dimension, shape, intensity, etc. There remains a gap in developing targeted methodology based on buildings types as well as buildings composition. In this study, we examine the urban fabric of two historical districts in China as case studies. By applying graph theory, we attempt to quantify the compositional patterns of urban fabric by setting plots and building typological units as nodes. This approach reveals differences in building types and urban fabric organization between the two cases. Our findings demonstrate that quantitative statistics and graph theory analysis, based on parameters such as building width, depth, and courtyard units, effectively uncover the patterns across different fabric compositions. Furthermore, metric intervals derived from graph theory indicators—such as node degree and depth values—provide a means to distinguish the unique characteristics of each case. This study highlights the potential of integrating morphological analysis with graph theory to advance the understanding of urban fabric complexity of historical area.

Keywords: Building types · Gragh theory · Plot patterns · Urban fabric

1 Introduction

Urban morphology creates a multi-level structure of urban form, offering a systematic framework to analyze how buildings and urban environments develop and change. Many scholars highlight the close link between building types and urban fabric. Building types are the basic units and framework of the urban system. The concept of Typomorphology integrates urban morphological analysis with typological evolution, providing a more comprehensive view of urban form's structure and changes (Moudon, 1986). For traditional Chinese urban fabric, much research has explored its composition and patterns from the perspective of building types (Whitehand and Gu, 2007; Chen and Romice 2009). However, existing studies mainly focus on the evolution of types and texture, without clearly describing the mathematical relationship between building types and fabric composition.

Y. Liu et al. (Eds.): CDRF 2025, *Transindividual Intelligence*, pp. 420–429, 2026.
https://doi.org/10.1007/978-981-92-0615-5_36

China's construction methods such as material use and bay width, as well as architectural hierarchy regulations, determine the existence of this "mathematical relationship." For example, the Yingzao Fashi (Treatise on Architectural Methods) regulated traditional timber architecture and modular systems, while the Lifang system influenced urban land division, and the Yufu system dictated building scales based on social and political status. Under the combined influences of ventilation and lighting needs, construction techniques, and institutional regulations, traditional residential clusters formed by enclosing courtyards with multiple single buildings, known as "courtyard houses", are common in most regions of China (Zhang et al., 2020). Additionally, climate, culture, and topography together shape the unique urban fabrics in different regions of China.

To reveal the morphological features of historical areas more precisely, many quantitative studies have emerged. However, these methods are less likely to capture the unique characteristics of traditional Chinese courtyard houses, such as "Jian" (bay), "Jin" (depth unit) and courtyard sequences, which are crucial for understanding Chinese historical urban fabric. Meanwhile, existing metrics focus on visible geometric features, but struggle to reflect differences caused by underlying topological properties and organizational logic. Graph theory, which represents complex relationships between entities, has shown potential in urban morphology studies. Therefore, this paper takes Nanjing and Datong as examples, using graph analysis to quantify the relationship between building types and urban fabric. It aims to reveal the organizational structure of historical fabric and compare how building types influence the fabric across different regions.

2 Literature Review

In the quantification of urban form, based on the integrated element hierarchy framework derived from Cozen and Caniggia's theories (Kropf, 2014), traditional fabric have generated over 140 metrics across different levels such as blocks, streets, plots, and buildings. These metrics describe the multi-scale morphological features of historical fabric, including dimension, shape, and intensity (Dibble et al., 2019; Fleischmann et al., 2021). At the Plot level, existing metrics can measure plot size and shape, but they still fall short in describing plot patterns. At the Building level, existing metrics mainly focus on quantifying individual buildings, but they fail to describe the relationships between buildings. In traditional Chinese courtyard houses, the relationship between yards and buildings is crucial for understanding urban fabric. At the Area level, while the existing metrics do account for yards. However, the yards in traditional Chinese courtyard houses are not isolated or residual "open spaces," but are intrinsically linked to building types. Current metrics can describe yard size and shape, but they do not capture the organizational relationship between yards and buildings. Additionally, Current quantitative metrics lack recognition and description of "building type" units. Most existing metrics focus on individual elements (such as buildings, plots, and blocks) without integrating building type units as an intermediate level.

Moreover, there are many metrics describing visible geometry, and research on topological relationships mostly concentrated on street (Hillier and Hanson 1984; Marshall, 2005). The uniqueness of Chinese building types necessitates expanding topological metrics to analyse relationships between urban elements. Therefore, geometric-based quantification methods show limitations in analyzing plot patterns, building types, and their relationships with fabric. In recent years, graph theory has been applied to urban morphology studies. By defining nodes and edges, it can describe building plan (Dong et al., 2022; Xiao and Liu, 2023), street network organization (Marshall, 2016), and regional structure (Esch et al., 2014). It has strong abstraction and spatial analysis capabilities. However, how to define nodes and edges of graph, and how to integrate graph theory with morphological metrics to create a quantification system that describes both structural form and specific building and fabric forms, still requires further explorationh.

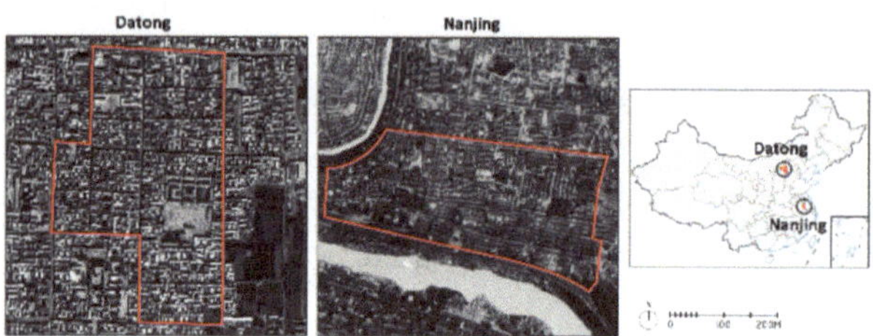

Fig. 1. Fabric satellite images of two research samples and the location of cities in China

3 Methodology

3.1 Traditional Fabric Diagram Based on Building Types

The study selects representative historical areas from Datong in northern China and Nanjing in southern China, namely Datong Gulou East Street and Nanjing Mendong area (Fig. 1). The data is based on vector maps suitable for statistical analysis, which are derived from cross-referencing satellite images and historical maps. These maps include information on streets, blocks, plots and buildings.

Traditional Chinese courtyard houses are characterized by buildings arranged around a central yard, with multiple courtyard units forming a building complex. Each courtyard unit is referred to as a "Jin". However, even within China, there are regional differences in describing architectural features. For example, regarding the counting of "Jin," in northern areas like Datong, it is based on the core yard (a standard courtyard enclosed by a main hall, south room, and east and west wing rooms), while in southern regions like Nanjing, it is based on the main building such as the main hall. Therefore, Therefore, this article establishes a unified method for dividing courtyard units, laying the foundation for subsequent fabric diagram and Graph node settings based on building types.

As shown in Fig. 2, for Datong, horizontal unit = n*standard Jin + m*side courtyard, vertical unit = n*regular Jin + m*First Jin. The depth composition of first Jin can be three cases: ①South room ②Forecourt ③South room + Forecourt. A regular Jin is defined as a courtyard unit which depth composed of a core yard + a main hall. For Nanjing, horizontal unit = n*Lu ("Lu" refers to a set of courtyards arranged along an axis in traditional courtyard houses in South region. The main axis usually has a strict layout of houses and courtyards, while the side Lu are relatively flexible). vertical unit = n*regular Jin + m*First Jin. The depth composition of an first Jin is only one case: ①South room.

Based on the above division of courtyard units, while setting the following statistical items: width, depth, and area of plot and regular Jin. It is used to supplement the differences in the fabric composition of the units with more tangible geometric dimensions.

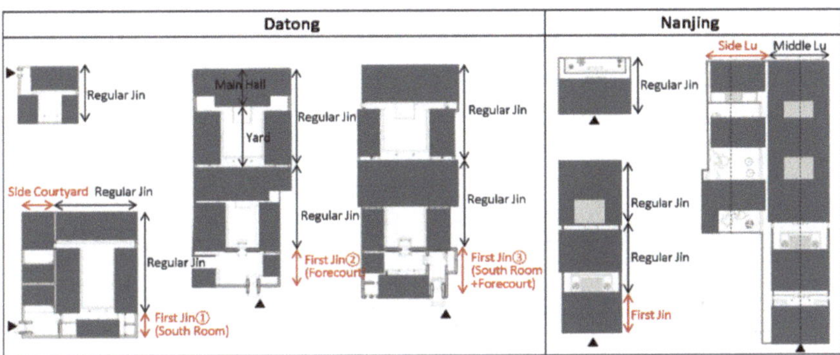

Fig. 2. Identification and unit division of traditional Chinese courtyard houses

3.2 Architectural Typological and Graph Setting

In spatial studies, basic spatial elements are typically regarded as "nodes," and the channels connecting these elements are "edges." The spatial topological structure formed by network can represent the spatial structure. Therefore, based on the analysis and setting of the courtyard type mentioned above, as shown in Fig. 3, the block boundaries are the main objects of study (usually, a rectangular block is divided into four boundaries: east, west, south, and north). The paths that can be directly reached in space are the edges, and graphs are constructed on two levels.

Layer 1 uses plots as nodes, divided into "○" traditional courtyard plots and "●" plots updated to modern buildings or unrecognizable as courtyard types (based on Sect. 3.1, if no regular Jin courtyard unit is identified, it is classified as a non-courtyard plot). Plots connect directly to block boundary nodes or via alley nodes based on the location of entrances relative to streets.

As an important unit of traditional Chinese courtyard houses, Jin has internal structural integrity and external connectivity. Therefore, Layer 2 defines the graph based on the courtyard unit characteristics discussed earlier, using each courtyard unit as a node to further deconstruct the "○" traditional courtyard plot from Layer 1. The new node types include Regular Jin, First Jin, and Side Courtyard. Here, "edges" not only represent the connection between plots and external streets via entrances but also signify the internal unit connections and path organization within courtyard residences.

For each block boundary under the two level settings, the following metrics are calculated: node degree, number of nodes and edges in the network starting from the block boundary, network density, and maximum depth value. The node degree counts the number of edges directly connected to the block boundary node, indicating the number of openings. The number of nodes and edges in the network represents the network's scale and the amount of space organized by the block boundary. Network density, measured by the formula $2E/(V(V-1))$, where E is the number of edges and V is the number of nodes, reflects the network's completeness and the tightness of spatial connections. The maximum depth value, which is the diameter of the network starting from the block boundary, indicates the level of spatial privacy as one moves deeper into the network. By combining these metrics with the inherent attributes of the block boundary, such as length and orientation, the effectiveness of graph theory in revealing the role of building types in fabric formation at different levels is assessed.

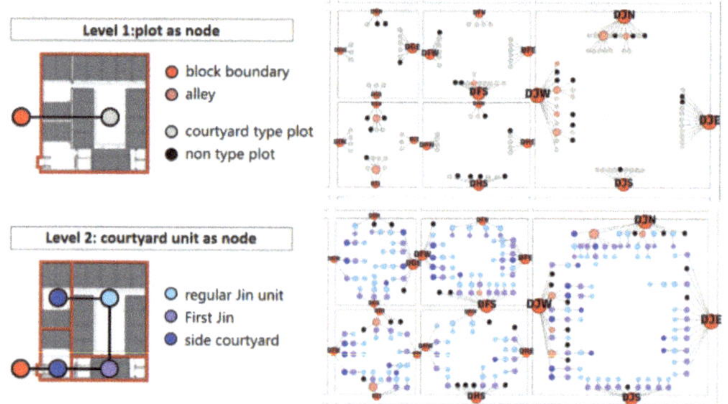

Fig. 3. Two level of graph settings and block network starting from the block boundary

4 Result

4.1 Urban Fabric Mapping and Data Statistics

Combining the 1958 "Datong Topographic Map" with the 1965 satellite map, the Datong study area covers 212,078 m^2 with 314 plots. Combining the 1937 Nanjing property map with the 1929 aerial photo, the Nanjing study area covers 161,484 m^2 with 385 plots. After identifying plots that conform to traditional courtyard types and removing those non courtyard type plots, Datong and Nanjing have 228 and 313 plot samples remaining (Fig. 4-a). Based on the building type settings discussed earlier, these plots are further divided into courtyard units (Fig. 4-b).

4.2 Dimensions Statistics and Comparative Analysis

Comparison at the plot level shows that in Datong, most plots have width of 14–26 m and depth of 15–30 m; in Nanjing, most plots have width of 2–20 m and depth of 8–38 m. The comparison reveals that: Nanjing plots have a smaller width but a wider

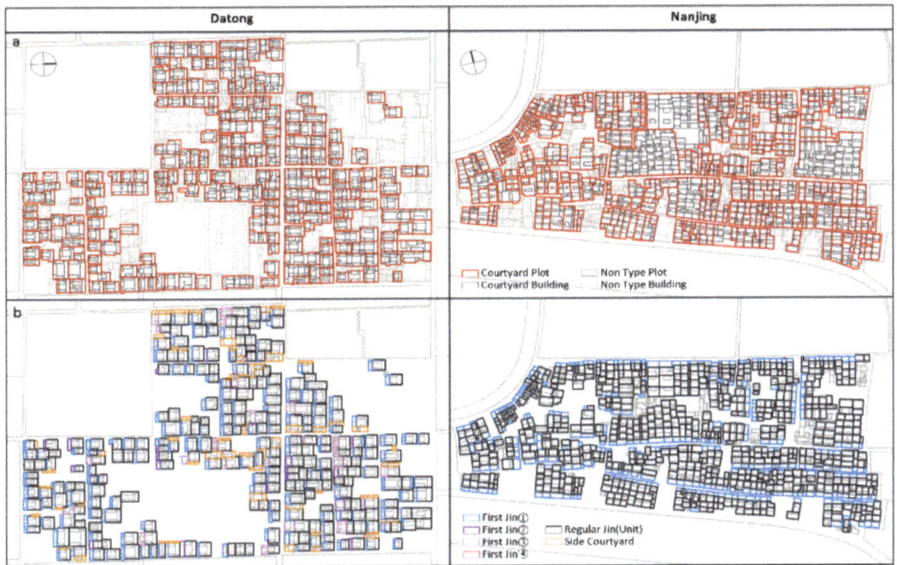

Fig. 4. Urban fabric mapping and courtyard unit diagram

range of depth, also have a smaller area. Comparison at the courtyard unit level shows that in Datong, most regular Jin courtyard units have width of 14–24 m and depth of 13–21 m. In Nanjing, most regular Jin courtyard units have width of 4–15 m and depth of 8–14 m. The comparison reveals that Datong's regular Jin are wider and deeper than those in Nanjing. Nanjing's regular Jin have a relatively consistent depth, while their width varies more significantly.

4.3 Graph Data Statistics and Comparative Analysis

Figure 5 shows the Graph diagrams of two cities at two levels, and shows the displays the network type distribution of Level 2, arranged by the direction and length of block boundaries, with the depth values of each node listed vertically.

(1) Comparison of block boundary length and node degree:

Node degree reflects the number of openings (plot and alley entrances) in a block boundary. As shown in table of Fig. 6, Datong has the lower average node degree, meaning fewer openings. However, the average ratio of length/node degree is similar, indicating that entrances are placed every 25–28 m along block boundaries in two cities. This suggests that the logic of the building type composition in two cities is similar, with differences mainly in block division. The bar chart in Fig. 6 shows the length of boundaries in different directions and the line chart shows the node degree. In Datong, east-west boundaries are shorter and have lower node degrees than north-south boundaries. In Nanjing, most east-west boundaries are short, with many having zero node degrees, indicating entrances are mostly on north-south. Nanjing's node degree trend

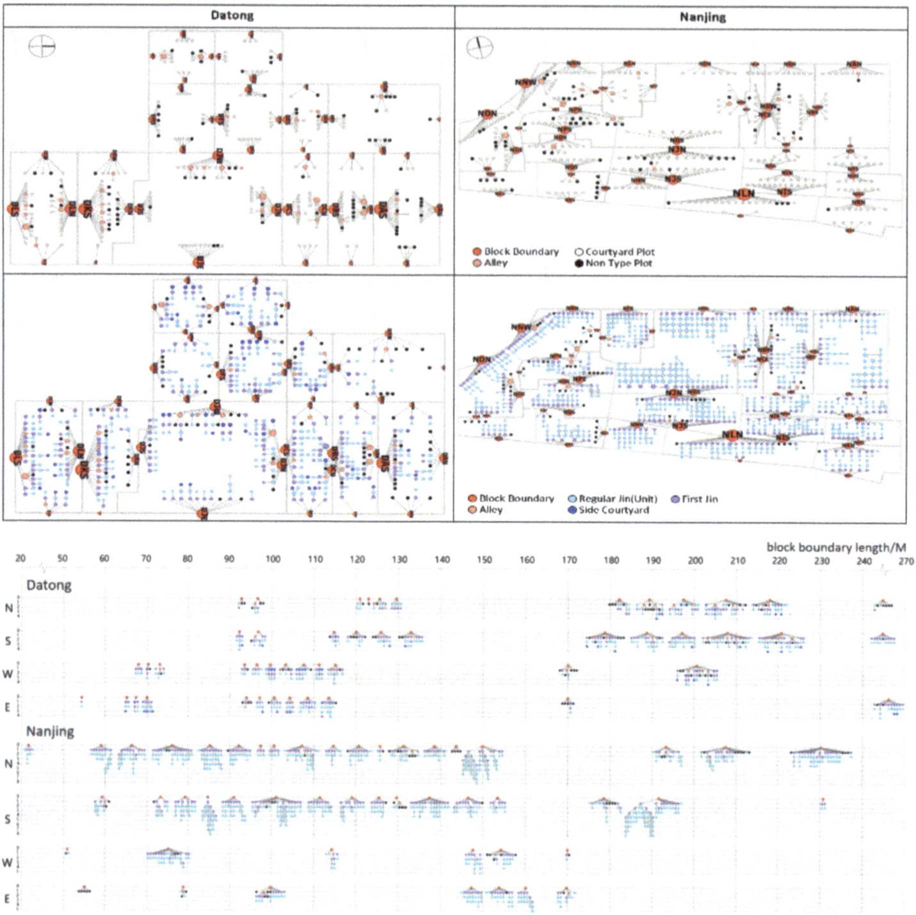

Fig. 5. Graph diagram of two cities of two levels; Network type distribution of Level 2

doesn't match length due to multi-Lu courtyards occupying widths but having only one entrance, lowering node degree (Fig. 7-b).

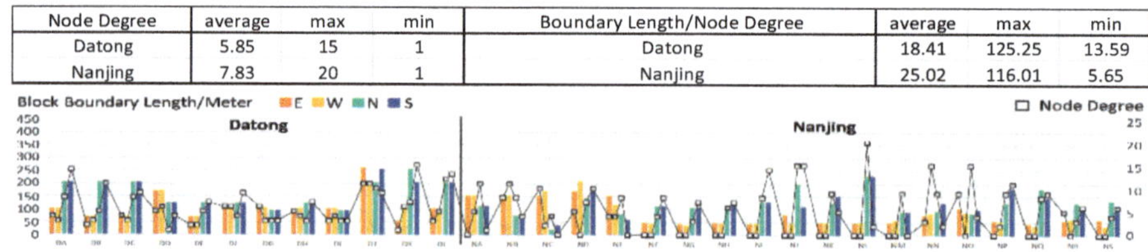

Node Degree	average	max	min	Boundary Length/Node Degree	average	max	min
Datong	5.85	15	1	Datong	18.41	125.25	13.59
Nanjing	7.83	20	1	Nanjing	25.02	116.01	5.65

Fig. 6. Comparative analysis of block boundary segment length and node degree

(2) Comparison of network nodes, edges, and density:

As shown in Table of Fig. 7, Level 1 represents the total connected plots, and Level 2 represents the total connected courtyard units. The ratio of Level 2 to Level 1 indicates the average number of courtyard units per plot, with Nanjing having a higher value. Network density measures the completeness and tightness of connections. With the same number of nodes, Nanjing's network density is higher due to its multi-Lu large courtyards, which add horizontal connections between Lu, increasing the number of edges. While Datong basically maintains vertical development. For example, Nanjing's NAS (Fig. 7-b) and Datong's DBS (Fig. 7-c) both have 28 nodes, but NAS has a higher network density (0.11) than DBS (0.07).

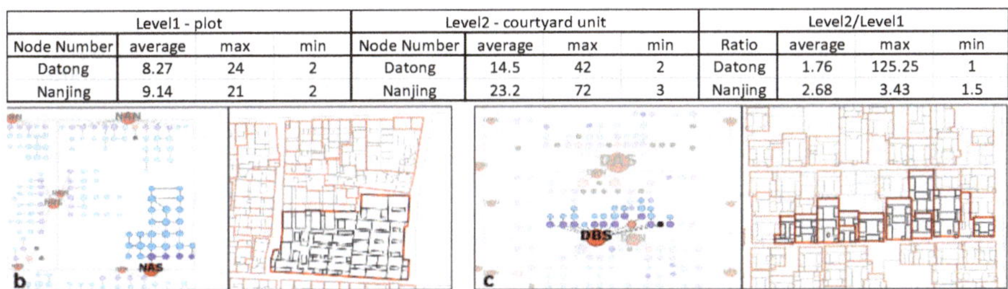

Level1 - plot				Level2 - courtyard unit				Level2/Level1			
Node Number	average	max	min	Node Number	average	max	min	Ratio	average	max	min
Datong	8.27	24	2	Datong	14.5	42	2	Datong	1.76	125.25	1
Nanjing	9.14	21	2	Nanjing	23.2	72	3	Nanjing	2.68	3.43	1.5

Fig. 7. a. Comparative of node number b. Boundary NAS (Nanjing) c. Boundary DBS (Datong)

(3) Comparative analysis of maximum depth values and node degree:

As shown in Fig. 8, Level 1's depth values vary only based on the alleys, divided into two categories: 1 and 2. The average value of Level 1 indicates the probability of alleys appearing in the fabric. Datong has the higher average maximum depth value, suggesting more alleys within the block to connect internal plots. Level 2, influenced by the multi-Lu and multi-Jin large courtyards in Nanjing. The superposition of horizontal and vertical hierarchical sequences has led to three peaks with a depth of 8 and the higher average maximum depth value.

From the scatter plot distribution at Level 2, the trend of node degree - maximum depth value generally shows a positive correlation. Datong is mainly distributed in the area with low node degree and low depth value, but there are also special spots with high node degree and low depth value. Nanjing has a wider range of depth-degree distribution, but there are special spots with low node degree and high depth value. These special spots in Nanjing usually appearing in large courtyards houses in the north-south direction block boundaries.

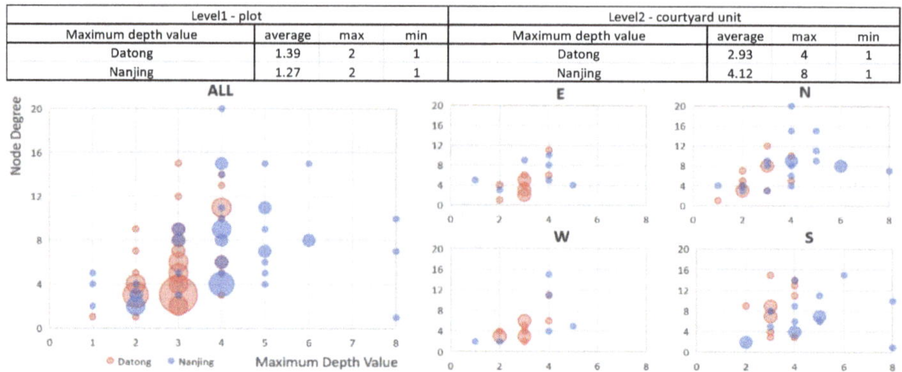

Level1 - plot				Level2 - courtyard unit			
Maximum depth value	average	max	min	Maximum depth value	average	max	min
Datong	1.39	2	1	Datong	2.93	4	1
Nanjing	1.27	2	1	Nanjing	4.12	8	1

Fig. 8. Comparative of maximum depth value and node degree (red Datong, blue Nanjing)

5 Conclusion and Discussion

The study obtained the following three conclusions:

1. Graph theory serves as an effective tool for analyzing the characteristics of urban tissue in historical areas. By defining courtyard units as nodes and their interconnections as edges, the number of nodes and their depth values—particularly those connected to edges along block perimeters—can reveal plot patterns and building type compositions. This approach provides a structured way to quantify and compare urban fabric configurations.
2. The courtyard unit in Chinese architecture serve as fundamental units of urban fabric and can be employed as a measurement scale. The width of a courtyard unit (often three bays) and the depth of a courtyard unit (one Jin), along with their specific dimensions, vary between the two cities studied, reflecting differences in cultural, climatic, and regional contexts.
3. Datong and Nanjing exhibit distinct distribution thresholds in graph depth values, particularly in relation to the node degree at block boundaries. These thresholds serve as effective indicators of differences in the morphological characteristics of urban fabric.

The study identifies the buildings as type units first, and divides the type units into courtyard units, and tries to express the characteristics of the urban fabric through the description of the relationship between the street-block-plot-courtyard units. The research was conducted based on the plot boundaries, but due to the lack of information, the location of entrances and exits, and the connection of courtyard units were inferred based on the existing building types, which may be inaccurate and need to be verified subsequently.

Acknowledgements. This research is funded by the Natural Science Foundation of Jiangsu Province (BK20242016), the JPI Urban Europe and the National Natural Science Foundation of China (NSFC-JPI_UE) (Grants 72361137008) and the National Natural Science Foundation of China (No.51708274).

References

Chen, F., Romice, O.: Preserving the cultural identity of Chinese cities in urban design through a typomorphological approach. Urban Des. Int. **14**(1), 36–54 (2009)

Dibble, J., et al.: On the origin of spaces: Morphometric foundations of urban form evolution. Environ. Plan. B: Urban Anal. City Sci. **46**(4), 707–730 (2019)

Dong, Y., Han, D., Trisciuoglio, M.: A graphical method of presenting property rights, building types, and residential behaviors: A case study of Xiaoxihu historic area, Nanjing. Front. Archit. Res. **11**(6), 1077–1091 (2022)

Esch, T., Marconcini, M., Marmanis, D.: Dimensioning urbanization—An advanced procedure for characterizing human settlement properties and patterns using spatial network analysis. Appl. Geogr. **55**, 212–228 (2014)

Fleischmann, M., Romice, O., Porta, S.: Measuring urban form: overcoming terminological inconsistencies for a quantitative and comprehensive morphologic analysis of cities. Environ. Plan. B: Urban Anal. City Sci. **48**(8), 2133–2150 (2021)

Hillier, B., Hanson, J.: The Social Logic of Space. Cambridge University Press, Cambridge (1984)

Kropf, K.: Ambiguity in the definition of built form. Urban Morphol. **18**(1), 41–57 (2014)

Marshall, S.: Streets and Patterns. Spon Press, London and New York (2005)

Marshall, S.: Line structure representation for road network analysis. J. Transp. Land Use. **9**(1), 29–64 (2016)

Moudon, A.V.: Built for Change: Neighborhood Architecture in San Francisco. MIT Press, Cambridge (1986)

Whitehand, J.W.R., Gu, K.: Extending the compass of plan analysis: A Chinese exploration. Urban Morphol. **11**(2), 91–109 (2007)

Xiao, X., Liu, Z.: Quantifying urban network transitions with evolution degree. Nexus Netw. J. **25**(Suppl 1), 471–480 (2023)

Zhang, Q., Hua, X., Huang, H.: Zhongguo Chuantong Minju Gangyao [Outline of Traditional Chinese Residential Buildings]. China Architecture & Building Press, Beijing (2020)

Urban Data Analytics, Modeling and Simulation

Pedestrian Trajectory Simulation on University Campus Using Deep Inverse Reinforcement Learning: A Case Study of Southeast University Wuxi Campus

Wang Guangjin, Huang Ruike, Li Xiang, and Li Li(✉)

Southeast University, Si-Pai-Lou 2#, Nanjing, Jiangsu Province, China
101012053@seu.edu.cn

Abstract. Campus users' daily activities rely on on-campus facilities, making campus environmental organization crucial to their quality of life. Consequently, campus space optimization is a key research focus. However, existing methods, using pedestrian flow simulation tools based on spatial topology or deep-learning-based prediction models, often neglect environmental factors' influence on behavior. Campus users' behavior is goal-oriented, and ignoring environmental elements may lead to design misjudgments. Inverse reinforcement learning(IRL) can model human-environment interactions in real-world settings, offering a novel approach to pedestrian trajectory simulation. Thus, this study uses Southeast University Wuxi Campus as a case study, collecting real pedestrian trajectories and environmental data. A deep maximum entropy IRL method constructs a pedestrian trajectory simulation model. After validation, the model effectively simulates behavior under environmental influences and quantifies their impact, providing a useful tool for campus space optimization.

Keywords: University Campus · Pedestrian Trajectory Simulation · MaxEnt IRL · Environmental Behavior

1 Introduction

University campus usage continuously adapts to changes in faculty and student population, discipline development, and urban environment [1, 2]. This dynamic adaptation creates a gap between actual and planned functional organization, making rational campus space development an ongoing iterative process. To meet campus structural optimization demands, pedestrian flow simulation methods identifying environmental factors offer valuable theoretical support.

In previous crowd flow simulation studies, some relied on spatial topological relationships for modeling and simulation [3], but their validity was questionable in intentional campus environments. Others used statistical methods but struggled to fully reveal environmental factors' driving mechanisms on behavior [4]. While some combined LSTM or GCN with attention mechanisms to predict pedestrian flow [5, 6], they often overlooked

Y. Liu et al. (Eds.): CDRF 2025, *Transindividual Intelligence*, pp. 433–442, 2026.
https://doi.org/10.1007/978-981-92-0615-5_37

pedestrian-environment relationships or focused on limited spatial nodes. Additionally, some used multi-agent models based on attraction and social forces, adjusting environmental factor parameters to make the simulated trajectories more realistic [7]. However, manual parameter adjustments compromised model objectivity. Therefore, this study introduces a deep maximum entropy inverse reinforcement learning (IRL) multi-agent pedestrian flow simulation algorithm. It leverages neural networks to learn complex human-environment interactions and map pedestrian movement to environmental factors.

Southeast University Wuxi Campus, fully operational since 2022, is an ideal case study due to its recent rapid academic expansion and discipline adjustments. In this context, this paper employs Wi-Fi probe technology suitable for the mesoscopic scale to collect real user trajectories within the campus as expert trajectories for agent learning. Through extensive training and comparison of real and simulated trajectories, a reliable simulation model is developed, and the model's reliability is tested against the campus's adjustment needs.

2 Data Collection and Pedestrian Flow Simulation Methods

This section mainly consists of two parts: data collection and pedestrian flow simulation. The acquisition of effective training data forms the foundation of pedestrian flow simulation, which involves the rationality of trajectory data collection and processing methods, as well as the extraction of environmental factor feature data. The pedestrian flow simulation section covers the construction of the algorithm, as well as the tuning and training of model parameters.

2.1 Collection and Preprocessing of Expert Trajectory Data

The application of Internet of Things (IoT) technology in trajectory collection effectively overcomes the limitations of traditional survey methods, such as resource waste and subjectivity. Currently, technologies such as GPS, Wi-Fi, and radar positioning are widely used for trajectory monitoring in urban environments. Among these, Wi-Fi probe technology is particularly well-suited for trajectory collection in mesoscopic-scale areas like university campuses [8].

2.1.1 Collection and Cleaning of Expert Trajectory Data

Based on the effective detection range of 50 m for Wi-Fi probes, this study deployed 19 monitoring devices at key transportation and public space nodes on the Southeast University Wuxi Campus (Fig. 1). The device layout was designed to ensure the uniqueness of the paths between detection points while minimizing the overlap of detection areas between different monitoring points. Over the course of two weeks, 28.87 million raw data points were collected.

However, the raw data contains a significant amount of noise, such as user data with only one record, data that fluctuates between 1 to 3 probes, and data with missing values. Additionally, due to the concentration of campus facilities, overlap in the detection ranges

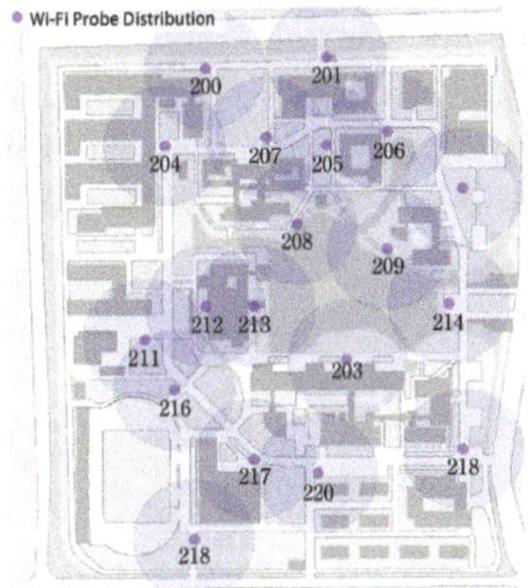

Fig. 1. The facility layout, including the placement of probes

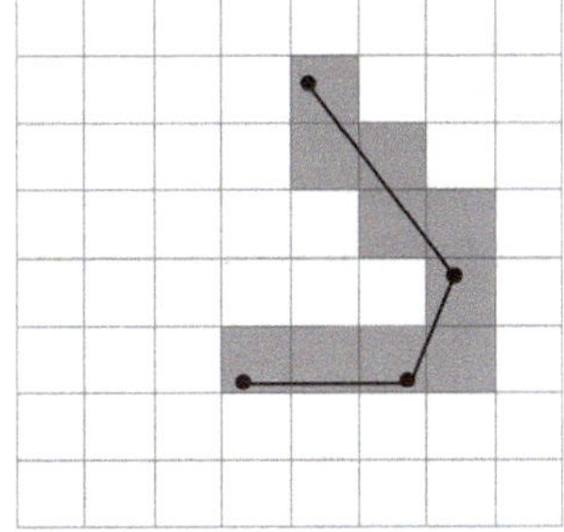

Fig. 2. The principle of trajectory mapping

of devices is inevitable, resulting in a large number of ping-pong data points that oscillate between multiple monitoring points. To obtain valid trajectory data, this study reasonably conducted data cleaning, trajectory reconstruction, and filtering progresses. Ultimately, over 260,000 valid trajectories were extracted, covering 2,306 users.

2.1.2 Grid Mapping and Splitting of Expert Trajectories

In the inverse reinforcement learning(IRL) environment, the trajectory of the agent consists of a sequence of discrete states (grid), such as [(100, 1), (101, 0), (101, 0), …], where the first integer in each tuple represents the state (grid cell) number and the second integer represents the action: 1 for right, 0 for staying, 2 for left, 3 for up, and 4 for down. To match the agent's trajectory, the expert trajectory must be converted to the same format by mapping it onto the corresponding grids (Fig. 2).

In the experiment, directly using the expert trajectories mapped onto grids for training yielded suboptimal results. However, splitting trajectories into multiple sub-trajectories significantly improved performance, as human behavior changes over time and goals are not always constant. Therefore, all trajectories were split into sub-trajectories, with an average length of 22 empirically determined.

2.2 Extraction of Environmental Feature Data

Through a systematic review of the literature, campus environmental factors were classified into eight major categories: buildings, squares, sidewalks, roads, vegetation, water bodies, parking lots, and other facilities. Given their spatial prominence and significance, buildings and sidewalks were further subdivided into two levels, yielding 41 total environmental factor types. Then environmental feature data was extracted based on these categories (Fig. 3) through the following steps: First, the campus master plan was rasterized into a 35×40 grid, balancing simulation accuracy (enhanced by finer grids) with computational power limitations. Next, the locations of each environmental factor were labeled, with the value of the grid cell where the environmental factor is located set to 1, and 0 for all other cells. To reflect the influence of environmental factors on surrounding areas, the feature value of all other grid cells was calculated based on their distance from the grid cell containing the environmental factor. For example, for a particular environmental factor, the value of a grid cell in its corresponding environmental feature grid is equal to the sum of the values of all the factor's grid cells, weighted by the inverse of their distances to the target grid cell.

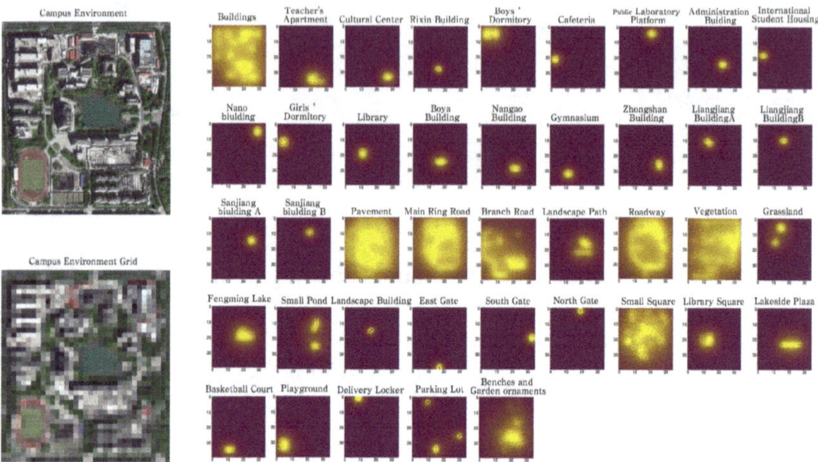

Fig. 3. Environmental feature network extraction

2.3 Pedestrian Flow Simulation Method

This study's pedestrian flow simulation method is based on maximum entropy inverse reinforcement learning (MaxEnt IRL). After obtaining expert trajectory data for IRL, the algorithm is constructed and integrated with environmental feature data. By mimicking expert decision-making and behavior, a pedestrian flow simulation modelperceiving the campus environment is developed.

2.3.1 Principle of MaxEnt IRL and Algorithm Construction

MaxEnt IRL builds upon on reinforcement learning (RL). A key distinction is that RL requires manually defining a reward function (which evaluates the feedback the agent receives after taking an action in the environment), and its resulting policy often needs subjective testing to meet desired goals. In contrast, IRL infers the reward mechanism by learning from expert data, essentially mimicking expert behavior. A challenge with IRL is that multiple policies can be learned from the same expert data. MaxEnt IRL addresses this by selecting the policy with the highest information entropy, ensuring an optimal solution. The concept of information entropy, introduced by Shannon in 1948 [9], is crucial here: higher entropy indicates the agent has captured more features from expert trajectories while still effectively fitting them. The mathematical expression for entropy, $H(X)$, is a law defined by the formula:

$$H(X) = -\sum_{i=1}^{n} P(x_i) \log P(x_i) \tag{1}$$

Where $P(x_i)$ is the probability of the random variable X taking on the i-th value. A detailed mathematical derivation and the principles of MaxEnt are provided in Suzuki's algorithm introduction [10], and will not be discussed in detail here.

The reward mechanism derived from MaxEnt IRL is represented by a neural network. Its input comprises environmental feature values, and its output is the reward for each state. It iteratively updates the reward neural network's parameters, allowing the agent's state visit frequency to gradually approach the expert trajectory's. The algorithm structure (Fig. 4) involves the following detailed steps: ①Calculate the frequency of expert visits to each state and the distribution of expert trajectory starting points. ②Extract environmental feature data and initialize the reward distribution for each state by randomly generating neural network parameters. ③Combine the state's transition probabilities and the discount factor γ to perform V-value (State Value) iteration until convergence, then use the V-values to derive the agent's policy. ④Forward computation: Obtain the agent's visit frequency to each state based on the policy. ⑤Combine the expert's trajectory starting point distribution to compute the difference in visit frequency between the agent and the expert at each state, then compute the gradient of the neural network parameters and update the parameters.

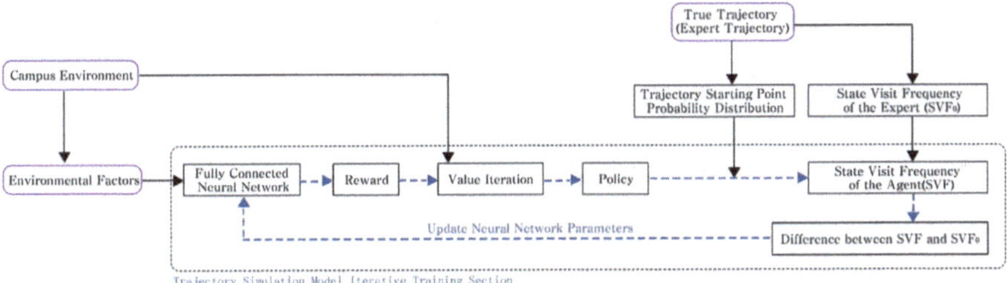

Fig. 4. Algorithm overall framework

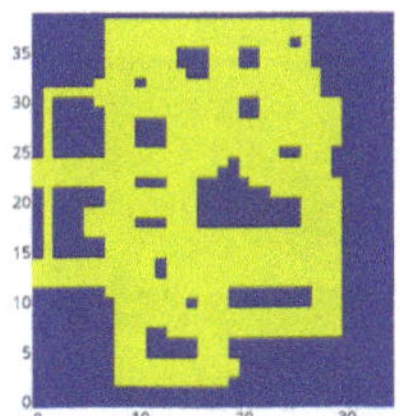

Fig. 5. Activated states (states visited by expert trajectories)

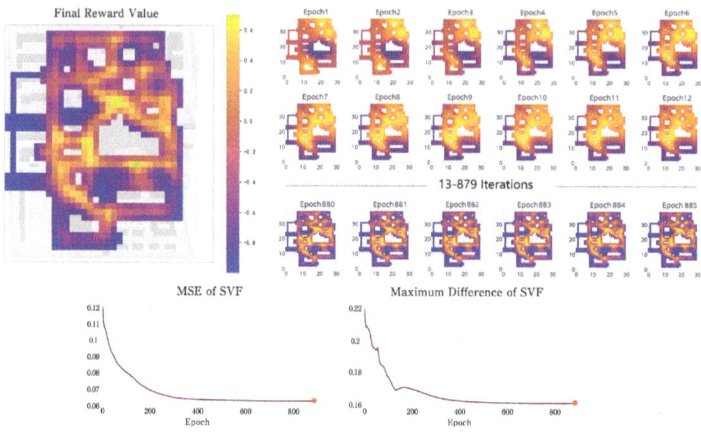

Fig. 6. Iteration process of model training

2.3.2 Model Training

The core of model training is the iterative optimization of the reward neural network's parameters, which requires setting hyperparameters. Following extensive simulation experiments and a grid search for hyperparameter tuning, the final configuration is as follows: a learning rate of 0.0001 and an L2 regularization coefficient of 0.5. The neural network consists of seven fully connected layers with dimensions (82, 164, 328, 328, 328, 164, 82), followed by a ReLU output layer. ELU activation is used between fully connected layers to prevent gradient vanishing and ReLU dead zones.

To avoid computational waste, the training environment is restricted to states visited by expert trajectories (Fig. 5). Additionally, to simulate the uncertainty in state transitions, the agent transitions based on reward value with 75% probability and randomly with 25%. In the V-value iteration, the discount factor γ for V-value iteration, which weights future rewards, is set to 0.95.

After 885 iterations in the formal training, both the mean squared error and the maximum difference in state visitation frequencies between the agent's and expert's trajectories steadily decreased, indicating that the agent effectively learned the characteristics of the expert trajectory. Additionally, the reward value iteration visualization (Fig. 6) clearly illustrates the gradual refinement of the reward values.

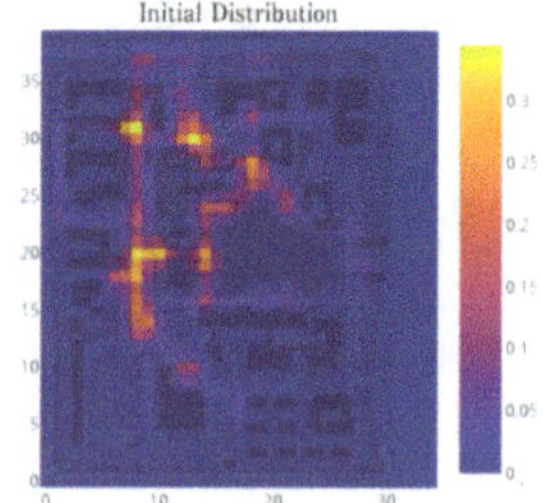

Fig. 7. Initial distribution (left)

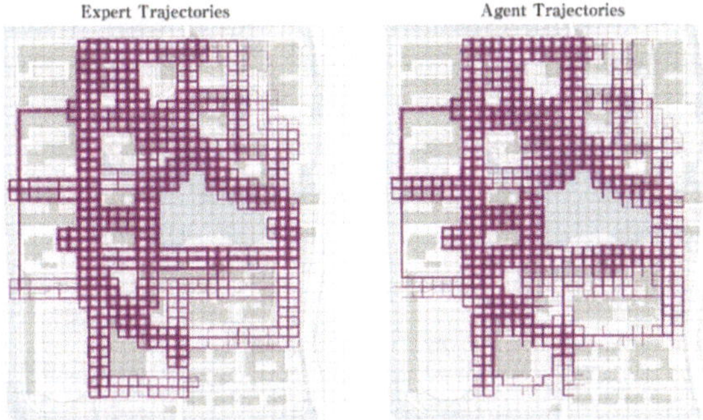

Fig. 8. Comparison of agent-simulated trajectories with expert trajectories (right)

3 Simulation Results and Applications

This chapter covers model validation and application. Pedestrian flow simulation is validated from three perspectives: ①Assessing state reward distribution (Fig. 6) alignment with campus conditions by incorporating trajectory starting state probability (Fig. 7); ②Evaluating the agent's learning performance by comparing its trajectory with the expert's (Fig. 8); ③And quantifying environmental factor attractiveness on pedestrian behavior by extensively sampling the reward neural network to extract reward values. Following this validation, the application section presents a case study on Zhongshan Courtyard's vitality at Southeast University Wuxi Campus, demonstrating the simulation tool's practical utility.

3.1 Analysis of Simulation Results

The preliminary analysis of the final reward distribution (Fig. 6) mainly reveals as follows: ①The highest reward values at Southeast University Wuxi Campus are located near the library, cafeteria, and administrative building, matching the highest foot traffic in the expert trajectory, which aligns with the campus layout: The library serves as the main study space, the cafeteria is a daily necessity, and the administrative building houses faculty, staff, and students. ②The foot traffic in the dormitory area is high, but the reward value here is not as high. This is not an anomaly, as the starting probability near the dormitories is quite high (Fig. 7). ③The small garden between Boya Building and South High Courtyard has low reward values and foot traffic, which aligns with reality. Although Boya Building is the only teaching facility, low traffic is due to the South High Courtyard's underutilization, leading post-class students to return to dormitories or visit more functional areas like the library. These findings indirectly validate the alignment of the model's reward distribution with the actual campus usage conditions.

In the comparison of the agent's and expert's trajectories (Fig. 8), line thickness and opacity represent foot traffic. Overall, the agent's trajectory features mostly aligns with the expert's, with differences in distribution observed only in localized areas. These discrepancies are due to two factors: limited precision in the environmental grid and features, hindering accurate foot traffic learning, and a 25% random transition probability, leading to deviations in decision-making.

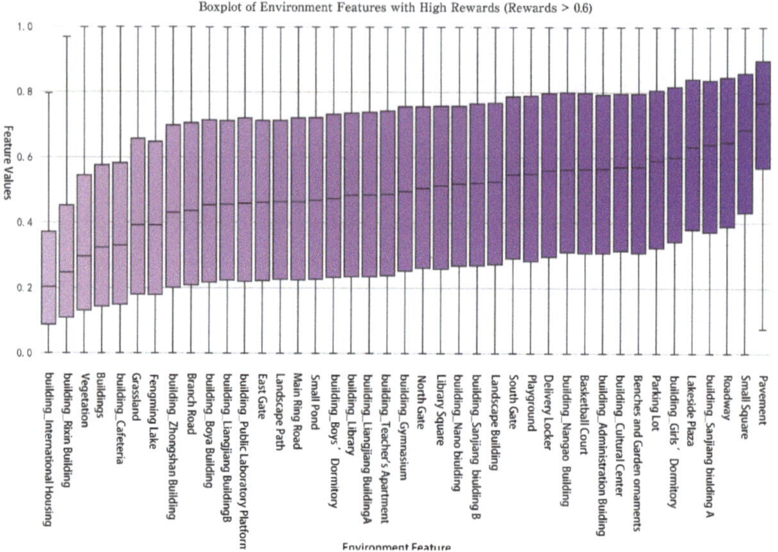

Fig. 9. Feature values of environmental factors

To estimate the feature values of environmental factors, 1 million samplings of the feature-value to reward-value mappings. In each round, 41 random floating-point numbers (0–1) representing 41 environmental factors, and the reward values computed by the neural network were recorded. The final dataset, filtered for overall reward values greater than 0.6, enabled analysis of individual factor feature values. Analysis (Fig. 9) shows that activity infrastructure like sidewalks, small squares, roadways, and lakeside squares have higher feature values, while dormitories, Rixin Building, vegetation, and buildings have lower values. Notably, office buildings in the northern campus, such as Sanjiang Yuan and Liangjiang Yuan, have higher feature values compared to those in the south, like Nanguo Yuan, Rixin Building, and Zhongshan Yuan. This may be due to lower usage and fewer facilities in the south. Anomalously, although the lawn facilities are located near the northern office and dormitory areas, their feature values are relatively low, possibly due to a lack of shelter from tall vegetation and seating.

3.2 Improving Campus Utilization Status with the Trained Model

The pedestrian flow simulation tool aims to provide timely feedback for campus designers and administrators. At Southeast University Wuxi Campus, a 200-square-meter café in the first-floor lobby of Zhongshan Courtyard has struggled with low vitality since its opening. Leveraging the pedestrian flow prediction model, this study attempts to enhance the area's spatial reward value and vitality by adjusting environmental factors. Specifically, based on the model's derived environmental attractiveness values, we propose adding functional areas similar to the northern office zone within Zhongshan Courtyard and reallocating some administrative functions from the Administrative Building, and adding seating and landscape to the small square near the café. Inputting these adjusted environmental factors into the simulation model showed significant increase in reward values and agent trajectories near Zhongshan Courtyard (Fig. 10). But these simulated outcomes currently lack post-modification empirical validation.

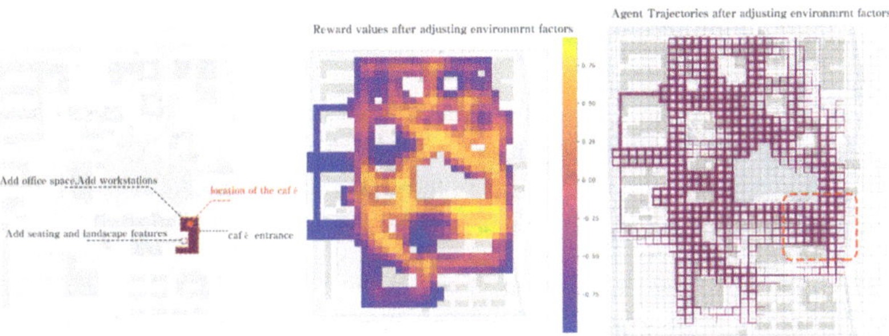

Fig. 10. Zhongshan courtyard café vitality enhancement plan and the post-renovation simulation results of campus reward values and trajectories

4 Summary and Discussion

This study proposes a pedestrian flow simulation model capable of deeply perceiving campus environmental factors based on deep maximum entropy inverse reinforcement learning, aiming to provide theoretical support and practical tools for campus space optimization and renovation. With Southeast University Wuxi Campus as a case study, this paper details the data processing workflow and parameter settings for model training, as well as its practical applications. The entire process is independent of Southeast University Wuxi Campus's specific conditions, thus offering general applicability to university campuses and providing practical value for campus environmental design and evaluation.

This study still has certain technical limitations. Firstly, due to computational constraints, the precision of the environmental grid division and environmental factors classification is limited, resulting in inaccuracies in the simulation of local areas. Secondly, the simulation only covers a two-week period, while pedestrian flow research in university campuses should be analyzed across different time periods. Future research could improve environmental data accuracy and account for temporal variations to enhance the model's accuracy and practicality.

References

1. Fu, W.: Enhancing university campus landscape design through regression analysis: Integrating ecological environmental protection. Soft. Comput. **27**, 16309–16329 (2023)

2. Iranmanesh, A., Mousavi, S.A.: City and campus: Exploring the distribution of socio-spatial activities of students of higher education institutes during the global pandemic. Cities. **128**, 103813 (2022)
3. Zhang, J., Senousi, A.M., Zhao, P., Law, S., Liu, X.: Exploring 3D spatial morphology using multilayered space syntax, network science and wi-fi log data. Urban Info. **2**, 1 (2023)
4. Hu, X., Ren, Y., Tan, Y., Shi, Y.: Research on the spatial and temporal dynamics of crowd activities in commercial streets and their relationship with formats—A case study of Lao men dong commercial street in Nanjing. Sustainability. **15**, 16838 (2023)
5. Feng, J., Yu, L., Ma, R.: AGCN-T: A traffic flow prediction model for spatial-temporal network dynamics. J. Adv. Transp. **2022**, 1–12 (2022)
6. Liu, C.H., et al.: Modeling citywide crowd flows using attentive convolutional LSTM. In: 2021 IEEE 37th International Conference on Data Engineering (ICDE), pp. 217–228. IEEE, Chania, Greece (2021)
7. Dijkstra, J., Jessurun, A.J.: Agent-based pedestrian activity simulation in shopping environments using a choice network approach. In: Wąs, J., Sirakoulis, G.C., Bandini, S. (eds.) Cellular Automata, pp. 680–687. Springer International Publishing, Cham (2014)
8. Wan, H., Wang, S., Du, J., Li, L., Zhang, J.: Trace analysis using wi-fi probe positioning and virtual reality for commercial building complex design. Autom. Constr. **153**, 104940 (2023)
9. Shannon, C.E.: A mathematical theory of communication. Bell Syst. Tech. J. **27**(3), 379–423 (1948)
10. Ziebart, B.D., Maas, A., Bagnell, J.A., Dey, A.K.: Maximum entropy inverse reinforcement learning. In: Proceedings of the 23rd AAAI Conference on Artificial Intelligence (AAAI'08), Chicago, Illinois, USA, vol. 8, pp. 1433–1438 (2008)

AeroboX: A Drone-Based Multi-Environmental Responsive Urban Facility in Hazards Detection and Prevention for the Elderly

Xiaomeng Zhang[1], Jiadong Liang[1], Yuji Huang[1], Wei Peng[1], Yu Yan[2(✉)], and Leiqing Xu[3]

[1] Architectural Association School of Architecture, London, UK
[2] Shenzhen University, Shenzhen, China
yimyvu@szu.edu.cn
[3] Tongji University, Shanghai, China

Abstract. Recent research has highlighted drones as the emerging application of assistive technology (AT) in age-friendly design. However, existing assistive drones are often limited to single-scenario applications such as locating or fall detection, which fall short in complex urban outdoor environments. This study proposes AeroboX, an Arduino-based drone prototype that integrates environmental sensors and an AI vision camera to assist the elderly in avoiding urban outdoor risks such as falls, infections, and traffic hazards. Based on a scenario-driven design research framework, the paper develops a drone-based prototype as a mobile urban facility that can continuously and autonomously interact with users and urban outdoor environment by performing functions such as traffic signal and obstacle recognition, hazard alarming, and sun exposure protection. The results of the feasibility test validate the prototype's effectiveness, providing new insights into drones as cognitive augmentation for the elderly to build an aging-friendly life circle and an inclusive city.

Keywords: Environmental Responsive · Sensor · Multi-sensory · Inclusive City

1 Introduction

The rapid development of smart technologies has been contributing significantly to assistive technologies (AT) for the elderly in enhancing their safety and independence. Recent research has highlighted drones as the emerging application of AT in age-friendly design due to their high mobility, manoeuvrability, and three degrees of freedom [1]. Previous assistive drones focused on medication delivery [2], fall detection [3], and health monitoring [4], utilizing technologies such as IoT, cloud computing, and sensors, as well as deep learning-based computer vision techniques such as object detection and semantic segmentation.

However, while existing systems effectively address specific needs like first-aid or healthcare such as therapeutic intervention [1], indoor individual's state monitoring [4], and medical delivery during pandemic [5], they operate in isolated scenarios and remain

Y. Liu et al. (Eds.): CDRF 2025, *Transindividual Intelligence*, pp. 443–455, 2026.
https://doi.org/10.1007/978-981-92-0615-5_38

limited in adapting to complex and dynamic urban environments. Moreover, the potential of drones to integrate multiple sensors for multi-modal interactions has yet to be fully explored [6]. Therefore, in the field of assistive drones, it is essential to consider a comprehensive and adaptable way to respond to multi-dimensional needs. In this study, we propose a drone-based multi-environmental responsive assistive urban facility for the elderly, which we defined as AeroboX (Fig. 1).

Based on our secondary research from the World Health Organization (WHO), National Institutes of Health (NIH), etc., there are five most common urban outdoor risks for the elderly: traffic accidents [7], falls [8], heatstroke [9], infection [10], and poor lighting related safety issues [11]. AeroboX is designed to detect these hazards and respond with visual, auditory, and environmental feedback. The system consists of five functional modules (Fig. 2), built on two hardware platforms:

- Arduino: Social Distance Monitoring Module, Automatic Sunshading Module, Automatic Lighting Module
- HuskyLens: Obstacle Recognition Module, Traffic Light Recognition Module

This paper proposes a multidisciplinary approach that embeds multi-sensory interaction into the drone for assisting the elderly in complex urban outdoor spaces. AeroboX demonstrates scenario innovation that drones serve as mobile urban facilities, contributing to an aging-friendly life circle [12] and inclusive city. Unlike previous systems targeting isolated scenarios such as indoor health care or medical delivery with single inputs and limited interactions, our solution converts multi-modal inputs: KNN-based object recognition on HuskyLens, environmental sensing (temperature, UV, light intensity) and human proximity detection on Arduino - and converts them into targeted outputs: visual and auditory alerts and automatic adjustments (auto shading and lighting). Its modularized architecture and robust integration allow for flexible adaptation to diverse urban environments, marking a paradigm shift from single-function device to holistic environmental coordinator.

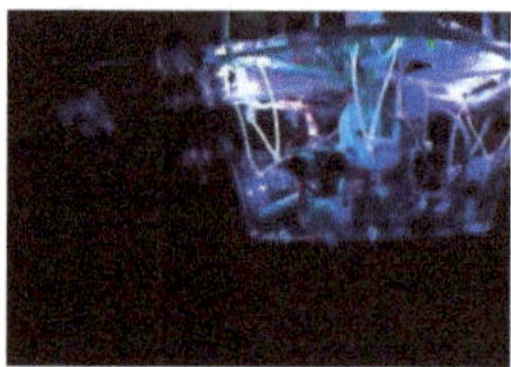

Fig. 1. AeroboX prototype

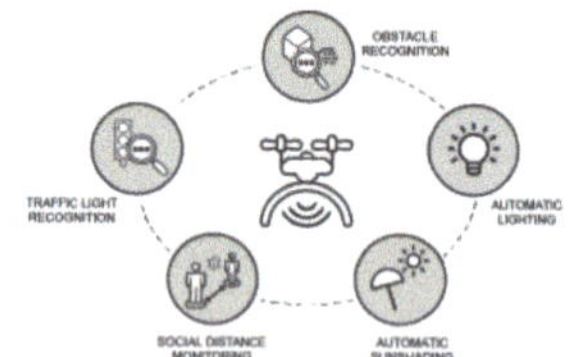

Fig. 2. Five modules of AeroboX

2 Related Work

Arduino-based Assistive Device Kulkarni et al. [13] and Nadikattu et al. [14] developed wearable devices for maintaining social distance during COVID-19. Based on the WHO's 1.5-m guideline [15], they used PIR sensors to detect vicinal violations and trigger audio alerts. This inspired our adoption of similar detection logic and thresholds. Beingolea et al. [16] developed the Smart Cap, which monitors UV index to alert users of high exposure; however, the study overlooks temperature-which can still lead to heatstroke even in low UV index. We address this limitation by integrating both UV and temperature sensors. Kilari et al. [17] developed an automatic light intensity control system using Arduino UNO and LDR to adjust lighting based on outdoor ambient light levels; yet it remains isolated from broader mobility-related safety concerns. Collectively, these studies inform the design concept of AeroboX by demonstrating the potential of Arduino-based systems to deliver real-time environmental feedback and enhance user safety through automated responses-principles that underpin AeroboX's integrated and scenario-driven approach. Building upon these foundations, this study extends the application scope through a modular design that supports multi-scenario application in complex outdoor settings.

Object Recognition Methods Advanced deep learning models such as YOLO, Haar Cascade, and SegFormer, as well as AI camara like HuskyLens, are widely used for object recognition. The 'Flying Guide Dog' built by Tan et al. [18], explored the combination of drone and street view semantic segmentation with SegFormer to identify walkable paths and traffic signals, helping the user to cross a street safely. Nascimento [3] presented 'QuadAALper', a drone that can autonomously navigate inside the house and recognize a person lying on the floor. In his work, the author used Haar Cascade in OpenCV for detecting fallen people. Additionally, HuskyLens, a more lightweight solution, has been used for face recognition with Arduino UNO in a medication delivery drone developed by Baloola et al. [5]. This method does not require any Internet connection and offer lower power consumption. Han et al. [19] developed a walking assistive device for visually impaired individuals to enhance their safety during mobility. They used HuskyLens for distinguishing obstacle types and provide voice information via a buzzer. Inspired by its Arduino compatibility, real-time feedback, and capacity of object recognition without external servers and cloud-computing, we chose HuskyLens as visual recognition tool for our device.

3 Materials and Methods

We adopt a design research framework consisting of demand analysis-prototyping-feasibility test (Fig. 3). This chapter introduces the prototyping part, including hardware development and software development.

3.1 Hardware Development

3.1.1 Microcontroller, Sensing Devices and Actuators

The core of the AeroboX system is Arduino UNO, a cost-effective open-source platform that allows for high customization by integrating with multiple sensors and actuators. The system is powered by a 5 V USB power supply.

To respond to multiple environmental factors, various sensors, actuators, and an AI camera are integrated into one system, which are given in Table 1.

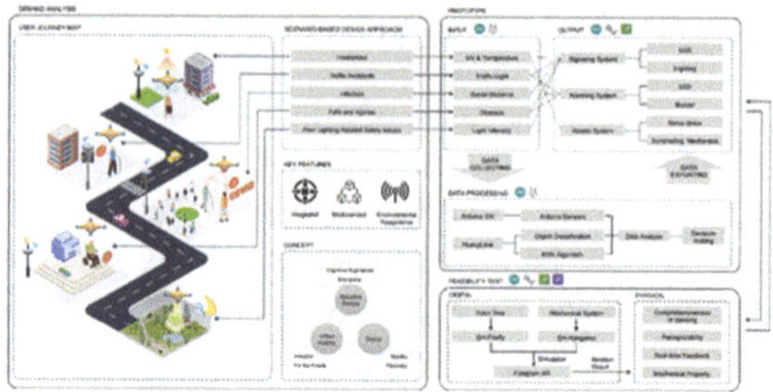

Fig. 3. Design research framework

Table 1. Sensors, microcontroller, and actuators of AeroboX

Qty.	Sensors	Qty.	Actuators
01	HuskyLens	01	Arduino UNO
03	HC-SR501 PIR sensor	01	5 V Power Supply
03	HC-SR04 ultrasonic sensor	01	LED
01	DHT11 temperature sensor	01	Buzzer
01	GUVA-S12SD UV sensor	01	Servo motor
01	GM55 LDR	01	CRM2212 KV1400 Propeller

3.1.2 Prototyping and Integration

The skeleton design of AeroboX was carefully considered to ensure effective integration of all components (Fig. 4), which adhere to the following principles:

- Cover a wide detection area for sensing from multiple angles.
- Accurately capture environmental conditions without being obstructed by other components.
- Effectively connected to other sensors or actuators.

The skeleton was designed and modelled in Rhinoceros with appropriate dimensions. The primary support structure and sunshading boards were fabricated in acrylic using laser cut, while joints were made of resin in SLA 3D printing to maintain precise dimensions and ensure structural integrity.

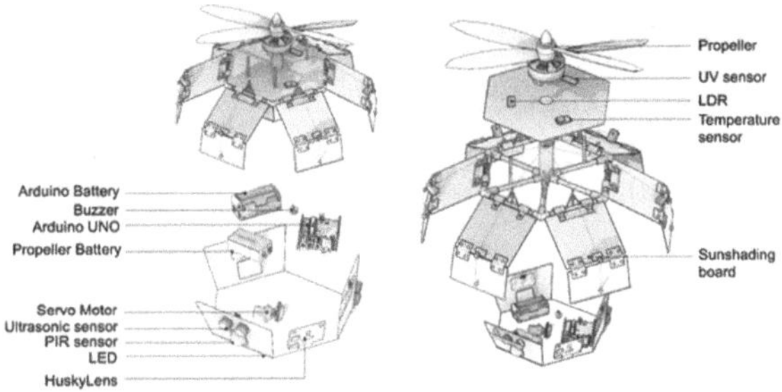

Fig. 4. Skeleton and components of AeroboX

3.1.3 Mechanical System Design

This section aims to design an automatically deployable sunshading structure. Two structural options—the rod system and board system (Fig. 5)—were simulated in Grasshopper using Kangaroo. Based on visual evaluation, the board system was selected for its integrity and structural stability. It consists of a servo motor, spool, strings, and sunshading boards. The servo motor controls the spool, and when activated by sensors detecting high UV or temperature, it rotates to pull the six strings attached to the sunshading boards to deploy as an umbrella. As shown in Fig. 4, the main tension point lies at the end of board B. When board B rises, it drives board A to rotate. After both drop 30°, board A stops while board B continues another 60°. Powered by the motor, the boards rise sequentially. A 2.5 cm spool radius enables 7 cm string travel with 180° rotation.

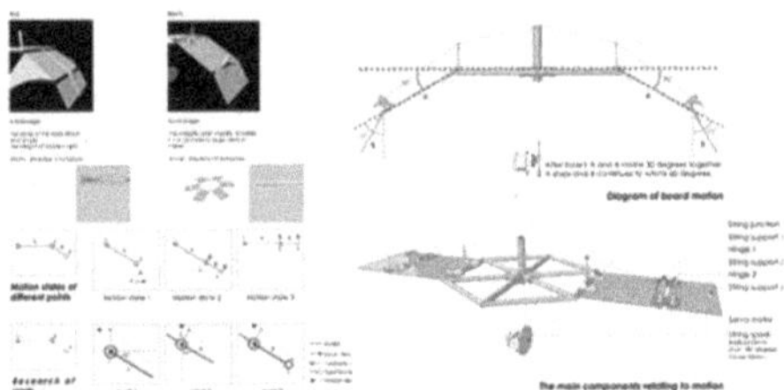

Fig. 5. Mechanical system for sunshading

3.2 Software Development

The software system of AeroboX consists of the following five modules, which correspond to five outdoor risks: traffic accidents, falls, infection, heatstroke, and poor lighting related safety issues (Fig. 6).

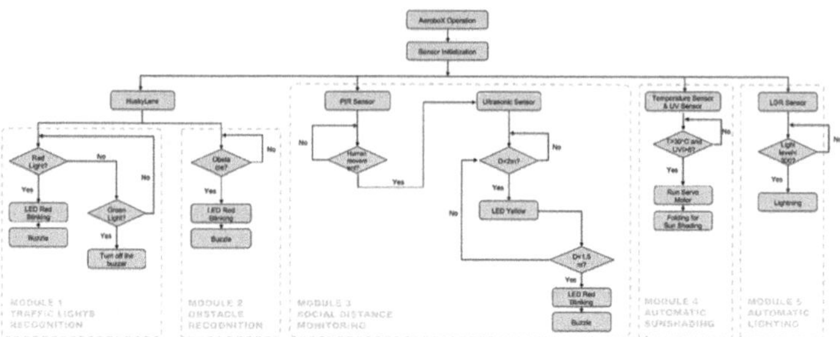

Fig. 6. AeroboX logic flowchart

Obstacle Recognition Module and Traffic Light Recognition Module We use HuskyLens to recognize obstacles that cause falls (including staircases, roadblocks, and curbstones) and traffic lights (red, green) based on the built-in KNN algorithm. Due to the intelligent traffic control system in our experiment location, yellow lights rarely appear in visual data and were therefore not included in the detection. When these obstacles or a red light is detected, the system triggers the buzzer for an audio alert and the LED turns red. The steps are as follows. (Fig. 7)

1. Sample Classification and Labeling. We distinguish between positive and negative samples to mitigate overfitting, and each sample is assigned a unique ID (Table 2).

Table 2. Positive and negative sample labeling

Positive sample					Negative sample	
Staircases (upstair view)	Traffic light (red)	Staircases (downstair view)	Roadblocks	Uneven ground (curbstone)	Traffic light (green)	Null class (flat ground)
ID 1	ID 2	ID 4	ID 5	ID 6	ID 3	ID7–11

2. Sample Images Collection. For each sample, we collect 50 images (the maximum supported amount), which cover various angles, materials, and lighting conditions.
3. Learning and Classification. The samples are processed using HuskyLens' object classification function, which is trained based on the built-in KNN algorithm. When completing, HuskyLens can recognize these samples in real world and transmit their ID numbers to Arduino.

4. Arduino Interaction. When the system detects positive samples, Arduino activates the buzzer to alert users to the presence of red traffic signal, staircases (upstair and downstair view), roadblocks and curbstones.

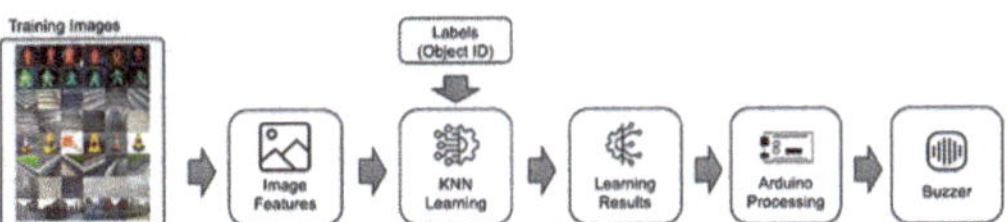

Fig. 7. HuskyLens training and output process

Social Distance Monitoring Module The system uses three pairs of PIR sensors and an ultrasonic sensor for 360° coverage. When the PIR sensor detects movement, the ultrasonic sensor begins measuring the distance between individuals. When the distance is 1.5 m–2 m, a yellow LED turns on to caution the user. If the distance decreases to less than 1.5 m, the system activates a red LED and sounds a buzzer to alert the user.

Automatic Sunshading Module In this module, the system integrates a temperature sensor and a GUVA-S12SD UV sensor. When the temperature exceeds 30 °C and UV Index reaches 6, the system automatically deploys a mechanical sunshade. For UV Index calculation, the Arduino system operates as follows:

1. Reading raw value from GUVA-S12SD UV sensor using analogRead(), which corresponds to an output voltage given by the formula:

$$V_{out} = \frac{raw\ value}{1023} \times 5 \tag{1}$$

2. Converting the Output Voltage to UV Index. The UV Index is calculated using the following formula:

$$\text{UV Index} = \frac{V_{out} - 1.0}{0.1} \tag{2}$$

where is the measured output voltage (ranging from 1.0 V to 3.3 V). This formula converts the voltage into a UV Index value ranging from 0 to 15.

3. Arduino checks the UV index and triggers the mechanical sunshade when the value exceeds 6.

Automatic Lighting Module The system includes a LDR to detect low-light condition. When the surrounding light level drops below 300, equivalent to less than 10 lux (considered poor visibility for human sight), the system automatically activates an LED light.

4 Feasibility Test and Results

We evaluated the performance of HuskyLens for obstacle recognition and traffic signal in real-world urban outdoor environments (Fig. 8). In this study, obstacles refer to three fall-related urban features: staircases, curbstones, and roadblocks. The core experiment focused on staircase detection across scenarios varying in two surface materials (rough granite, smooth tile), four lighting conditions (daytime: diffused light/shadow; nighttime: artificial light/no light), and two viewpoints (upstairs/downstairs), with 16 samples per condition to assess recognition accuracy. Figure 9 illustrates the results of the staircase recognition evaluation. The overall accuracy refers to the correct triggering of the buzzer, independent of object class confusion between stair views or curbstones, as it reflects the system's primary function: detecting potential tripping hazards and issuing alerts. Night illumination significantly improved overall accuracy, with smooth tile stairs achieving higher (up to 75%) than rough granite stairs (53.3%). In contrast, under diffuse daylight, rough granite outperformed smooth tiles, with 6.7% and 7.2% higher accuracy for upstair and downstair views. However, shadows reduced each texture's overall accuracy by 17–37%. Besides, misclassification between upstair and downstair view, staircases and curbstones also occurred under all conditions: rough granite showed higher misclassification rates as curbstones under both lighting conditions-up to 20% in daylight and 40% at night, while smooth tiles remained largely distinguishable, with only 6.3% misclassified as curbstones at night. Apart from staircases evaluation, under diffuse daylight, the overall accuracy for curbstone recognition was 90.0% (9% misclassified as down-view stairs), while roadblocks recognition reached 93.7%. In terms of traffic signal, under diffuse daylight, it achieved 80% success in recognizing red/green lights with correct ID assignment, though accuracy dropped when red-colored moving objects appeared nearby.

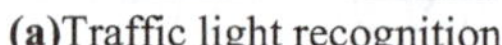

(a) Traffic light recognition

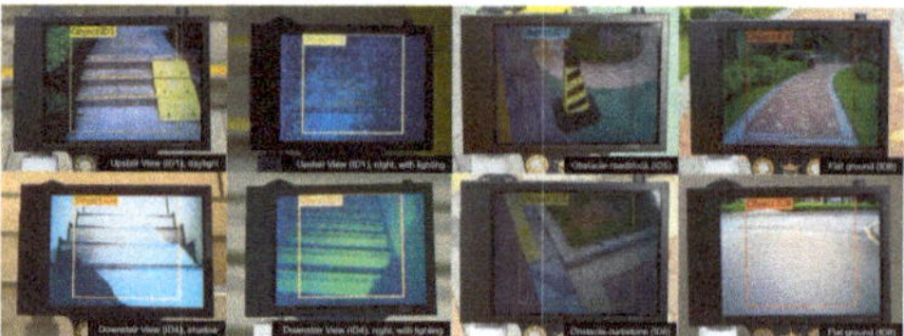

(b) Obstacles and Negative sample recognition

Fig. 8. HuskyLens recognition results

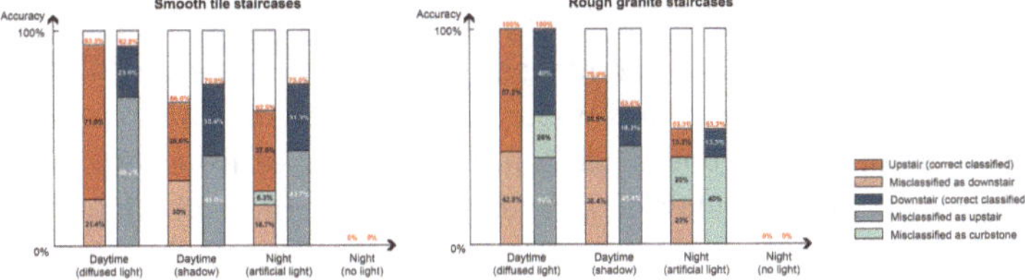

Fig. 9. Staircase recognition evaluation between smooth tile and rough granite under different light conditions and viewpoints

We also invited a focus group of 10 tester to test the performance of Arduino sensors and actuators in real-world. For the Automatic Sunshading Module, we used Kangaroo and Firefly in Grasshopper with Fologram AR to optimize spool radius and string length. After six iterations, the final prototype (with spool radius 2.5 cm) folded in 1.1 s under simulated conditions, showing high efficiency. In the Social Distance Module, LED and buzzer alerts were triggered within 0.1 s when a tester approached at 0.9 m. Eight out of ten testers found the visual and audio signals clear and intuitive (Fig. 9). In the Automatic Lighting Module, lights activated instantly as ambient brightness decreased (Fig. 10). We also conducted flight tests. The results indicated that the system demonstrated limited flight stability, with difficulties in accurately control the flight direction.

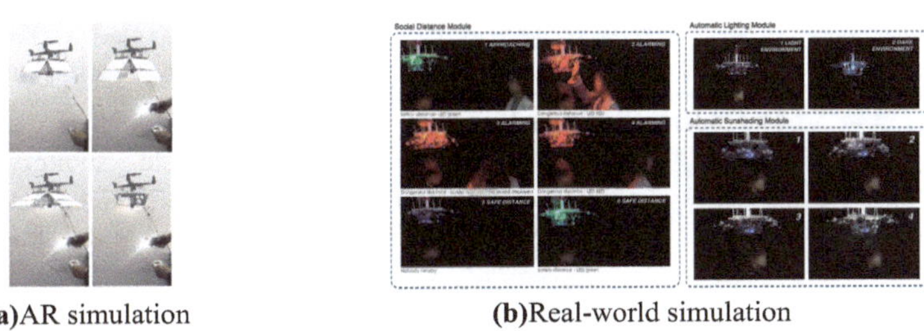

(a)AR simulation **(b)**Real-world simulation

Fig. 10. Feasibility test of Arduino-based scenarios in AR and real-world simulation

5 Discussion and Future Work

The AeroboX prototype demonstrates robust multi-sensory integration, effective mechanical design, and real-time environmental responsiveness, validated by feasibility tests. As shown in Table 3, compared to existing assistive drones that focus on single functions like emotion recognition or fall detection, AeroboX uniquely combines traffic and obstacle detection, multi-environmental sensing, proximity monitoring, and multi-modal feedbacks into a modular, adaptive system. Also, its automatic lighting significantly enhanced HuskyLens' obstacle recognition performance at night.

Despite these strengths, AeroboX remains limited in visual recognition accuracy, flight stability, and long-term evaluation. The results of HuskyLens experiments demonstrate that it is highly influenced by lighting, shadows, and surface texture. At night, artificial illumination enhanced edge clarity, particularly on smooth, reflective tiles, while rough granite scattered light and reduced contrast, which results in lower recognition accuracy for rough surfaces than smooth surfaces at night. Under daylight, specular reflections on tiles hindered contour detection, lowering distinguishability compared to rough granite, and shadows significantly impaired recognition by introducing abrupt brightness shifts, false edges, and texture loss, especially on rough granite; misclassification between stairs and curbstones occurs more frequently on rough surfaces, primarily due to their similar material properties. Also, KNN model's lack of spatial depth awareness led to frequent confusion between upstair and downstair views, which share similar 2D features. Overall, HuskyLens effectively detects objects with distinct visual contrast (especially clear shape boundaries and colors) due to KNN's reliance on feature matching, but lacks generalization for objects with similar materials, high sensitivity to lighting conditions, or ambiguous visual boundaries. Future work may consider adopting Maixduino-a Kendryte K210-based, Arduino-compatible, but more flexible edge computing AI camera that allows broader visual recognition models like YOLOX-Tiny and CNN, which would enable the construction of an extensive dataset for multi-category object and material recognition. This study also faces challenges in flight stability due to payload and equipment limitations. Future work could adopt a quadrotor drone instead of counter propeller to improve stability. Additionally, no longitudinal user study was conducted due to time constraints. Future research should include long-term evaluations with diverse users to evaluate sustained effectiveness and user experience.

System/Device	Targeted users	Scenario	Functional scope	User interaction
AeroboX	The elderly	Outdoor urban hazards detection and prevention	Integrated (traffic light and obstacle response, sun protection, distance monitoring, lighting)	Visual and auditory alerts, automatic lighting and shading
Nascimento et al. [3]	The elderly	Indoor navigation and fall detection	Single-function	Autonomous operation with smartphone-based control
Martínez et al. [4]	The elderly	Indoor individual state monitoring	Single-function	Facial emotion recognition
Ascensao et al. [1]	Individuals with autism	Dance Movement Therapy Intervention	Single-function	Visual and motion-based interaction with adaptive control
Tan et al. [18]	The visually impaired	Outdoor urban street guidance	Multiple (walkable path detection, traffic light recognition)	Voice feedback and haptic guidance

(continued)

(*continued*)

System/Device	Targeted users	Scenario	Functional scope	User interaction
Baloola et al. [5]	Patients during COVID-19	Medical delivery in pandemic	Single-function	Facial recognition and IoT guide landing

Table 3. Comparative analysis of AeroboX and other assistive drone systems

System/Device	Targeted users	Scenario	Functional scope	User interaction
AeroboX	The elderly	Outdoor urban hazards detection and prevention	Integrated (traffic light and obstacle response, sun protection, distance monitoring, lighting)	Visual and auditory alerts, automatic lighting and shading
Nascimento et al.[3]	The elderly	Indoor navigation and fall detection	Single-function	Autonomous operation with smartphone-based control
Martínez et al.[4]	The elderly	Indoor individual state monitoring	Single-function	Facial emotion recognition
Ascensao et al. [1]	Individuals with autism	Dance Movement Therapy Intervention	Single-function	Visual and motion-based interaction with adaptive control
Tan et al.[18]	The visually impaired	Outdoor urban street guidance	Multiple (walkable path detection, traffic light recognition)	Voice feedback and haptic guidance
Baloola et al.[5]	Patients during COVID-19	Medical delivery in pandemic	Single-function	Facial recognition and IoT guide landing

6 Conclusion

In conclusion, this paper proposes AeroboX, a drone-based multi-environmental responsive device as mobile urban facility that aims to assist the elderly in avoiding potential urban outdoor risks. The KNN-based HuskyLens AI camera enables real-time obstacle recognition, with performance influenced by light condition, shadows, and material texture, highlighting its strengths and limitations in complex urban environments. Compared to existing assistive devices that are often limited to single-function applications, AeroboX allows for a more comprehensive solution to various outdoor risks. The AeroboX prototype demonstrates robust integration of multi-sensory interaction, effective mechanical design, and real-time environmental responsiveness, providing new sights into drones serve as mobile urban facility to build an aging-friendly life circle and inclusive city.

Acknowledgements. This work was supported by the National Key R&D Program of China (Grant No. 2024YFF0618400), under the project "Research on Key Technology for Detection and Evaluation of Living Environment and Rehabilitation Assistive Device for Elderly People with Disability and Dementia."

References

1. Ascensao, T., Jamshidnejad, A.: Autonomous socially assistive drones performing personalized dance movement therapy: An adaptive fuzzy-logic-based control approach for interaction with humans. IEEE Access. **10**, 15746–15770 (2022)

2. Kim, S.J., Lim, G.J., Cho, J., Côté, M.J.: Drone-aided healthcare services for patients with chronic diseases in rural areas. J. Intell. Robot. Syst. **88**(1), 163–180 (2017)
3. Nascimento, R.M.G.: QuadAALper-The Ambient Assisted Living Quadcopter. Universidade do Porto, Porto (2015)
4. Martínez, A., Belmonte, L.M., García, A.S., Fernández-Caballero, A., Morales, R.: Facial emotion recognition from an unmanned flying social robot for home care of dependent people. Electronics. **10**(7), 868 (2021)
5. Baloola, M.O., Ibrahim, F., Mohktar, M.S.: Optimization of medication delivery drone with IoT-guidance landing system based on direction and intensity of light. Sensors. **22**(11), 4272 (2022)
6. Baer, M., Tilliette, M.-A., Jeleff, A., Ozguler, A., Loeb, T.: Assisting older people: From robots to drones. Geron. **13**(1), 57–58 (2014)
7. Yee, W.Y., Cameron, P.A., Bailey, M.J.: Road traffic injuries in the elderly. Emerg. Med. J. **23**(1), 42–46 (2006)
8. Centers for Disease Control and Prevention. Facts about Falls. Older Adult Fall Prevention. https://www.cdc.gov/falls/data-research/facts-stats/index.html. (2024)
9. World Health Organization. Heat and Health . World Health Organization. https://www.who.int/news-room/fact-sheets/detail/climate-change-heat-and-health. (2024)
10. Mueller, A.L., McNamara, M.S., Sinclair, D.A.: Why does COVID-19 disproportionately affect older people? Aging. **12**(10), 9959–9981 (2020)
11. Dev, M.K., Black, A.A., Cuda, D., Wood, J.M.: Low light exposure and physical activity in older adults with and without age-related macular degeneration. Transl. Vis. Sci. Technol. **11**(3), 21 (2022)
12. Wu, X.A., Xu, L.Q.: Research on the form of community life circle planning: discussion on the scope of age-friendly life circle and walking friendly life circle. J. Hum. Settl. W. China. **36**, 74–82 (2021)
13. Kulkarni, M.D., Alfatmi, K., Deshmukh, N.S.: Social distancing using IoT approach. J. Electr. Syst. Inf. Technol. **8**(1), 15 (2021)
14. Nadikattu, R.R., Mohammad, S.M., Whig, P.: Novel economical social distancing smart device for Covid19. Int. J. Electr. Eng. Technol. **11**(4), 204–217 (2020)
15. World Health Organization: Operational considerations for case management of COVID-19 in health facility and community. Interim guid., Pediatr. Med. Rodz. **16**(1), 27–32 (2020)
16. Beingolea, J.R., Zea-Vargas, M.A., Huallpa, R., Vilca, X., Bolivar, R., Rendulich, J.: Assistive devices: Technology development for the visually impaired. Designs. **5**(4), 75 (2021)
17. Kilari, G., Mohammed, R., Jayaraman, R.: Automatic light intensity control using arduino UNO and LDR. In: 2020 International Conference on Communication and Signal Processing (ICCSP). IEEE (2020)
18. Tan, H., Chen, C., Luo, X., Zhang, J., Seibold, C., Yang, K., et al.: Flying guide dog: Walkable path discovery for the visually impaired utilizing drones and transformer-based semantic segmentation. In: 2021 IEEE International Conference on Robotics and Biomimetics (ROBIO). IEEE (2021)
19. Han, J.H., Yoon, I., Kim, H.S., Jeong, Y.B., Maeng, J.H., Park, J., et al.: Mobility support with intelligent obstacle detection for enhanced safety. Optics. **5**(4), 434–444 (2024)

A Multimodal Large Language Model to Support Participatory Urban Village Renovation

Daxu Wei(✉) and Christiane M. Herr

School of Design, Southern University of Science and Technology, Shenzhen, China
weidx2025@mail.sustech.edu.cn, cmherr@sustech.edu.cn

Abstract. This study examines the potential application of multimodal large language model (MLLM) in the participatory design process of urban village renovation, with a focus on enhancing public participation and optimizing design proposals. In the context of rapid urbanization in China, urban villages face severe challenges, including aging infrastructure and overcrowding, making their revitalization a critical task in urban renewal. Current renovation projects predominantly rely on expert- and government-led decision-making, often resulting in a mismatch between design proposals and residents' needs. This research proposes a method that integrates residents' feedback through multimodal inputs, including images and text, to generate design solutions that align with their preferences and the local environment. The study was conducted in Dakan Village, Shenzhen, where a participatory design approach was employed to collect and analyze residents' feedback, using GPT Vision and Latent Diffusion Model (LDM) to generate and refine design proposals. The results show that the integration of multimodal data facilitated residents' expression of design needs, narrowing the gap between design proposals and actual needs. The study also demonstrates that MLLM technology, combined with Low-Rank Adaptation (LoRA) and ControlNet fine-tuning and prompt engineering, can generate designs that are more aligned with residents' expectations. However, the research highlights limitations, including the need for further refinement of the designs and the challenge of model interpretability. The findings suggest that while MLLM provide significant support in participatory design, professional expertise remains essential in final design adjustments, and future studies should focus on improving model transparency and feasibility for broader applications in urban renewal projects.

Keywords: Urban Village Renovation · Participatory Design · Multimodal Large Language Model · Urban Regeneration

1 Introduction

In the rapid urbanisation process in China, urban villages face significant challenges as a unique area of urban transformation (Zhao et al., 2021). Due to China's land rights policies, the government expanded cities by requisitioning farmland around urban areas, which led to villages being surrounded by urban development, ultimately forming the

Y. Liu et al. (Eds.): CDRF 2025, *Transindividual Intelligence*, pp. 456–465, 2026.
https://doi.org/10.1007/978-981-92-0615-5_39

present-day urban villages (Wu et al., 2013). As a result, urban villages in large cities are often areas inhabited by low-income groups, facing issues such as aging buildings, outdated infrastructure, and poor environmental conditions (Fig. 1). The revitalisation and transformation of urban villages have become key tasks in the urban renewal of China's metropolitan areas, aimed at improving the quality of living environments in older areas, enhancing infrastructure, and promoting social cohesion and inclusivity (Wolfe, 2017).

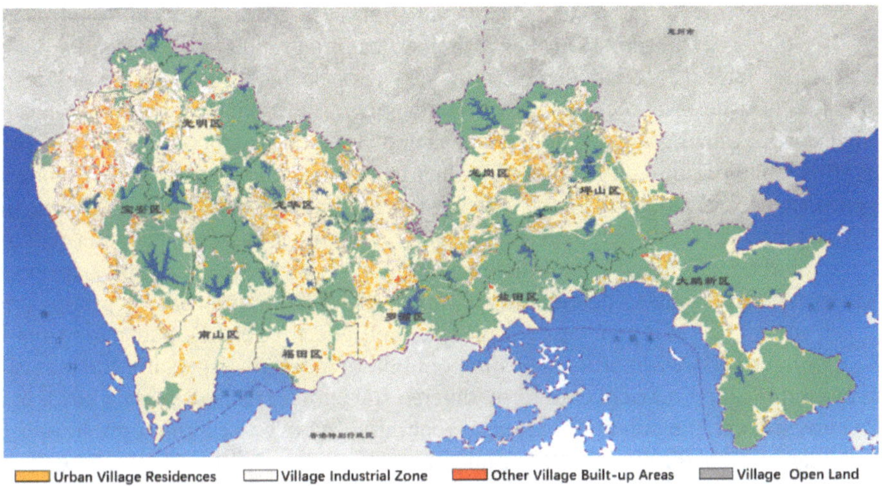

Fig. 1. Distribution of urban villages in Shenzhen (Shenzhen Municipal Planning and Land Resources Commission, 2018)

Due to the high professional threshold of urban renewal, most urban village transformation projects currently rely on a top-down decision-making model, led by governments and real estate developers (Li et al., 2020). This has led to a mismatch between design proposals and the actual needs of residents (Zhuang et al., 2017). In this context, public participation has been introduced into the decision-making process of urban village transformation, with the aim of reconciling conflicts of interest among different stakeholders (Zhou, 2014). Participatory design provides stakeholders with an opportunity to actively engage in the design process (Halskov & Hansen, 2015), combining their lived experiences with the technical and design knowledge of professional designers (Kensing & Blomberg, 1998; Warr & O'Neill, 2004). However, existing forms of public participation are mainly limited to simple surveys or hearings (Simonsen & Robertson, 2012), and some scholars argue that this form of participation is merely "symbolic" and does not truly safeguard the rights of residents (Xu & Lin, 2019). This situation leaves urban village residents at a disadvantage due to their lack of professional knowledge or difficulties in expressing their views, while the disagreements among multiple stakeholders further intensify the controversy surrounding transformation plans (Scolobig, 2016; Chaskin et al., 2012).

With the widespread application of artificial intelligence technologies, MLLM such as GPT Vision (Wu et al., 2024; Yang et al., 2023) and LDM (Rombach et al., 2022) have gradually been applied to architectural design (Li et al., 2024; Korzynski et al., 2023). MLLM effectively integrates visual and linguistic information, demonstrating strong capabilities in tasks such as image generation, content analysis, visual question answering, and cross-modal reasoning (Yin et al., 2024). They also offer more efficient and equitable feedback integration approaches through various input methods, including text and images (Wang et al., 2024). For instance, Bratteteig and Verne (2018) explored how artificial intelligence influences the development of participatory design techniques, analysing the characteristics of AI and machine learning, and discussing the differences in user participation in AI system design integration compared to traditional design methods. Stigberg et al. (2024) further examined the integration of participatory design with artificial intelligence, focusing on how AI can be incorporated into the design process and how design processes affect AI design, while emphasising mutual learning among users, designers, and developers. Quan and Lee (2025) explored the possibility of integrating ChatGPT into public participation planning support systems (PPSS) through a hypothetical case study in Seoul, developing a ChatGPT-based web platform that allows stakeholders to continuously engage and realise more inclusive decision-making, enhancing public participation in urban regeneration projects. Xu et al. (2024) proposed an AI-enhanced multimodal collaborative design (AI-MCD) framework integrating large language model (LLM), generative artificial intelligence (GenAI), and mixed reality (MR), aiming to address inefficiencies in collaborative design, especially those arising from communication barriers caused by diverse stakeholder backgrounds. Guridi et al. (2024) investigated the potential of image-generating AI in co-designing public spaces, finding that by transforming public opinions into visual concepts, which can enhance design diversity, improve participation, and increase efficiency.

Although existing studies have explored the application of AI technology in participatory design, little research has been conducted on how MLLM can play a role in the participatory design of urban village transformation, particularly in integrating and analysing residents' opinions and optimising transformation plans. This study aims to utilise MLLM to support public participation in urban village transformation projects, promoting communication and collaboration among residents, designers, and governments, and proposing a participatory design method that facilitates the expression of intentions in different modalities. The pilot study is conducted in Dakan Urban Village, Shenzhen. First, residents' feedback on transformation plans is collected, including text descriptions and images in various forms. Specific design intention images are generated using LDM (Esser et al. 2021; Rombach et al., 2022), and GPT Vision is employed to analyse these images and extract key elements of residents' preferences. These elements are then organised and used to train a LoRA model to fine-tune LDM. Finally, the LoRA model is combined with ControlNet in LDM to generate specific design proposals.

2 Methodology

The study began by researching the pilot area and identifying the participating residents, then collected feedback from residents on the design intent through a questionnaire and used GPT Vision to understand the primary images, followed by applying LoRA to fine-tune the LDM, and finally validated the generated design results through a case study (Fig. 2).

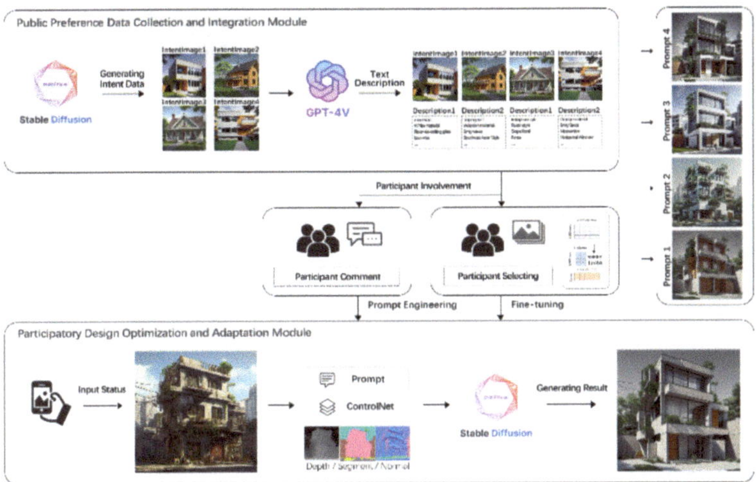

Fig. 2. Design system flowchart

2.1 Research Area and Participants

This study selected Dakan Village in Shenzhen as the pilot research area to examine the feasibility of the proposed methodology due to its typicality and representativeness in urban village renovation. Dakan Village is located in Nanshan District, Shenzhen, facing issues such as aging infrastructure, insufficient public space, and overcrowded living conditions. It covers an area of 8.9 km^2, with a population of over 50,000 and approximately 1,485 buildings. The choice of this area as the research sample not only reflects common problems in urban village renovation but also provides a reference for the renovation of other similar areas. For the purpose of this study, Building No. 32 on Dakan Village Commercial Street was selected as the research object for architectural renovation design. This area primarily consists of antiquated low-rise residential buildings with a lack of public space, low greening rates, and poor spatial quality. Additionally, nearby public squares and green spaces were selected as research objects for public space and landscape renovation, aiming to cover various types of urban renewal. The stakeholders involved in the participatory design of the study included 11 residents living in the building or nearby public spaces, as they are familiar with the area and are direct beneficiaries of the renovation plan, making their participation high (Fig. 3).

2.2 Data Collection and Processing

To ensure that residents could express their design intentions intuitively, the study employed LDM to generate 30 images of architectural exteriors, public facility layouts, and landscaping designs to establish a visual library, including modern glass curtain-wall buildings, traditional Chinese buildings, minimalist style buildings, as well as public spaces such as sports plazas, waterfront squares, children's playgrounds, and landscape designs like Chinese gardens and modern gardens. Based on these images, we used GPT Vision for few-shot learning to extract key design elements such as architectural style,

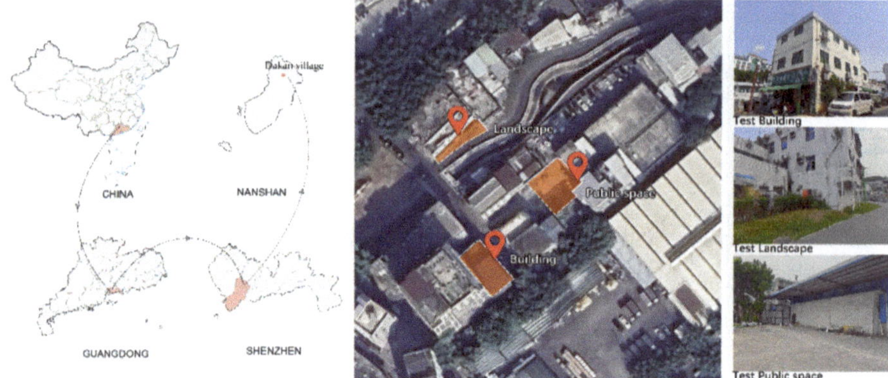

Fig. 3. Research area and status

color, layout, and materials, generating corresponding text descriptions and forming a dataset of 30 image-text pairs (Fig. 4). The semi-structured questionnaire was created, including design style preference questions for architecture, public space, and landscaping, as well as supplementary suggestions for the renovation plan. Residents selected 19 images from the picture library through an online questionnaire and provided brief text descriptions explaining why they chose those images or what specific adjustments they hoped for in the design. Ultimately, 11 valid responses were collected, including the design intention images selected by residents and supplementary text descriptions, the keywords from the text descriptions of residents were then extracted for the next stage of prompt engineering to guide the generation results.

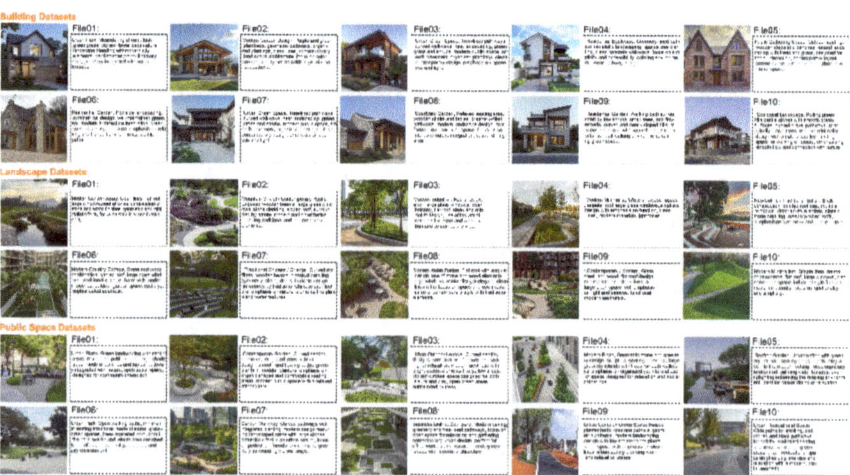

Fig. 4. Design intent images and descriptions

2.3 Model Fine-Tuning

To generate images that align with residents' design intentions, we fine-tuned the LDM model using LoRA (Low-Rank Adaptation), which enables efficient fine-tuning by adjusting only a small portion of the pre-trained model parameters, resulting in low resource requirements (Hu et al., 2021). We used the images selected by residents and paired them with the simplified text labels to form a dataset of 19 image-text pairs. During training, we employed the AdamW 8-bit optimizer, with a network dimension of 32. The training was conducted on a single RTX 4080 Super GPU.

2.4 Case Study

We first took photographs of the current state of Building No. 32 on Dakan Village Commercial Street and its surrounding squares and landscapes, and input these into LDM, loading the LoRA model. Subsequently, the design results were fine-tuned through ControlNet's semantic and depth maps and guided by the residents' design intention prompts. The experimental results showed that in different urban village renovation scenarios, the adjusted prompts effectively guided the model to generate design images that met residents' expectations (Fig. 5). For example, when inputting "a modern-style building with a simple and elegant appearance, incorporating some Lingnan architectural features" as prompt into Output Building02, the model generated an image showcasing a modern-style building façade, featuring a Lingnan-style roof and wooden material cladding, in line with the residents' description. This design not only reflected residents' preferences but also fully considered the constraints of the actual environment, ensuring the feasibility and practicality of the design.

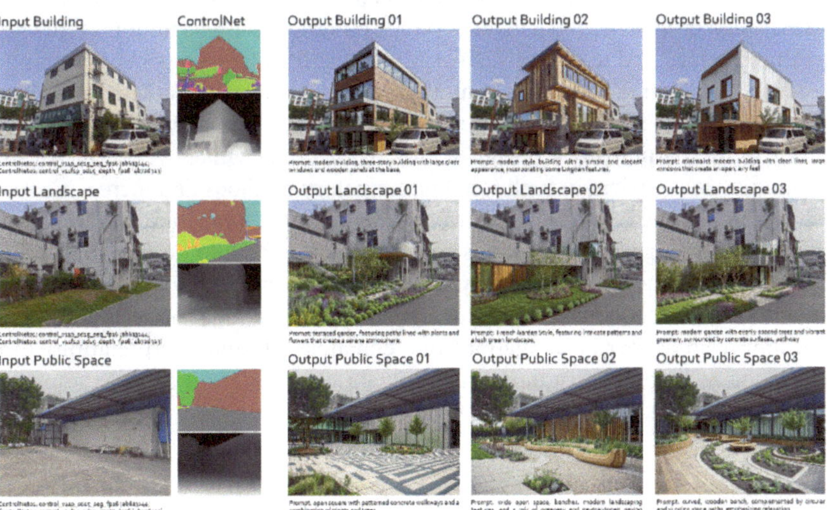

Fig. 5. Input parameters and generated results

3 Results and Analyses

To evaluate the generated design outcomes, we conducted a dual-version questionnaire survey assessing the design images. The expert version was completed by seven professionals in urban planning and architectural design, focusing on three key metrics: generated image quality, design rationality, and implementation feasibility. The public version involved eleven urban village residents, primarily evaluating image quality, design outcomes, and expectation alignment. The results from the expert ratings showed an image quality score of 8.0, design rationality of 7.1, and implementation feasibility of 7.5. The public version yielded image quality ratings of 8.2, design evaluation scores of 8.1, and expectation alignment ratings of 9.0.

The findings demonstrate that through fine-tuning, the MLLM can effectively generate design proposals that align with residents' expectations. Overall, most generated solutions satisfy fundamental design requirements while adhering to the basic principles of urban village renewal. However, certain design proposals still exhibit irrational elements - for instance, in "Landscape Proposal 01", the large ramp design results in significant waste of usable space, and the road system suffers from unclear path indications. This suggests that although the model can generally produce expected design solutions, its heuristic generation approach still contains uncertainties and requires further optimization through integration with professional expertise to better align with practical design knowledge and functional requirements.

4 Discussion

This study aims to explore the application of MLLM in the participatory design process of urban village renewal, with a particular focus on how it integrates residents' preferences, optimises design proposals, and enhances public participation. The findings indicate that, by combining image and text inputs, MLLM can facilitate residents in expressing their design needs, generating proposals that align closely with their intentions. Compared to related studies in existing literature, this research further expands the application of MLLM in the participatory design scenarios of urban village renewal, addressing the communication barriers in traditional participatory methods and increasing resident involvement in the renovation plans. However, there are still some limitations in the study, particularly as some generated designs may not fully adhere to design conventions or environmental constraints. Additionally, the lack of interpretability in MLLM continues to pose challenges in understanding the design decision-making process. Future research could benefit from deeper investigations into enhancing design interpretability through the integration of participatory inputs and domain expertise, as well as examining how expanded public engagement datasets might influence participatory outcomes.

5 Conclusion

This study applies MLLM technology to urban village renewal, thereby expanding the scope of its application in the field of participatory design. The research demonstrates that MLLM allows non-architectural residents to express design needs and preferences more conveniently and intuitively, overcoming the communication barriers present in traditional participatory methods and enhancing resident involvement in the renovation plans. By combining images and text, residents' feedback becomes more diverse, and the integration of multimodal data allows for a more comprehensive, objective, and dynamic collection of feedback, narrowing the gap between design proposals and actual needs. Despite the issue of model interpretability. Future research can address the lack of professionalism in AI-generated results by incorporating rule-based control conditions, further improving the model's effectiveness while also enhancing its interpretability.

References

Bratteteig, T., Verne, G.: Does AI make PD obsolete? Exploring challenges from artificial intelligence to participatory design. In: Proceedings of the 15th Participatory Design Conference: Short Papers, Situated Actions, Workshops and Tutorial, vol. 2, pp. 1–5 (2018). https://doi.org/10.1145/3210604.3210646

Chaskin, R., Khare, A., Joseph, M.: Participation, deliberation, and decision making: The dynamics of inclusion and exclusion in mixed-income developments. Urban Aff. Rev. **48**(6), 863–906 (2012). https://doi.org/10.1177/1078087412450151

Esser, P., Rombach, R., Ommer, B.: Taming transformers for high-resolution image synthesis. In: Proceedings of the IEEE/CVF Conference on Computer Vision and Pattern Recognition (CVPR), pp. 12873–12883 (2021) https://openaccess.thecvf.com/content/CVPR2021/html/Esser_Taming_Transformers_for_High-Resolution_Image_Synthesis_CVPR_2021_paper.html?ref=//githubhelp.com

Guridi, J.A., et al.: Image generative AI to design public spaces: A reflection of how AI could improve the co-design of public parks. Digit. Gov.: Res. Pract. **6**, 3656588 (2024). https://doi.org/10.1145/3656588

Halskov, K., Hansen, N.B.: The diversity of participatory design research practice at PDC 2002–2012. Int. J. Hum.-Comput. Stud. **74**, 81–92 (2015). https://doi.org/10.1016/j.ijhcs.2014.09.003

Hu, E. J., et al. LoRA: Low-Rank Adaptation of Large Language Models. (2021). http://arxiv.org/abs/2106.09685

Jiang, Y., Mohabir, N., Ma, R., Wu, L., Chen, M.: Whose village? Stakeholder interests in the urban renewal of Hubei Old Village in Shenzhen. Land Use Policy. **91**, 104411 (2020). https://doi.org/10.1016/j.landusepol.2019.104411

Kensing, F., Blomberg, J.: Participatory design: Issues and concerns. Comput. Support. Coop. Work. **7**(2), 167–185 (1998). https://doi.org/10.1023/A:1008689307411

Korzynski, P., Mazurek, G., Krzypkowska, P., Kurasinski, A.: Artificial intelligence prompt engineering as a new digital competence: Analysis of generative AI technologies such as ChatGPT. Entrep. Bus. Econ. Rev. **11**(3), 25–37 (2023). https://doi.org/10.15678/EBER.2023.110302

Li, C., Zhang, T., Du, X., Zhang, Y., & Xie, H. Generative AI Models for Different Steps in Architectural Design: A Literature Review. (2024). https://doi.org/10.48550/arxiv.2404.01335

Quan, S.J., Lee, S.: Enhancing participatory planning with ChatGPT-assisted planning support systems: A hypothetical case study in Seoul. Int. J. Urban Sci. **29**(1), 89–122 (2025). https://doi.org/10.1080/12265934.2025.2462823

Rombach, R., Blattmann, A., Lorenz, D., Esser, P., Ommer, B.: High-resolution image synthesis with latent diffusion models. In: Proceedings of the IEEE/CVF Conference on Computer Vision and Pattern Recognition (CVPR), pp. 10684–10695 (2022) https://openaccess.thecvf.com/content/CVPR2022/html/Rombach_High-Resolution_Image_Synthesis_With_Latent_Difusion_Models_CVPR_2022_paper.html

Simonsen, J., Robertson, T. (eds.): Routledge International Handbook of Participatory Design. Routledge (2012). https://doi.org/10.4324/9780203108543

Scolobig, A.: Stakeholder perspectives on barriers to landslide risk governance. Nat. Hazards. **81**(1), 27–43 (2016). https://doi.org/10.1007/s11069-015-1787-6

Stigberg, S.K., Carcani, K., Joshi, S.G., Bratteteig, T.: Participatory design meets artificial intelligence: Co-imagining mutual learning of AI technologies and designing with AI tools. In: Adjunct Proceedings of the 2024 Nordic Conference on Human-Computer Interaction, pp. 1–3 (2024). https://doi.org/10.1145/3677045.3685471

Shenzhen Municipal Planning and Land Resources Commission: Shenzhen urban village (old village) overall planning (2018–2025) [深圳市城中村(旧村)总体规划 (2018–2025)]. Shenzhen Government, Shenzhen (2018)

Wang, Y., et al. Exploring the Reasoning Abilities of Multimodal Large Language Models (MLLM): A Comprehensive Survey on Emerging Trends in Multimodal Reasoning. (2024). http://arxiv.org/abs/2401.06805

Wu, F., Zhang, F., Webster, C.: Informality and the development and demolition of urban villages in the Chinese peri-urban area. Urban Stud. **50**(10), 1919–1934 (2013). https://doi.org/10.1177/0042098012466600

Wu, W., Yao, H., Zhang, M., Song, Y., Ouyang, W., & Wang, J. GPT4Vis: What can GPT-4 do for Zero-Shot Visual Recognition? (2024). https://doi.org/10.48550/arXiv.2311.15732

Xu, S., Wei, Y., Zheng, P., Zhang, J., Yu, C.: LLM enabled generative collaborative design in a mixed reality environment. J. Manuf. Syst. **74**, 703–715 (2024). https://doi.org/10.1016/j.jmsy.2024.04.030

Xu, Z., Lin, G.C.: Participatory urban redevelopment in Chinese cities amid accelerated urbanization: Symbolic urban governance in globalizing Shanghai. J. Urban Aff. **41**(6), 756–775 (2019). https://doi.org/10.1080/07352166.2018.1536420

Yang, Z., et al. The Dawn of LMMs: Preliminary Explorations with GPT-4V(ision). (2023). https://doi.org/10.48550/arXiv.2309.17421

Zhao, Y., An, N., Chen, H., Tao, W.: Politics of urban renewal: An anatomy of the conflicting discourses on the renovation of China's urban village. Cities. **111**, 103075 (2021). https://doi.org/10.1016/j.cities.2020.103075

Zhuang, T., Qian, Q.K., Visscher, H.J., Elsinga, M.G.: Stakeholders' expectations in urban renewal projects in China: A key step towards sustainability. Sustainability. **9**(9), 1640 (2017). https://doi.org/10.3390/su9091640

Zhou, Z.: Towards a collaborative approach? Investigating the regeneration of urban villages in Guangzhou, China. Habitat Int. **44**, 297–305 (2014). https://doi.org/10.1016/j.habitatint.2014.07.011

Unraveling the Unseen Gaps: The Differential Influence of Panoramic Videos and Traditional Visual Media on Landscape Perception

Yanrong Zhu, Yanting Shen, and Jiawei Yao(✉)

Tongji University, 1239 Siping Road, 200092 Shanghai, China
jiawei.yao@tongji.edu.cn

Abstract. This study investigates the information interaction between individuals and environments through controlled experiments comparing the mechanisms of influencing visual perception of static photographs, conventional fixed-perspective videos, and panoramic videos, providing guidance for media selection in diverse landscape visual preference experiment scenarios. All the three methods introduce measurement errors, with panoramic videos exhibiting the highest scene fidelity while static photographs show the lowest. Mean score reveals that video-based media elicit more positive visual responses than static images. Variance shows non-panoramic media lose perspective information, with regular videos at 56.51% and photos at 57.74% of panoramic video variance. Correlation coefficient quantification further confirms that static photographs distort subjects' perception of landscape elements like walls and roads. Although panoramic videos employing autostereoscopic 3D projection technology deliver coherent, 360° immersive visuals that accurately convey spatial information and simulate real environments, practical experimental efficiency considerations necessitate tailored media selection based on specific spatial typologies being simulated.

Keywords: Image perception · Visual assessment · Landscape preference · Panoramic projection · Interactive evaluation

1 Introduction

Human-environment interaction constitutes a dynamic process of material and informational exchange, wherein individual perception and behavior are co-shaped by ambient environmental characteristics [1, 2]. This transindividual paradigm underscores the critical role of accurate perception data acquisition in building intelligent human-environment networks [3].

Visual aesthetic quality, as an emergent property of human-environment interaction [4], serves as both a key indicator of habitat quality and a focal point for studying the laws of body-environment interaction. While simulated environments are widely used in visual perception research, their efficacy remains constrained by technological limitations in replicating authentic spatial experiences [5].

Y. Liu et al. (Eds.): CDRF 2025, *Transindividual Intelligence*, pp. 466–475, 2026.
https://doi.org/10.1007/978-981-92-0615-5_40

Contemporary digital technologies offer new possibilities through enhanced spatial data acquisition and intuitive human-machine interfaces. Immersive Virtual Reality makes scene simulating more accurate [6]. Within this framework, machines transcend their conventional roles as mere capturers and decoders of human perceptual inputs; they function as corporeal extensions of human agents, augmenting their capacity to assimilate and interpret ambient environmental information with heightened precision. But landscape preference assessments predominantly rely on traditional photographs [7] or static videos [8], despite documented limitations in conveying multidimensional spatial information [9].

Emerging visualization techniques like panoramic projection lack quantitative validation of their superiority over traditional media, leaving two critical questions unanswered: the extent to which they improve environmental perception accuracy and how cost-benefit considerations should guide media selection. This study addresses these gaps through controlled experiments comparing general static photographs, general videos, and panoramic videos, quantifying media-specific information loss in landscape simulation and establish evidence-based guidelines for cost-effective media selection.

2 Previous Research

2.1 Scene Simulation in Landscape Visual Preference Experiments

Visual preference arises from the dynamic interplay between environmental perception and cognitive interpretation, shaping an individual's perceptual schema [1]. Landscape visual preference reflects "the landscape's capacity to elicit aesthetic satisfaction" (Jacques, 1980) [10], providing critical insights into human perception mechanisms. Due to practical constraints, simulated environments are often necessary for landscape preference studies—for example, the Kaplan "Content Recognition Method" employs photo sampling and statistical analysis to decode environmental perceptions [11]. However, technological mediation introduces scene reproduction errors, with variance magnitudes dependent on the presentation media and the design of human-computer interaction mode.

2.2 Evolution of Presentation Medium: From 2D Static Pictures to Immersive Dynamic Panoramas

According to Gibson's theory of visual perception, image media can be categorized into static and dynamic types with the latter providing a closer approximation to natural visual experience [12]. While static images remain predominant in Chinese landscape studies [7], their one-sided presentation with restricted viewpoints compromises scene originality and the reliability of their preferred data is questionable. Conventional videos partially address this limitation through motion continuity but still suffer from perspective constraints [8]. Panoramic videos offer a solution by providing 360° coherent immersive visual experiences [13], implemented through two primary approaches: recorded panoramas (presented via web platforms [14] or VR headsets [15]) and modeled panoramas (computer-generated virtual environments [16]). Although comparative studies have demonstrated comparable scene simulation efficacy between recorded and modeled panoramas (Xu et al., 2022) [17], the "recorded panoramic roaming" technology used in these experiments is still not a continuous spatial sequence. The effectiveness of real-scene panoramic video in visual preference studies remains to be validated.

2.3 Principles of Human-Computer Interaction and Data Fidelity

In an ideal state, human-computer interaction (HCI) should be an intuitive exchange without secondary conversion, without hardware devices such as keys and touchscreens outside the body, nor the human body actively processing and expressing the perceived information [18].

In the application of visual perception research, virtual reality technology has optimized HCI experience. The three-dimensional interaction mode has been improved in both information input and output aspects: more visual devices focusing on embodiment are adopted, and intuitive postural interaction methods such as gestures and galvanic skin responses, to enhance the fidelity of perceived information transmission. However, in practical use, current virtual reality devices will affect individual visual perception due to interference from other senses. For example, when presenting virtual panoramic scenes, head-mounted VR devices often cause a sense of pressure on the head, affecting the user's physical feelings and further interfering with visual perception and evaluation. In contrast, non-wearable autostereoscopic 3D panoramic projection devices can reduce such interference, but they have higher requirements for the experimental site.

3 Experiments Methods and Procedures

The controlled experiment investigated visual perception differences across three presentation formats (static photographs, conventional videos, and panoramic videos), pioneering a cost-effective and user-friendly imaging and data collection paradigm based on real-scene panoramic videos. (Fig. 1).

All the image data are captured from real scenes. Considering HCI, all the subjects used intuitive gestures to express the visual evaluation scores. The comparison of the imaging methods and HCI media of the three groups of experiments is shown in Fig. 2. 18samples were collected from each video group, for a total of 1,710 online questionnaires for general photographs.

3.1 Data Processing

The real-time scores were synchronized with the actual paths using a frame-sampling and image recognition approach. Specifically, numerical scores were extracted from screen recordings at a frequency of 1 frame per second (1 Hz), obtaining ratings corresponding to each second of footage along every path. The collected data underwent normalization before calculating mean scores and variances for each frame. Finally, semantic segmentation technology was employed to extract and analyze landscape element from video frames, followed by computation of their correlation coefficients with the corresponding ratings (Fig. 3).

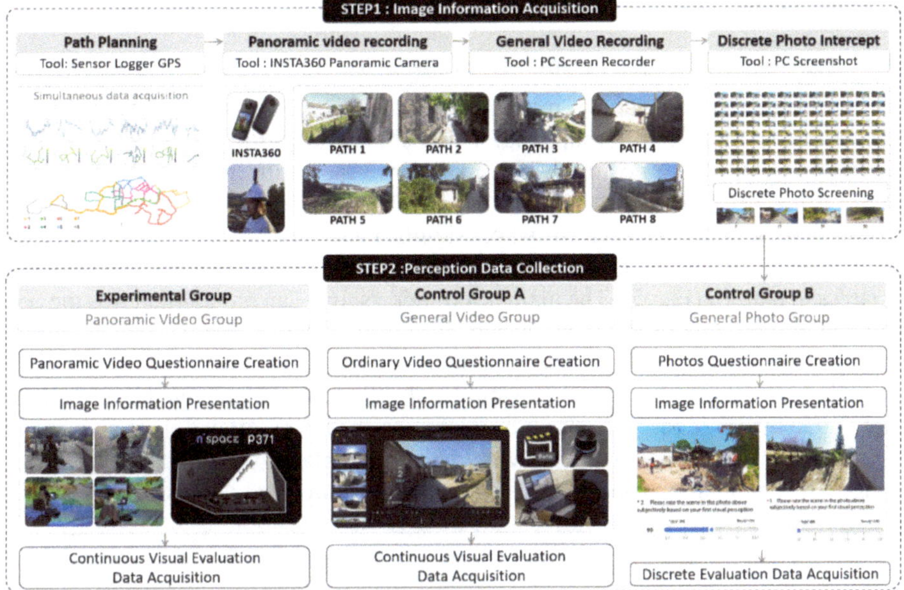

Fig. 1. Control experiment flow chart

	Medium of Presentation (all electronic images)	Presentation Tools	Human-Computer Interaction(HCI) Medium					
			Information Input Stage		Information Output Stage		Information Feedback Stage	
			Hardware	Software	Hardware	Software	Hardware	Software
Panoramic Video Group	360° Panoramic Moving Image	Naked Eye Panorama Projector (nspace p371, five-sided projector)	Enclosed Hexahedral Projection Room (4m×4m×3.9m)	nsapce	Slide Interactive Knob	EV Recording	15-inch Electronic Display	System Value Display
General Video Group	Fixed-Perspective Motion Image	Laptop Computer (DELL, 15 inches)	15-inch Electronic Display	Insta360 Player	Slide Interactive Knob	EV Recording	15-inch Electronic Display	System Value Display
General Photo Group	Fixed-Perspective Static Image	Smartphone/Computer	Electronic Display	Wenjuanxing	Touch Screen /Mouse	Wenjuanxing	Electronic Display	Wenjuanxing

Fig. 2. Comparative analysis of human-computer interaction modes for the three groups4.Data analysis and experimental results

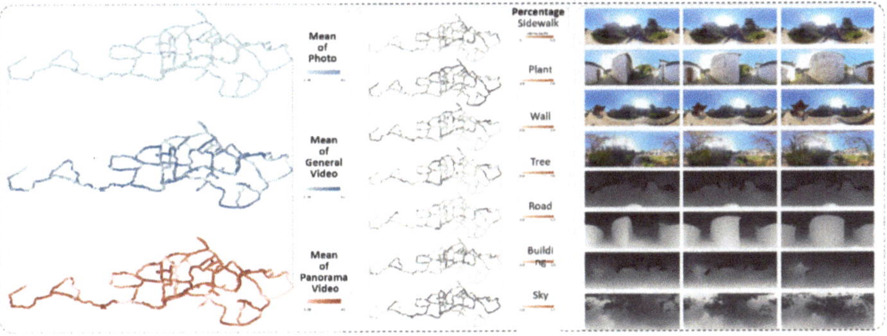

Fig. 3. Data processing analysis charts

3.2 Data Analysis

3.2.1 Mean Score Analysis: Dynamic vs. Static Media

The value and the fluctuations of mean scores of the video groups were overall higher than that of the photograph group (Fig. 4), separately suggesting that moving images typically provide a more positive visual experience than still images, and areas outside the artificially filtered nodes causing large visual difference were lost in the static screenshots to be displayed. In areas where significant numerical stratification appeared, general video group scored highest in multiple places. For example, for the node within 445–465 s in path 10 (Fig. 5), The main information showed in photograph was the open space, while the scene of the video group shifted from the dimmer space of a narrow alleyway to the open space with relatively better lighting and views. The transition and contrast between the scenes in videos would make the subjects feel that the landscape is "better" than the static screenshots alone. However, the lower left and right views of the node have a more cluttered landscape with rubbish and plastic sheds, which can only be shown in the panoramic video. So, this case also illustrates that the loss of information due to the restricted viewpoints affects the final visual evaluation results.

3.2.2 Variance Analysis: As a Fidelity Metric

Overall variance of the photo group is the lowest, while the panoramic video group's highest. Hypothesized reasons include: People tend to have a convergent preference for screened, representative photographs; the greater richness of the visual information in the panoramic video would lead to greater variance; and the suitability of the panoramic equipment for some subjects also led to variance.

Videos convey the positive visual experience of walking along a linear landscape space such as a river (Fig. 6) better than photographs. But cluttered elements like modern commercial on the side of the road were only presented in the panoramic video group, which had the highest variance. The quantification of variance indicates the highest spatial information fidelity of panoramic videos. If ideally panoramic video would show image information from the full viewpoint of the actual scene, then comparing the mean of the variances would also reflect the loss of information due to the missing viewpoints in the remaining two control groups. The average variance of the general video group was calculated to be 56.51% of that of the panoramic video group and 57.74% of that of the photo group.

3.2.3 Comparative Analysis of Correlation

The correlation coefficients (Fig. 7) between landscape elements and aesthetic ratings demonstrate that dynamic imaging significantly amplifies perceptual differentiation of elements, with panoramic videos providing more comprehensive representation of elements like walls and roads, intensifying their visual impacts. Overall, the absolute correlation coefficients in the photo group were predominantly below 0.15, while those in video groups exceeded 0.35, indicating dynamic media's magnification effect on element-specific preferences. Specifically, the "wall" element showed consistent positive correlations across groups, with panoramic and conventional video groups exhibiting

458.74% and 461.97% higher coefficients than photos respectively, suggesting dynamic presentation enhances positive engagement with architectural features. Conversely, the "road" element predominantly showed negative correlations, with absolute coefficient values in panoramic videos exceeding photos by 817.97%, conventional videos by 562.94%, and panoramic videos surpassing conventional videos by 380.89% ($p < 0.01$ for all comparisons). These differences reveal that road elements exert stronger visual influences in dynamic contexts, particularly when fully visualized in panoramic formats, which uniquely convey their complete spatial presence.

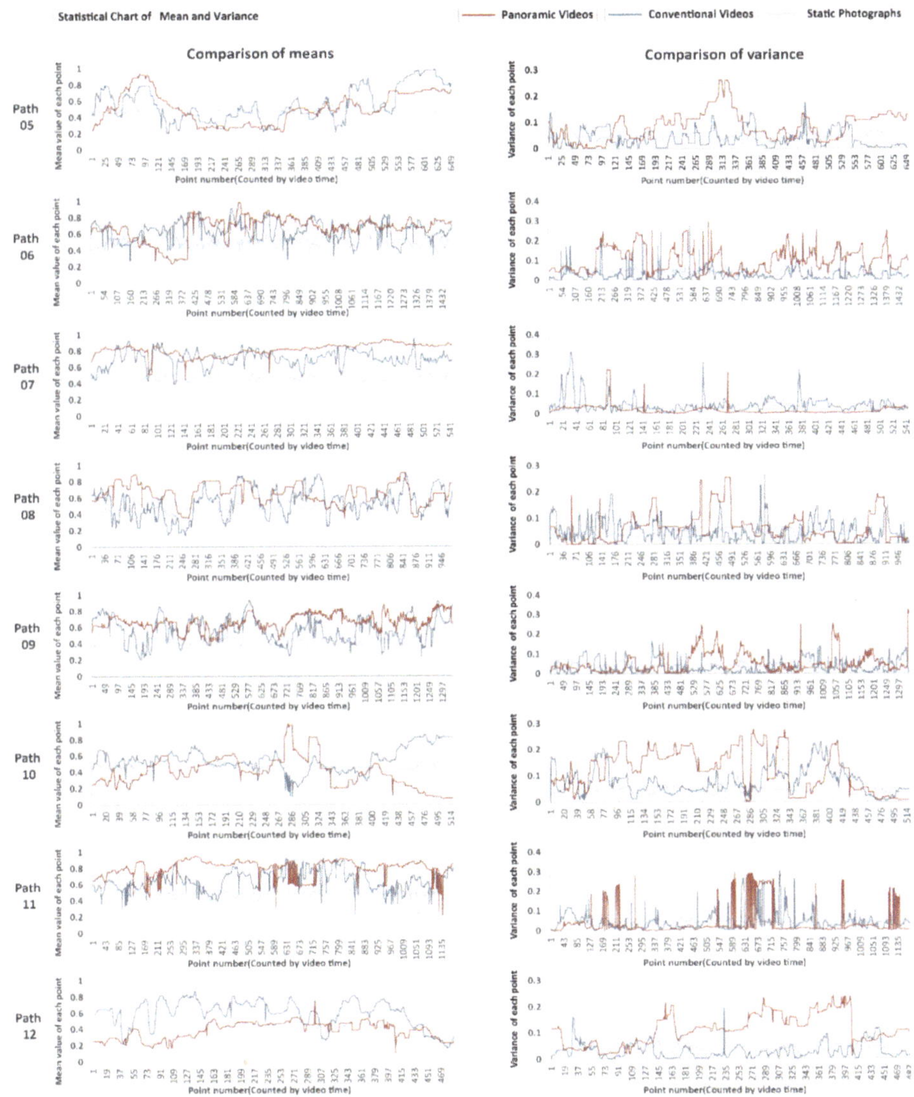

Fig. 4. Statistical chart of the mean and variance of the points within the 8 paths

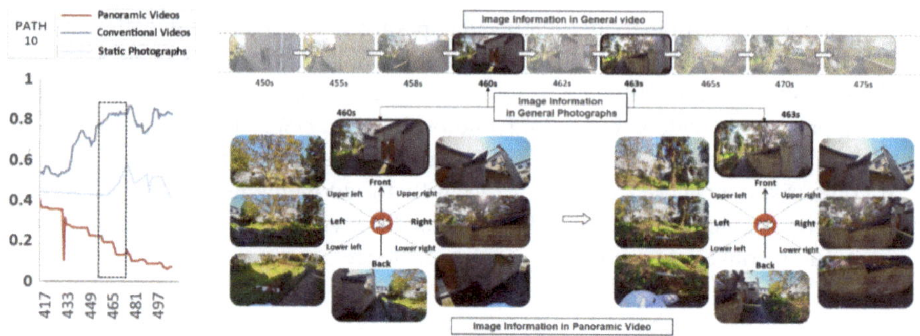

Fig. 5. Analysis of differences in mean values: Comparison of image information across three presentation media (take a node in path 10 as an example)

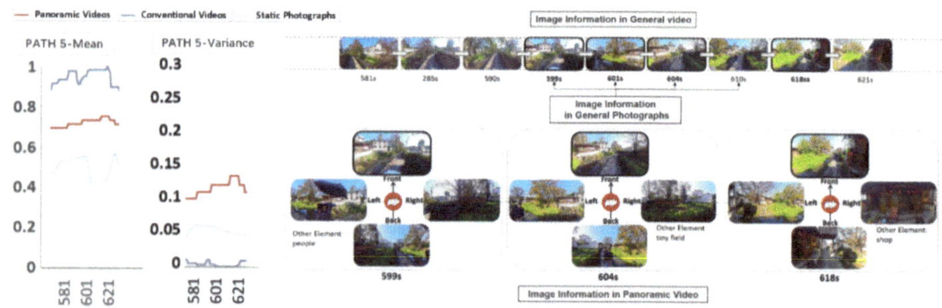

Fig. 6. Analysis of differences in variance values: comparison of image information across three presentation media (take a clips in path 5 as an example)

Path		wall	building	sky	tree	road	sidewalk	plant
5	mean_pic	0.11	0.09	-0.07	-0.05	-0.16	-0.04	0.01
	mean_mp4	0.33	0.17	-0.27	0.08	-0.39	-0.17	0.42
	mean_pa	0.32	0.33	-0.25	-0.10	-0.53	-0.01	0.20
6	mean_pic	-0.01	-0.01	0.01	0.00	0.01	0.00	-0.01
	mean_mp4	0.23	-0.11	0.01	0.17	-0.12	-0.12	0.15
	mean_pa	0.13	-0.03	0.07	0.09	-0.04	0.04	-0.10
7	mean_pic	0.06	0.00	-0.04	0.02	-0.06	0.06	0.01
	mean_mp4	-0.01	0.07	0.02	0.01	-0.17	0.02	0.01
	mean_pa	0.23	-0.30	0.22	0.29	-0.01	-0.21	0.03
8	mean_pic	0.03	-0.04	0.04	0.00	0.01	0.01	-0.01
	mean_mp4	0.07	-0.03	-0.18	0.25	0.01	-0.01	0.02
	mean_pa	0.25	-0.14	0.01	0.17	-0.21	-0.10	0.09
9	mean_pic	-0.02	0.04	-0.03	-0.01	-0.01	-0.04	-0.01
	mean_mp4	0.18	-0.36	0.19	0.27	0.04	0.06	0.30
	mean_pa	0.00	-0.13	0.09	0.01	0.14	0.10	0.00
10	mean_pic	0.09	-0.12	0.33	0.11	0.07	0.04	-0.01
	mean_mp4	0.16	-0.05	-0.37	0.33	-0.54	0.02	0.30
	mean_pa	-0.03	-0.22	0.54	-0.14	0.50	-0.14	-0.26
11	mean_pic	0.03	0.06	-0.05	-0.01	-0.04	0.04	-0.03
	mean_mp4	0.10	-0.08	0.16	-0.09	-0.13	0.01	0.02
	mean_pa	0.14	-0.11	0.09	0.03	-0.16	0.09	-0.18
12	mean_pic	0.01	0.05	-0.05	-0.01	0.02	0.01	0.00
	mean_mp4	0.10	0.26	-0.08	-0.23	-0.32	0.00	0.17
	mean_pa	0.15	-0.08	0.06	0.06	-0.32	-0.11	0.32

Fig. 7. Table of correlation analysis for each path based on semantic segmentation

4 Conclusion and Discussion

4.1 Error Control Framework in Visual Perception Experiments

The study identified key challenges in translating visual perception into quantifiable data. While gesture-based scoring (e.g., knob adjustment, touchscreen sliding) improved intuitiveness compared to verbal or textual responses, hardware limitations introduced new biases. For instance, delayed feedback from interactive knobs and inconsistent gesture-to-score mappings caused rating discrepancies for similar scenes. Although unified interaction designs (real-time display for video groups, touchscreen simulation for photo groups) enhanced cross-group comparability, they could not fully eliminate subjective cognitive processing during scoring. Physiological monitoring devices (e.g., EEG, EDA) were excluded to avoid physical discomfort, but this trade-off limited direct access to unconscious perceptual signals.

Three interdependent factors determine perceptual data accuracy in landscape studies. Scene fidelity is affected by media - specific information loss. Embodied immersion requires multi - sensory engagement—visual simulation alone can't represent real - world cognitive integration. Signal fidelity is distorted during conversion to evaluative info and external expression. Optimizing human - computer interaction, like using gesture - based interaction instead of rating scales or verbal descriptions, can minimize errors.

4.2 Recommendations and Strategies

In conclusion, panoramic videos have the highest scene fidelity among the three imaging methods, while ordinary photos the lowest. Yet all introduce errors to experimental results and necessitate situation-based adjustment in application. Panoramic projection systems have relatively high requirements for video recording equipment and processing equipment [19]. Ordinary photos are suitable for large-scale landscape preliminaries like different park styles or mountain ranges due to their capacity for simple screening and obtaining broad visual preference statistics. Ordinary videos are apt for landscapes with linear space such as streets or riverbanks. Panoramic videos are ideal for detailed, localized complex landscapes or interaction studies despite cost.

Therefore, in visual preference studies, real-world panoramic video technology based on panoramic projection holds significant advantages due to its continuity in imaging, full-view coverage, and realism. These strengths make it particularly suitable for experiments involving sequential spatial environments composed of relatively small-scale and highly detailed scene nodes. Specific application scenarios include: investigating how landscape elements in narrow alleys or linear streets with complex local features influence perceived visual quality, or exploring how specific visual elements such as greenery or wayfinding cues affect human behavior in urban public space segments like atriums or open corridors, where the density of visual information is typically higher.

Furthermore, this technology offers more precise and reliable technical support for further design optimization and control. It can be used, for example, to determine locations for partial spatial renovations in existing environments, or to accurately simulate and conduct "immersive evaluations" of design proposals before implementation. Given that panoramic projection based on real-world video involves directly recording and projecting actual scenes, it avoids the time-consuming and costly 3D modeling required in model-based VR panoramic walkthrough technologies. Prospectively, the continued development of panoramic live-streaming and projection technologies will provide stronger technical support for the real-time reconstruction of spatial environments, enhancing both environmental fidelity and the efficiency of visual preference data collection.

4.3 Limitations and Future Directions

Besides sample size disparity that may magnify individual errors, the experiments still have shortcomings in feedback interface design. And separation of score and image interfaces in panoramic viewing caused operational confusion. The adding of Integrated AR overlays or haptic feedback knobs may enhance control precision. What's more, the absence of auditory cues also negatively interfered with visual evaluation, and future experiments could incorporate multimodal stimuli, such as soundscape, olfactory cues, to better simulate real-world perception. Additionally, future experiments could add dimensions of VR-rendered panoramas and panoramic photographs, forming a more complete comparison.

References

1. Huang, J., Li, Y., Liang, J., et al.: Research on flow simulation and tourism routes optimization based on visual preference: A case study of Gulangyu Island. Tour. Trib. **39**, 85–100 (2024)
2. Celis Bueno, C., Schettini, C.: Transindividual affect: Gilbert Simondon's contribution to a Posthumanist theory of emotions. Emot. Rev. **14**, 121–131 (2022)
3. Liu J, Luo D, Fu X et al (2023) Design strategy of multimodal perception system for smart environment. In: Marques G, González-Briones A (eds) Internet of Things for Smart Environments. Springer International Publishing, Cham, pp. 93–115
4. Daniel, T.C.: Whither scenic beauty? Visual landscape quality assessment in the 21st century. Landsc. Urban Plan. **54**, 267–281 (2001)
5. Dong, S., Ma, J., Xin, W.: Research progress and trend of landscape visual evaluation-knowledge atlas analysis based on CiteSpace. J G UNIV: Nat. Sci. Ed. **41**(5), 1–13 (2023)
6. Zhao, M., Ding, C., Crossley, T.: Integration of EEG and deep learning on design decision-making: A data-driven study of perception in immersive virtual architectural environments. In: 28th International Conference on Computer-Aided Architectural Design Research in Asia, CAADRIA 2023 (2023)
7. Xu, T., Xu, C., Lan, S., et al.: Study on the characteristics of public preference of urban visual landscape. Chin. Landsc. Archit. **39**(11), 15–21 (2023)
8. Chen, H., Wang, M., Zhang, Z.: Research on rural landscape preference based on TikTok short video content and user comments. Int. J. Environ. Res. Public Health. **19**, 10115 (2022)
9. Bishop, I.D., Rohrmann, B.: Subjective responses to simulated and real environments: A comparison. Landsc. Urban Plan. **65**, 261–277 (2003)
10. Tang, X., Wang, X.: Landscape visual environment assessment (LVEA): Concept, origin and development. J. Shanghai Jiaotong UNIV (Agric. Sci.). **25**, 173–179 (2007)
11. Patrick, A.: Visual preference research: An approach to understanding landscape perception. Chin. Landsc. Archit. **29**, 22–26 (2013)
12. Gibson, J.J.: The ecological approach to the visual perception of pictures. Leonardo. **11**, 227–235 (1978)
13. Guo, M., Zhang, J., Yang, Z., et al.: An empirical study on the response of university students to viewing autumn secondary forest phytocommunities landscape via virtual reality in Northeast China. Ecol. Indic. **158**, 111450 (2024)
14. Shen, H., Aziz, N.F., Lv, X.: Using 360-degree panoramic technology to explore the mechanisms underlying the influence of landscape features on visual landscape quality in traditional villages. Eco. Inform. **86**, 103036 (2025)

15. Yu, N., Lin, Y., Gao, T., et al.: The relationship between landscape preference and the restoration effect of different types of park green spaces. Landsc. Archit. Acad. Journa. **40**, 54–62 (2023)
16. Xu, L., Meng, R., Huang, S., et al.: Healing oriented street design: Experimental explorations via virtual reality. Urban Plan. Int. **34**, 38–45 (2019)
17. Xu, J., Zhu, X., Wang, S., et al.: Perception evaluation of visual quality of public spatial environment in shopping centers: A validity study of subjective evaluation in simulated scenes. New Archit. **3**, 21–26 (2022)
18. Zhang, F., Dai, G., Peng, X.: A survey on human-computer interaction in virtual reality. Sci. Sin. (Inf.). **46**, 1711–1736 (2016)
19. Survey on QoE Evaluation of Panoramic Video: Survey on QoE evaluation of panoramic video. J. Signal Process. **38**, 1831–1842 (2022)

Enhancing Graph Machine Learning Workflows with Visual Programming for Urban Design

Luis Felipe Palomares Avena[1(✉)], Bruno Miguel Zarrabe Ricoy[2], Hetal Bharwani[2], and Angelos Chronis[2]

[1] Tecnologico de Monterrey, Eugenio Garza Sada 2501, 64849 Monterrey, Mexico
luisfelipe.palomares@tec.mx

[2] Institute for Advanced Architecture of Catalonia, Pujades 102, 08005 Barcelona, Spain

Abstract. Graph Machine Learning (GML) has emerged as a transformative tool for analyzing complex urban environments, enabling the discovery of hidden spatial patterns and their associated features. By integrating graph theory with deep learning, GML facilitates advanced spatial modeling and analysis of multi-dimensional data, often surpassing the capabilities of traditional machine learning models. However, the inherent complexity of GML workflows, often reliant on advanced coding skills and fragmented toolchains, presents significant barriers to adoption among urban planners and architects. This study addresses these challenges and provides a comprehensive overview of GML within architecture and urban design, highlighting the importance of facilitating human interaction with AI systems. By leveraging Grasshopper's visual programming platform, this research proposes a simplified, accessible workflow that aims to democratize predictive urban analysis tools, enhancing spatial analysis and real-time predictive data visualization for GML processes.

Keywords: Graph machine learning · Urban networks · Graph neural networks · Visual programming

1 Introduction

Graph Machine Learning (GML) has emerged as a transformative technique in architecture and urban analysis, enabling the discovery of hidden spatial patterns and their associated features [11]. GML provides a versatile framework for studying spatial patterns that correlate with most aspects of city life and for identifying factors that could influence accessibility, mobility, public space, and even safety. These capabilities empower urban planners to target possible interventions more precisely, potentially mitigating future urban challenges. The ability to visualize and interpret spatial relationships through graph structures could be critical for uncovering patterns, communicating findings, and making informed urban design decisions. Despite its potential, GML often remains inaccessible to urban planners and architects due to its reliance on advanced coding skills and

†Tecnologico de Monterrey, School of Architecture, Art and Design, Ave. Eugenio Garza Sada 2501, Monterrey, N.L., México, 64849.

Y. Liu et al. (Eds.): CDRF 2025, *Transindividual Intelligence*, pp. 476–490, 2026.
https://doi.org/10.1007/978-981-92-0615-5_41

familiarity with machine learning methodologies. Programming languages like Python, while powerful, present a steep learning curve for many professionals and students without formal technical training. Traditional ML implementations, while effective, are time-consuming and demand intricate coding knowledge to build, test, and debug. Furthermore, modern urban planning projects frequently involve complex datasets and multi-step processes across various platforms, requiring efficient computational workflows. These technical barriers prevent many users from leveraging data-driven analysis and predictive approaches to inform urban design decisions.

This paper addresses the accessibility challenge by introducing a visual programming workflow that enhances and simplifies the GML process—from data preparation to model training. By integrating Grasshopper with simplified GML methodologies, this research aims to democratize predictive urban analysis tools, enabling urban planners to efficiently process spatial data and make informed design decisions without extensive coding expertise.

2 Background and State-of-the-Art

While integrating graphs with deep learning is a relatively new area of study, the analysis of graphs themselves has been an established field for quite some time [13]. Graph-based spatial analysis within architecture and urban design was first introduced by Bill Hillier in the late 1970 as Space Syntax Theory [5]. Representing urban data with graphs can capture the intrinsic topological information and knowledge in the data, as well as many techniques developed to analyze the urban graph data [14].

2.1 Graphs and Machine Learning

Graph theory facilitates the understanding of intricate relationships within networks. When combined with deep learning, it becomes vital in understanding the multidimensional data, allowing for advanced spatial modeling and analysis [8]. Traditional machine learning models built for tabular data often fail to capture the complex interactions within an urban setting. In contrast, graph structures naturally represent spatial relationships, enabling models to identify subtle patterns that go beyond mere numerical proximity [12]. Alymani et al. highlights the challenges of traditional ML methods, emphasizing their limitations in handling the relational nature of urban data and the advantages offered by GML [1]. Some studies implementing GML have even been shown to outperform traditional ML techniques [7]. Graphs can contain more information from different domains, because they are not grid-based, and do not only represent information about elements (nodes, pixels, etc.) but also the relationships between nodes. This is specifically helpful in architecture and urban design where we define space as an assemblage of physical elements which are all somehow connected [2]. Kipf makes emphasis on the use of Graph Neural Networks as a way to model the complexities associated with urban datasets [9]. Additionally, integrating distance into the edges aligns with core urban planning principles, emphasizing the significance of proximity and connectivity in spatial relationships. By incorporating Euclidean and Dijkstra-calculated distances, the graph representation captures both geometric precision and practical travel considerations, enhancing its realism [12].

2.2 Visual Programming for Accessibility in Machine Learning

While at the moment, there isn't a dedicated tool or plugin to implement GML within Grasshopper, the current process involves integrating a variety of digital tools that usually incorporate python libraries leading to disconnected and fragmented processes. A typical workflow involves the use of Topologic in connection with visual programming tools, such as Dynamo, Sverchok, and Grasshopper, integrating the DGL Python library [8]. Topologic enables hierarchical and topological representations of cities, architectural spaces, buildings and artefacts through non-manifold topology (NMT), providing lightweight representations [6]. A more common approach involves using Python machine learning algorithms directly within Google Colab. While this may appear to be a more convenient solution, it often presents challenges such as limited immediate feedback, as well as restricted interaction and visualization of data.

3 Methodology

The following exploration was designed to explore the feasibility of running a complete graph machine learning (GML) methodology directly within Grasshopper (GH) environment and ecosystem, taking advantage of Python 3 capabilities inside Rhino 8. The focus was to determine if complex machine learning workflows could be integrated into Grasshopper's visual programming environment, enabling enhanced data manipulation, dataset analysis, prediction and integrated visualization without leaving the platform. The methodological framework was structured into three distinct sections:

1. The creation of GH components to integrate GML capabilities.
2. The implementation of a complete GML process within the context of a real-world project.
3. A comparative analysis of the workflow between two platforms that can host GML processes: Google Colab and Grasshopper.

3.1 GML within Grasshopper

The integration into Grasshopper offers users the benefit of its visual scripting interface, making spatial, geometry, and graph-related data analysis more intuitive. By adapting existing Python code to fit Grasshopper's structure, the experiment sought to simplify the overall process, re-arranging the input structure and workflow that could streamline these operations for a broad user base. The main strategy involved breaking down the complete GML workflow into smaller components that could function within Grasshopper's modular structure.

A total of 31 unique GH components were developed (freely accessible at https://github.com/luisFPA/safeNet), encompassing a range of functionalities essential for spatial and topological analysis. These scripts integrated tasks including attaching data to GML processes, visualizing graphs and graph inversions to better understand network connectivity and structural relationships and performing clustering analyses with K-means and DBSCAN algorithms using the scikit-learn library. The components also enabled advanced data manipulation tasks such as DataFrame handling, as well as the construction of NetworkX and DGL graphs, which were necessary for processing data in graph neural networks through a node classification component. Node classification is one of the most common and widely utilized tasks on graph data, requiring a model to predict the true category for each node [10]. Finally visualizing the results as graphs directly on the Rhino viewport and charts plots generated through Grasshopper and Matplotlib.

3.2 Case study and application in Mexico City

The second part of the methodology involved applying GML processes to a real-world urban project to evaluate its effectiveness and usability. The methodology was implemented in Mexico City as a case study, focusing on a predictive analysis based on crime against women (Fig. 1). Urban safety was recognized as an increasingly urgent concern as cities grow in complexity, requiring proactive interventions to reduce crime through artificial intelligence strategies, particularly against vulnerable populations such as women.

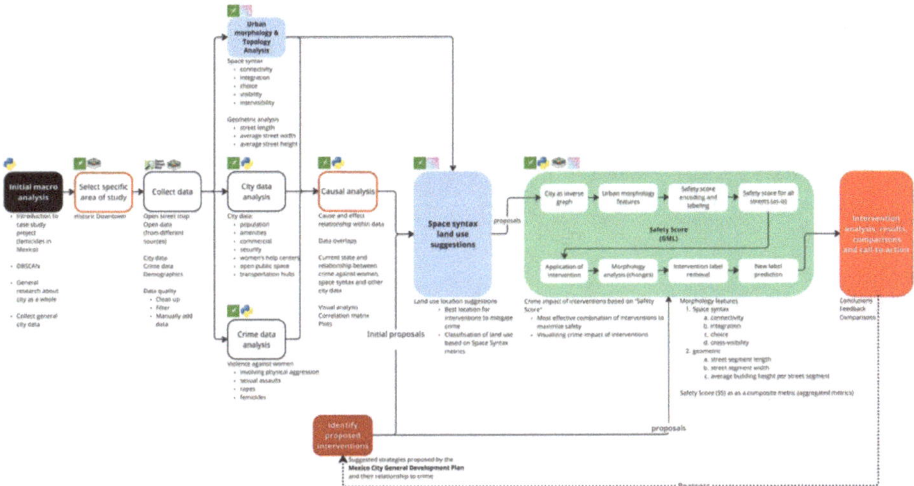

Fig. 1 Complete urban analysis and GML methodology applied to the Mexico City project.

The process began with identifying proposed interventions expected to contribute to crime reduction, which were derived from a comprehensive analysis of strategies outlined in the Mexico City General Development Plan (Fig. 2). The next phase focused on using space syntax metrics to generate land use suggestions that would optimize the effectiveness of these interventions. Central to the methodology was the construction and application of a safety score (Fig. 3), where the urban environment was represented as an inverse graph with street midpoints as nodes and connections as edges. The model integrated space syntax metrics (connectivity, integration, and choice), a cross-visibility metric reflecting building surveillance from across a street, as well as geometric features such as street length, width, and building height to capture the topological and morphological characteristics of the city. These factors fed into the safety score, which served as a baseline to assess potential land use interventions. After establishing the initial safety score, the proposed interventions were applied to the model to evaluate their impact on urban morphology. Once the interventions were incorporated, the existing safety labels on each street segment were removed, preparing the model for a new prediction (Fig. 4). Subsequently, a graph machine learning model was employed to predict new safety scores for each street segment. In the final phase, a comprehensive analysis was conducted to evaluate the effectiveness of the interventions in enhancing safety. Visualization tools were used to illustrate the spatial impacts, helping identify the most effective strategies. This approach provided guidance for targeted urban interventions aimed at improving public safety, particularly in reducing crimes against women.

Fig. 2 Proposed interventions within Mexico City's Historic Center based on Mexico City General Development Plan.

3.3 Comparative Analysis

The following approach focuses on comparing Grasshopper and Python within Google Colab for Graph Machine Learning (GML) workflows, emphasizing the processes, tools, and overall efficiency, based on the observed steps as the new workflow and tools were applied to the Mexico City case study. This comparison aimed to systematically evaluate and measure specific variables to assess their user interface, methodology and usability, among other observed factors. Instead of relying solely on computational time, which would be highly dependent on hardware performance, this comparison prioritizes the steps required to achieve a given task according to the following testing criteria:

- Installation and setup.
- Workflow complexity.

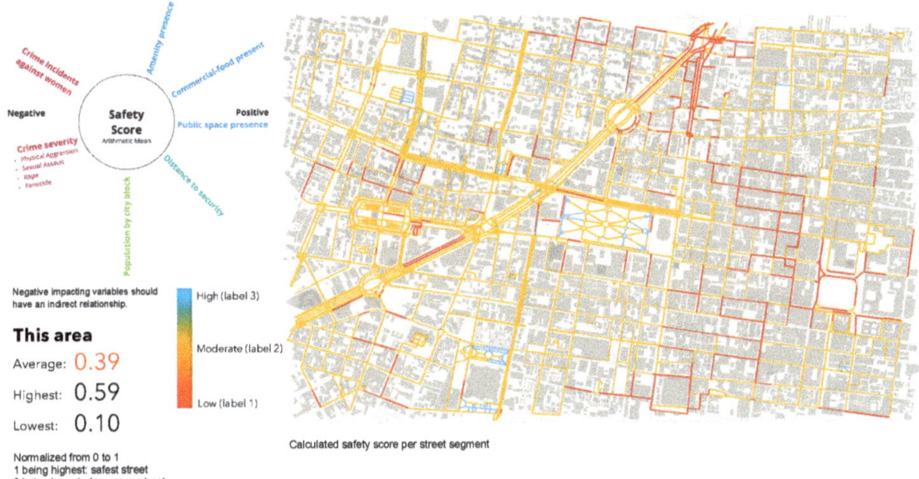

Fig. 3 Visualization of Safety Score analysis of Mexico City's Historic Center.

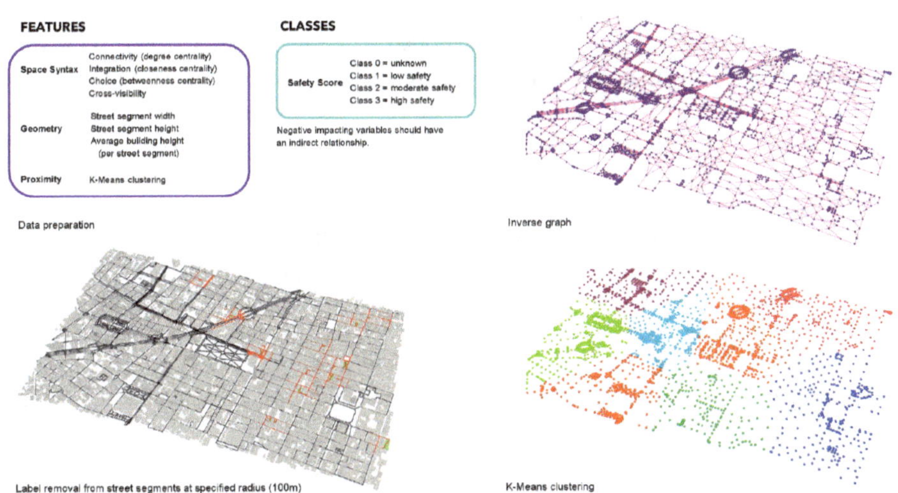

Fig. 4 Graph preparation for GML tasks involving node classification. Counter-clockwise from upper left corner: Features and classes to predict, node removal, K-Means clustering, and inverse graph, all within the Rhino-Grasshopper interface.

- Data management.
- Geometric capabilities.
- Visualization and user interaction.

4 Results

This section presents the results of applying Graph Machine Learning (GML) methodology in Grasshopper. It analyzes the model's performance in predicting crime risk against women in Mexico City, evaluates its effectiveness, and compares the approach with Google Colab to highlight differences in methodology and outcomes.

4.1 Grasshopper Methodology

Figure 5 outlines the reworked and simplified graph machine learning workflow enabled by Grasshopper used for data manipulation, encoding, feature extraction, graph construction, model training, and evaluation stages. The process begins with the feature extraction phase, where raw data is processed to assign values to nodes or edges in the graph through geometric manipulation within Grasshopper. For edge prediction, the graph can be inverted so that edges become nodes and nodes become links between them. This way the same node classification algorithm can be performed. Distinct types of raw data are normalized and combined to create a comprehensive set of features. Additionally, geometric data (such as points or segments) undergoes K-means clustering, followed by one-hot encoding, to enrich the feature set with spatial clusters that may correlate with input data. Next, in the graph construction phase, the normalized features and classes are assigned to nodes and edges within a NetworkX graph. This graph is then converted into a DGL (Deep Graph Library) format to facilitate compatibility with graph neural network (GNN) models. The NetworkX graph can also be exported in CSV format for further analysis, while the DGL graph is saved in binary format (BIN) for direct input into the GNN. The GNN model training step node classification, which enables the model to predict labels for nodes based on spatial and topological relationships within the graph. This step encompasses three primary stages: preparation, learning, and an iterative evaluation process to refine classifications and address missing class labels. Finally, the evaluation phase includes both visual (plots) and quantitative (performance scores) analyses. These evaluations assess the model's accuracy in classifying nodes (or edges), contributing to an understanding of spatial relationships and risk distribution across the city being analyzed.

4.2 Case Study Model Performance

The model demonstrated strong performance, exhibiting a commendable balance between precision and recall; however, there remains potential for further enhancement, particularly in terms of precision. The predictions predominantly aligned with the second label, corresponding to a moderate safety score. In evaluating the proposed intervention sites, we found that six of these sites resulted in an improved safety score. Conversely, four sites exhibited at least one street where the safety score declined, while five sites remained unchanged. Additionally, two sites were not assessed due to errors encountered during the analysis. The accompanying graphs, shown in Fig. 6, illustrate both the incomplete and complete safety score distributions, effectively highlighting the locations of the assigned labels. These visualizations provide critical insights into the impact of proposed interventions on urban safety dynamics. The analysis revealed that

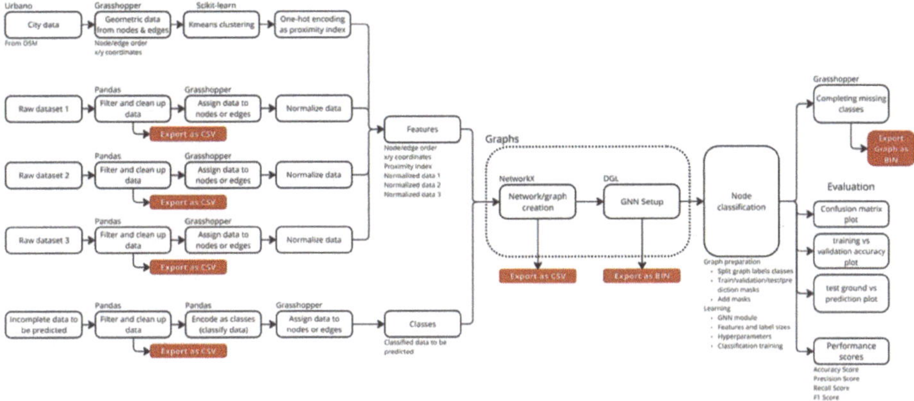

Fig. 5 Graph Machine Learning process and methodology through visual programming. Example shows three sets of data as features.

certain geometric and morphological adjustments, such as creating new pedestrian pathways, had a complementary effect on the success of public space interventions. These findings highlight the importance of considering both spatial configuration and physical design when planning safety-enhancing interventions. The study also identified a critical threshold for prediction accuracy. It was observed that beyond a 300-m radius from the intervention site, the reliability of the predicted safety scores began to decline. This suggests that the influence of public space interventions on safety is spatially limited and diminishes with distance.

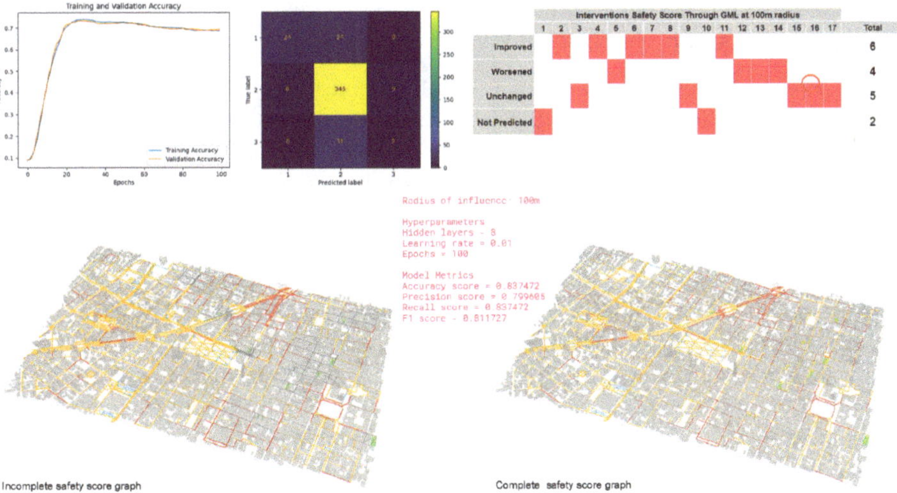

Fig. 6 Results of GML process implementation in Mexico City's crime issues.

4.3 Comparative: Steps and Focus Areas

The following section presents the findings from the application of Graph Machine Learning (GML) in Grasshopper and Google Colab for the Mexico City case study as a comparative between the two platforms.

Installation and Setup. The setup process for running GML varies between Grasshopper and Python in Google Colab, with each platform relying heavily on external libraries to extend functionality. Both platforms use core libraries such as DGL, NetworkX, and Scikit-learn for GML (Table 1). However, Grasshopper requires setting up a new Python environment to implement GML capabilities and makes use of task-specific plugins like Urbano [4] and DeCodingSpaces [3] for spatial data and space syntax analysis. In contrast, Google Colab allows for a more straightforward setup through a simple!pip command, integrating libraries like OSMNX for city data, Momepy for morphological analysis, Shapely for geometric manipulation, and Matplotlib for data visualization, offering robust tools for spatial and urban data management.

Workflow Complexity. Workflow complexity and execution steps differ significantly between Grasshopper and Python in Google Colab, largely due to their respective approaches to modularity and visualization of algorithms. Google Colab offers modularity by dividing code into individual cells, which allows flexibility for iterative coding but can complicate workflows due to the non-linear execution of code. This makes it harder to manage multiple interconnected modules.

Grasshopper simplifies workflows with its modular, node-based components that integrate Python algorithms into individual visual frameworks. This modularity reduces the number of steps required throughout the process, as multiple algorithms can be smoothly integrated into a single component allowing for easier execution, particularly for users with limited coding experience. However, Grasshopper is limited to node classification tasks, requiring graph inversion when edge classification is needed.

Dataset Processing. Python in Google Colab demonstrates the capability to process large datasets efficiently, such as those representing the entire city of Mexico City. These datasets encompass a significant number of nodes and spatial features, showcasing Colab's ability to handle extensive and complex urban data. Google Colab's flexibility and computational resources make it ideal for large-scale analyses that involve extensive spatial and tabular data. It supports robust workflows that can dynamically scale as datasets grow in size and complexity. Its reliance on cloud-based resources reduces hardware constraints, enabling the processing of large-scale datasets with ease.

Grasshopper, in contrast, is limited in its capacity to handle large datasets. For instance, when using OpenStreetMap (OSM) data, the platform enforces a cap of 50,000 nodes, restricting the area of analysis to approximately 6,000,000 square meters (e.g., a region measuring 2,000 m x 3,000 m). This limitation makes Grasshopper more suitable for small-scale, detailed studies rather than citywide investigations. However, it struggles to scale up when processing extensive tabular data or performing large-scale spatial computations.

Geometric Operations in Attribute Assignment. A crucial step within the GML process involves the encoding of data to specific nodes or edges within the graph as attributes.

The process of adding attributes to nodes and edges within the graph, such as selecting the closest nodes or creating bounding boxes with points inside, is inherently a geometric operation. Grasshopper excels in such tasks due to its geometry-first design. Its tools

Table 1 List of Python libraries needed to run a complete Graph Machine Learning process comparing their use within Google Colab and Grasshopper platforms

Library	Google Colab	Grasshopper	Purpose
DGL	Install through Pip	Install v2.0.0 through Pip	Builds and trains graph neural networks.
GeoPandas	Pre-installed	Not needed	Extends Pandas to handle geospatial data.
Google.colab Drive import	Pre-installed	Not needed	Mounts Google Drive to current cloud-based coding environment.
Matplotlib	Pre-installed	Install v3.9.0 through Pip	Creates static, animated and interactive visualizations.
MomePy	Install through Pip	Not needed	Urban morphology analysis for spatial and network-based metrics.
NetworkX	Pre-installed	Install v3.2.1 through Pip	Creates, analyzes and visualizes complex networks.
Node2Vec	Install through Pip	Not needed	Algorithm for learning feature representations of graph nodes.
NumPy	Pre-installed	Install v2.0.2 through Pip	Package for numerical computing in Python.
OSMNX	Install through Pip	Not needed	Extracts, processes, and analyzes OpenStreetMap data.
Pandas	Pre-installed	Install v2.2.2 through Pip	Data manipulation and analysis library for structured data.

(*continued*)

Table 1 (*continued*)

Library	Google Colab	Grasshopper	Purpose
Phik	Install through Pip	Could not be installed	Measures correlation between categorical and numerical data.
Scikit-learn	Pre-installed	Install v1.5.1 through Pip	Machine learning library for classification, regression, and clustering.
Shapely	Pre-installed	Not needed	Geometric operations and manipulation of planar spatial objects.
Torch	Pre-installed	Install v2.2.1 through Pip	Deep learning framework optimized for tensors and GPUs.

and components are optimized for handling spatial data, making it easier to perform these operations interactively and efficiently. In Google Colab, these operations can be replicated but would typically require custom scripts or the use of multiple libraries.

Visualization. Visualization and user interaction play crucial roles in computational design and analysis, particularly when working within urban settings. Grasshopper is specifically designed to prioritize interactive, geometric outputs, enabling users to explore results in real time. This feature is particularly valuable for design processes that require iterative refinement, as it allows for immediate visual feedback and adjustments. Grasshopper's emphasis on visualizing data interactively and instantaneously makes it especially accessible for beginners, who can easily manipulate and understand the geometry they are working with without needing to delve deeply into coding or complex algorithms by creating the graph directly on the Rhino viewport. In the study, graphical outputs, including training versus validation accuracy plots and confusion matrices, were generated and displayed directly within the Grasshopper interface. Charts depicting training versus validation accuracy and confusion matrices were generated directly within the Grasshopper interface using the Matplotlib library. Additionally, evaluation metrics—including accuracy score, precision score, recall score, and F1 score—were computed through a dedicated standalone component to provide a comprehensive assessment of model performance and complement the visualized results.

5 Discussion

This paper demonstrates the feasibility of adapting a complete GML methodology and its associated algorithms to the Grasshopper ecosystem. While challenges such as file size limitations, library compatibility, and advanced graph plotting persist, the coupling of parametric modeling with graph-based spatial analysis offers distinct advantages. These include seamless integration of analytical results into the design process and the capacity to generate and evaluate multiple design iterations efficiently, highlighting Grasshopper capability as a prototyping tool [3]. Grasshopper's unique blend of robust geometry manipulation, real-time feedback, and visual programming capabilities positions it as an ideal platform for integrating graph machine learning into spatial analysis, offering an interactive framework that fosters human-AI collaboration. The comparative analysis highlights Grasshopper's strengths in usability and geometric manipulation, particularly when attaching spatial features to nodes and edges. In contrast, Google Colab offers scalability, an advanced library ecosystem, and the capacity to handle large datasets. The designed workflow combines multiple algorithms within individual modules, simplifying tasks for users with limited scripting expertise thus reducing the number of steps required within the GML process. Through the proposed methodology, Grasshopper's components streamline workflows by visually integrating geometric and topological data, reducing the complexity of feature extraction, ease in geometric data encoding, and graph neural network model training [1, 9].

The created toolset aims to broaden its scope beyond node classification to address a wider range of GML tasks. These include link prediction, which focuses on determining the likelihood of an edge between arbitrary nodes, as demonstrated in frameworks like DGL. Additionally, graph classification and regression tasks can assign labels or predict values for entire graphs, while neighbor sampling methods enable efficient selection of node subsets during training. The incorporation of three-dimensional graphs into the toolset would also represent a significant enhancement. While two-dimensional graphs suffice for many urban design applications, some scenarios—such as volumetric spatial analysis and multilevel architectural environments—require the complexity of 3D graphs to accurately capture spatial relationships. This capability extends the toolset's applicability to a broader range of architectural and urban challenges.

6 Conclusion

Our work builds upon previous GML foundational studies and platforms while addressing significant gaps in the current body of research, emphasizing the accessibility and usability of GML tools, thus promoting a more inclusive adoption of data-driven urban analysis. By embedding advanced GML workflows within an intuitive, visual programming environment, we intend to contribute to a paradigm shift in how urban data is analyzed and utilized in design processes. This integration offers a user-friendly workflow that facilitates access to sophisticated machine learning tools and data analysis that fosters broader engagement with computational methods in urban planning and architecture. Our methodology enables more nuanced and actionable insights into urban environments, ultimately supporting more informed and effective design strategies, advancing the accessibility of AI tools that open new avenues for exploring the complex interplay between urban form and function, as well as human-machine interaction.

Beyond professional practice, the proposed tool and workflow holds significant value in educational settings. By integrating GML into a widely adopted architectural tool, we intend to lower the entry barrier for architecture and urban design students, as well as non-programmers to explore advanced spatial modeling and analysis. The visual and interactive nature of Grasshopper enhances engagement, offering immediate feedback that improves understanding and application of complex algorithms and data. This makes it a practical tool for teaching data-driven workflows based on GML fundamentals, enabling hands-on learning through real-world examples. To further support this integration, short courses and workshops can be developed around the tool, covering key topics such as graph theory, GML tasks, feature assignment and spatial analysis. Our research contributes to the evolving intersection of computational design, urban planning, and architectural education by making sophisticated analytical tools more accessible and impactful.

Future research could explore promising directions, such as the adoption of 3D graph representations for volumetric spatial analysis [7] and the incorporation of advanced graph machine learning tasks, including link prediction and graph classification. These advancements have the potential to significantly enhance the accessibility and applicability of GML in addressing intricate urban challenges. The capabilities of GML offer novel opportunities for the analysis of complex urban networks, contributing to the paradigm shift toward data-informed decision-making. This democratization of computational and AI-driven tools empowers architects and urban planners to tackle multifaceted issues inherent in cities. By integrating sophisticated spatial modeling and analysis with graph machine learning, users can derive actionable insights into building and urban topologies, streamline design processes, and sustain leadership in architectural and urban innovation [8]. The toolset's versatility opens doors to other architectural domains, including building performance analysis, optimization of structural networks, and exploration of interior spatial configurations. This adaptability highlights the relevance of GML techniques to diverse architectural contexts. It establishes a foundation for a tool that merges visual programming strengths with coding-based GML workflows. By addressing current limitations and refining the methodology, the development of a fully integrated plugin for Grasshopper becomes achievable. Such a tool would not only simplify complex GML processes even further, but also make them accessible to a broader audience, paving the way for predictive analysis and fostering data-informed decision-making across various fields.

References

1. Alymani, A., Jabi, W., Corcoran, P.: Graph machine learning classification using architectural 3D topological models. SIMULATION. **99**(11), 1117–1131 (2023). https://doi.org/10.1177/00375497221105894
2. Bauscher, E., Dai, A., Elshani, D., Wortmann, T.: Learning and generating spatial concepts of modernist architecture via graph machine learning. In: Proceedings of the 29th International Conference of the Association for Computer Aided Architectural Design Research in Asia (CAADRIA), vol. 1, pp. 159–168. The Association for Computer-Aided Architectural Design

Research in Asia (CAADRIA), Hong Kong (2024). https://doi.org/10.52842/conf.caadria.2024.1.159
3. Bielik, M., Schneider, S., König, R.: Parametric urban patterns: Exploring and integrating graph-based spatial properties in parametric urban modelling. Des. Tool Dev. **1**, 701–708 (2012). https://doi.org/10.52842/conf.ecaade.2012.1.701
4. Dogan, T., Saraf, N.: Urbano: A new tool to promote mobility-aware urban design, active transportation modeling and access analysis for amenities and public transport. In: Proceedings of the Symposium on Simulation for Architecture and Urban Design. Delft, Netherlands (2018). https://doi.org/10.22360/SimAUD.2018.SimAUD.028
5. Hillier, B., Hanson, J.: The Social Logic of Space. Cambridge University Press (1984). https://doi.org/10.1017/CBO9780511597237
6. Jabi, W., Aish, R., Lannon, S., Chatzivasileiadi, A., Wardhana, N.M.: Topologic—A toolkit for spatial and topological modelling. In: Computing for a Better Tomorrow, Proceedings of the 36th eCAADe Conference, vol. 2, pp. 449–458 (2018). https://doi.org/10.52842/conf.ecaade.2018.2.449
7. Jabi, W., Alymani, A.: Graph Machine Learning using 3D Topological Models. Society for Computer Simulation International (SCS) (2020)
8. Jabi, W., Leon, D.A., Alymani, A., Pourali Behzad, S., Salamoun, M.: Exploring building topology through graph machine learning. In: Crawford, A., Diniz, N., Beckett, R., Vanucchi, J., Swackhamer, M. (eds.) CADIA 2023: Habits of the Anthropocene: Scarcity and Abundance in a Post-Material Economy, vol. 3, pp. 24–32. Association for Computer Aided Design in Architecture (ACADIA), Denver, USA (2023) https://papers.cumincad.org/cgi-bin/works/paper/acadia23_v3_179
9. Kipf, T.N., Welling, M.: Semi-supervised classification with graph convolutional networks arXiv preprint https://arxiv.org/abs/1609.02907. (2017). https://doi.org/10.48550/arXiv.1609.02907
10. Node Classification with DGL. Deep Graph Library. Last accessed Retrieved 7 Jan 2025. (n.d.). https://docs.dgl.ai/tutorials/blitz/1_introduction.html#sphx-glr-tutorials-blitz-1-introduction-py
11. Pagani, A., Mehrotra, A., Musolesi, M.: Graph input representations for machine learning applications in urban network analysis. Environ. Plan. B. **48**(4), 741–758 (2021). https://doi.org/10.1177/2399808319892599
12. Sideris, N., Bardis, G., Voulodimos, A., Miaoulis, G., Ghazanfarpour, D.: Enhancing urban data analysis: leveraging graph-based convolutional neural networks for a visual semantic decision support system. Sensors. **24**(4), 1335 (2024). https://doi.org/10.3390/s24041335
13. Wang, M., et al.: Deep graph library: A graph-centric, highly-performant package for graph neural networks. arXiv preprint https://arxiv.org/abs/1909.01315. 2019. https://doi.org/10.48550/arXiv.1909.01315
14. Li, Y., Zhou, X., Pan, M.: Graph neural networks in urban intelligence. In: Graph Neural Networks: Foundations, Frontiers, and Applications, pp. 579–593. Springer Nature Singapore, ingapore (2022)

CityFlow: An Advanced Case-based Design System Integrated with Expert Workflow and LLM Retriever

Kai Hu[1(✉)], Ariel Noyman[2], Adrian Mora-Carrero[3], Parfait Atchade-Adelomou[2], Luis Alonso-Pastor[2], Markus ElKatsha[2], Yan Zhang[2], Yubo Liu[1], and Kent Larson[2]

[1] School of Architecture, South China University of Technology, Guangzhou, China
arhukai@mail.scut.edu.cn

[2] MIT Media Lab, Massachusetts Institute of Technology, Cambridge, MA, USA

[3] Escuela Técnica Superior de Ingenieros Informáticos, Universidad Politécnica de Madrid, Campus Montegancedo, Madrid, Spain

Abstract. Urban environments present increasingly complex challenges, which demand innovative, interdisciplinary solutions. This paper introduces a novel AI-enhanced methods-as-case framework, designed to support decision-making in urban design. Addressing the limitations of traditional rule-based and case-based design paradigms, our framework combines these approaches to enhance case retrieval and interpretable revision processes. The resulting CityFlow platform provides a user-friendly, low-code environment for data analysis, urban evaluation, and simulation, fostering collaboration and knowledge sharing among diverse users. By integrating LLMs and a knowledge graph, CityFlow facilitates rapid prototyping and iterative refinement of urban design workflows. We discuss the platform's current capabilities, aligned with early stages of AI advancement (L1-L2), and outline future directions towards more autonomous AI assistance (L3), ultimately redefining human-computer interaction in urban science and paving the way for more sustainable and collaborative urban design and research.

Keywords: Artificial intelligence · Computer-aided design · City sciences · Large language model · Case-based design

1 Introduction

Urban environments are becoming increasingly complex, facing significant challenges in areas such as sustainability, mobility, public health, and equity (Noyman 2022). Urban design problems are often considered "wicked problems" (Rittel and Webber 1973), meaning they lack definitive solutions, have unclear criteria for success, and involve conflicting stakeholder values. Addressing these issues requires innovative, interdisciplinary approaches and collaborative research to build more sustainable and liveable cities.

Y. Liu et al. (Eds.): CDRF 2025, *Transindividual Intelligence*, pp. 491–501, 2026.
https://doi.org/10.1007/978-981-92-0615-5_42

A model is a simplified representation of a real-world system, used to understand, predict, or control it by focusing on key elements and relationships. However, as Michael Batty notes, no single model can fully address the complexity of a city problem. Instead, a multi-modeling approach — utilizing multiple models to represent the diverse values and urban dynamics — is necessary (Batty 2021).

How to build an intelligent multi-modeling system to support decision-making in complex urban environments? Historically, Computer-aided design (CAD) methods have relied on two primary paradigms: rule-based design (RBD) and case-based design (CBD). RBD is simple and efficient for clear problems, but struggles with complexity and change. CBD handles complexity well by learning from past examples, but can be inaccurate, inflexible, and difficult to implement.

This paper proposes an AI-enhanced framework by combining CBD and RBD methods. A platform named CityFlow is built based on the framework. With an integrated AI assistant, CityFlow enables users to search and create diverse workflows regarding data analysis, urban evaluation, and simulation, making it easier to discover, share, and refine innovative approaches for urban problem-solving.

2 Literature Review

The research on design methods and systems can be traced back to the 1960s, when the pioneers, such as Bruce Archer and John Chris Jones, started the Design Method Movement (DMM) (Langrish 2016). They aimed to make design more systematic by using rules and procedural systems. While the DMM declined in the 1970s, its focus on structured methods helped shape the development of CAD. RBD methods, like Shape Grammar and Space Syntax, provide powerful tools for architects and urban planners. However, some designers dislike these approaches, finding them overly rigid and uncomfortable with the idea of reducing complex human behavior and experiences into logical frameworks (Cross 1993).

Unlike RBD, CBD systems leverage past experiences and knowledge to tackle new challenges. One of the earliest examples was Archie-II, which built a database of buildings by analyzing their floor plans and usage across various factors (Domeshek and Kolodner 1993). Other CBD tools used in architecture include CADRE (Dave et al. 1995), FABEL (Voss 2014), and Architable (Wei 2013). Early CBD tools were primarily focused on retrieving design cases, lacking true intelligence and automation in adapting and adjusting historical cases.

In recent years, with the rapid development of deep learning, many researchers have developed new CBD tools with deep learning. For example, Huang and Zheng first applied Generative Adversarial Networks (GANs) to the recognition and automatic generation of apartment building floor plans (Huang and Zheng 2018). Liu et al. expanded the application of GANs to the generation of university campuses and Jiangnan gardens (Liu et al. 2021, 2022). Steven Jige Quan proposed an Urban-GAN CBD framework to generate urban forms in the Manhattan area (Quan 2022). While powerful, the "black-box generation" of deep learning makes these CBD tools less interpretable, and it is difficult for designers to further adjust the generated results.

As summarized in Table 1. RBD excels with clear rules and is easy to use, but struggles with ambiguity, while CBD thrives on complex problems but lacks explainability and adaptability.

Table 1 Comparison between Rule-based & Case-based design systems

	Rule-based design system	Case-based design system
Complexity Handling	Weak; struggles with ambiguous problems	Strong; supports fuzzy matching
Flexibility	Low; rules hard to update	High; adaptable through learning new cases
Usability	High; clear rules, easy to implement	Low; poor explainability, difficult to identify/adapt cases
Adaptability	High; applicable across scenarios with well-defined rules	Low; past cases often fail in new scenarios

3 Methodology

To address these limitations, we introduce an LLM-enhanced Methods-as-Cases Framework by combining CBD and RBD methods. As an old Chinese saying goes, give people fish, you feed them for a day, teach them how to fish, and you feed them for a lifetime. Unlike conventional CBD systems that store static outcomes, the proposed framework stores rule-based design algorithms as cases. Additionally, by incorporating LLM-powered retrieval and coding assistance, this framework enhances CBD by improving the accuracy of case retrieval and supporting intelligent modification of design algorithms.

As shown in Fig. 1, this framework follows a case-based reasoning process, comprising four key steps: Retrieve, Revise, Reuse, and Retain.

- **Retrieve:** The **LLM Retriever** finds appropriate design algorithms based on natural language inputs from the database.
- **Revise:** Modify and update existing design algorithms according to new design requirements with the assistance of the **LLM Coder**.
- **Reuse:** Utilize the **Containerized Interpreter** to execute newly created design algorithms and generate new solutions for specific design problems.
- **Retain:** Store the new design algorithms and their associated design problems in the **Knowledge Graph (KG) Database** for future retrieval and reuse.

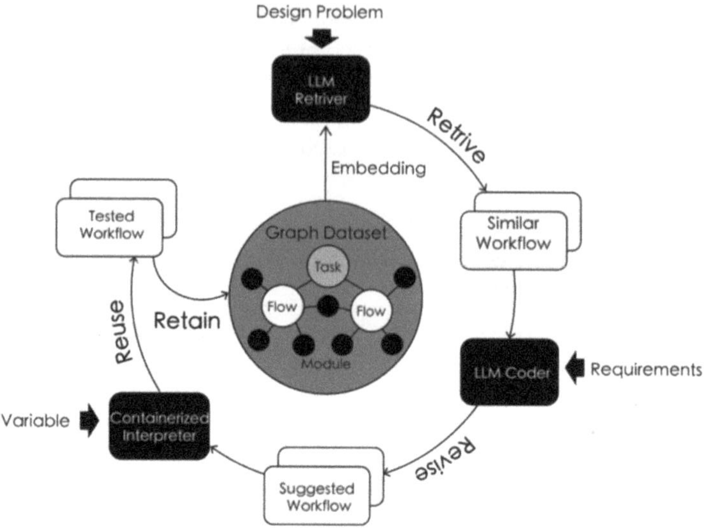

Fig. 1 The LLM-enhanced Methods-as-Cases Framework

3.1 Textual Embedded Semantic Knowledge Graph (KG)

Our framework focuses on storing design methods rather than the outcomes. To achieve this, Directed Acyclic Graphs (DAGs) are used to organize the design processes (See Fig. 2). Thinking of each node within the DAG as a specific operation, it contains both the code to execute and a clear description of its purpose. These nodes are linked together, representing a chain of both reasoning and action, where the output of one node directly feeds into the next. This structure, similar to Grasshopper, makes the reasoning process easy to understand and operate by Designers and LLMs. To make this information useful for searching, we convert all information into vector embeddings using the all-MiniLM-L6-v2 model. This allows us to perform semantic searching, finding relevant design methods based on their meaning.

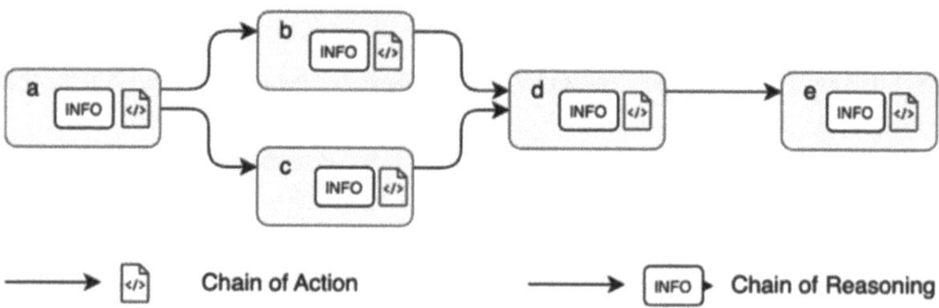

Fig. 2 The proposed DAG structure includes both the chain of reasoning and the chain of action.

3.2 LLM-enhanced Retriever

Design tasks often involve ambiguous requests – for example, the client may need a "colorful black." Standard search methods, like keyword or even semantic search, struggle with these kinds of requests because they don't understand the underlying intent. To address this, we've developed a new retrieval method that combines LLMs with KG. (See Fig. 3 for an overview of the process.)

Our method works as follows: when a design request is received, an LLM first breaks it down into clearer and more specific sub-questions. Next, a semantic search is used to identify relevant entities within the KG based on these refined questions. To broaden the search, we expand these initial entities by exploring their relationships to other nodes in the KG – the extent of this expansion is controlled by a defined 'search depth'. Finally, another LLM analyzes the retrieved information, scoring it based on its relevance to the refined questions. This allows the system to provide a more accurate answer with informed recommendations.

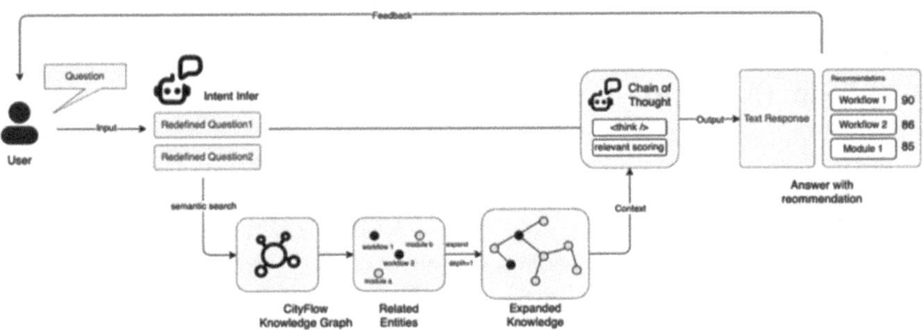

Fig. 3 LLM-enhanced Retrieve method based on intent inference and Graph RAG

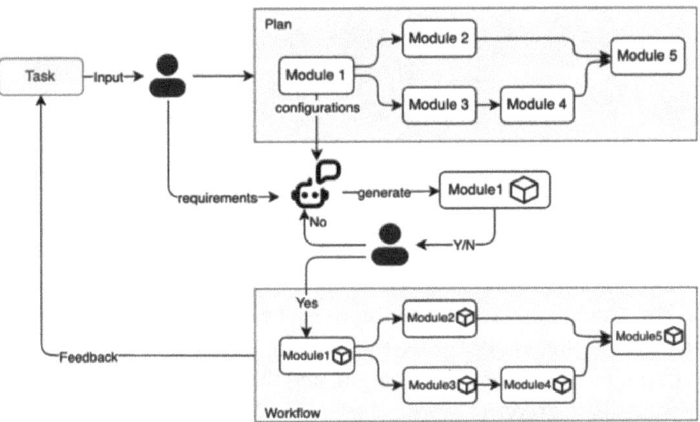

Fig. 4 Human-AI Collaborative Revise Process

3.3 Human-AI Collaborative Revise Process

The conventional CBD systems often struggle with revising an existing case. The "black-box" generation of deep learning tools makes it difficult for designers to understand the reasoning process. To address this issue, our proposed framework introduces a more flexible revision process by empowering users to actively shape the work process with the help of an AI module assistant.

The collaboration process between the user and the module assistant is illustrated in Fig. 4. The user acts as the task planner and decision-maker, responsible for designing the overall structure of the workflow. The module assistant, on the other hand, serves as a collaborator, allowing users to use natural language to request the assistant to write or modify modules or to suggest optimizations. This involves a multi-turn collaboration where the user decides whether to adopt the assistant's suggestions, edits, and refines the generated code, and ultimately completes the workflow construction. This approach effectively helps designers unfamiliar with programming to build complex workflows, while also enabling designers and researchers with programming expertise to increase efficiency and quickly test and validate problem-solving approaches.

4 CityFlow Platform

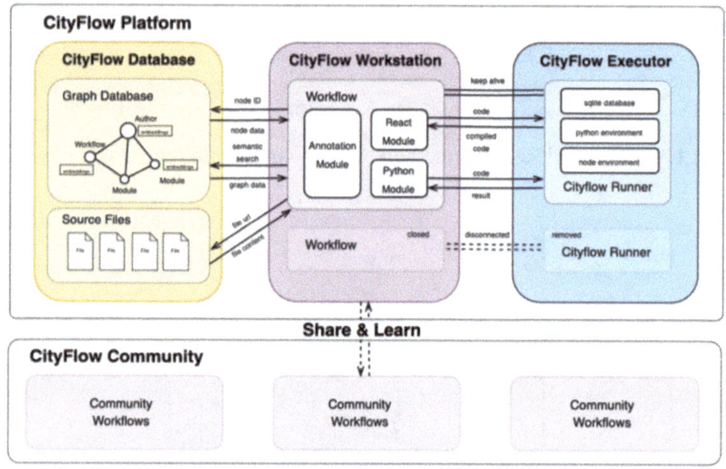

Fig. 5 The system architecture of CityFlow Platform

Based on the framework mentioned above, we built a platform named CityFlow. As shown in Fig. 5, this platform comprises three core components: the CityFlow Knowledge Graph, the CityFlow Workstation, and the CityFlow Executor. The CityFlow Knowledge Graph is a graph database that stores workflow information in the form of nodes and edges. The CityFlow Workstation serves as the primary user interface, providing an easy-to-use, modular, graphical programming interface. The CityFlow Executor is responsible for managing the execution of workflow computation logic.

Workflows designed in the CityFlow platform can be shared and published within the CityFlow community, fostering collaboration and communication among urban design researchers.

CityFlow offers a collaborative platform for users from diverse backgrounds, aiming to build smarter, more sustainable urban communities:

4.1 A Teaching Tool for Urban Educators

CityFlow has been applied in undergraduate education as an experimental course for fourth-year students. This course aims to familiarize students with methods for urban environment assessment through AI collaboration. As shown in Fig. 6(a), the teaching resources have been organized by the CityFlow KG digitally. A chatbot built upon the CityFlow KG and the Deepseek-R1 model can help students answer related learning questions and provide personalized learning pathways. Furthermore, with the assistance of the CityFlow AI Coder, students build agent-based models to simulate pedestrian behaviors and analyze the relationship between campus spatial layout and pedestrian activity. The CityFlow platform provides educators with an interactive teaching tool, enhancing the scientific rigor, practical application, and cutting-edge techniques of teaching, preparing students for the challenges of future architectural and urban development.

4.2 A Low-Code Environment for Urban Designers

Figure 6(b) showcases a workflow developed within the CityFlow platform, demonstrating how designers can efficiently assess accessibility in the urban environment. CityFlow utilizes a user-friendly, drag-and-drop interface that allows designers to visually connect different data processing steps. This approach leverages reusable modules, built using either Python or React, to handle tasks such as reading diverse urban datasets (e.g., openstreet maps, points of interest), performing spatial analysis to calculate urban metrics, and generating visualizations to communicate findings. By assembling these pre-built modules into a complete workflow, the designer can significantly streamline the assessment process, reducing repetitive tasks and accelerating the identification of design issues within the urban environment.

4.3 An AI Coder for Module Builder

For module builders, CityFlow incorporates an AI-powered coder that assists users in custom Python and JavaScript modules. As shown in Fig. 6(c), the AI Coder intelligently interprets user requests, translating them into functional code snippets aligned with CityFlow module structures. This empowers users to rapidly prototype and test new functionalities, such as custom data processing pipelines or interactive visualizations, without needing to be expert programmers. The system also provides suggestions and explanations, guiding users through the coding process and promoting learning. This significantly reduces development time and allows users with limited programming experience to contribute to CityFlow's capabilities.

4.4 An Open Platform for Urban Researchers

CityFlow integrates AI-powered search engines into community workflows, creating a web-based platform that fosters collaboration among urban scientists. It enables users to easily search for and share research findings, workflows, and case studies, contributing to a continuously evolving database of urban design solutions.

Figure 6(d) demonstrates the publishing page of a traffic simulation model in the CityFlow platform. Cityflow allows researchers to clearly showcase their methodologies and results, making it easier for others to reproduce findings and build upon existing work. It drives iterative innovation in urban science by facilitating the exchange of knowledge and best practices.

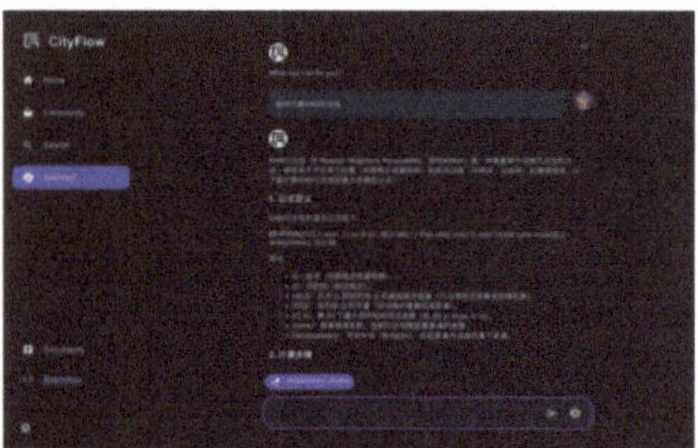

(a)The integration of the CilyFlow KG and Deepseek offers students improved comprehension of urban assessment methods.

(b)The visual programming interface in CityFlow help designerscreate workHow to evaluate amenity accessbility in Cambridge.

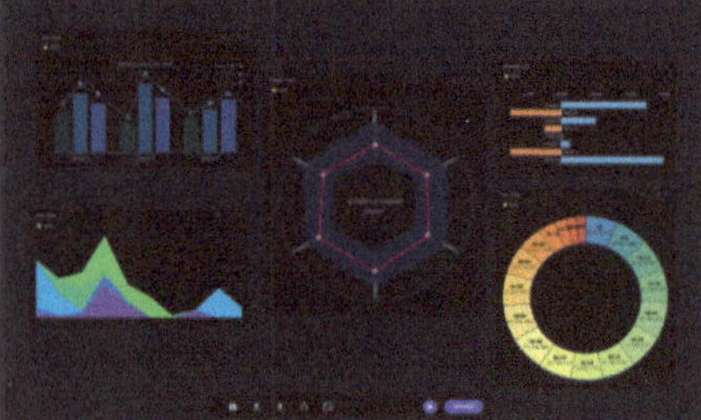

(c) The data visualization modules built by the Al Coder in CityFlowplatform.

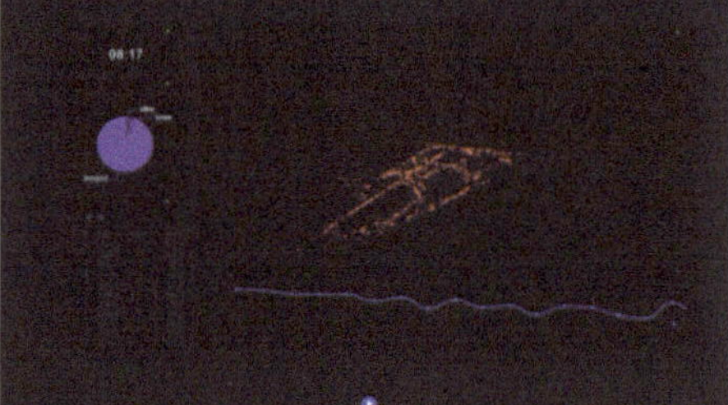

(d) A pedestrian simulation model of MIT campus published by researchers in the CityFlow platfomn.

Fig. 6 The user interface of CityFlow Platform

5 Discussion

Science fiction films like Her and Iron Man have long envisioned AI assistants capable of understanding and assisting humans. Now, rapid advancements in AI, particularly LLMs, are making these scenarios a reality, promising to revolutionize how we interact with computers. OpenAI, as one of the top AI companies, has outlined five stages from AI to AGI (L1-L5). As of the writing of this paper, the most advanced LLMs have essentially reached the L2 level, but have not yet fully evolved to L3. This paper further extends the discussion about how these different levels of AI (L1-L3) can be integrated with the CityFlow platform, creating a new human-computer interaction (HCI) paradigm. Here's a breakdown:

- **L1 Dialogue Assistant:** Acts as an intelligent search engine, answering questions and guiding users.
- **L2 Collaborative Assistant:** Functions as an experienced programmer, helping translate design ideas into code and workflows.
- **L3 Task Assistant:** Operates as an independent design assistant, autonomously completing tasks and generating solutions based on given objectives.

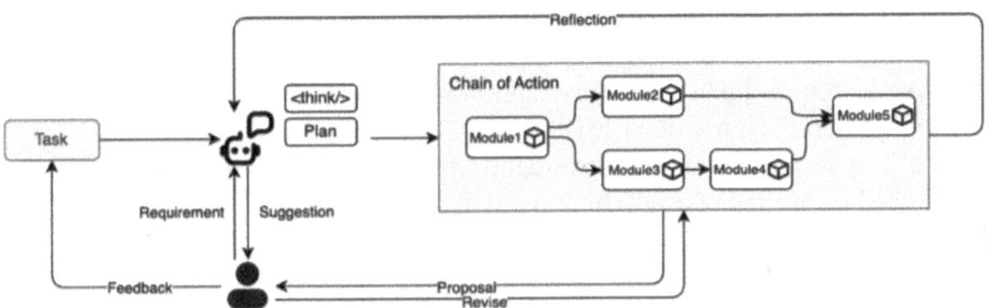

Fig. 7 In the L3 level, the user is the decision-maker, and the AI assistant is the actor.

As shown in Figs. 2 and 3, the CityFlow platform has essentially implemented L1 and L2 levels of HCI. While the AI assistant can significantly enhance human efficiency, a substantial amount of work still requires human involvement. As illustrated in Fig. 7, unlike previous "black-box" generation based on deep learning, L3-level HCI interaction intuitively demonstrates the AI assistant's "thinking" process and "acting" process. Simultaneously, users can take over the AI assistant's work at any time, adjust its working process, or propose new requirements, prompting the AI assistant to make modifications. This deep collaborative mode greatly enhances the operability and interpretability of generative AI, allowing designers or researchers to better focus on the problem itself, grasp the primary contradictions, and leverage the advantages of AI to improve work efficiency.

6 Conclusion

Overall, this paper proposed a novel framework to build an intelligent multi-modeling system for decision-making support in complex urban environments. The resulting CityFlow platform, built upon this framework, has been applied in education and research projects. By integrating LLMs and a user-friendly, low-code environment, CityFlow empowers users of varying technical expertise to rapidly prototype, analyze, and share solutions for complex urban challenges. The ongoing work towards the next level of AI integration promises even greater autonomy and efficiency, redefining the HCI paradigm in urban science and paving the way for more sustainable and collaborative urban design and research.

References

Batty, M.: Multiple models. Environment and Planning B: Urban Analytics and City Science. **48**(8), 2129–2132 (2021). https://doi.org/10.1177/23998083211051139

Cross, N.: A history of design methodology. In: de Vries, M.J., Cross, N., Grant, D.P. (eds.) Design Methodology and Relationships with Science, pp. 15–27. Springer, Netherlands (1993). https://doi.org/10.1007/978-94-015-8220-9_2

Dave, B., et al.: Case-based spatial design reasoning. In: Haton, J.P., Keane, M., Manago, M. (eds.) Advances in Case-Based Reasoning. EWCBR 1994. Lecture Notes in Computer Science, vol. 984, pp. 198–210. Springer, Berlin, Heidelberg (1995)

Domeshek, E., Kolodner, J.: Using the points of large cases. AI Edam-Artificial Intelligence for Engineering Design Analysis and Manufacturing. **7**(2), 87–96 (1993)

Huang, W., Zheng, H.: Architectural drawings recognition and generation through machine learning. In: Proceedings of the 38th Annual Conference of the Association for Computer Aided Design in Architecture, pp. 18–20. Association for Computer Aided Design in Architecture (ACADIA), Mexico City, Mexico (2018)

Langrish, J.Z.: The design methods movement from optimism to Darwinism. In: Lloyd, P., Bohemia, E. (eds.) Future Focused Thinking - DRS International Conference. Brighton, United Kingdom (2016)

Liu, Y. et al.: Exploration on machine learning layout generation of Chinese private garden in Southern Yangtze. In: In: Proceedings of the 2021 DigitalFUTURES: The 3rd International Conference on Computational Design and Robotic Fabrication (CDRF 2021) 3, pp. 35–44 (2022)

Liu, Y., Luo, Y., Deng, Q., Zhou, X.: Exploration of campus layout based on generative adversarial network: Discussing the significance of small amount sample learning for architecture. In: Proceedings of the 2020 DigitalFUTURES: The 2nd International Conference on Computational Design and Robotic Fabrication (CDRF 2020), pp. 169–178 (2021)

Noyman, A.: CityScope: An Urban Modeling and Simulation Platform. Massachusetts Institute of Technology (2022)

Quan, S.J.: Urban-GAN: An artificial intelligence-aided computation system for plural urban design. Environment and Planning B: Urban Analytics and City Science. **49**(9), 2500–2515 (2022)

Rittel, H.W., Webber, M.M.: Dilemmas in a general theory of planning. Policy. Sci. **4**(2), 155–169 (1973)

Voss, A.: Case design specialists in FABEL. In: Issues and Applications of Case-Based Reasoning to Design, pp. 301–335. Psychology Press (2014)

Wei, L.K.: Research on Architectural Space Retrieval and Generation Based on CBR and HTML5. Tianjin University (2013)

BY NC ND

Research on Noise Distribution in Residential Areas Based on Urban Section and Profile Graph: A Case Study of Nanjing

Suyi Shen[1], Ziyu Tong[1(✉)], and XiaoDong Lu[2]

[1] School of Architecture and Urban Planning, Nanjing University, 210093 Nanjing, China
shensuyi@smail.nju.edu.cn, tzy@nju.edu.cn

[2] School of Architecture and Fine Art, Dalian University of Technology, 116081 Dalian, China
lxd3721@dlut.edu.cn

Abstract. With the acceleration of urbanization and the increase in population density, noise pollution has become a major factor adversely affecting residents' quality of life. The correlation between noise and urban form has been widely confirmed. However, existing studies have mainly focused on noise distribution along a single dimension, failing to capture its continuous variation in three-dimensional (3D) space. This study proposes a method combining urban sectional analysis and noise simulation, using typical residential samples from Nanjing and applying *CadnaA* for noise simulation, generating sectional noise distribution graphs to analyze noise variation in 3D space. The findings show that building layout and height significantly influence noise propagation and attenuation, resulting in noticeable variations in noise intensity across different areas and building components. This method offers a new perspective for understanding the complex effects of urban morphology on noise propagation and provides theoretical support for optimizing urban noise management to improve residents' quality of life.

Keywords: Noise distribution · Residential form · Urban section · Profile graph

1 Introduction

With the acceleration of urbanization and the increase in population density, urban noise pollution has become an increasingly severe issue, negatively affecting residents' quality of life and their physical and mental health. Reports indicate that in 2023, around 5.7 million noise complaints were filed in China, reflecting an increase of 1.2 million compared to the previous year, with megacities like Shanghai and Beijing accounting for 30% of the total complaints. Among these, complaints about social life noise account for 68.4% [1]. Numerous studies have shown that urban noise propagation is closely related to urban forms—such as streets, roads, buildings and land uses [2]. Understanding how different urban forms influence noise propagation, has become a significant topic in urban morphology and acoustics research.

At the macro scale, noise distribution is mainly influenced by factors such as urban density, functional zoning, and street network density; at the micro scale, factors such

Y. Liu et al. (Eds.): CDRF 2025, *Transindividual Intelligence*, pp. 502–511, 2026.
https://doi.org/10.1007/978-981-92-0615-5_43

as building volume, orientation, number of stories, spatial arrangement, and the location and geometric features of open spaces directly affect noise propagation [3]. Current studies analyzing the influence of micro-scale urban morphology on noise propagation through noise maps generally fall into three main categories. The first category focuses on horizontal noise distribution in different urban forms. For example, Zhou et al. analyzed the traffic noise distribution and its influencing factors in four types of residential areas through planar noise maps [4]. The second category investigates vertical noise distribution based on building heights. For instance, Oliveira and Silva explored the influence of urban form on facade noise variations by presenting noise distribution on different floors through vertical noise maps [5]. The third category separately analyzes horizontal and vertical noise data to identify the characteristics of noise distribution in both directions under different urban forms. For example, Pan Ning analyzed horizontal and vertical noise maps to assess the effects of street network forms and building layouts at different block scales on traffic noise [6]. Despite revealing noise characteristics in different dimensions, these analyses fail to comprehensively capture the continuous spatial variation of noise.

To address this gap, this study proposes a noise analysis method based on urban section. By setting a series of parallel profiles within residential areas, continuous slices representing the spatial characteristics of the area are obtained. These slices integrate horizontal and vertical noise data to construct a series of sectional noise distribution maps, offering a clearer depiction of continuous variation of noise in space. This method aims to provide a new tool for understanding the complex relationship between noise propagation and urban morphology at the block scale, while supporting urban noise management.

2 Methodology

2.1 Integrating Urban Sectional Analysis and Noise Simulation: A New Methodology

This study integrates urban sectional analysis with noise simulation to develop a new noise assessment model. Urban section provides a multidimensional perspective by presenting the spatial arrangement of elements such as buildings, streets, and open spaces in both horizontal and vertical dimensions. It illustrates how noise is distributed and continuously varies across space. This method provides a more intuitive demonstration of noise propagation path, showing how noise levels change as sound travels from the source to different locations. It effectively captures the influence of different urban forms on noise propagation and attenuation, while also enhancing understanding of residents' noise perception under various scenarios.

The study uses *CadnaA* software for noise simulation, predicting noise intensity at various locations and heights. Sectional noise distribution maps are then generated to reflect noise perception at different spatial positions. The simulation is set up as follows:

1. Noise Sources Setup: Major noise sources (e.g., road traffic volume, vehicle speed) are determined based on actual traffic data.

2. Building and Street Parameters Setup: data of buildings are imported to simulate the propagation and attenuation of noise within the urban space.
3. Noise Receiver Location Setup: Noise receiver points are located in both horizontal and vertical dimensions to reflect how people perceive noise at different positions and heights. Horizontally, the points are placed at the average standing height of a person (about 1.5 m) to simulate outdoor exposure. Vertically, they are positioned at the center of windows (typically the midpoint of the floor) to reflect how indoor residents are affected by external noise.

2.2 Generation of Sectional Line Diagrams and Area Division

To present the spatial variation of noise distribution in residential areas, this study constructs continuous noise distribution maps using a series of parallel sectional slices. These slices are distributed at regular intervals along a specific direction, forming continuous profiles that reflect the spatial characteristics of the area.

As shown in Fig. 1, which illustrates the sectional line diagram and area division, each slice is subdivided into five parts to distinguish the different facades of buildings and external spaces:

(I) Open space near the noise source: This area is closest to the noise source and experiences the highest levels of direct noise exposure.
(II) Facade facing the noise source: The primary interface that initially receives and reflects the noise.
(III) Roof area: Since this study focuses on areas directly affecting residents, the roof, typically sparsely populated, is not equipped with receiver points.
(IV) Facade facing away from the noise source: The secondary interface where noise is attenuated after being blocked by the front facade.
(V) Open space far from the noise source: Located farther from the noise source, this area typically experiences lower noise levels than the others.

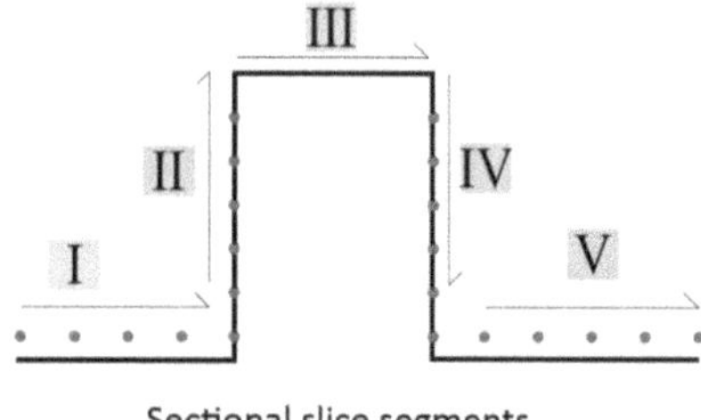

Fig. 1 Diagram of sectional line and area division

2.3 Expansion of Sectional Line Diagrams and Visualization of Sectional Noise Distribution

To better illustrate the continuous variation of noise across different spatial locations, this study horizontally expands the sectional line diagrams and visualizes the simulated noise data. This process generates sectional noise distribution maps.

Figure 2(a) shows the expanded sectional line diagram based on Fig. 1, where the noise distribution points of each region (I, II, III, IV, V) are sequentially arranged along the X axis, clearly illustrating their spatial relationships. Simulated data are integrated to simultaneously present dynamic noise changes along both horizontal and vertical dimensions within the expanded sectional diagram. Figure 2(b) illustrates the variation in noise intensity at different receiver points. To better illustrate the trend of noise intensity changes with spatial location, the distances between receiver points were standardized. The X-axis represents the positions of the receiver points, starting from the road noise source, with equal spacing between adjacent points. The Y-axis corresponds to the noise levels at these points. Vertical lines indicate key turning points of the building or site, while I, II, III, IV, and V denote five regions exhibiting distinct noise propagation and attenuation patterns.

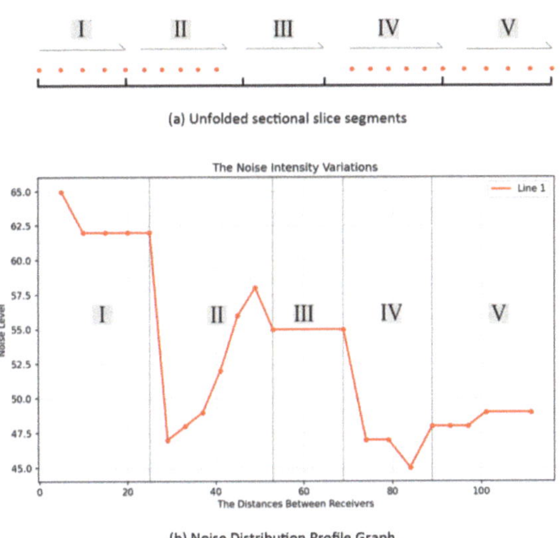

Fig. 2 Diagram of the sectional noise distribution visualization process

By visualizing sectional noise distribution maps, this study integrates separate two-dimensional data into a continuous line graph that presents dynamic noise changes along both horizontal and vertical dimensions. This provides a more intuitive display of the continuous noise variation across space.

2.4 Quantitative Analysis of Sectional Noise Distribution

This study performs a quantitative analysis of spatial noise variation using the noise distribution maps. The analysis relies on the noise fluctuation frequency (f), which reflects the periodic variation of noise at different locations. Frequency is calculated by measuring the distance between consecutive peaks (period, T) and taking its reciprocal. The period is defined as the distance from one peak to the next, and the frequency indicates the number of fluctuations per unit distance.

The period (T, m) is calculated using the following formula:

$$T = \frac{D}{N} \tag{1}$$

where D is the distance between peaks; N is the number of points between peaks

The frequency (f, Hz) is calculated as:

$$f = \frac{1}{D} \tag{2}$$

By calculating the fluctuation frequency of each noise distribution line, the spatial noise distribution can be quantitatively described, providing data to better understand the relationship between noise fluctuations and spatial arrangement.

3 Case Study

3.1 Study Examples

This study selects typical urban patterns derived from a large number of residential samples in Nanjing, a medium-sized Chinese city with typical urbanization characteristics. The selected patterns reflect common features of Nanjing's urban structure, including various building forms, block layouts, and spatial organization. These features are essential for understanding how different urban forms affect noise propagation and distribution.

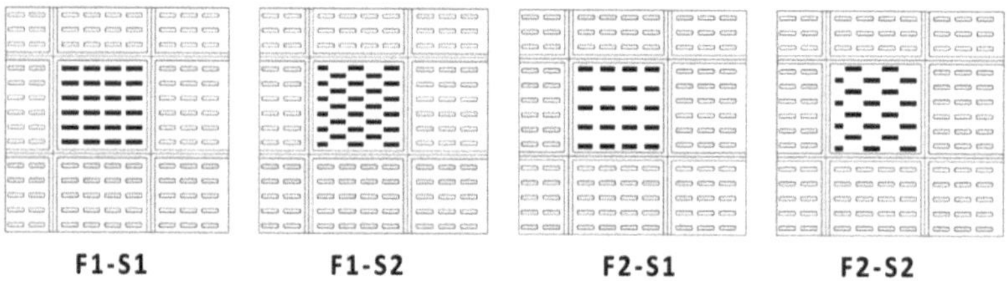

Fig. 3 Residential patterns

Noise in low-rise residential areas is mainly influenced by traffic noise at ground level and the surrounding environment, with noise propagation typically being more uniform and less variable. In contrast, mid- and high-rise buildings have more complex noise paths due to factors like reflection, diffraction, and multi-path propagation. Therefore, this study focuses on mid- and high-rise buildings, selecting the F1 series (8–11 stories) and the F2 series (11–18 stories) as representative samples (Fig. 3). In terms of building layout, the S1 series samples use a parallel array layout, where buildings are evenly distributed along the street. The S2 series samples use a staggered array layout, where buildings are offset relative to each other, creating a more complex spatial arrangement.

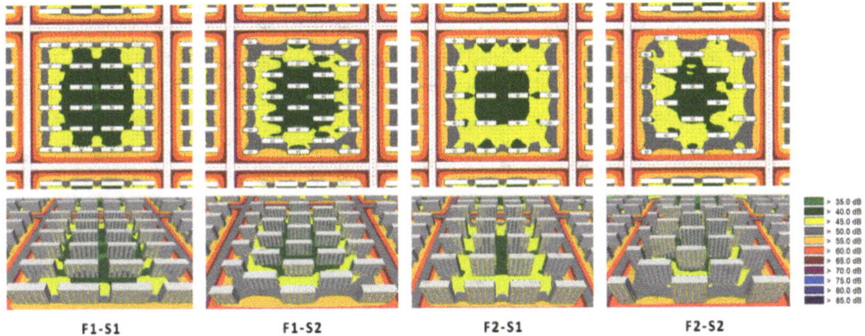

Fig. 4 Noise maps

3.2 Simulation Results: Traditional 2D Noise maps

As shown in Fig. 4, the simulation results are represented by color variations, which indicate the noise intensity across different regions. The S1 series samples, with a parallel array layout, exhibit more uniform noise propagation. In contrast, the S2 series exhibit a more uneven noise distribution, potentially due to reflection and diffraction effects caused by the staggered arrangement of buildings.

Traditional noise maps display noise data only in a single dimension and cannot simultaneously represent noise data in both horizontal and vertical directions within a single map. Consequently, it is difficult to visually capture the continuous variation of noise in space.

3.3 Sectional Noise Distribution Graphs

This study employs a noise distribution analysis method based on urban section, which integrates horizontal and vertical data to generate sectional noise distribution graphs for typical residential areas in Nanjing. This method analyzes noise variation across different spatial locations and heights. The results indicate that noise intensity is influenced by the complex interactions within urban form, exhibiting significant variations across different regions and building components.

3.3.1 Periodic Fluctuation Analysis of Noise Distribution

Periodic fluctuation analysis is conducted on the sectional noise distribution lines to observe the variation cycles of noise data and calculate the fluctuation frequency (f). These frequencies represent the rate of noise fluctuations along each line, with higher frequencies indicating more frequent and intense fluctuations, suggesting more complex changes that could be influenced by external factors. Lower frequencies indicate more sparse and stable fluctuations with smaller changes, leading to relatively lower noise pollution.

The noise fluctuation characteristics differ for parallel array (F1-S1, F2-S1) and staggered array (F1-S2, F2-S2) buildings.

In the F1-S1 area (Fig. 5), the noise fluctuation frequencies range from 0.50 Hz to 0.60 Hz (Table 1). Line 1 and Line 4 have the lowest frequency of 0.5 Hz, showing slower noise fluctuations with longer periods and gradual intensity changes. In contrast, Line 2 has a frequency of 0.57 Hz, indicating more frequent fluctuations. Line 3 has the highest frequency of 0.60 Hz, showing the fastest noise variation with denser fluctuations patterns.

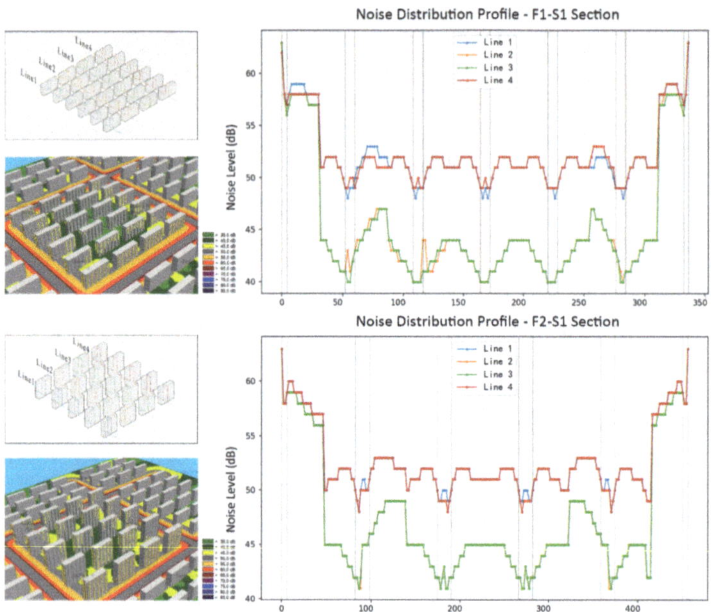

Fig. 5 Noise distribution profile graph - parallel array

Table 1 Frequency of noise fluctuations (Hz) - parallel array

	Line1	Line2	Line3	Line4
F1-S1	0.50	0.57	0.60	0.50
F1-S2	0.43	0.44	0.43	0.38

In the F2-S1 area (Fig. 5), frequencies range from 0.32 Hz to 0.63 Hz (Table 1). Line 1 has the lowest frequency of 0.32 Hz, showing slow noise changes. Lines 2, 4, and 5 have frequencies of 0.49 Hz, 0.47 Hz, and 0.46 Hz, with more frequent fluctuations. Line 3 has the highest frequency of 0.63 Hz, indicating the most frequent and intense fluctuations.

In the F1-S2 area (Fig. 6), frequencies range from 0.38 Hz to 0.44 Hz (Table. 2). Line 4 has the lowest frequency of 0.38 Hz, showing slower fluctuations. Lines 1 and 3 have frequencies of 0.43 Hz, with moderate fluctuation speed. Line 2 has a slightly higher frequency of 0.44 Hz, with slightly more frequent fluctuations.

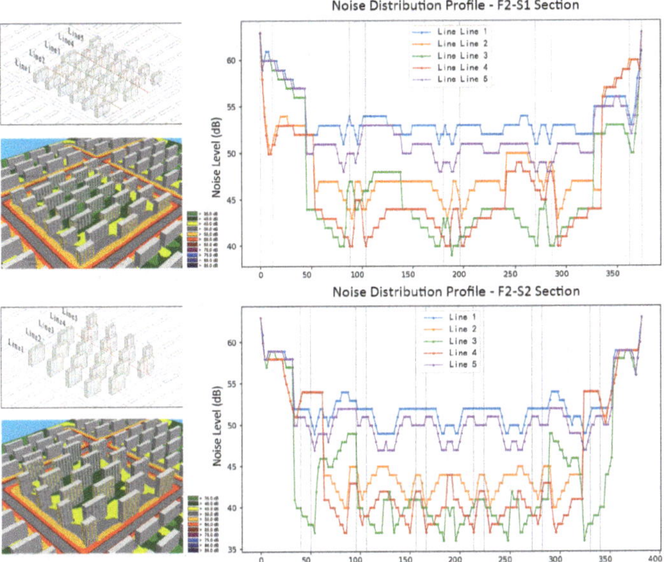

Fig. 6 Noise distribution profile graph - staggered array

Table 2 Frequency of noise fluctuations (Hz) - staggered array

	Line1	Line2	Line3	Line4	Line5
F2-S1	0.32	0.49	0.63	0.47	0.46
F2-S2	0.34	0.42	0.47	0.46	0.35

In the F2-S2 area (Fig. 6), frequencies range from 0.34 Hz to 0.47 Hz (Table. 2). Lines 1 and 5 have the lowest frequencies of 0.34 Hz and 0.35 Hz, showing slower fluctuations. Line 2 has a slightly higher frequency of 0.42 Hz, with slightly more frequent fluctuations. Lines 3 and 4 have higher frequencies of 0.46 Hz and 0.47 Hz, showing more frequent fluctuations.

3.3.2 Impact of Building Layout and Floor Height on Noise Fluctuations

Noise fluctuation frequencies vary across regions, revealing the complex impact of building layout, floor height, and other factors on noise propagation. Specifically, in parallel array buildings, the F1-S1 shows higher noise fluctuation frequencies, while the F2-S1 has a broader frequency range with greater variation, indicating more complex noise fluctuations. In contrast, in staggered array buildings, noise fluctuation frequencies are generally lower and more stable, especially in the F1-S2, where the frequency variation is smaller, suggesting more uniform noise fluctuations.

A comparison of the F1 and F2 series reveals that difference in floor height significantly affects noise propagation. In the F1 series (8–11 stories), noise fluctuations are relatively stable with lower frequencies, and the variation remains uniform, regardless of the layout type (parallel or staggered). In contrast, in the F2 series (11–18 stories), the increased height leads to more complex noise propagation paths, with enhanced reflection, diffraction, and refraction effects, resulting in more frequent and intense fluctuations.

A comparison of the S1 and S2 series shows that the S2 series exhibits lower and more restricted noise fluctuation frequencies, suggesting that the staggered layout (S2) does not necessarily increase noise fluctuations but may contribute to their stability. Although the staggered layout is more complex and typically enhances reflection and diffraction effects, it may reduce fluctuation frequencies by increasing the propagation path length and allowing for more attenuation space. In contrast, the parallel layout (S1) is more regular, with more direct noise propagation paths, leading to higher fluctuation frequencies.

4 Discussion and Conclusion

This study proposes a novel noise distribution analysis method based on urban profiles, providing insights into the complex relationship between block-scale urban form and noise propagation. It also offers theoretical support for the optimization of noise management.

Compared to traditional noise maps, which fail to capture noise changes in both horizontal and vertical dimensions, this study constructs continuous noise distribution maps, depicting dynamic noise changes along both axes and offering a clearer view of spatial noise variation. It also allows for a clearer differentiation of noise variation between different building components (e.g., facades, open spaces).

However, this research has some limitations. First, the sample types are limited to typical urban patterns and do not encompass all potential residential layouts. Future studies could extend this method to a broader range of urban forms and noise source scenarios. Second, the analysis primarily relies on simulation data, instead of incorporating subjective perception surveys. Future research might combine surveys or field measurements to better assess the impact of noise on residents' lives, thereby further validating and improving the method's applicability.

References

1. Ministry of Ecology and Environment of the People's Republic of China: China environment noise prevention and control annual report. https://www.mee.gov.cn/hjzl/sthjzk/hjzywr (2024). Aaccessed 30 Jan 2025
2. Yildirim, Y., Arefi, M.: Seeking the nexus between building acoustics and urban form: A systematic review. Curr. Pollut. Rep. **9**, 198–212 (2023) https://doi.org/10.1007/s40726-023-00250-1
3. Berghauser Pont, M., Forssén, J., Haeger-Eugensson, M., Gustafson, A., Achberger, C., Rosholm, N.: Using urban form to increase the capacity of cities to manage noise and air quality. Urban Morphol. **27**(1), 51–70 (2023) https://doi.org/10.51347/UM27.0003

4. Zhou, Z., Kang, J., Zou, Z., Wang, H.: Analysis of traffic noise distribution and influence factors in Chinese urban residential blocks. Environ. Plan. B Urban Anal. City Sci. **44**(3), 570–587 (2017) https://doi.org/10.1177/0265813516647733
5. Oliveira, M.F., Silva, L.T., Pedro, J.B.: How urban noise can be influenced by the urban form. In: Advances in Biology, Bioengineering and Environment, pp. 31–36, 2010
6. Pan, N: The influence of urban morphology on the spatial distribution of traffic noise at block scale. Master's thesis, Dalian University of Technology, Dalian, China (2017)

Research on the Relationship between Urban Spatial Form and Street Attractiveness and Design Suggestions Based on Multi-source Big Data and Deep Learning Methods —— A Case Study of Representative Blocks in Hong Kong

Wei Wenjiao(✉) and Zhang Yu

School of Architecture, Southeast University, Nanjing, China
1324740563@qq.com

Abstract. Taking Hong Kong's representative blocks as the research object, this study explores the relationship between various types of urban spatial form indicators and street attractiveness by establishing a quantitative analysis model based on multi-source big data. The study is based on web data crawling, using LDM model and sliding window method to determine the scope of the experiment, constructing a map of Hong Kong's gastronomy, and calculating the attractiveness of each block as the dependent variable of the model. Using SDNA, spatial statistics of POI data, semantic segmentation and other methods, a macro and meso-micro scale urban form evaluation index system is constructed, and each index is used as the independent variable. The results show that (1) accessibility, population density and street attractiveness have a non-linear relationship, and functional mixing degree is positively correlated with attractiveness (2). Among the streetscape indicators, green view index and closure have an inverted U-shaped relationship with block attractiveness, while walkability and imageability are basically positively correlated with attractiveness (3). With regard to the streets where shops with different consumption characteristics gather, we propose shop location and street design suggestions to enhance the attractiveness of the blocks from the perspective of their macroscopic characteristics and interface perceptions.

Keywords: Streetscapes · Business vitality · Multi-source big data · Computer vision techniques · SDNA

1 Introduction

In recent years, with the stock transformation of the national urban space development paradigm and the upgrading and transformation of the social consumption structure, the attractiveness evaluation of the commercial street pedestrian environment, as the core carrier of the urban public space, has become an important basis for measuring the

Y. Liu et al. (Eds.): CDRF 2025, *Transindividual Intelligence*, pp. 512–522, 2026.
https://doi.org/10.1007/978-981-92-0615-5_44

quality of the city. The current technological revolution in society is deeply integrated with public health demands, and the physical commercial space enhances consumer stickiness through environmental aesthetics and scene narratives, and its attractiveness evaluation indexes are reconstructed due to the health and safety paradigm in the post epidemic era [1], and in summary the study of the attractiveness of the commercial walking environment has become a general trend.

With the development of the times and technological innovation, domestic and foreign scholars are increasingly rich in research in this area. For the exploration of a quantitative analysis system for walkable blocks, Shields et al. [2] propose to divide the indicators into two categories: urban community-level and street-level; for the exploration of streetscape environmental perception, Ki et al. [3] proposed that street greening, pavements, enclosure, diameter/height ratio, obstacles and visual complexity are pedestrian indices to be calculated; for an exploration of computer vision techniques, Zhen Wei et al. [4] used the Segnet model to study the age-friendliness of street environments, Raveena Marasinghe et al. [5] have systematically summarised the application of CV techniques in urban planning; for an exploration of the relationship between street environment and capitalisation efficiency, Hahm, Yoon, & Choi [6] suggest that high quality streetscape promotes consumption. Based on previous studies, we innovatively take "urban spatial pattern indicators and street attractiveness" as the entry point, analyze the data from multiple sources, develop a quantitative evaluation index system, explore the relationship between variables, and provide scientific support for how to improve the attractiveness of streets.

2 Methods and Data

2.1 Research Framework

This paper constructs a research framework on the relationship between urban spatial form indicators and street attractiveness in Hong Kong (Fig. 1), and uses data crawling, LDM modelling, SDNA, spatial statistics of POI data, and semantic segmentation to build a quantitative analysis model based on multi-source big data to explore the relationship between variables, and to further propose corresponding recommendations for shop location and street design. Catering accounts for a large proportion of local businesses, and to reduce multi-factor interference, this study focuses on restaurants, cafes and bars, using their vibrancy to represent the commercial vitality of the block.

2.2 Multi-source Data

2.2.1 POI Data

Use Geopandas library and Osmnx library to get the POI data of Openstreetmap website, define its categories as 'restaurant', 'cafe', 'fast_food', 'food_court', 'bar' and 'pub' six categories, and each POI data includes attribute information such as gastronomy name, latitude and longitude and gastronomy type.

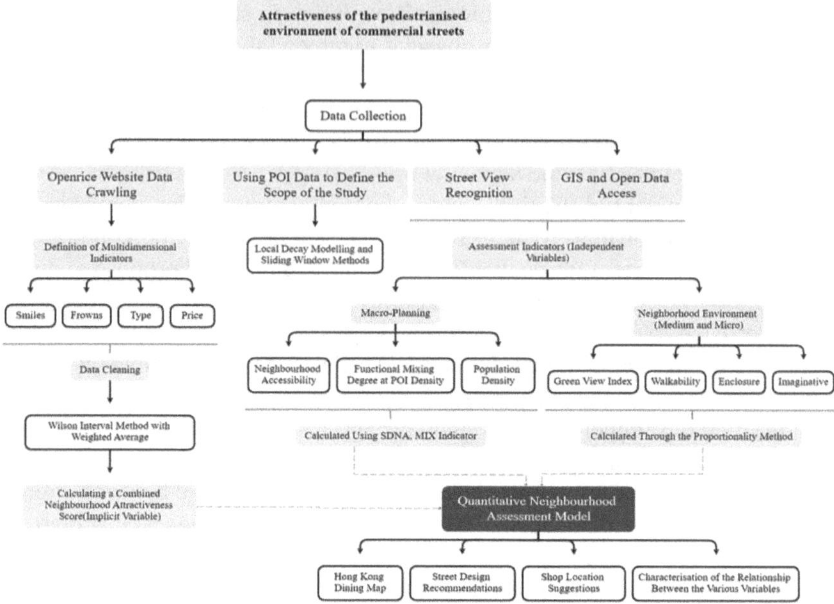

Fig. 1 Research flowchart

2.2.2 Openrice Website Data and Population Density Data

Openrice (https://www.openrice.com/zh/hongkong) is the most popular catering website in Hong Kong with more comprehensive information on catering data.Census and Statistics Department (https://www.censtatd.gov.hk/), this website provides more complete demographic statistics, including the number of people and the area of each sub-district in Hong Kong.

2.2.3 SVI Data Sources

We chose the Chinese map service provider Baidu map platform (https://lbsyun.Baidu.com/) to access the SVI resources through its API interface. The OSM road network was processed in ArcGIS by cropping projection and densification, and a sampling point file (.csv) was generated with a spacing of 40 m. This file was used with python code to obtain sample point panoramic mapping from SVI resources, which was processed into a usable streetscape. Later, manual photo screening was performed to clean up invalid distorted images to create a streetscape dataset.

2.3 Methodologies

2.3.1 Local Decay Modelling (LDM) with Sliding Window Approach

Firstly, data preparation. Crawl the data of Hong Kong catering POI with Python, total 5,283 data, all the line data and surface data in it were unified to be processed as multi-point data, and the total 7,953 data were aggregated. In order to ensure the accuracy of the research results, data need to be cleaned and pre-processed to exclude the data with missing or irregular format. After processing, 7,036 data were finally obtained and exported as .CSV files.

A local decay model (LDM) was used to analyse the density of restaurant clusters in Hong Kong in order to identify the specific study area. The formula is as follows, for each point i, its density contribution can be expressed as:

$$D_i(x, y) = \frac{1}{2\pi h^2}\exp\left(-\frac{d_i^2}{2h^2}\right) \tag{1.1}$$

where $D_i(x,y)$ is the density contribution of point i at position (x,y);d_i is the distance from point i to position (x,y); h is the bandwidth, used to control the rate of attenuation.

The density contributions of all points are summed to get the total density at position (x,y) with the following equation:

$$D(x, y) = \sum_{i=1}^{n} D_i(x, y) \tag{1.2}$$

The latitude and longitude of Hong Kong are used as the overall study area, and the latitude and longitude corresponding to the catering POI points are read. Define the Gaussian kernel function, determine the local decay model (LDM), set the bandwidth parameter to 0.05, and plot a 3D surface map of the decay situation, with the X-axis and Y-axis denoting the latitude and longitude, respectively, and the Z-axis denoting the density of Hong Kong's catering agglomeration (Fig. 2). A sliding window is implemented using scipy.ndimage.maximum_filter to find the local maximum of the catering agglomeration. The size of the sliding window is set to 5. The 95% quantile of the density value is calculated as the threshold for filtering the high density peak points, the density peak points are extracted and visualized (Fig. 2), and suitable blocks within one kilometre around each peak point are selected to delineate the catering agglomeration area.

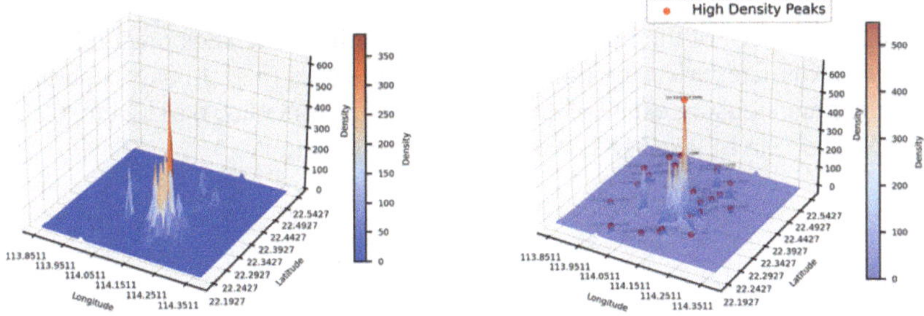

Fig. 2 Catering cluster density and peak density points in Hong Kong

2.3.2 Commercial Gastronomy Consumption Characteristics Identification and Block Attractiveness Scoring

Openrice is the most popular restaurant information platform in Hong Kong. We use Scrapy crawler framework to capture the relevant data (.csv file) of each restaurant, including restaurant name, address, price, type, good and bad reviews, etc. After cleaning and validating the data, we draw a map of Hong Kong's restaurants (Fig. 4) and extract the consumption characteristics of the blocks.

The Wilson interval method was used to calculate the number of positive reviews (smiles) and the number of negative reviews (frowns) of the catering shops to get the comprehensive score of each shop. The z in the formula is the quantile of normal distribution, z = 1.96 at 95% confidence level. Formula:

$$aggregate_{score} = \frac{p + \frac{z^2}{2n} - z\sqrt{\frac{p(1-p)}{n} + \frac{z^2}{4n^2}}}{1 + \frac{z^2}{n}} \tag{2.1}$$

Included among these p = smiles/(smiles+frowns), n = smiles + frowns, z is the z-value corresponding to the confidence level (e.g. z = 1.96 at the 95% confidence level). A weighted average of the composite scores was used to calculate the total customer satisfaction (block attractiveness) of block catering outlets.

2.3.3 Macrometric Assessment of the Built Environment

Commonly used indicators were selected to analyse block urban form, including block accessibility, functional mix, and population density.

Calculate accessibility values for each block: Obtain the block road network information from OpenStreetMap website and clean up the road network data by spatial selection, attribute filtering, topological rules and data editing. In SDNA, set the radius index to 1000, 2000, ..., N, calculate the important parameter MHD (Mean Hybrid Distance)-used to express centrality, equivalent to Mean Depth in spatial syntax-for each road segment. The overall accessibility of each block is obtained by averaging the above data.

Calculate the value of the mixed-use index for each block: The crawled POI data for the study area was categorised by function, e.g. commercial (e.g. shops, restaurants); residential (e.g. flats, housing estates); entertainment (e.g. cinemas, parks), etc. Calculated using the Mixed Use Index (MXI) with the following formula:

$$MXI = -\sum_{i=1}^{n}\left(\frac{p_i}{P} \cdot \ln\left(\frac{p_i}{P}\right)\right) \tag{3.1}$$

where p_i is the number of POIs in class i;P is the total number of POIs in the block;n is the number of categories of POIs.

According to the formula, process the data in ArcGIS so that each POI data corresponds to the block parcel, count how many POI data points (i.e., P-value) there are in a parcel, export the chart to EXCEL, count the number of various types of POIs in each parcel (Pi-value), and thus calculate the ratio of various types of POIs within each parcel, Pi/P, and substitute it into the above formula to get the various parcel MXI value, and take the average value to get the overall MXI value of the block.

Population density analysis: Check the demographic statistics of the blocks at the Census and Statistics Department of the Hong Kong Government (https://www.censtatd.gov.hk/) for subsequent analytical studies.

2.3.4 Semantic Segmentation of Street View Images

The street scene dataset is processed for semantic segmentation using the fcn_resnet101 model. FCN_ResNet101 uses ResNet-101 as a backbone network to extract image features and recover spatial resolution through FCN upsampling to achieve pixel-level prediction. Different categories of elements - containing roads, plants, sky, buildings, cars, etc. - are marked with different colours (Fig. 3), and the segmented image is output to quantify the relative share of each streetscape constituent of the image, which is visualised in ArcGIS (Fig. 4).

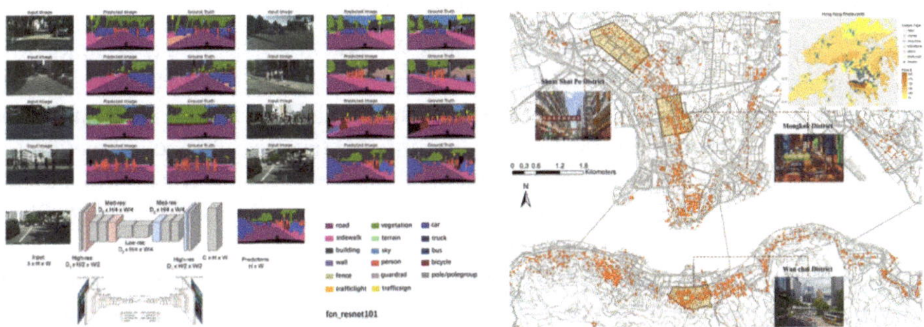

Fig. 3 SVI semantic segmentation

Fig. 4 Hong Kong dining map

Street perception assessment evaluates the impact of street environment on people's perception through quantitative indicators, which makes urban streetscape research towards data and science. We selected several commonly used indicators, including green view index, closure, walkability, and imageability, to construct a streetscape evaluation system at the meso-micro level by studying the results of previous researchers. The calculation method is as follows (Table 1).

Table 1 Description of indicators.

Evaluation indicators	Qualitative definition	Calculation formula
Green View Index	Percentage of area of greenery (trees, shrubs, lawns, etc.) visible to pedestrians in street space	$C_1 = P_{plants} + P_{terrain}$
Enclosure	The extent to which street space is enclosed by physical elements (e.g., buildings, plants, fences, etc.)	$C_2 = \frac{P_{building}+P_{plants}+P_{wall}}{P_{road}+P_{sidewalk}+P_{terrain}}$

(continued)

Table 1 (*continued*)

Evaluation indicators	Qualitative definition	Calculation formula
Walkability	The combined ability of the street environment to provide safety, convenience, comfort and enjoyment for pedestrians	$C_3 = \frac{P_{sidewalk}+P_{fence}}{P_{road}}$
Imageability	The overall perceptual qualities of a streetscape conveyed through visual, spatial, and cultural symbols.	$C_4 = P_{building} + P_{sign}$

Notes: C_i (i = 1, 2, 3, 4) represents the objective equations of four perceptions; P represents the percentage of an element in the overall image.

3 Case Studies

3.1 Calculation of Urban Macro-scale Indicators

The Local Decay Model (LDM) and the Sliding Window Method were used to identify the study blocks as Mong Kok, Sham Shui Po and Wan Chai (Fig. 4), and the specific process is described in Sect. 2.3.1. After data crawling and score calculation, the specific process is described in Sect. 2.3.2, the three categories of block consumption characteristics, namely price, business model and business format, were extracted and attractiveness scores of the blocks were calculated, with Mong Kok block at 0.826, Sham Shui Po at 0.715 and Wan Chai at 0.803.

At the macro level, the above three blocks were analysed in terms of accessibility, functional mix and population density, and the specific analysis process is shown in 2.3.3, the results of which are as follows: the Mong Kok block has an accessibility composite score of 1,110.5051, an MXI value of 1.8803, and a population density of 44,150/km^2, the Sham Shui Po block has an accessibility composite score of 970.9413, an MXI value of 1.6937 and a population density of 47,190/km^2, and the Wan Chai block had an accessibility composite score of 822.5997, an MXI value of 1.5803 and a population density of 15,630/km^2.

3.2 Calculation of Urban Meso-micro-scale Indicators

Through semantic segmentation of the streetscape image, the process of which is described in Sect. 2.3.4, the green view index, closure, walkability and imageability of the three blocks are calculated and visually represented (Fig. 5).

4 Discussion and Conclusions

4.1 Relationship between Urban Macro-morphology and Block Commercial Attractiveness

Comparing the data of the three blocks, it can be found that Mong Kok has the highest attractiveness score, followed by Wan Chai and Sham Shui Po; for accessibility, Mong Kok has the highest score, followed by Sham Shui Po and Wan Chai; and for population density, Sham Shui Po has the highest score, followed by Mong Kok and Wan Chai (see Table 2); suggesting that there is a non-linear relationship between accessibility, population density and attractiveness of the street, and that ease of access to a certain place and the concentration of people in that place don't necessarily determine the perception of the streetscape and commercial vibrancy of the place. In terms of MXI, Mong Kok has the highest value, followed by Wan Chai and Sham Shui Po, showing that MXI value is positively related to attractiveness.

Table 2 Comparison of accessibility and functional mix in three blocks

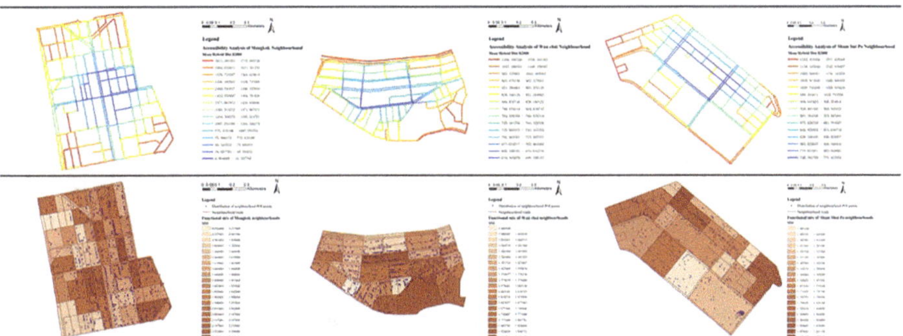

Combined with the consumption characteristics, we can put forward a general catering site selection proposal. High-attractiveness blocks have a variety of consumption levels and a wide range of shop prices, with the majority of boutique chain shops (e.g. Caprice) and branded independent shops providing multi-functional services and high-quality environment; medium-high-attractiveness blocks have a high concentration of fast-consumption chain shops (e.g. McDonald's) and community restaurants, which are convenient, cost-effective and multi-functional for leisure, office and party; medium-attractiveness blocks have a high concentration of snack vendors and individually owned independent shops due to the general low consumption level and single surrounding business in the area.

4.2 Relationship between Street Interface Characteristics and Block Attractiveness

4.2.1 Analysis of Streetscape Indicators

The distribution of the streetscape indicators is shown in Fig. 5 and analysed below.

In terms of closure, the Mong Kok block is the highest, such as Tung Choi Street and Nathan Road, which are narrow in scale, with numerous signboards and a high degree of façade continuity; the Sham Shui Po block is the second highest, such as Kau Kong Street, which has a high density of buildings, but is mostly made up of old houses, with small derelict open spaces in certain areas; and the Wan Chai block is the lowest, with the main street of the block significantly wider than that in the other two blocks, and with relatively high openness in redeveloped areas such as Lei Tung Street. Combined with the attractiveness score, the Mong Kok block scores the highest and has a high degree of closure. It is evident that a certain degree of closure can promote commercial vibrancy, and the design can be designed to prolong the dwell time of pedestrians through measures such as improving the continuity of the interface of the shops. The Sham Shui Po block also has a high level of enclosure, but its block attractiveness is the lowest, suggesting that enclosure is not the main factor determining its attractiveness or that excessive enclosure in this block has a negative impact. In summary, there is an inverted U-shaped relationship between enclosure and block attractiveness, with too high a level leading to a sense of oppression and too low a level making the space less enclosed, with appropriate values providing a positive impact on block attractiveness.

In terms of Green View Index, Wan Chai Block has the highest because of the presence of large green spaces such as Wan Chai Park and the density of street trees on both sides of streets such as Queen's Road East; Mong Kok block has a lower GVI because of its predominantly commercial and residential functions, high density of buildings, and congested roads; and Sham Shui Po Block has the lowest GVI because of its predominantly old blocks, poor street environment, and its predominantly small-scale parks and sparse street trees. Combining the attractiveness scores of the three blocks, the Mong Kok block has the highest score but a lower GVI, indicating that GVI is not the main factor determining the attractiveness of the block or that too high a GVI has a negative impact. In summary, street attractiveness has a roughly inverted U-shape relationship with green visibility, too much greenery will obscure commercial signboards and increase maintenance costs, too little greenery will affect the streetscape perception, and appropriate values provide positive impacts on block attractiveness.

In terms of walkability, the Mong Kok block is the highest, forming a high-density pedestrian network such as the Tung Choi Street - Fa Yuen Street - Sai Yeung Choi Street South triangle, with a high commercial density and where shopping on foot has become a major way of life. The Wan Chai block is the next highest, with good pedestrian facilities and good accessibility. The Sham Shui Po block is the lowest, with ageing buildings and utilities in the block, chaotic traffic, and the presence of light goods vehicles occupying the road for loading and unloading goods, resulting in pedestrians being forced to detour around the motorway. Combined with the attractiveness score analysis, walkability has a roughly positive relationship with block attractiveness.

In terms of imageability, the Mong Kok block is the highest, with its typical street characteristics - cyberpunk, such as Nathan Road, Sai Yeung Choi Street South and other sections of the road gathering more than 2,000 neon signs; the Wan Chai area is the second highest, with its dichotomy of old and new buildings, with old buildings such as Wo Cheong Pawnshop Riding Mansions becoming landmarks and historical and cultural symbols, and new buildings such as the Hong Kong Hopewell Centre, etc., representing the development of the times and an international image. Sham Shui Po is the lowest, with many buildings having aged facades, a dilapidated image, a shortage of peripheral services and unclear signage. Combined with the analysis of the attractiveness scores, there is a roughly positive relationship between the imageability of each block and the attractiveness of the block.

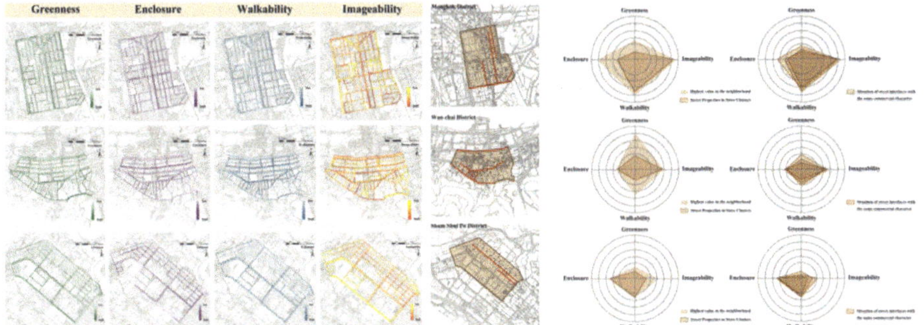

Fig. 5 Distribution of perceptual qualities

Fig. 6 Radar map analysis of streetscape metrics

4.2.2 Shop Location and Street Design Recommendations

Streets where popular shops are clustered are identified by selecting shops with more than 500 reviews in Openrice in three blocks. Radar charts were used to analyse the streetscape indicators in the clustered areas (Fig. 6). Shops preferred by consumers in Mong Kok blocks are clustered around Dundas Street, Kwong Wah Street, Fa Yuen Street North and Tung Choi Street area, which are mostly block businesses with medium average price, single format and dominated by independent shops. Compared with other streets, they are characterised by high walkability and imageability, and average green view index and closure. Shops preferred by consumers in the Wan Chai block are mostly clustered in Johnston Road, Queen's Road East and Hennessy Road, and are mostly block businesses with higher average prices, a wide range of formats, and a mix of stand-alone and chain shops, which are characterised by low green view index, high imageability and closure, and average walkability when compared with other streets. Sham Shui Po blocks with shops preferred by consumers are mostly clustered in Yuanzhou Street and Fuk Wing Street, and are mostly block businesses with low average prices, a single format, and dominated by independent shops, with low green view index, high closure and walkability, and an average imageability when compared to other streets.

In summary, design recommendations can be made for streets with different consumption types, while clarifying the suitable block siting environment for different types of shops to further promote the attractiveness of the block. For the average price of medium, single-format stand-alone shops, the street design needs to emphasise walking comfort and clearer signage; for the average price of higher average price, a rich mix of stand-alone shops and chain shops, the emphasis needs to be on the openness of the walking space, clear and complete signage and street facilities; for the average price of low average price, a single-format stand-alone shops, need to focus on the reasonable degree of spatial openness and walking comfort.

4.3 Summary and Outlook

This study takes typical commercial districts in Hong Kong as empirical cases, constructs a quantitative analysis model based on multi-source big data, and systematically investigates the correlation mechanisms between urban spatial form indicators and commercial attractiveness. Data-driven store location optimization strategies and streetscape design recommendations are then proposed. Compared with traditional survey methods, this approach demonstrates significant advantages in data collection efficiency and analytical accuracy. However, the study still has the following limitations: First, data availability constraints restrict the regression-based statistical modeling of causal relationships between independent and dependent variables; second, the current methodology has not yet been validated for generalizability in other urban contexts, which will be an important direction for future research.

References

1. Chen, Y., Zhu, M., Zhou, Q., Qiao, Y.: Research on spatiotemporal differentiation and influence mechanism of urban resilience in China based on MGWR model. Int. J. Environ. Res. Public Health. **18**(3), 1056 (2021)
2. Shields, R., Gomes da Silva, E.J., Lima e Lima, T., Osorio, N.: Walkability: A review of trends. J. Urbanism. **16**(1), 19–41 (2023). https://doi.org/10.1080/17549175.2021.1936601
3. Ki, D., Lieu, S., Chen, Z., Lee, S.: A novel walkability index using Google Street View and deep learning. SSRN. (2022), 4097441
4. Wei, Z., Cao, K., Kwan, M.-P., Jiang, Y., Feng, Q.: Measuring the age-friendliness of streets' walking environment using multi-source big data: A case study in Shanghai, China. Cities. **148**, 104829 (2024). https://doi.org/10.1016/j.cities.2024.104829
5. Marasinghe, R., Yigitcanlar, T., Mayere, S., Washington, T., Limb, M.: Computer vision applications for urban planning: A systematic review of opportunities and constraints. Sustain. Cities Soc. **100**, 105047 (2024). https://doi.org/10.1016/j.scs.2023.105047
6. Hahm, Y., Yoon, H., Choi, Y.: The effect of built environments on the walking and shopping behaviors of pedestrians: A study with GPS experiment in Sinchon retail district in Seoul, South Korea. Cities. **89**, 1–13 (2019)

Exploring Generative Design for the Renovation of University Research Buildings to Facilitate Academic Information Dissemination

Cai Siyu[1], Deng Qiaoming[1](✉), and Liu Yubo[2]

[1] School of Architecture, South China University of Technology, Guangzhou 510640, China
dengqm@scut.edu.cn

[2] State Key Laboratory of Subtropical Building and Urban Science, Guangzhou, China

Abstract. This study explores the renovation of research buildings at South China University of Technology using a multi-objective genetic algorithm (MOGA) for generative design. The process optimizes building area, visual accessibility, information capacity, and minimizes outdoor thermal radiation. Initial issues like inaccessible pathways and small rooms were resolved through algorithm refinement, while the second optimization round improved environmental adaptability. The final design meets key performance indicators, with manual adjustments enhancing local functionality. This research highlights the potential of intelligent algorithms in collaborative design and provides insights for university research building renovation.

Keywords: Information Behavior · MOGA · Generative Design · Interdisciplinary Communication · Transindividual Collaboration · Research Buildings

1 Introduction

As economic and technological development accelerates, universities play a key role in fostering interdisciplinary innovation. However, traditional departmental layouts hinder academic information exchange. Optimizing campus spaces can help overcome this challenge.

Inspired by MIT's Endless Corridor and the Free University of Berlin, this study explores how low-rise, high-density research buildings enhance information flow and collaboration. Focusing on the renovation of research buildings at South China University of Technology, it applies a generative design approach using a Multi-Objective Genetic Algorithm (MOGA) to optimize spatial configurations.

By translating academic information behaviors into spatial metrics, the algorithm generates and refines multiple design iterations. The final plan, combining algorithmic optimization and manual adjustments, balances space efficiency with improved information flow (Fig. 1).

Y. Liu et al. (Eds.): CDRF 2025, *Transindividual Intelligence*, pp. 523–534, 2026.
https://doi.org/10.1007/978-981-92-0615-5_45

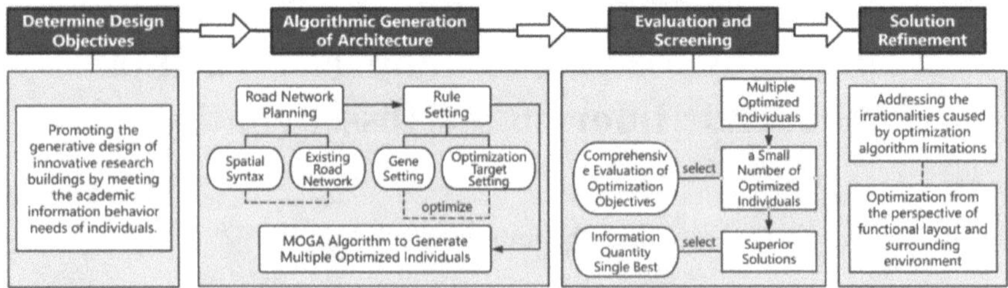

Fig. 1. Generative design and evaluation process based on the MOGA algorithm

2 Quantitative Metrics for the Relationship Between Physical Space and Innovation

Sociology and architectural studies establish the "environmental space – interactive communication – innovation" framework, emphasizing communication as key to quantifying innovation. MIT's Thomas J. Allen (1977) demonstrated a negative correlation between distance and communication frequency, shaping research building layouts [1]. In the 1990s, UCL's Bill Hillier linked spatial structure to team interaction [2], while Christine Thaler (2005) found that higher spatial integration fosters new interactions [3]. Jean D. Wineman (University of Michigan) further showed that shorter distances between faculty offices increase collaboration [4].

Campus surveys indicate faculty and students mainly interact with familiar individuals in public spaces, limiting interdisciplinary engagement. Academic information acts as a bridge across disciplines, as seen in MIT's "Endless Corridor" and the Free University of Berlin's exhibition areas, which enhance academic exchange [5].

This study quantifies "people encountering academic information" for MOGA optimization. Information acquisition time ties to outdoor thermal radiation—comfortable spaces extend exposure. Visual accessibility enhances information filtering. Corridor length increases serendipitous encounters, while building area determines total information capacity. Embedding these spatial metrics into the optimization process advances transindividual intelligent collaborative design, offering insights for future university research building renovations.

3 Generative Design Based on the MOGA Algorithm

3.1 MOGA Algorithm

The Multi-Objective Genetic Algorithm (MOGA) optimizes complex design problems by simulating natural selection. In urban planning and architectural generative design, it manages interdependent goals efficiently. For example, Milad Showkatbakhsh and Mohammed Makki used a genetic algorithm with feedback mechanisms to regulate urban form through five genetic variables [6]. This study applies the Wallacei plugin on Rhino's Grasshopper to optimize university research building design.

3.2 Site Conditions

3.2.1 Overview of the Site

This study focuses on the Wushan Campus's primary teaching and experimental area at South China University of Technology for generative design. Covering 45,960 m^2, it includes 150 classrooms and labs, surrounded by research institutions and landmarks like West Lake and historic red-brick buildings. The renovation must balance functionality with cultural and environmental integration (Fig. 2).

Fig. 2. Schematic diagram of site location

3.2.2 Road Network Design

The site borders residential areas to the west and landmarks like West Lake Garden Hotel and Yifu Humanities Center to the south. Traditional red-brick buildings to the east attract high foot traffic, making pedestrian flow key. Using Depthmap for space syntax analysis, the design establishes an east-west main axis and three north-south secondary axes, forming a pedestrian grid to optimize connectivity and integrate with the surroundings (Fig. 3).

The road network analysis used a space syntax-based Choice model with an 800 m radius ($\approx$10-min walk) to assess traversability. By integrating integration and choice metrics, the model quantifies centrality and accessibility, guiding optimization design (Table 1).

3.3 Gene and Optimization Objective Setup

In the MOGA framework, this study creates a gene pool for architectural generation and defines optimization objectives through parameter exploration. Adjusting gene combinations balances objectives, avoiding extreme design biases for coordinated outcomes.

In this study, all genetic algorithm experiments use the following parameters to ensure reproducibility and adequate convergence (Table 2).

Fig. 3. Site analysis and preliminary setup

Table 1. Quantitative records of road network design space syntax

Original Road Network	Road Network Setting One	Road Network Setting Two
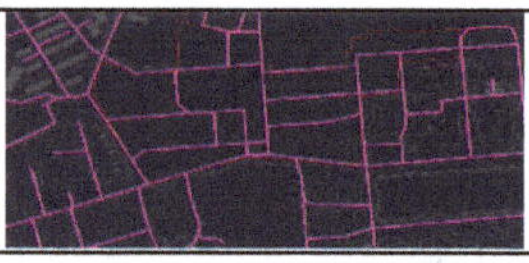		
NACH 800 = 18.65	NACH 800 = 20.24	NACH 800 = 21.18

These parameter values were chosen based on preliminary tuning experiments (Population ∈ [50, 150], Generations ∈ [30, 70], Crossover ∈ [0.6, 0.9], Mutation ∈ [0.05, 0.2]), ensuring a diverse and convergent Pareto front within 50 generations.

3.3.1 Gene Pool Setup

Genes are classified into four types: building height, room depth, mid-rise configuration, and mass excavation. Table 3 details their parameter ranges and design rationale.

Before algorithm integration, the site's road network was subdivided using a 9 m × 9 m modular grid for smaller units, with block heights set at 0 m or 4 m. Larger units used a windmill-shaped subdivision, controlling vertical mass generation through parameters like building height, room depth, and mass excavation. The scheme applies a layered generation strategy, refining genetic parameters to meet optimization goals. Room depth aligns with standard lab dimensions, mid-rise heights balance efficiency and floor area

Table 2. Algorithm parameter settings

Parameter	Symbol	Value	Description
Population size	Population	100	Number of candidate solutions per generation
Number of generations	Generations	50	Total evolutionary iterations
Crossover rate	Crossover	0.8	Probability of gene recombination between parents
Mutation rate	Mutation	0.1	Probability of random gene mutation in an individual
Selection strategy	–	Tournament selection	Each round selects the best individuals for the next generation
Convergence criterion	–	None	Fixed iterations used to facilitate performance comparison

Table 3. Composition and parameter ranges of the gene pool

Gene	Gene pool	Design rationale
Building height	South side: 4–12 m; North side: 8–16 m	Ensures a comprehensive balance among regional sunlight, landscape, and building mass.
Room depth	East-West: 6–9 m; North-South: 9–12 m	Meets the spatial functional requirements and ensures adequate natural ventilation.
Mid-rise configuration	Height: 20–28 m	Provides the necessary supplement of high-rise buildings within the site while controlling overall density.
Mass excavation	Excavation area ratio: $<35\%$ of the total volume	Enhances the openness of the building space while ensuring a certain floor area ratio.

ratio, and mass excavation enhances permeability while maintaining environmental harmony. Building height is capped at four stories to integrate with traditional architecture and lakeside areas (Fig. 4).

3.3.2 Optimization Objectives

This study defines four key optimization objectives:

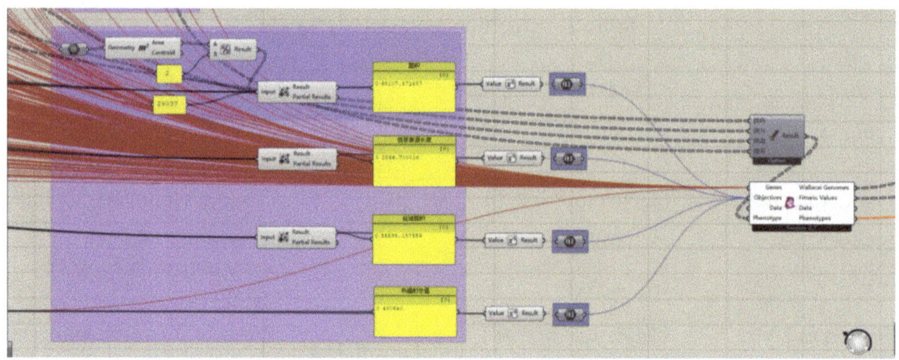

Fig. 4. MOGA algorithm localized cell diagram

1. Maximizing Information Flow

   Quantified by the total length of building edges connected to corridors, reflecting the layout's support for academic interaction. Longer corridors foster spontaneous knowledge exchange and interdisciplinary communication.
2. Minimizing Outdoor Heat Radiation

   Using the Ladybug plugin, heat radiation is optimized by adjusting building morphology and excavation ratios, reducing heat exposure and enhancing pedestrian dwell time for better academic interaction.
3. Maximizing Visual Connectivity

   Visibility analysis calculates connectivity, optimizing spatial permeability for efficient interdisciplinary information retrieval. Clear sightlines and broad views improve information absorption and selection.
4. Maximizing Building Area

   Building area ensures efficient land use and accommodates academic content. Larger buildings with more rooms increase information capacity (Table 4).

Table 4. Composition and parameter range of the gene pool

Optimization objective	Quantitative indicator	Extreme case
Maximizing information Accessibility	Total length of building edges adjacent to corridors	Excessively long buildings, insufficient excavation
Minimizing outdoor heat radiation	Total heat radiation calculated using Ladybug	Buildings too low, excessive excavation
Maximizing visual accessibility	Total visual accessibility calculated through isovist analysis	Buildings too short, excessive excavation
Maximizing usable building area	Total building area	Buildings too tall, insufficient excavation

The dynamic balance between optimization objectives is maintained through competing objectives and genetic control, preventing extremes. For example, maximizing building area offsets excessive excavation from prioritizing visual accessibility. Genetic control adjusts excavation ratios and building height to balance information accessibility and minimize heat radiation, ensuring coordinated outcomes. These objectives enable collaborative multi-agent design.

To understand how our four optimization objectives interact—and where trade-offs arise—we perform a pairwise analysis and summarize the findings in Table 5.

Table 5. Correlation and trade-off network among optimization objectives

Objectives	Correlation/trade-off	Implication
Building area vs. corridor length	Positive correlation	Enlarging total floor area often allows longer, more serpentine corridors, boosting information flow but increasing construction cost.
Building area vs. heat comfort score	Negative trade-off	More area typically means greater façade exposure and heavier thermal load, reducing heat comfort unless mitigated by shading or insulation.
Building area vs. visual field	Positive correlation	Larger floorplates afford wider sightlines and more vantage points, improving visibility metrics.
Corridor length vs. heat comfort score	Negative trade-off	Longer corridors create more glazed or open interfaces that admit solar gains, worsening heat radiation metrics.
Corridor length vs. visual field	Positive correlation	Extended, winding corridors increase opportunities for vistas and "serendipitous encounters," enhancing visual field area.
Heat comfort score vs. visual field	Mild trade-off	Maximizing daylighted views (visual field) often involves large windows or open façades, which can elevate solar heat gain.

By explicitly articulating these interdependencies, the paper offers practitioners a clear framework for navigating the multi-criteria landscape of research-building renovation.

3.4 Evaluation and Selection

The MOGA algorithm generates multiple optimized solutions with quantitative values for the four objectives (Fig. 5). To find the optimal solution, we rank the indicators and select representative solutions. The one with the highest information accessibility value is chosen (Fig. 6).

A rigorous evaluation system ensures reliability, preventing inaccessible spaces or uneven layouts. A multi-dimensional control strategy balances information accessibility and heat radiation. Further screening eliminates poor layout schemes, identifying a solution that performs well across all criteria: information accessibility, thermal environment, visual accessibility, and building area.

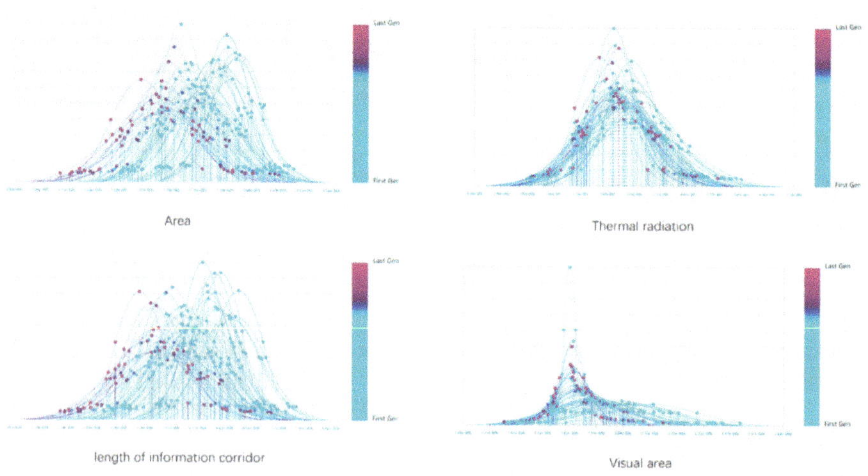

Fig. 5. Line graph of optimization objective values for each generated individual.

The experimental results show that the multi-objective genetic algorithm solutions meet optimization goals, achieving spatial balance and design alignment. This confirms the method's applicability for multi-objective optimization in architectural design, providing a scientific evaluation framework and practical reference for generative design.

3.5 Generative Design Application

This study applied architectural generative design to the SCUT Wushan campus's core teaching and research area.

In the first stage (Scheme 1), issues like blocked paths, small rooms, and insufficient building area emerged due to inadequate path widths, limited genetic pool conditions, and poorly set area goals. Refinements were made by adjusting road widths, genetic conditions, and area targets.

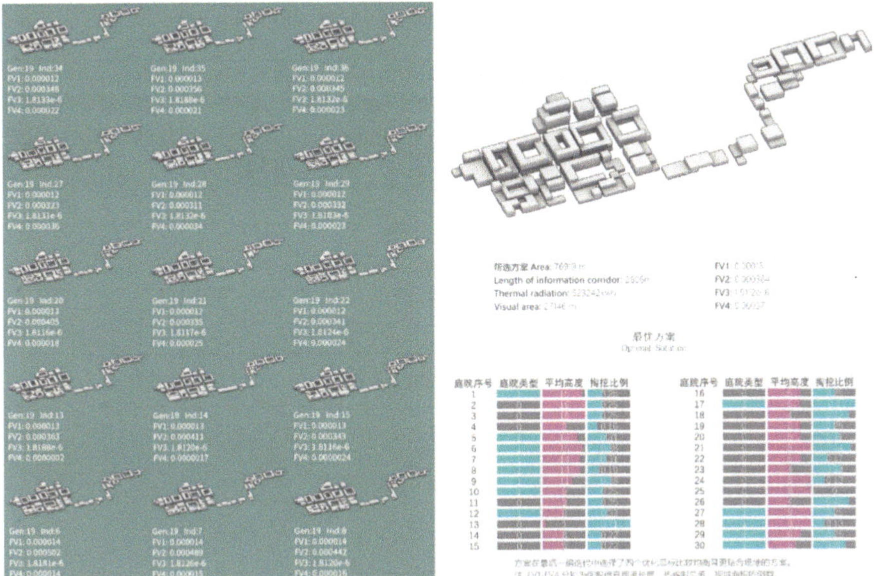

Fig. 6. Display of the small number of optimized solutions selected in the experiment (left) and the final optimized solution (top right), along with part of their genetic sequences (bottom right).

In the second stage, spatial improvements were made, but integration with surrounding structures needed work. Some buildings obstructed traditional structures, and low-efficiency single-story buildings near West Lake persisted. Adjustments to excavation, height controls, and building genes were applied, with a height limit near West Lake.

Minor issues like insufficient excavation near West Lake and corridor uniformity remained. Manual adjustments improved local conditions and functionality (Fig. 7). The results show that iterative optimization with manual intervention effectively balances function and integration (Table 6 and Fig. 8).

Table 6. Comparison of generated solutions and result analysis

Scheme	Scheme 1	Scheme 2	Scheme 3
Generated Solution			
Building Area (m²)	75,340	77,219	76,919
Information Corridor Length (m)	2,542	2,610	2,606
Outdoor Heat Radiation (kW·h)	508,300	518,900	523,242
Visual Field Area (m²)	26,890	27,020	27,146
Radar Chart			

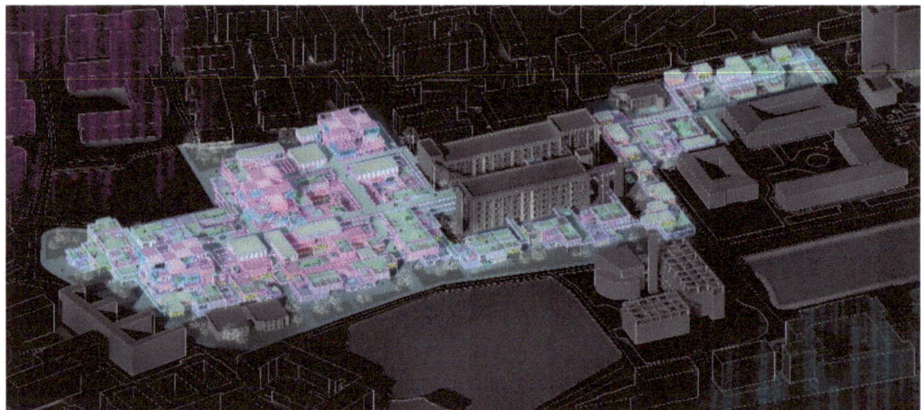

Fig. 7. Results display

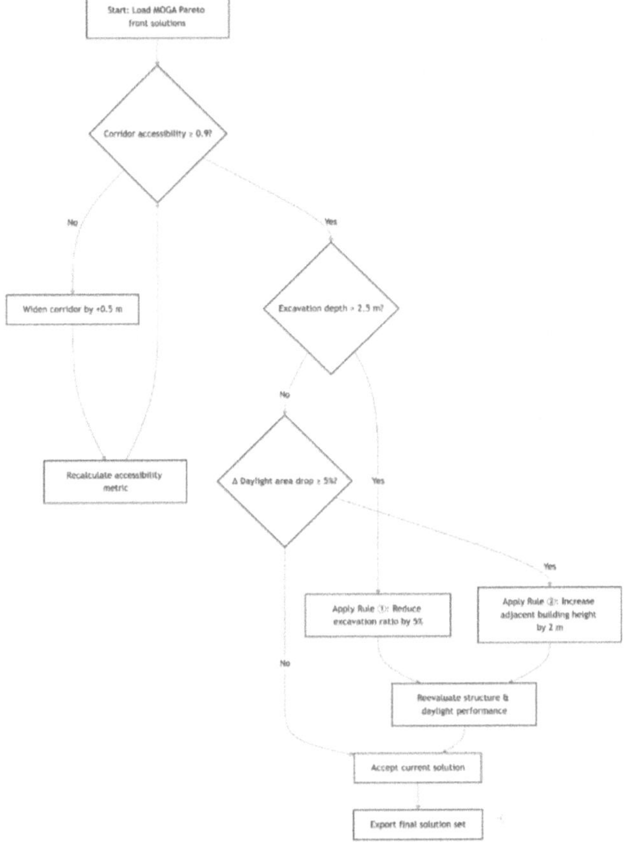

Fig. 8. Manual post-processing decision flowchart

4 Conclusion

This study applies a multi-objective genetic algorithm to optimize architectural generative design, focusing on improving academic information flow in university research buildings. Through iterative refinement of algorithm rules—such as gene pool constraints and optimization settings—the spatial organization and functionality were enhanced. Adjusting control node interactions achieved a balance between functional performance and site adaptability.

The results show optimized solutions meeting ideal metrics in building area, information corridor length, and visual connectivity, with manual interventions improving local functionality. This research demonstrates the potential of multi-objective genetic algorithms in generative design for university research buildings, offering a methodological reference for intelligent, site-specific design. Integrating intelligent algorithms with manual adjustments offers a novel approach for research building innovation.

References

1. Liu, Y.B., Tang, Y.Y., Deng, Q.M.: Review of spatial quantification research in Europe and America oriented by scientific innovation. Contemp. Archit. **07**, 43–47 (2020)
2. Allent, J.: Managing the Flow of Technology: Technology Transfer and the Dissemination of Technological Information within the R&D Organisation. The Massachusetts Institute of Technology, Cambridge, MA (1977)
3. Penn, A., Hillier, B.: The social potential of buildings: spatial structure and the innovative milieu in scientific research laboratories. In: Proceedings of Corporate Space and Architecture International Symposium, Ministère de l'équipement du logement et des transports, Paris, pp. 38–43 (1992)
4. Sailer, K., Thomas, M.: Correspondence and non-correspondence: using office accommodation to calculate an organisation's propensity for new ideas. In: Proceedings of the 12th International Space Syntax Symposium, International Space Syntax Symposium (2019). http://www.12sssbeijing.com/proceedings/download.php?lang=cn&class1=165. Accessed 15 Jan 2021
5. Deng, Q.M., Liu, Y.B., Ji, M.: Campus planning and design for interdisciplinary exchange and cooperation. Era Archit. **02**, 30–35 (2021). https://doi.org/10.13717/j.cnki.ta.2021.02.006
6. Showkatbakhsh, M., Makki, M.: Application of homeostatic principles within evolutionary design processes: adaptive urban tissues. J. Comput. Des. Eng. **7**(1), 1–17 (2020)

Simulation and Green Solution for Urban Crime: Application of Large Language Model and Agent-Based Model

Lu Chen[4(✉)], Wenzhen Jia[1], Peixu Sun[2], Zelin Liu[2], and Yuan Lai[3(✉)]

[1] School of Landscape Architecture, Beijing Forestry University, Beijing 100083, China
[2] China Architecture Design & Research Group, Beijing 100044, China
[3] Department of Urban and Rural Planning, School of Architecture, Tsinghua University, Haidian District, Beijing 100044, China
yuanlai@tsinghua.edu.cn
[4] Department of Landscape Architecture, School of Architecture, Tsinghua University, Haidian District, Beijing 100084, China
chenlu24@mails.tsinghua.edu.cn

Abstract. Urban environmental safety is critical for sustainable city development, yet the complex mechanisms underlying crime patterns pose significant research challenges. This study analyzes 12,893 criminal judgments (PCJs) from three Chinese megacities over three years using Large Language Models (LLM), developing a crime pattern simulation model that integrates Interpretable Machine Learning (IML) with Agent-Based Modeling (ABM). Then, three urban renewal scenarios were tested. The result shows that: (1) Socioeconomic and built-environment factors account for 33.3%–44.5% of outdoor crime spatial variation; (2) Nighttime lighting emerges as a key predictor, while canopy cover exhibits threshold effects on specific crime types; (3) Green regeneration scenarios demonstrate significantly lower crime rates. The proposed multidimensional framework provides actionable spatial planning insights and policy recommendations for safer cities.

Keywords: Urban Crime · Paper of Criminal Judgment · China's Mega-cities · Interpretable Machine Learning · Agent-based Model

1 Introduction

Urban safety has emerged as a critical indicator for assessing social resilience (Grimaldi et al. 2023). Urban crime, a major risk factor for urban safety, is significantly correlated with the urban environment, especially in mega-cities (He et al. 2020; Shi et al. 2019). The analysis and simulation of crime patterns in mega-cities can provide reliable support for safe city planning, which is a key aspect of smart governance in mega-cities.

Theoretical frameworks in environmental criminology, such as Crime Prevention Through Environmental Design (CPTED), provide foundational support for urban safety research. With initial urban crime research focusing on the mutually reinforcing relationship between street vitality and crime prevention, recent studies have further focused

Y. Liu et al. (Eds.): CDRF 2025, *Transindividual Intelligence*, pp. 535–544, 2026.
https://doi.org/10.1007/978-981-92-0615-5_46

on the more complex associations and mechanisms between urban spatial features and crime patterns (Chen et al. 2024). In addition to socio-economic factors, green space has also been recognized as a key factor in the spatial kind of criminal behavior (Abdillah et al. 2024). Thus, the complex association of urban crime with urban spatial features requires the integration of multiple dimensions of variables, which is a key challenge in realizing crime simulation on a large urban scale. Another challenge that supports the simulation of crime patterns is the collection of crime distribution, such data remain scarce and difficult to obtain especially in China's mega-cities,. While publicly available PCJ provides detailed case-level information, its internal information is scattered and difficult to be standardized.

This study aims to address the two major challenges mentioned above in order to elucidate the mechanisms between urban environment and crime, and simulate crime patterns under different old urban areas development scenarios

2 Material and Methodology

The research framework is structured across three levels (Fig. 1): (1) Dataset Establishment, (2) Mechanism Analysis, and (3) Multi-scenario Simulation.

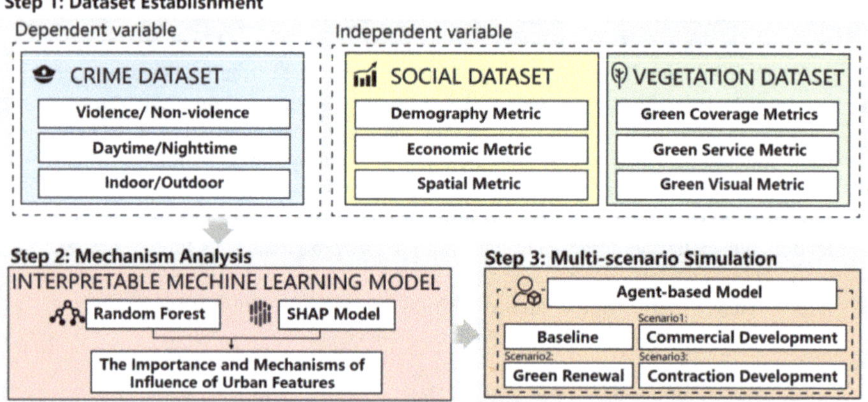

Fig. 1. The framework of this paper

This paper focuses on crime patterns in China's Mega-cities with populations exceeding 10 million, targeting three representative cities: Beijing, Chongqing, and Guangzhou. Large Language Model (LLM), Interpretable Machine Learning (IML) and agent-based modeling (ABM) serve as the key technologies to address the aforementioned challenges.

2.1 Dataset Establishment Supported by LLM

To mitigate confounding effects from non-urbanized areas, the study focuses on urban core districts and nearby suburban areas, using neighborhood boundaries as the smallest unit for analysis. Twelve socio-economic and spatial environmental features were collected from various sources (Gao, J. et al. 2023; Gong et al. 2020). Then, PCJs sourced from the China Judgments Online platform were parsed with LLM to interpret the urban crime pattern. The types of crimes are limited to those related to the features of urban space. Detailed charge classifications presented in Table 1.

A corpus of 12,893 PCJs were processed through the GLM-4-FlashX LLM (developed by ZhiPu AI), selected for its optimized performance in vertical task processing with rapid response capabilities (72.14 tokens/s generation speed) The resultant spatial-temporal crime database contains critical analysis dimensions: temporal classification (daytime/nighttime); crime type (violence/non-violence); spatial typology (indoor/outdoor). With the combination of three dichotomous parameters, the criminal types are divided into eight categories. (Fig. 2). Model reliability was verified through manual parsing of 200 judgment samples (96.5% alignment)."

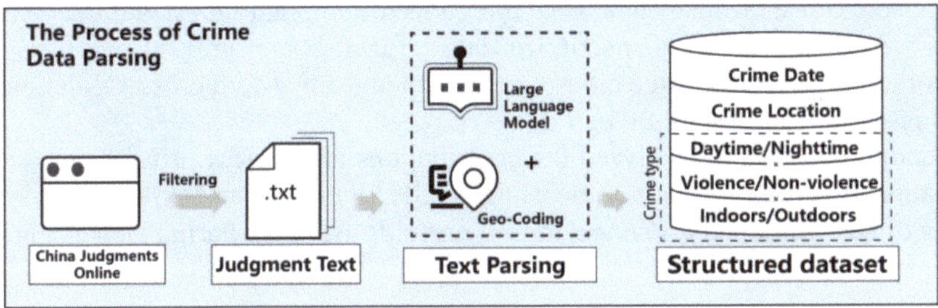

Fig. 2. The process of crime data parsing

Then, crime data undergo geocoding process through the Amap API. to derive precise latitude-longitude coordinates. PCJs exhibiting geographical obfuscation in crime location descriptions were systematically excluded to maintain analytical validity. Finally, all data is compiled in GIS software, aggregated to the neighborhood scale through spatial joining.

Table 1. The criminal charges in PCJs

Classification	Criminal charges	Total number
Violation of Citizens' Personal Rights	Intentional Homicide	14
	Intentional Injury	732
	Rape	24
	Forcible Indecent Assault	36
	Negligent Homicide	19
Violation of Property	Theft	4310
	Robbery	41
	Snatching	17

(*continued*)

Table 1. (*continued*)

Classification	Criminal charges	Total number
Endangering Public Safety	Intentional Destruction of Property	54
	Dangerous Driving	7007
	Traffic Violation	639

2.2 Mechanism Analysis by IML

The Random Forest (RF) model was used to explore the relationship between 12 urban environmental features and neighborhood crime rates. Crime rate was defined as the ratio of total crimes to the neighborhood population (per 10,000 people) during the study period. Since crime events were categorized into eight types, eight separate RF models were constructed. To maximize data utilization and prevent overfitting, 547 neighborhoods across the three cities were modeled using 4-fold cross-validation, with a training-to-validation set ratio of 7:3.

In addition, the study analyzed the contributions of different urban features to the explanation of various types of crime using the SHAP method and explored the specific impacts of key factors on crime and their thresholds using the Partial Dependency Plot (PDP) method.

2.3 Multi-scenario Simulation Based on ABM

ABM enables the simulation of social systems and complex socio-spatial phenomena. The study combines the IML results and the urban road network to establish a three-way "potential offender-potential victim-urban environment" ABM, and applies the model to the central city of Beijing, predicting the changes of the urban crime pattern under three different old urban areas development scenarios (Fig. 3). The model is a highly generalized virtual world of urban crime, which presupposes the distribution of potential offenders and potential victims based on the population of different neighborhoods. When a potential offender and a potential victim meet at a certain point according to the rule-based movement, the decision of whether a crime occurs is based on the Agent Crime index of the neighborhood where they meet. The Agent Crime index is calculated as follows:

$$Agent\ Crime\ index = \frac{\text{number of crimes in neighborhoods (Predicted by RF)}}{\text{number of encounters in ABM}}$$

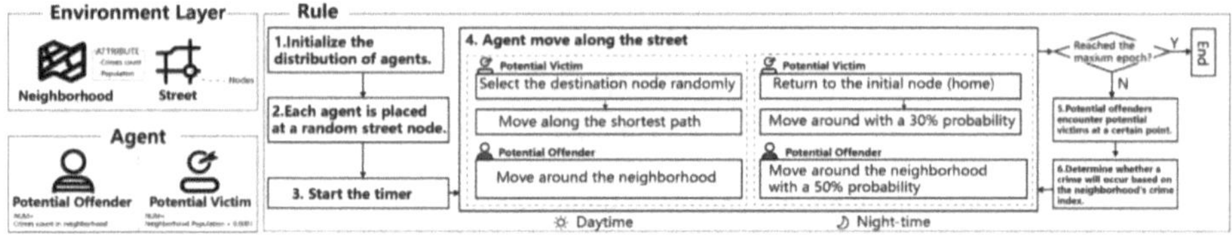

Fig. 3. Framework of ABM

Under different development scenarios of the city, the Agent Crime index of neighborhoods will change correspondingly, thus changing the security pattern of the city. Based on the potential development paths of China's mega-city, we assume the following three development patterns for the old urban areas of Beijing (Table 2), then conduct twenty rounds of simulation in the virtual world. The simulated crime locations represent the urban crime patterns under different scenarios.

Table 2. Development patterns of cities in multi-scenario simulations

	Description	Parameter change settings in neighborhoods
Baseline	—	—
Commercial Development	Commercialization of old urban areas and further strengthening of core areas	**Old urban areas with top 20% of nighttime lights**: population + 1%, nighttime lighting + 15%, POI density + 10% **Old urban areas with bottom 20% of nighttime lights**: population − 1%, nighttime lights − 15%, POI density − 10%
Green Renewal	Renovation of vegetation in old urban areas and creation of micro-parks by utilizing fragmented spaces	**All old urban areas**: park area per capita + 5%, green visibility rate + 10%, canopy cover + 10%
Contraction Development	Inhibition of further construction development in core areas and decentralization of urban construction to peripheral areas	**Old urban areas with top 20% of nighttime lights**: population − 1%, nighttime lights − 15%, POI density − 10% **Old urban areas with bottom 20% of nighttime lights**: population + 1%, nighttime lights + 15%, POI density + 10%,canopy cover − 10%

3 Result

The crime location distribution and dataset construction results for three cities are shown in Fig. 4, with 7,251 crime locations identified. All cities exhibit spatial clustering of crimes rather than random distribution.

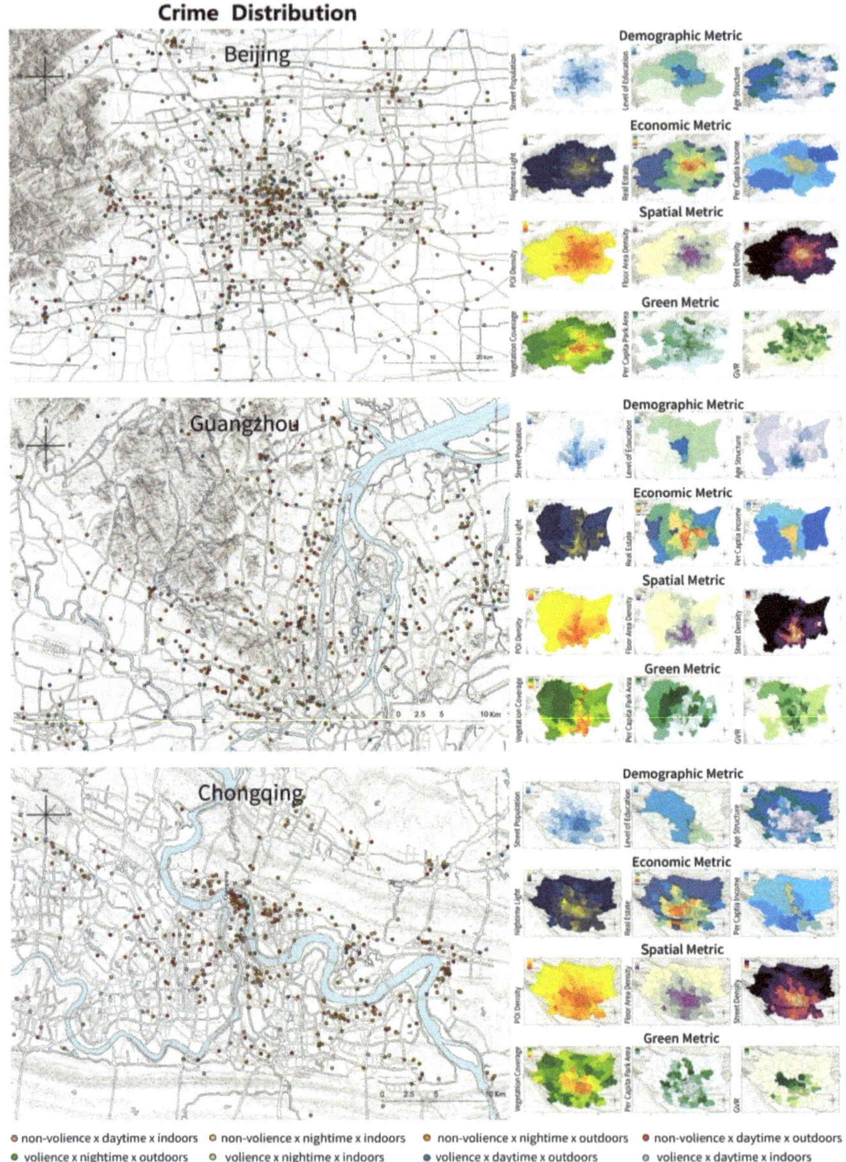

Fig. 4. Urban features dataset

The RF models show that the 12 neighborhood features proposed have certain explanatory contribution for outdoor crimes. The model for violent nighttime outdoor crimes performs best, explaining 44.5% of crime distribution variation. Other outdoor crime models have R-squared values around 0.33–0.34 (Fig. 5).

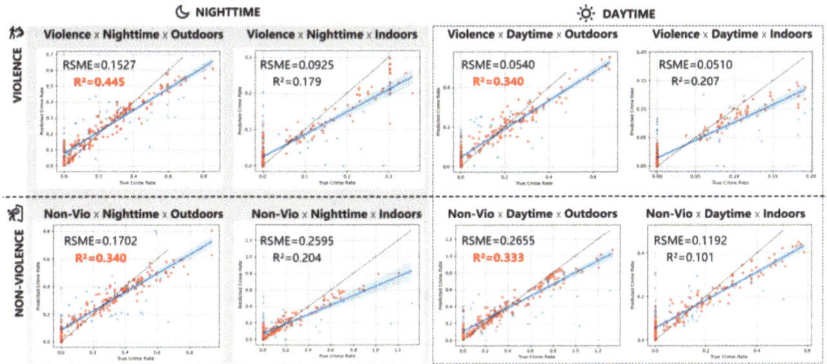

Fig. 5. The performance of the eight RF models

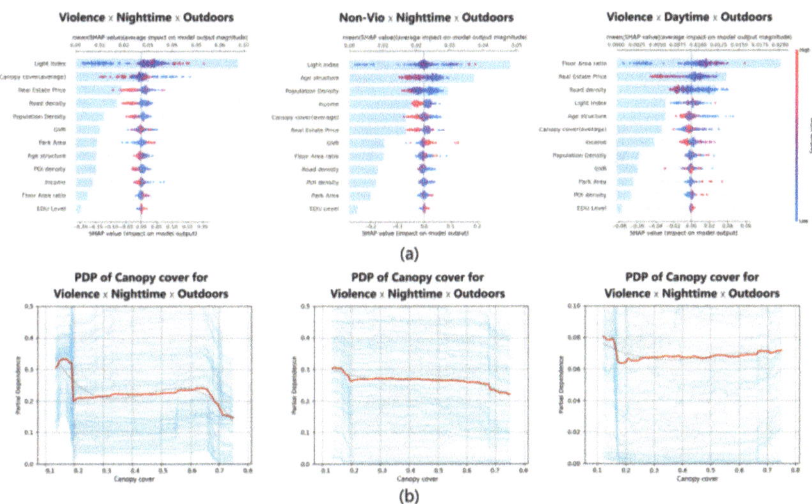

Fig. 6. Contributions of different factors to RF models. (a) Explanatory contributions of urban features for different types of crime. (b) PDP for canopy cover attribute

IML analysis was conducted on three RF that showed good prediction performance (Fig. 6). The results show that: for the violent nighttime outdoor crime model, the most significant contributing feature is the nighttime light index, followed by canopy cover, property prices, and road density. According to the partial dependence plot results, canopy cover is generally negatively correlated with crime, with a clear threshold around 20%. This means that when the average tree canopy cover of a block exceeds 20%, the nighttime outdoor violent crime in the block will significantly decrease.

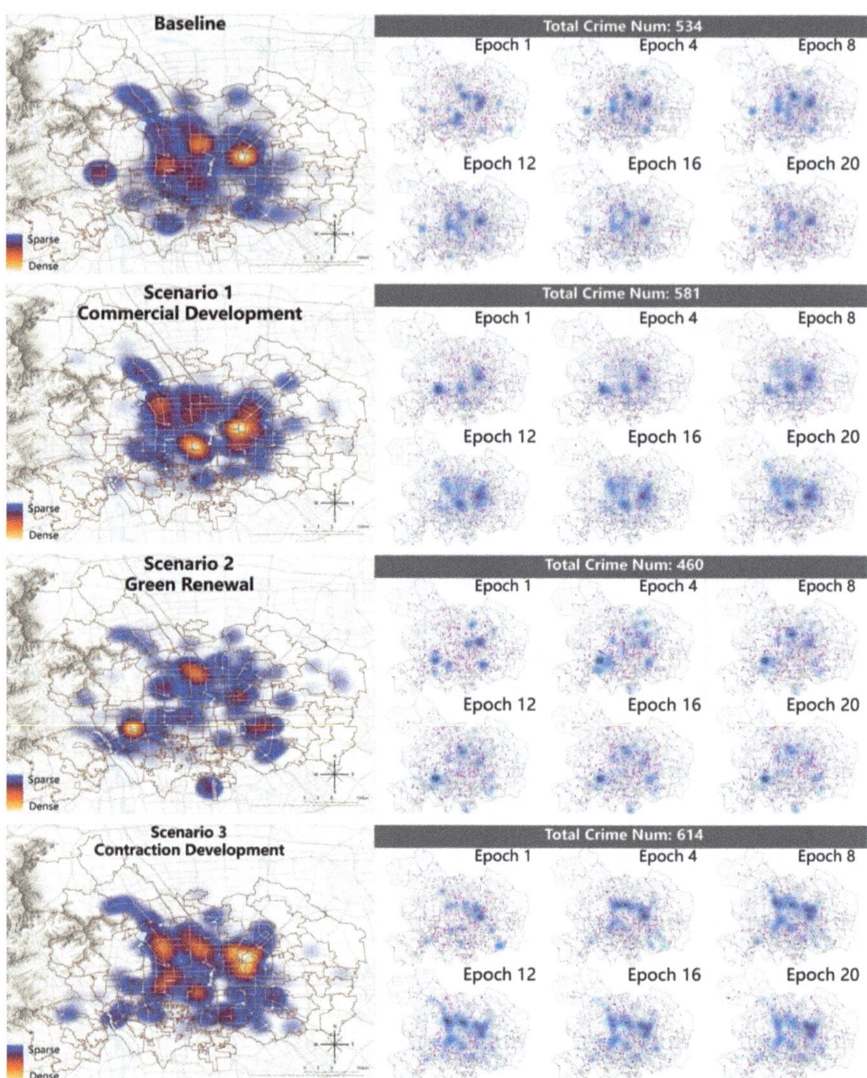

Fig. 7. The ABM prediction of crime pattern in different development scenarios of Beijing's old urban areas.

For daytime outdoor violent crimes, the most significant features are property price and floor area ratio, followed by road density, nighttime lighting, age structure, and canopy cover. Canopy cover has a weak but noticeable threshold around 17%. When canopy cover exceeds this threshold, crime rates decrease. Further increases in canopy cover do not significantly change the crime pattern. It should be noted that the result doesn't prove a direct causal link between green space and crime.

The ABM was constructed using the IML results as rule parameters (Fig. 7). The results show that, compared with the baseline scenario, the commercial development approach in the old urban area will make crime locations more concentrated and lead to an increase in the total number of crimes. Under the contraction development scenario, the number of crimes will rise significantly and form several discrete clusters around the old urban area. However, green renewal of the old urban area can disperse crime locations outside the old urban area and reduce the total number of crimes (by about 13.8% in this simulation).

4 Discussion and Conclusion

The key findings of this paper are: (1) The urban environment can influence crime patterns, twelve features proposed in this study explain 33.3%–44.5% of the variation in outdoor crime distribution, but have weaker explanatory power for indoor crime; (2) Although the model does not prove direct causality, it reveals that canopy cover significantly affects the distribution patterns of certain crimes, providing a new perspective for the development of urban green spaces; (3) The simulation results show that commercial and contraction development scenarios both lead to an increase in total urban crime. Contraction development also disperses crime clusters, making urban security management more challenging. In contrast, green renewal of the old urban area can mitigate these effects by reducing crime rates and preventing large-scale crime concentration. However, this study has limitations: it did not compare crime patterns across different cities or account for significant urban characteristics such as topography and climate.

The research is supported by two key technologies: large language models (LLMs) and agent-based modeling (ABM). LLMs facilitate structured data extraction from textual sources, enabling efficient parsing of crime locations and integration with GIS-based socio-economic and environmental datasets, thereby minimizing manual processing. ABM, on the other hand, simulates micro-level individual behaviors to predict macro-level system changes. ABM can also incorporate spatial patterns of cities into the model by integrating GIS data, which makes it a huge and yet untapped potential in the analysis of urban social relations. The key to hindering the realization of this potential lies in the setting of agent rules. The countermeasure in this study is to build an IML model based on the multi-dimensional data and use the predictive results as the parameters of ABM multi-scenario simulation. This innovative combination method compensates for the inadequacy of the IML in dealing with the the spatial vector information on the one hand, and improves the accuracy and credibility of the ABM output results on the other hand. It is worth noting that, along with the enhancement of LLM's in-depth analysis ability, whether it is possible to directly interpret crime mechanisms and define agent rules and parameters through the analysis of extensive PCJs is a direction worthy of further exploration. However, it is also necessary to be alert to the uncontrollability and limited interpretability of LLM outputs, so further excavation of urban crime mechanisms and causality is still very critical.

Acknowledgements. This research was funded by the National Natural Science Foundation of China (#72274101), the National Key Research and Development Program of China (#2022YFC3800603), and the Tsinghua-Toyota Joint Research Fund.

References

Abdillah, Widianingsih, I., Buchari, R.A., Nurasa, H.: From urban crime areas to urban resilience: Lessons learned from Bandung city, Indonesia. Cogent Soc. Sci. **10**(1), 2369171 (2024). https://doi.org/10.1080/23311886.2024.2369171

Chen, J., Li, H., Luo, S., Su, D., Zang, T., Kinoshita, T.: Exploring the complex association between urban form and crime: Evidence from 1,486 U.S. counties. J. Urban Manag. **13**(3), 482–496 (2024). https://doi.org/10.1016/j.jum.2024.05.008

Gao, J., et al.: China regional 250 m normalized difference vegetation index data set (2000–2023). National Tibetan Plateau/third pole environment data center (2023). https://doi.org/10.11888/Terre.tpdc.300328

Gong, P., et al.: Mapping essential urban land use categories in China (EULUC-China): Preliminary results for 2018. Sci. Bull. **65**(3), 182–187 (2020). https://doi.org/10.1016/j.scib.2019.12.007

Grimaldi, M., Coppola, F., Fasolino, I.: A crime risk-based approach for urban planning. A methodological proposal. Land Use Policy. **126**, 106510 (2023). https://doi.org/10.1016/j.landusepol.2022.106510

He, Z., Deng, M., Xie, Z., Wu, L., Chen, Z., Pei, T.: Discovering the joint influence of urban facilities on crime occurrence using spatial co-location pattern mining. Cities. **99**, 102612 (2020). https://doi.org/10.1016/j.cities.2020.102612

Shi, Y., Hu, X., Yang, H., Yang, J.: Format compositions and spatial distribution characteristics of central districts in megacities: People's square and Shanghai's Lujiazui district. Urban Des. Int. **24**(4), 260–270 (2019). https://doi.org/10.1057/s41289-018-0077-9

Author Index

Y. Liu et al. (Eds.): CDRF 2025, *Transindividual Intelligence*, pp. 545–547, 2026.
https://doi.org/10.1007/978-981-92-0615-5

Zeitfracht Medien GmbH
Ferdinand-Jühlke-Straße 7
99095 Erfurt, Deutschland
produktsicherheit@kolibri360.de